16th International Conference on Nanomaterials - Research and Application (NANOCON 2024)

Brno, Czech Republic
16-18 October 2024

ISBN: 979-8-3313-2321-9

Printed from e-media with permission by:

Curran Associates, Inc.
57 Morehouse Lane
Red Hook, NY 12571

Some format issues inherent in the e-media version may also appear in this print version.

Printed with permission by Curran Associates, Inc. (2026)

For permission requests, please contact Tanger Ltd
at the address below.

Tanger Ltd
Keltickova 62
710 00 Slezka Ostrava
Czech Republic

www.tanger.cz

Additional copies of this publication are available from:

Curran Associates, Inc.
57 Morehouse Lane
Red Hook, NY 12571 USA
Phone: 845-758-0400
Fax: 845-758-2633
Email: curran@proceedings.com
Web: www.proceedings.com

TANGER Ltd.
The Czech Society for New Materials and Technologies
Regional Centre of Advanced Technologies and Materials

2024 NANOCON®

16th International Conference on Nanomaterials - Research & Application

CONFERENCE PROCEEDINGS

October 16 - 18, 2024
Brno, Czech Republic, EU

ISSN 2694-930X

NANOCON 2024 **Conference Proceedings**

Different Authors

October 16-18, 2024 Orea Congress Hotel Brno, Czech Republic, EU

Issued by: TANGER Ltd., Keltickova 62, 710 00 Ostrava, Czech Republic, EU

Edition: 1st Edition, 2025

SESSION A - NANOMATERIALS FOR ELECTRONIC, MAGNETIC AND OPTIC APPLICATIONS. CARBON NANOSTRUCTURES, QUANTUM DOTS

SESSION B - INDUSTRIAL AND ENVIRONMENTAL APPLICATIONS OF NANOMATERIALS

SESSION C - BIONANOTECHNOLOGY, NANOMATERIALS IN MEDICINE

SESSION A
NANOMATERIALS FOR ELECTRONIC, MAGNETIC AND OPTIC APPLICATIONS. CARBON NANOSTRUCTURES, QUANTUM DOTS

Chairmen

Prof. Dr. Eduard HULICIUS Institute of Physics of the CAS, Prague, Czech Republic, EU

Assoc. Prof. Dr. Martin KALBAC J. Heyrovsky Institute of Physical Chemistry of the CAS, Prague, Czech Republic, EU

USE OF CARBON NANOSTRUCTURES FOR SENSOR PREPARATION

[1]Jaroslav HOLZEL, [1]Jan SLANY, [1]Karel KLOUDA, [1]Petra ROUPCOVA

[1]*VSB - Technical University of Ostrava, Ostrava, Czech Republic, EU,* jan.slany@vsb.cz

https://doi.org/10.37904/nanocon.2024.4984

Abstract

Graphene oxide is a carbonaceous material that has unique physicochemical properties (specific surface area, pollution resistance, conductivity, high mechanical strength) and can be produced in large quantities with relative ease. It is a material formed by the oxidation of graphene, which can be further modified with, for example, ascorbic acid to form reduced graphene oxide (rGO), or with hydrogen peroxide to form holey graphene oxide (HGO). The application of graphene oxide and its modifications in processes leads to improved sensitivity, electrocatalytic behavior and reduced fouling. Thus, a kampron, which is an electronic component with graphene oxide derivatives fabricated using 3D printing technology, has been experimentally prepared with the advantage of applications in various fields – physical, chemical and biological. Thus, in several stages, the action in the physical domain (pressure), chemical domain (gases and vapors) and biological domain (gases exhaled from the lungs) will be investigated. The changes caused by the investigated influences on the sensors are reflected by a change in the voltage on the sensor. This paper focuses on the first stage of the research, i.e. the fabrication and verification of reactivity of sensors printed in a single layer, which are similar to conventional comb or spiral sensors. The next stages of the research will focus on the investigation of sensors arranged in multiple layers, with the final stage involving the preparation of sensors with complex spatial shapes, while higher sensitivity is also expected. Carbon materials in the form of filaments will be used for 3D printing.

Keywords: Graphene oxide, sensors, kampron, 3D printing

1. INTRODUCTION

Graphene oxide and materials based on it are widely used materials mainly because of their excellent optical, mechanical and electrical properties, which can be used for a wide variety of applications. Attention is currently focused primarily on research into the industrial synthesis of graphene, with an emphasis on premium quality, while at the same time establishing a cheap and simple processing method to promote practical and industrial applications. [1] The materials are generally synthesized by long-established processes (Hummers method), whereby the graphite powder is oxidized to form graphene oxide (GO). From GO, it is possible to prepare, for example, rGO, i.e. reduced GO, by reaction with ascorbic acid, or HGO (holey graphene oxide, also referred to as perforated graphene), which can be prepared by the action of hydrogen peroxide. [2] The GO modifications differ quite significantly in properties from the original GO, for example in terms of electrical conductivity, mechanical strength or hydrophilic nature of the structures. [3] Reduced graphene oxide is a graphene material that is formed by reducing the number of oxygen groups of GO. After such reduction, the most resistant oxygen groups remain on the edges of GO. rGO has different electrical properties than graphene oxide, in particular conductivity. This is why the purpose is to take advantage of possible changes in conductivity due to the action of different sensor reagents. Holed graphene oxide (HGO) is then a graphene material with a large number of holes and pores in its basal plane. The last material is graphitic carbon nitride (g-C_3N_4), which appears to be promising for photocatalytic technologies (water splitting), for applications in supercapacitors (high charge density), or as a flame retardant. It is this material that was used for printing the sensors tested in this work, where the material consisted of a PP filament with 10% g-C_3N_4. Thus, in summary, all of these materials can be used, for example, to purify wastewater from drugs, drugs, metabolites, microorganisms, pollutants, dyes, radioactive isotopes and heavy metals [4], and for biomedical applications such as magnetic field-controlled nanobots - targeted delivery of drugs, antimicrobials, interaction with molecules, or possibly cancer diagnosis. [5 - 8] The last application can be for the preparation of sensors and biosensors, as they are active cells working due to changes in electrical properties during electrochemical processes (physical applications of GO in the form of sensors) [9, 10].

Sensors of all types are proven to improve the quality of life in homes and industry. Electrodes are the most important part of sensors, as they enable data reception and analysis. The most commonly used materials for the development of electrodes are gold, silver, aluminum and carbon, in the form of graphene can be described as a frequently chosen option due to its excellent electrical and crystal properties. In force sensors, for example, the signal response, limit of detection, maximum sensing range and potential reproducibility of the response are all of interest - and it is these properties that are of high importance in graphene oxide. Electrochemical sensors then take advantage of the relatively large specific surface area when working with biomolecules, where the speed of transport and small band gaps cause an interaction between the electrode and the substance being analysed. The undeniable advantage of this material is also its low environmental burden, making it a popular material compared to the other metals mentioned. [11, 12] The purpose of this paper is then to verify whether materials from the graphene family are indeed suitable for the preparation of sensors and for which purposes these sensors can be used.

2. EXPERIMENTAL PART

The following section briefly describes the materials used and the experimental procedure.

2.1 Materials and methods

Comprehensive research on the reactive properties of graphene derivatives (GO, rGO, HGO, etc.) requires the development of a stable platform for future sensors. The preparation of these modifications is, however, protected by a patent and for this reason, only the term "kampron" will be used, which is a collective term for electronic components and devices with graphene derivatives created using 3D printing technology while being cheap, fast, reliable and repeatable technology with great potential for the development of sensors in various application areas - physical, chemical and biological. It is important to note that this term has not been used in history before and is solely a designation created by the authors. The Kampron is capable of detecting different types of reactions and interactions both in the physical (measurement of physical quantities, such as force, pressure, temperature, etc.) and chemical (gases and vapors detection) domains. Biological measurements, i.e., measurements related to the physiological functions of organisms, will also be investigated. For this reason, the research has been divided into three stages, with the first stage focused on the production and research of reactivity sensors that are similar to conventional comb or spiral sensors and are printed in a single layer, the second stage is then dedicated to the production and research of reactivity sensors that are comb (spiral) but arranged in multiple layers, and the essence of the last stage is the production and research of reactivity sensors that will have more complex spatial shapes. Thus, higher sensitivity is also expected.

The following section then focuses on the results of the experiments performed in the first stage of the research - i.e. measuring the reactivity of GO derivatives using a kampron printed in a single layer. The following materials were used for the experiments - PLA (polylactid acid) conductive filament, PETG filament (supplier PRUSA Research) and GO, prepared according to the procedure of Hummer W.S. - Offerman R.E. 195. The following equipment was then used for the experiments - a cell for testing the kampron with gas inlet and outlet, an Arduino UNO microcontroller with Data logger shield for SD cards and a SHITO force gauge, type FS-500N.

First, the kampron was prepared by printing the electrodes for sensing the electrical voltage (resistance or conductance measurement) from PLA material with 10% graphite. The electrodes are on a non-conductive surface (PETG material). GO nanoparticles (100 nm on average) in an aqueous suspension were deposited on the electrodes, which formed an active layer at 40 - 50°C after 12 hours of drying, see **Figure 1**.

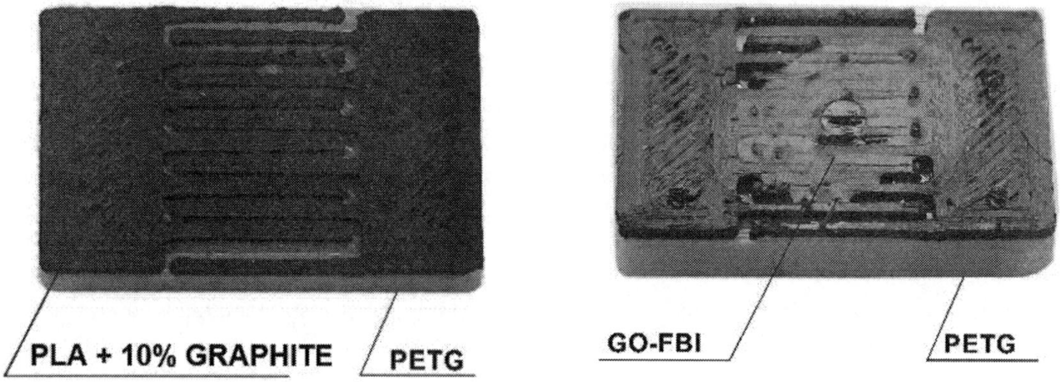

Figure 1 Kampron electrodes on insulator (left), GO on electrodes (right)

The kampron was then placed in a closed cell (**Figure 2**) with inlet and outlet for gases and vapors for reactivity measurements. The electrodes were connected to a voltage source produced by a microcontroller that simultaneously detects changes and stores the measured data on an SD card.

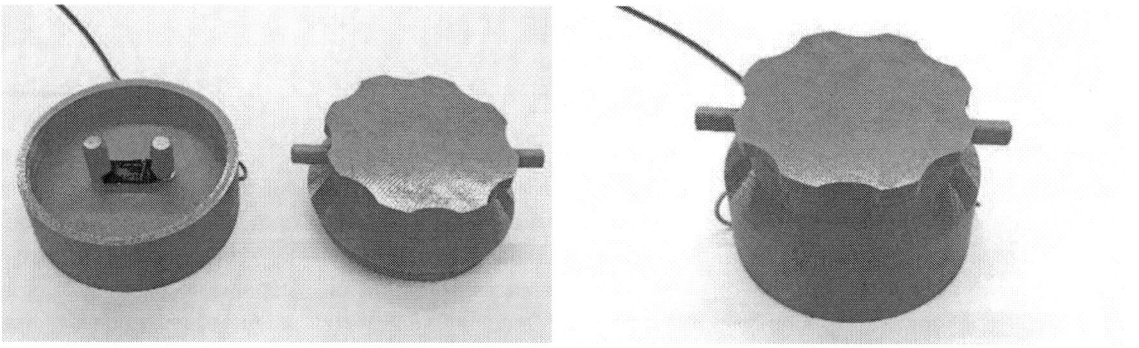

Figure 2 Open cell with kampron (left), closed cell (right)

2.2 Results

The measurement itself was performed using a microcontroller that contains an analog-to-digital converter that reads the changing voltage on the kampron and converts it to numbers between 0 and 1023 (it is 10-bit - it distinguishes 210 possible values). The maximum voltage on the microcontroller pin is 5 V. This value corresponds to 1023. For example, the numerical value 260 corresponds to a voltage of 1.27 V. To better plot the data in the graph, a converter was used. **Table 1** shows the conditions under which the measurements were performed and the following graphs show the results of physical, chemical and biological measurements. Chemical measurements were conducted using a 2-liter air tank, which was inflated with atmospheric air using a compressor set to a pressure of 1.2 bar. A material soaked in acetone or gasoline was placed in the intake where, due to the pressure, the vapors came into contact with the kampron. For the biological measurements, human exhalation lasting 20 seconds was used, which was then gradually atmospherically ventilated for 30 seconds.

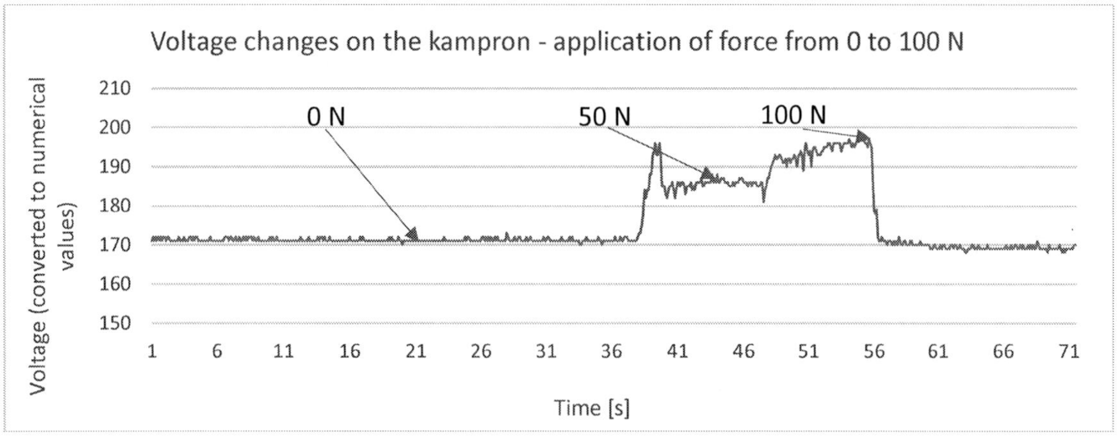

Figure 3 Voltage changes while testing force changes

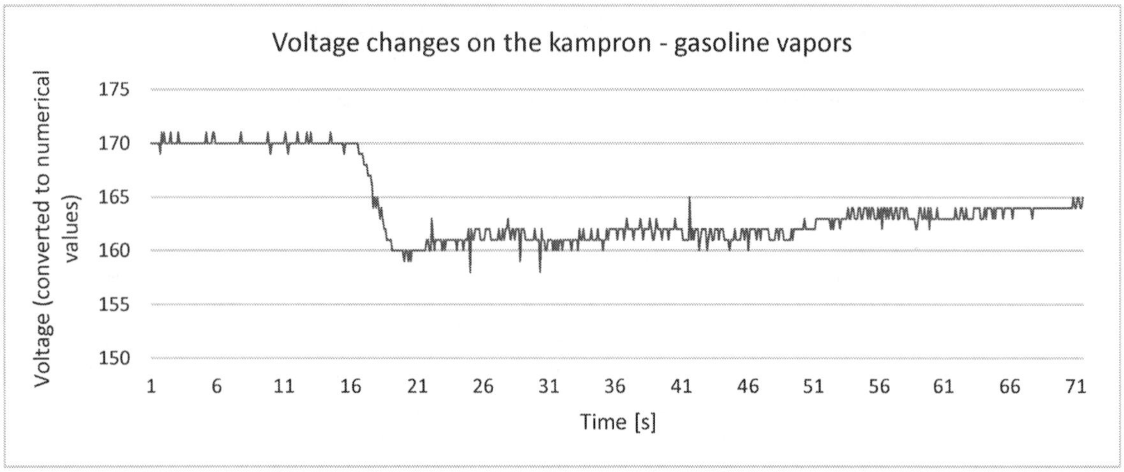

Figure 4 Voltage changes while testing vapors (gasoline)

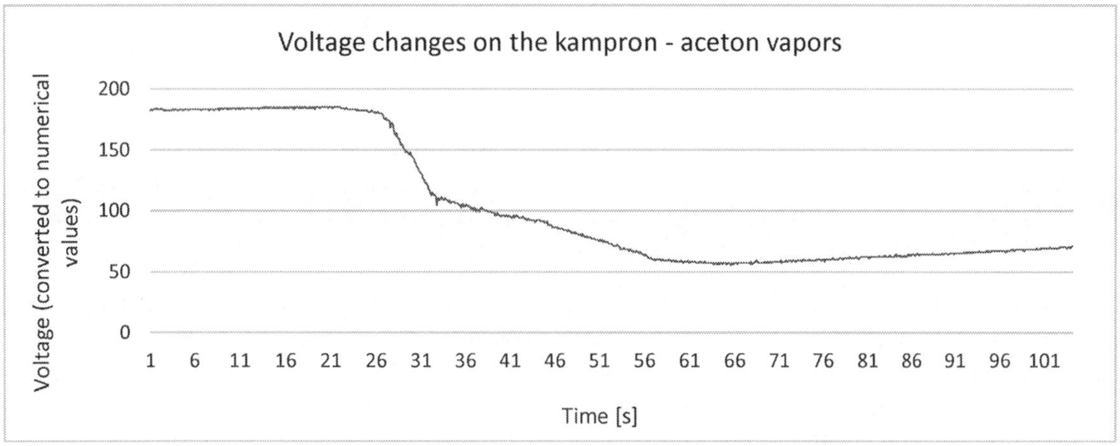

Figure 5 Voltage changes while testing vapors (aceton)

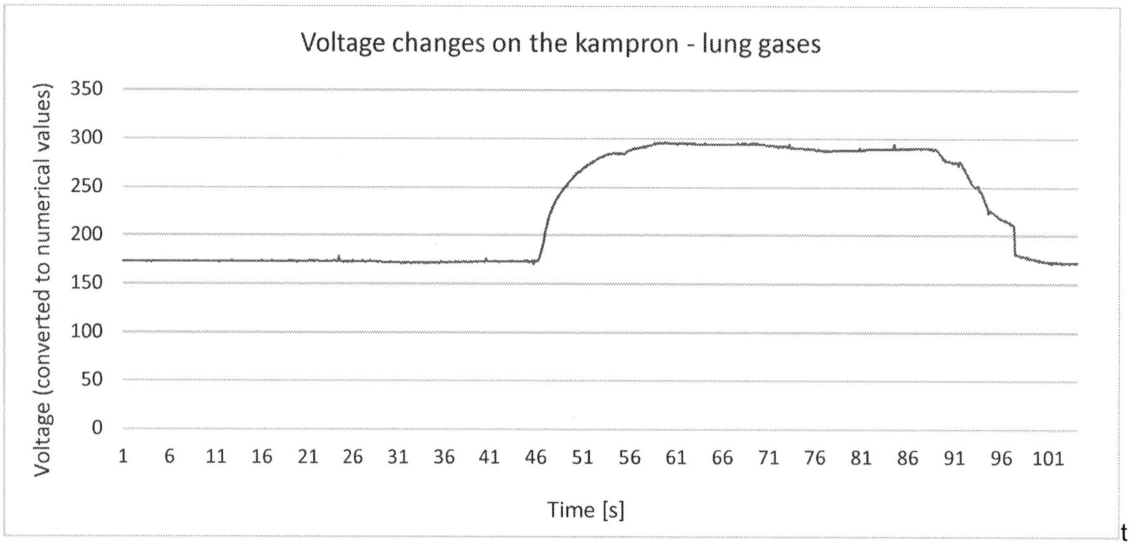

Figure 6 Voltage changes while testing moist gases from the lungs

The table below summarizes the experimental values obtained from the measurements.

Table 1 Experimental conditions and values obtained by measurements

Experimental conditions	Pressure measurement (physical measurement)	Measurement of flammable vapors (chemical measurement)	Measurement of flammable vapors (chemical measurement)	Measurement of gas mixtures (biological measurement)
Kampron material	Single layer 10x16x2.6 mm kampron, PLA + 10% GO electrodes	Single layer 10x16x2.6 mm kampron, PLA + 10% GO electrodes	Single layer 10x16x2.6 mm kampron, PLA + 10% GO electrodes	Single layer 10x16x2.6 mm kampron, PLA + 10% GO electrodes
Temperature [°C]	22 (laboratory and sample)	22 (laboratory and sample)	22 (laboratory and sample)	22 (laboratory) 34 (sample)
Pressure [hPa]	999	974	970	970
Humidity [%]	37	33	30	35
Subject of testing	Force applied to the kampron 0 - 50 - 100 N	Medical gasoline (hydrogenated gasoline)	Aceton	A person exhaled gases at 87% humidity
Voltage [numerical value]	171 - 185 - 197	170 - 160	182 - 55	173 - 295
Voltage [V]	0.84 – 0.9 – 0.96	0.83 – 0.78	0.89 – 0.27	0.85 – 1.44

As can be seen from the figures, kampron has demonstrated, as a basis for further research, that convincing results were already obtained in the first stage when measuring GO reactivity. It is assumed that the measurable voltage changes are the result of interactions of the reactants at the contact points of GO with the electrodes. This assumption will be developed in the next stages of the research. The tested method can be considered promising (it is cheap, fast and modifiable), which makes it applicable in future physical applications (sensors, reactors, etc.).

3. CONCLUSION

It has been experimentally verified that the use of GO and its modifications has considerable potential also in the field of sensors, where physical, chemical and biological measurements of stress changes after exposure of the material to various influences have been performed. In the verification, a material called kampron was chosen (the name was assigned by the authors, it is not a previously known material), where single layer 10x16x2.6 mm kampron, PLA + 10% GO electrodes were used. Thus, it can now be stated that the kampron prepared in the first stage of the research (single layer), in conjunction with the electrodes prepared from GO, met the expectations in terms of results. Attention will now be focused on the second and third stages, whereby multilayered helical structures and subsequently much more complex shapes will be tested. The materials used appear to be quite reliable due to their low cost, ease of preparation - and hence reproducibility, which is a key aspect in the experimental validation of the suitability of the materials.

REFERENCES

[1] WANG, H., YUAN, X., ZENG, G., WU, Y., LIU, Y., JIANG, Q., GU, S. Three Dimensional graphene based materials: synthesis and applications from energy storage and conversion to electrochemical sensor and environmental remediation. Adv. Colloid Interface Sci., 2015, Vol. 221, pp. 41-59. Available from: https://doi.org/10.1016/j.cis.2015.04.005

[2] AMIR FAIZ, M.S., CHE AZURAHANIM, C.A., RABA'AH, S.A., RUZNIZA, M.Z. Low cost and green approach in the reduction of graphene oxide (GO) using palm oil leaves extract for potential in industrial applications. Results Phys. 2020, Vol 16, 102954. Available from: https://doi.org/10.1016/j.rinp.2020.102954

[3] IBRAHIM, A., KLOPOCINSKA, A., HORVAT, K., ABDEL HAMID, Z. Graphene-based nanocomposites: synthesis, mechanical properties, and characterizations. Polym. (basel), 2021, Vol. 13. Available from: https://doi.org/10.3390/polym13172869

[4] JI, F., WANG, B., ZHANG, L. Light-Triggered Catalytic Performance Enhancement Using Magnetic Nanomotor Ensembles. Research, 2020, 6380794. Available from: https://doi.org/10.34133/2020/6380794

[5] IACOVACCI, V., RICOTTI, L., SIGNORE, G., VISTOLI, F., SINIBALDI, E., MENCIASSI, A. International Conference on Robotics and Automation (ICRA). 2019, pp 2495-2501. Available from: https://doi.org/10.1109/ICRA.2019.8794322

[6] SRIVASTAVA, S., MEDINA-SANCHEZ, M., KOCH, B., SCHMIDT, O. G. Medibots: Dual-Action Biogenic Microdaggers for Single-Cell Surgery and Drug Release. Adv. Mater. 2016, Vol. 28, pp 832-837. Available from: https://doi.org/10.1002/adma.201504327

[7] ALAPAN, Y., BOZUYUK, U., ERKOC, P., KARACAKOL, A. C., SITTI, M. Multifunctional Surface Microrollers for Targeted Cargo Delivery in Physiological Blood Flow. Sci. Robot. 2020, Vol. 5. Available from: https://doi.org/10.1126/scirobotics.aba5726

[8] VILLA, K., KREJCOVA, L., NOVOTNY, F., HEGER, Z., SOFER, Z., PUMERA, M. Cooperative Multifunctional Self-Propelled Paramagnetic Microrobots with Chemical Handles for Cell Manipulation and Drug Delivery. Adv. Funct. Mater. 2018, Vol. 28, 1804343. Available from: https://doi.org/10.1002/adfm.201804343

[9] ZHANG, Y., YUAN, K., ZHANG, L. Micro/Nanomachines: from Functionalization to Sensing and Removal. Adv. Mater. Technol. 2019, Vol. 4, 1800636. Available from: https://doi.org/10.1002/admt.201800636

[10] JURADO-SANCHEZ, B., PACHECO, M., ROJO, J., ESCARPA, A. Magnetocatalytic Graphene Quantum Dots Janus Micromotors for Bacterial Endotoxin Detection. Angew. Chem. Int. Ed. 2017, Vol. 56, pp 6957-6961. Available from: https://doi.org/10.1002/anie.201701396

[11] PUMERA, M. Graphene in biosensing. Mater. Today. 2011, Vol. 14, pp 308-315. Available from: https://doi.org/10.1016/S1369-7021(11)70160-2

[12] NAG, A., MITRA, A., MUKHOPADHYAY, S. CH. Graphene and its sensor-based applications: A review. Sensors and Actuators A:Physical. 2018, Vol. 270, pp 177-194. Available from: https://doi.org/10.1016/j.sna.2017.12.028

PEROVSKITE $CSPBBR_3$ NANOCRYSTALS WITH OLEIC LIGANDS GRAFTED ON ZNO SURFACES: DFT STUDY OF ELECTRONIC PROPERTIES

[1]Ghulam ABBAS, [1]Egor UKRAINTSEV, [1]Bohuslav REZEK

[1]Faculty of Electrical Engineering, Czech Technical University in Prague, Prague, Czech Republic, EU, abbasghu@fel.cvut.cz

https://doi.org/10.37904/nanocon.2024.4982

Abstract

Semiconductor $CsPbBr_3$ (CPB) perovskite materials exhibit narrow band gap (about 2.5 eV) and possess very fast (hundreds of picoseconds) and strong free exciton luminescence (green at the nanoscale). CPB combination with ZnO microrods can further enhance photoluminescence. In this work the impact of ZnO polar and non-polar surface terminations on the properties of ZnO system with CPB nanocrystals passivated by oleic ligands is investigated. Calculations of structure and electronic properties by density functional theory (DFT) methods are compared on polar surfaces (Zn-face (0001), O-face $(000\bar{1})$), and non-polar surfaces ($(10\bar{1}0)$ and $(11\bar{2}0)$) before and after grafting CPB crystal with oleic ligands (OA, OAm). We find that on Zn-face ZnO surface the attachment of CPB with ligands causes increase of the band gap whereas on the other three ZnO surfaces the attachment of CPB with ligands caused the decrease of the band gap. The position of Fermi level with respect to HOMO and LUMO is also changing, indicating possible charge transfer and doping effects.

Keywords: Perovskite, ligands, ZnO surfaces, density functional theory

1. INTRODUCTION

Semiconductor CPB perovskite have been regarded as exciting materials in recent years for owing exciting structural, electronic and optical properties, low cast fabrication and high power conversion efficiency [1-3]. This unprecedentedly high power conversion efficiency has led towards high absorption coefficient in visible light range, high photoluminescence quantum yields, low rate of electron-hole pair recombination, high bipolar charge carrier mobility, tunable emission wavelength, balanced charge mobility (electron-hole), high carrier lifetimes, narrow full width at half maxima, and high diffusion lengths [4]. CPB is promising material for high-energy radiation detection and optoelectronic applications due to its large carrier diffusion length, high carrier mobility and photophysical properties [5]. Unfortunately, poor stability of CPB hindered its potential applications [6]. To improve the stability of CPB generally different ligands or coating are applied. The oleic ligands (OA, OAm) generate surface passivation, prevent aggregation, enhance dispersion that is useful to enhance the optical properties and most importantly they provide environmental protection to CPB crystals [6]. $CsPbBr_3$ can be produced by changing the ratio of Cs to Pb, and the required phase of the crystal can be controlled by using suitable surface ligands or antisolvent [7]. To synthesize pure $CsPbBr_3$ or $Cs_4PbBr_6/CsPbBr_3$ perovskite composites, ligand-assisted supersaturated recrystallization procedure is used. In this procedure, Oleic acid (OA) and oleylamine (OAm) ligands are added to achieve evolution from $CsPbBr_3$ and intense green light emissive Cs_4PbBr_6/ OA/OAm to Cs_4PbBr_6 [7]. By introducing impurities into perovskite materials and applying computational engineering the optical and electronic properties can be altered, making them a suitable choice for different applications including solar cells [8]. Various strategies are being implemented for surface passivation of CPB through ligand exchange, doping, silica shells and introducing polymer encapsulations to tweak properties for different applications [9].

Recently, zinc oxide (ZnO) has gained valuable attention in different electronic applications where its wide band gap 3.37 eV and superior semiconductor properties provide required advantages [10]. The band gap, antioxidant and antimicrobial properties of zinc oxide can be tailored by doping of different elements [11]. Further, the surface chemistry and chemical stability of ZnO offers a platform for various molecular interactions at molecular and atomic levels [12]. The transparency and availability of ZnO offers its applicability in real-word applications [10]. Direct and wide band gap, well known optical properties, low toxicity, high biocompatibility and luminescence property of ZnO defines it extremely useful in various biomedical, heating and sensing applications [13-15]. It is important to mention that the same material can form different structures on the ZnO substrates depending on the surface terminations and orientations of the surface dipoles [16]. Therefore, it is important to study the impact of ZnO surface terminations on the properties of ZnO system with semiconductor perovskite CPB-like nanocrystals accommodated with oleic ligands to understand the properties of ZnO/CPB nanocrystals. In the current study, we aimed to elucidate the electronic interaction and properties of four different polar and non-polar ZnO surface orientations with and without CPB nanocrystals containing oleic ligands by theoretical calculations.

2. MATERIALS AND METHODS

The interaction of ZnO surface facets with CPB nanocrystals with oleic-ligands at atomic level has been investigated by implementing DFT method in Quantum ATK software. For this purpose, polar surfaces (Zn-face (0001) and O-face $(000\bar{1})$) and non-polar surfaces ($(10\bar{1}0)$ and $(11\bar{2}0)$ facets) were considered separately interacting with CPB slab containing oleic ligands (OA, OAm). To study the impact of ZnO surface terminations on the properties of the system, only one typical CPB slab with one orientation and one arrangement of oleic ligands was used while only the ZnO surface orientations were changed.

For computation of electronic properties of the proposed model, Linear Combination of Atomic Orbital was used as basis function. In this basis function, the generalized gradient approximation and the Perdew–Burke–Ernzerhof exchange–correlation functional were utilized for calculations. The DFT calculations were performed for eleven structures: for three unit cells of ZnO, CPB and CPB with ligands as shown in **Figures 1(a, d, g)**, for four different facets of ZnO, and for four structures, in which ligands were attached with CPB crystals and CPB crystals were placed on ZnO facets.

The ZnO surfaces and CPB slabs were prepared using cleave function in Quantum ATK. Interface builder function available in Quantum ATK was applied to define the bonding between ZnO surfaces, CPB slab and oleic ligands by design. Further, the oleic ligands (OA, OAm) were placed vertically with respect to CPB slab. A vacuum of 20 Å was imposed along the Z-axis to interrupt the lattice periodicity which eventually avoid nonphysical interaction of the wave functions. All systems under consideration were assumed to be periodic in A and B direction. Further the boundary conditions, Neumann to the left and Dirichlet to right along the C direction of the system were applied.

All surfaces of ZnO, CPB slabs and oleic ligands (OA, OAm) were relaxed (optimized) to their minimum energy levels to obtain the stable configuration before merging them together to run the final optimization process. Band structures and density of states (DOS) were calculated for those structures. Changes in electronic properties of the system were analyzed.

3. RESULTS AND DISCUSSION

The resulting optimized ZnO and CPB structures are shown in **Figure 1**. **Figure 1(b)** represents the band structure of optimized ZnO unit cell. This ZnO unit cell exhibits the direct band gap of 1.02 eV. The Fermi level (E_F) at the center confirms the intrinsic behavior with a direct band gap [5]. Similarly, **Figure 1(e)** represents the band structure of optimized CPB unit cell. The optimized CPB unit cell has a direct band gap of 2.75 eV. The value for the band gap obtained in **Figure 1(b)** was verified by DOS analysis of ZnO unit cell and the exact value of the band gap was calculated in **Figure 1(c)**. Similarly, the value of the band gap obtained for optimized CPB unit cell in **Figure 1(e)** was verified by DOS analysis in **Figure 1(f)**. The results obtained for the case of CPB unit cell are in good comparison to the literature [2]. We calculated the band gap for ZnO unit cell at 3.70 eV using also semi empirical approach (results not shown here) and that value is also close to the literature [5]. However, the semi empirical approach does not support the calculations for the combined ZnO/CPB system.

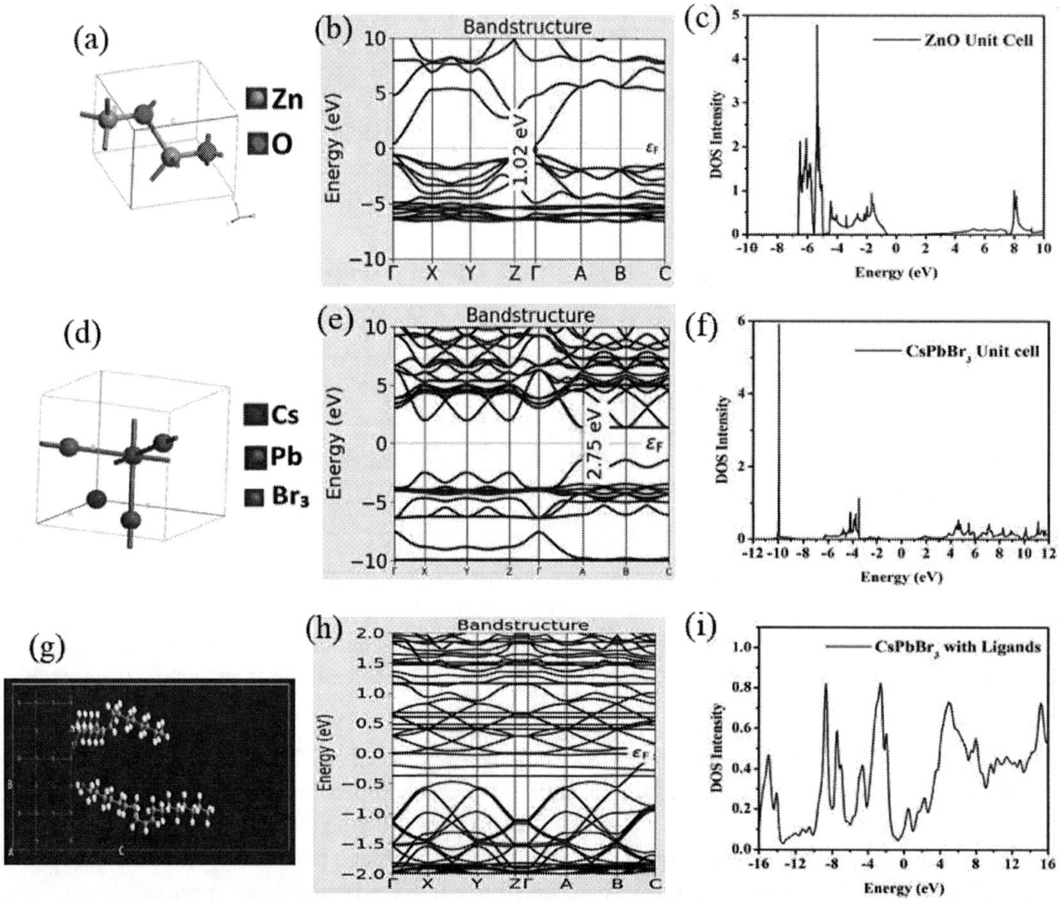

Figure 1 (a) Atomistic model, (b) band structure and (c) DOS analysis of optimized ZnO structure. (d) Atomistic model, (e) band structure and (f) DOS analysis of optimized CPB structure. (g) Atomistic model, (h) band structure and (i) DOS analysis of CPB with oleic ligands (OA/OAm). All calculated using DFT simulations.

Further, the interaction of CPB with oleic ligands (OA/OAm), as shown in **Figure 1(g)**, was considered and by band structure and DOS analysis the band gap of the system was found to be 0.0118 eV as shown in **Figures 1(h, i)**. These results show that when the oleic ligands (OA/OAm) are attached with CPB slab they significantly modify the electronic structure and band gap of CPB. When oleic ligands (OA/OAm) are attached with the surface of CPB possible local polarizations at the surface of oleic ligands (OA/ OAm) and CPB slab affect the electronic properties resulting in the modification of the band gap.

Figure 2 shows the optimized geometry of CPB slab containing oleic ligands (OA/OAm) and grafted on polar and non-polar ZnO slabs. Before further processing, the geometries in **Figure 2** were optimized to obtain the most stable configuration with the lowest energy states. **Figure 3** summarizes the results of simulations for

ZnO polar and non-polar facets ((0001), $(000\bar{1})$, $(10\bar{1}0)$ and $(11\bar{2}0)$) with and without CPB slab containing oleic ligands (OA/OAm). The DOS analysis was used to calculate the total unoccupied positions available for electrons at each energy level and assess the extent of contribution of each part of system towards the formation of HOMO and LUMO [17]. The area on the left with negative energy values represents the HOMO, while the region on the right with positive energy values represents the LUMO. The region of the band gap is present between these two regions in the center of the x-axis. The energy levels of HOMO and LUMO play a crucial role in the process of charge transfer between different parts of a system. A small energy gap between HOMO and LUMO facilitates the excitation and transfer of charge.

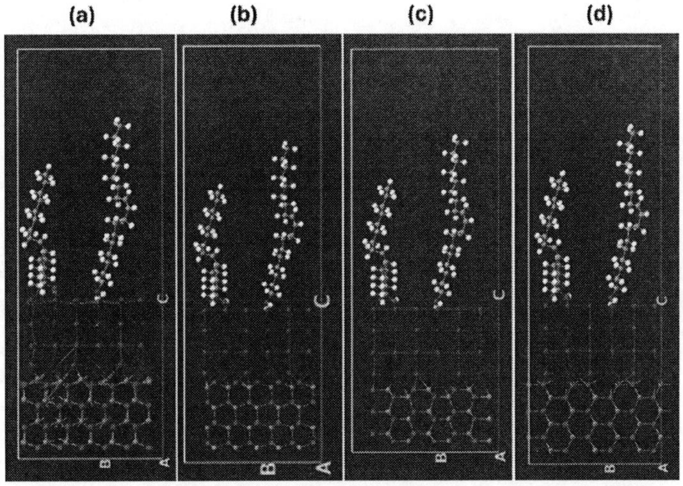

Figure 2 Optimized structural model of oleic ligands (OA, OAm) on CPB grafted on polar (a) Zn-face (0001), (b) O-face $(000\bar{1})$, and non-polar (c) $(10\bar{1}0)$ and (d) $(11\bar{2}0)$ ZnO facets.

Figure 3(a) shows that the band gap for the Zn-face polar slab (0001) is 1.06 eV. As a result of the interaction between the Zn-face polar slab (0001) with CPB slab containing oleic ligands (OA, OAm), the calculations show the value of the band gap increases and reaches 2.09 eV (**Figure 3(b)**). For the second case, the other polar face ZnO $(000\bar{1})$ was considered and after DOS analysis the band gap 1.78 eV was calculated (**Figure 3(c)**). When the CPB slab with oleic ligands was placed the results show the decrease in band gap from 1.78 eV to 0.25 eV (**Figure 3(d)**). Similarly, **Figures 3(e-f)** shows that the band gap for the case of ZnO $(10\bar{1}0)$ was modified from 2.29 eV to 1.98 eV and in **Figures 3(g-h)** for the case of ZnO $(11\bar{2}0)$ the band gap was modified from 4.01 eV to 1.99 eV.

It can be noticed that the placement of CPB slab with oleic ligands on the O-face ZnO resulted in decreasing the band gap significantly which can also result in smooth charge transfer between these moieties which is beneficial for better conductivity. Moreover, the results revealed that the attachment of CPB slab containing oleic ligands (OA, OAm) on Zn-face polar surface (0001) and slight increase of the bandgap is not improving the conductivity. Similar results were achieved using this similar approach on ZnO $(10\bar{1}0)$ and ZnO $(11\bar{2}0)$ facets.

The results in **Figure 3** demonstrate that it is possible to tune the band structure and DOS of the system. The electronic properties of polar and non-polar surfaces are modified in different manners [5]. In our case, all ZnO facets were tuned from their semiconductive nature to more conductive nature when ligands decorated CPB slab was placed on them except for the Zn-face ZnO (0001) surface where the wider band gap was observed. On the other hand, the $(000\bar{1})$ surface ZnO was tuned from 1.78 eV to 0.25 eV indicating drastic reduction in the band gap. These results conclude that the electronic properties of different ZnO facets can be tuned by introducing the CPB slabs passivated by oleic ligands (OA, OAm).

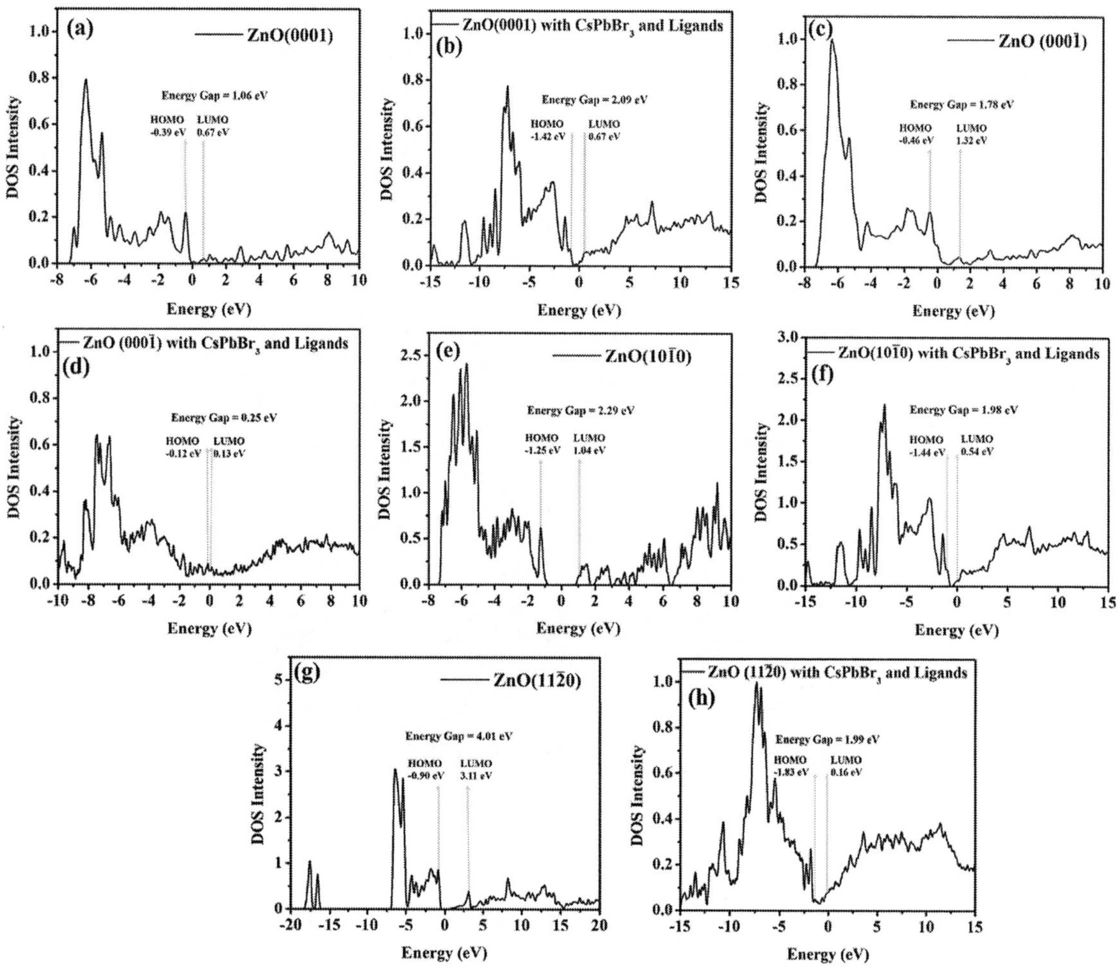

Figure 3 DOS analysis of polar (a) Zn-face (0001), (c) O-face $(000\bar{1})$, and non-polar (e) $(10\bar{1}0)$, (g) $(11\bar{2}0)$ ZnO facets standalone and after grafting CPB with oleic ligands (OA, OAm) on polar (b) Zn-face (0001), (d) O-face $(000\bar{1})$, and non-polar (f) $(10\bar{1}0)$, (h) $(11\bar{2}0)$ ZnO facets.

The flexible surface chemistry of CPB provides the option to dope them with different materials [18] for better conductivity for photovoltaic applications. Similarly, ZnO can be doped with different materials to modify its properties according to the requirements, i.e., Er-doped ZnO can be used for charge trapping process [19]. The luminescence and defects creation can be controlled by doping the ZnO nano and micro-rod with Er as well [20]. Similarly, molybdenum doped ZnO rods can be used to modify the luminescence and charge trapping phenomena of ZnO [21]. Our results show that merging CPB (with oleic ligands) and ZnO is other way to modify the electronic properties for different applications like scintillators or energy harvesting.

4. CONCLUSION

In this study the impact of ZnO surface orientation and atomic terminations on the properties of ZnO interface with semiconductor perovskite CPB nanocrystals passivated by common oleic ligands was investigated by theoretical DFT-based computations to understand the properties of ZnO/CPB interface. The results showed that the electronic properties of the ZnO/CPB system are modified in different ways depending on the ZnO surface orientation. On Zn-face ZnO surface the attachment of CPB with ligands caused the increase of the band gap from 1.06 eV to 2.09 eV. On the other three ZnO surfaces the attachment of CPB with ligands caused the decrease of the band gap. Fermi level position and conductivity of the systems was also varied. Therefore, the electronic properties of ZnO/CPB nanocrystals can be tuned for practical applications in scintillators or energy harvesting. In particular the decrease in the energy band gap between and increase in conductivity can facilitate the excitation charge transfer in the ZnO/CPB system.

ACKNOWLEDGEMENTS

This research was supported by the Grant Agency of the Czech Republic project 24-12872S and by the Ministry of Education Youth and Sports project CZ.02.01.01/00/22_008/0004617 (EcoStor) under the Operational Program Johannes Amos Comenius, call Excellent Research. Assistance of Hadi Hematian with computational settings and procedures is kindly acknowledged.

REFERENCES

[1] YI, Z., LADI, N. H., SHAI, X., LI, H., SHEN, Y. and WANG, M. Will organic–inorganic hybrid halide lead perovskites be eliminated from optoelectronic applications? *Nanoscale Adv.* 2019, vol. 1, pp. 1276–89.

[2] CHAILOO, Z., NOSHAD, H. and DOROUDI, A. Ab-initio investigation of the structural, electronic and optical properties of $CsPbBr_3$ perovskite under pressure using the DFT-1/2 method. *Physica B: Condensed Matter.* 2024, vol. 686, 415999.

[3] JOSHI, H., THAPA, R. K., LAREF, A., SUKKABOT, W., PACHUAU, L., VANCHHAWNG, L., GRIMA-GALLARDO, P., MUSA SAAD H-E, M. and RAI, D. P. Electronic and optical properties of cubic bulk and ultrathin surface [001] slab of $CsPbBr_3$. *Surfaces and Interfaces.* 2022, vol. 30, 101829.

[4] EZZELDIEN, M., AL-QAISI, S., ALROWAILI, Z. A., ALZAID, M., MASKAR, E., ES-SMAIRI, A., VU, T. V. and RAI, D. P. Electronic and optical properties of bulk and surface of $CsPbBr_3$ inorganic halide perovskite a first principles DFT 1/2 approach. *Sci Rep.* 2021, vol. 11, 20622.

[5] KHAN, T., BARANETS, S., GARTIA, M. R., WANG, J. and SHARMA, J. Mapping and Characterization of Local Structures of $CsPbBr_3$. *ACS Omega.* 2024, vol. 9, pp. 35789–97.

[6] LIU, Y., LI, Y., HU, X., WEI, C., XU, B., LENG, J., MIAO, H., ZENG, H. and LI, X. Ligands for $CsPbBr_3$ perovskite quantum dots: The stronger the better? *Chemical Engineering Journal.* 2023, vol. 453, 139904.

[7] LIANG, W., LI, T., ZHU, C. and GUO, L. Ligand-mediated evolution from CsPbBr3 to $Cs4PbBr_6/CsPbBr_3$ perovskite composites with intense green emission. *Optik.* 2022, vol. 267, 169705.

[8] MALEKA, P., DIMA, R., TSHWANE, D., NTWAEABORWA, O. and MAPHANGA, R. Phase Separation of Br-Doped $CsPbI_3$: A Combined Cluster Expansion, Monte Carlo, and DFT Study. *Molecules.* 2023, vol. 29, no.1, 92.

[9] C. G., S., JYOTHI, M. S. and BALAKRISHNA, R. G. Stabilization of $CsPbBr_3$ quantum dots for photocatalysis, imaging and optical sensing in water and biological medium: a review. *J. Mater. Chem. C.* 2022, vol. 10, pp. 6935–56.

[10] HEMATIAN, H., UKRAINTSEV, E. and REZEK, B. Strong Structural and Electronic Binding of Bovine Serum Albumin to ZnO via Specific Amino Acid Residues and Zinc Atoms. *ChemPhysChem.* 2022, vol. 23, e202100639.

[11] YADESA, D., GUYASA, J. N. and BEYENE, T. T. Effect of Codoping Zinc Oxide Nanoparticles with Sulfur and Nitrogen on Its Energy Bandgap, Antioxidant Properties, and Antibacterial Activity. *Advances in Materials Science and Engineering.* 2024, vol. 2024, pp. 1–11.

[12] KRÓL-GÓRNIAK, A., RAILEAN, V., POMASTOWSKI, P., PŁOCIŃSKI, T., GLOC, M., DOBRUCKA, R., KURZYDŁOWSKI, K. J. and BUSZEWSKI, B. Comprehensive study upon physicochemical properties of bio-ZnO NCs. *Sci Rep.* 2023, vol. 13, 587.

[13] PROCHÁZKOVÁ, L., VANĚČEK, V., ČUBA, V., PJATKAN, R., MARTINEZ-TURTOS, R., JAKUBEC, I., BURYI, M., OMELKOV, S., AUFFRAY, E., LECOQ, P., MIHÓKOVÁ, E. and NIKL, M. Core–shell ZnO:Ga-SiO_2 nanocrystals: limiting particle agglomeration and increasing luminescence via surface defect passivation. *RSC Adv.* 2019, vol. 9, pp. 28946–52.

[14] ALVARADO, J. A., GONZALEZ, G. S. A. and CAO, G. Rapid and efficient mineralization of methylene blue dye (above 95 % in 20 min) catalyzed by ZnO nanoparticles under sunlight. *Materials Today Sustainability.* 2024, vol. 27, 100883.

[15] GODSE, P. R., KADAM, S. A., NIMBALKAR, T. M., JADHAV, Y. M., JADHAO, Y. B., MA, Y.-R. and PATIL, V. B. Ultra-responsive and highly sensitive 1D ZnO nanotubes for detecting perilous low levels of NO_2 gas. *Mater. Adv.* 2024, vol. 5, pp. 2826–40.

[16] UKRAINTSEV, E., HEMATIAN, H. and REZEK, B. Polarization Controlled Assembly of Ultrathin Thiorphan Nanostructures on ZnO Surface Facets. *Langmuir*. 2023, vol. 39, pp. 1764–74.

[17] WAQAS, M., HADIA, N. M. A., SHAWKY, A. M., MAHMOOD, R. F., ESSID, M., ALOUI, Z., ALATAWI, N. S., IQBAL, J. and KHERA, R. A. Theoretical framework for achieving high Voc in non-fused non-fullerene terthiophene-based end-capped modified derivatives for potential applications in organic photovoltaics. *RSC Adv.* 2023, vol. 13, pp. 7535–53.

[18] SHAHZAD, M. K., HUSSAIN, S., ASHRAF, G. A., KHAN, M. R., TIRTH, V., ALQAHTANI, H., ALGAHTANI, A., AL-MUGHANAM, T., KHALIL, A. and AZEEM, W. Computational investigation on physical properties of lead based perovskite $RPbBr_3$ (R = Cs, Hg, and Ga) materials for photovoltaic applications. *Sci Rep.* 2024, vol. 14, 19692.

[19] BURYI, M., BABIN, V., REMEŠ, Z. and MIČOVÁ, J. Charge trapping processes in hydrothermally grown Er-doped ZnO. *Radiation Measurements*. 2022, vol. 150, 106700.

[20] BURYI, M., REMEŠ, Z., BABIN, V., ARTEMENKO, A., CHERTOPALOV, S. and MIČOVÁ, J. Cold plasma treatment of ZnO:Er nano- and microrods: The effect on luminescence and defects creation. *Journal of Alloys and Compounds*. 2022, vol. 895, 162671.

[21] BURYI, M., BABIN, V., ARTEMENKO, A., REMEŠ, Z., DĚCKÁ, K. and MIČOVÁ, J. Hydrothermally grown ZnO:Mo nanorods exposed to X-ray: Luminescence and charge trapping phenomena. *Applied Surface Science*. 2022, vol. 585, 152682.

INFLUENCE OF ER DOPING LEVEL AND POST-GROWTH PLASMA TREATMENT ON LUMINESCENCE PROPERTIES OF ZNO NANO- AND MICRORODS

[1]Maksym BURYI, [1]Shelja SHARMA, [2]Zdeněk REMEŠ, [3]Júlia MIČOVÁ

[1]Institute of Plasma Physics of the Czech Academy of Sciences, Prague, Czech Republic, EU, buryi@ipp.cas.cz, sharma@ipp.cas.cz

[2]Institute of Physics of the Czech Academy of Sciences, Prague, Czech Republic, EU, remes@fzu.cz

[3]Institute of Chemistry, Slovak Academy of Sciences, Bratislava, Slovakia, EU, Julia.Micova@savba.sk

https://doi.org/10.37904/nanocon.2024.4983

Abstract

Paper deals with the important question of the influence of plasma treatment on the luminescence properties of the ZnO nano- and microrods especially after the annealing in air at 700 °C. The annealing in air at 700 °C significantly suppresses excitonic luminescence (it is practically vanquished). However, hydrogen plasma treatment makes the excitonic luminescence to raise up again surpassing the initial level of intensity in the as grown sample. The influence of Er content on this effect has been studied. Temperature dependence of the excitonic band has been measured as well on example of the Er doping level at 0.25%. Interestingly, hydrogen plasma treatment applied after the annealing in air at 700 °C had strong influence on the excitonic intensity of the photon energy distribution as a function of temperature. Moreover, the effect of annealing in air and subsequent hydrogen plasma treatment on the allowed infrared 1.54 μm erbium transition is quite opposite. Exposure to hydrogen plasma leads after the annealing in air at 700 °C to the about double reduction of the 1.54 μm erbium transition. This phenomenon practically does not depend on Er content.

Keywords: ZnO nano- and microrods, plasma treatment, photoluminescence, excitonic band, defects-related band

1. INTRODUCTION

Zinc oxide is one of the most studied materials all over the world, still gaining attention due to its magnificent properties including wide direct bandgap (3.37 eV), high thermal conductivity (50 W/mK) and high exciton binding energy (60 meV at 300 K) [1–3]. Moreover, some other properties give a credit to ZnO creating an added value. These are anticorrosion and antimicrobial abilities [4–7]. Altogether, all of these makes zinc oxide, especially, at the nanoscale a promising material for various applications such as scintillators and nanomedicine [8–11], photovoltaics and photocatalysis [12–14] and many others.

Er supports growth of ZnO in the form of microrods. The larger the Er content (in the sequence 0.05 – 1%) the larger the microrods [15]. This and advanced experimental and theoretical studies led to the conclusion that erbium appears in the form of the Er_xO_y quantum dots serving as nucleation seeds [16].

The luminescence of the ZnO host is composed of the excitonic emission (having maximum at about 3.24 eV) and emission of native defects. The latter is observed within the 1-2 eV spectral region [17,18]. The typical defects in the ZnO nano- and microrods grown using hydrothermal growth are the neutral zinc vacancies (V_{Zn} at about 1.8-2 eV) and zinc dangling bonds (Zn_{DB} at about 2.4 eV) [15]. However, luminescence properties of the hydrothermally grown ZnO are strongly affectable by Er doping.

The red as well as the green bands increase upon the annealing in air at elevated temperatures up to 500 °C. Further increase of the annealing temperatures led to the slight decrease of the intensity of these bands [15].

Hydrogen plasma treatment results in the enhancement of the exciton-related luminescence band intensity. At the same time the VZn band intensity is decreasing. These processes depended on the Er doping level. The existence of energy transfer between the VZn and zinc interstitial was observed [19].

However, the influence of subsequent air annealing and hydrogen plasma treatment on the luminescence properties of Er doped ZnO nano- and microrods is unknown. This is the aim of the present work

2. EXPERIMENTAL

All of the analytical-grade reagents were used exactly as supplied, requiring no additional purification. Erbium(III) nitrate pentahydrate $Er(NO_3)_3{\cdot}5H_2O$ was obtained from Sigma-Aldrich, while zinc nitrate hexahydrate ($Zn(NO_3)_2{\cdot}6H_2O$) and hexamethylenetetramine (HMTA, $C_6H_{12}N_4$) were acquired from Slavus. A system called So-Safe Water Technologies, which has a conductivity of 0.20 $\mu S{\cdot}cm^{-1}$ at 25 °C, was used to purify deionized water. The hydrothermal growth method was used to create undoped and Er-doped ZnO nanorods with nominal compositions ZnO:Er (0.05, 0.25, 1%). First, using a magnetic stirrer, 200 ml of deionized water was used to dissolve the equivalent stoichiometric quantities of $Zn(NO_3)_2{\cdot}6H_2O$ and $Er(NO_3)_3{\cdot}5H_2O$, respectively. A Whatman 2 filter was used to filter the prepared solutions. For fifteen minutes, the filtered solutions were mixed and agitated at 400 rpm. In the final suspensions, the ZnO:Er(0.05, 0.25, 1%) had a nominal concentration of 25 mM. A 250 ml aqueous solution of HMTA at a concentration of 25 mM was then made. The solution was filtered through Whatman 2 straight to the $Zn(NO_3)_2{\cdot}6H_2O$ and $Er(NO_3)_3{\cdot}5H_2O$ mixing solutions after being stirred for 15 minutes. After stirring the reaction mixture, nanorods were hydrothermally grown by heating it to 90°C for three hours. After being separated and purified (by removing any remaining salts), the produced nanorods were centrifuged for 15 minutes at 18,000 rpm (RCF: 23,542 ×g) and then rinsed five times with deionized water. The samples were then lyophilized.

A homemade spectrometer containing the following components was used to study the characteristics of photoluminescence (PL): (a) a pulsed ultraviolet (UV) light emitting diode (LED) with a wavelength of 340 nm and a power of 1 mW; (b) a narrow band pass optical filter; (c) a Peltier cooled photomultiplier that is sensitive in the 355-800 nm spectral range (2 nm spectral resolution); (d) a spectrally calibrated double gratings monochromator SPEX 1672; (e) long-pass filters; (f) a lock-in amplifier referenced to the LED frequency (307 Hz).

Luminescence spectra are presented in the energy scale by using the Jacobian correction to convert from wavelength scale.

3. RESULTS AND DISCUSSIONS

UV-visible PL spectra measured in the ZnO:Er(0.05, 0.25 and 1%) samples annealed in air art 700 °C are shown in **Figure 1A**. They are composed of at least three contributions/bands appearing at approximately 1.82 eV, 2.4 eV and 3.25 eV. The origin of these bands is neutral zinc vacancy (V_{Zn}), zinc dangling bonds (Zn_{DB}) and excitonic emission (En), respectively [15]. It is known from previous studies [15] that the V_{Zn} in the as grown ZnO:Er(Mo) samples has the tendency to increase upon the annealing temperature reaching maximum intensity in the range 500-700 °C (the intensity is doubled as compared to the as grown samples). Moreover, annealing in air is known to strongly reduce or even extinguish the excitonic emission.

The intensity of the V_{Zn}, Zn_{DB} and En bands depends on the Er doping level. The intensity of the V_{Zn} and Zn_{DB} bands remains unchanged in the ZnO:Er(0.05, 0.25%) samples whereas it is about twice as large in the ZnO:Er(1%) sample. In contrast, the weak En band in the ZnO:Er(0.05%) is almost totally suppressed in the ZnO:Er(0.25, 1%). These are expected tendencies reported in [15]. They serve as a reference to the same samples annealed in air at 700 °C and subsequently treated by cold Ar + 5%H_2 plasma (see Experimental). The corresponding PL spectra are shown in **Figure 1B**.

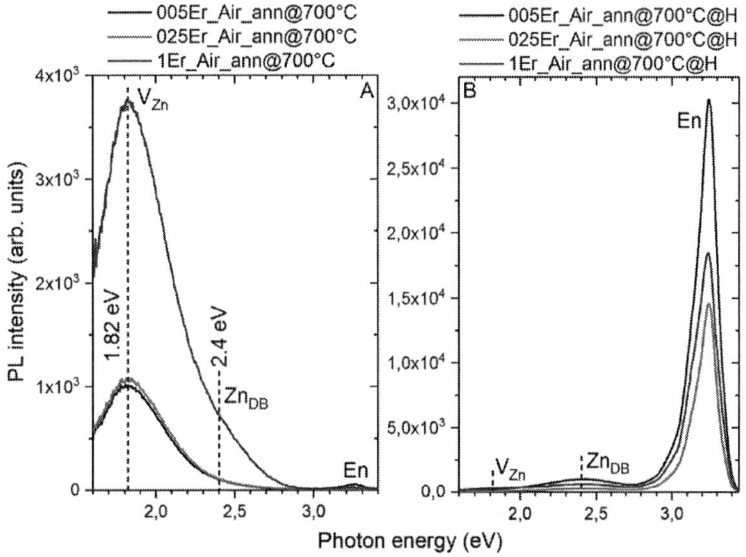

Figure 1 A - UV-visible PL spectra measured in the samples of ZnO:Er(0.05, 0.25 and 1%) annealed in air at 700 ℃ as also indicated in the legend. B - UV-visible PL spectra measured in the samples of ZnO:Er(0.05, 0.25 and 1%) annealed in air art 700 ℃ and subsequently hydrogen plasma treated as also indicated in the legend. En, ZnDB and VZn indicate specific emission bands originating from the excitonic luminescence zinc dangling bonds and neutral zinc vacancy [15].

They exhibit different trends as compared to the only annealed samples. The V_{Zn} band appeared about one order of magnitude suppressed in the ZnO:Er(0.25, 1%) while it was only fivefold weaker in the ZnO:Er(0.05%) after the plasma treatment. Oppositely, the Zn_{DB} band was about one order of magnitude and triple increased in the ZnO:Er(0.05%) and ZnO:Er(0.25%), respectively, while it remained unchanged in the ZnO:Er(1%) after the plasma treatment. This has never been observed before in the as grown samples of ZnO:Er(0.05, 0.25, 1%) nano- and microrods [15].

The En band became visible and very strong after the plasma treatment. The strongest it was in the ZnO:Er(0.05%). It is about twice as weak in the ZnO:Er(0.25,1%) as compared to the ZnO:Er(0.05%) sample.

All of these indicate strong influences of the annealing in air followed by plasma treatment as well as erbium doping. Recently, Er_xO_y-like structures (quantum dots, QD) have been reported to serve as the nucleation seeds for the ZnO nano- and microrods growth [16]. Er is a donor and thus can contribute its electron through the ZnO/Er_xO_y boundary to ZnO. This electron can change the charge balance appearing at a neutral V_{Zn} making it negative. The larger the Er content, the larger the number of donor electrons. Note that it has previously been found that at doping eve of 1% only about one half of the initial erbium remains in the ZnO material. Ar + 5%H_2 plasma is a reductive medium and injects electrons into the structure as well. Therefore, the V_{Zn} band exhibits the observed trends (**Figure 1B**). At the same time, hydrogen should interact with oxygen removing it thus contributing to the Zn_{DB} and oxygen vacancies (V_O) creation. Note that the hydrogen interacting with the surface of ZnO is not only in the form of a molecule (H_2) but it is atomic (H) and ionic (H^+). Therefore, one may expect high reactivity, much higher than in the case of the annealing in the reductive atmosphere. Oxygen is also building element for the Er_xO_y QDs. The lower the amount of these QDs the stronger the effect of hydrogen on the ZnO structure. The influence of the hydrogen plasma on the Er_xO_y QDs in this way can also be proven by measuring infrared PL of the main 1.54 μm Er^{3+} transition in all the samples. Note, that the existence of a shielding effect in ZnO:Er has been discussed in [15].

The IR PL spectra (the main 1.54 μm Er^{3+} transitions) of the samples studied (only annealed in air at 700 ℃ and also subsequently Ar + 5%H_2 plasma treated) are shown in **Figure 2**.

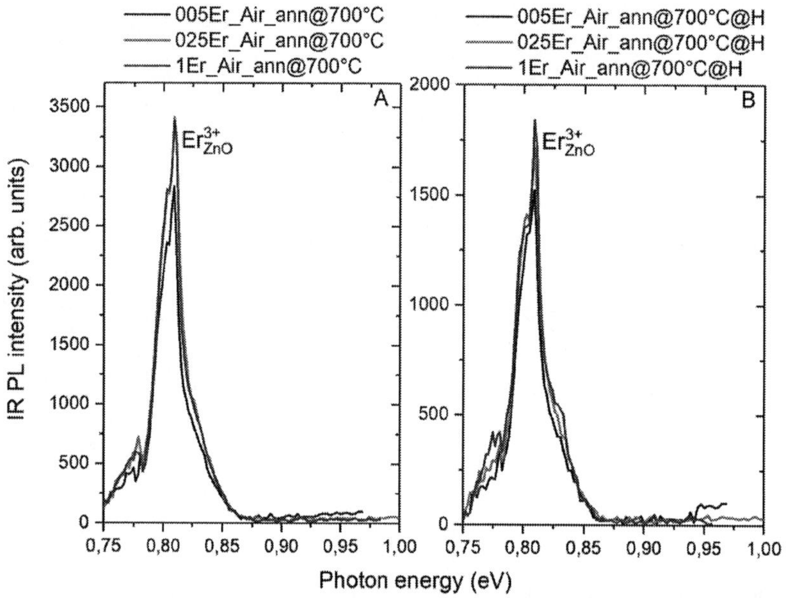

Figure 2 A - IR PL spectra measured in the samples of ZnO:Er(0.05, 0.25 and 1%) annealed in air at 700 ℃ as also indicated in the legend. B - IR PL spectra measured in the samples of ZnO:Er(0.05, 0.25 and 1%) annealed in air art 700 ℃ and subsequently hydrogen plasma treated as also indicated in the legend.

Naturally, the intensity of the main 1.54 µm Er^{3+} transitions is increasing but this trend is very weak in both only annealed and annealed and plasma treated samples. Due to this one may expect energy transfer or upconversion. Note that the maximum energy of the main Er transition is about 0.81 eV. Doubled it corresponds to 1.62 eV – the very beginning of the broad V_{Zn}. Unfortunately, this range is unreachable in the apparatus used (see Experimental). Therefore, the upconversion effect will be studied separately.

Ar + 5%H_2 plasma treatment resulted in the double decrease of the main 1.54 µm Er^{3+} transitions confirming the oxygen removal from the Er_xO_y QDs. Partial reduction of Er^{3+} to metal Er as well as Er^{2+} creation cannot be excluded in this case (see [16]).

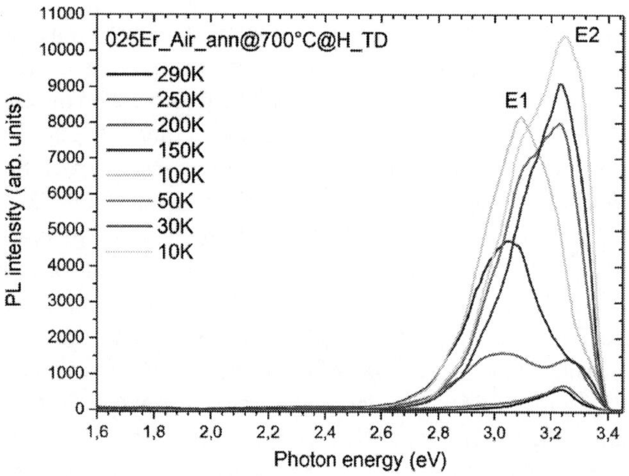

Figure 3 Temperature dependence of the En band measured in the ZnO:Er(0.25%) sample annealed in air at 700 ℃ with the subsequent plasma hydrogenation. E1,2 are specific contributions.

To get better insight into the effect of the plasma hydrogenation on the En band, it has been measured at low temperatures. The corresponding spectra are shown in **Figure 3** on example of the ZnO:Er(0.25%) where the intensity of this band was the weakest among the samples.

It can be seen that the spectrum is composed of at least two complex bands appearing at about 3 eV (E1) and 3.24 eV (E2). The second one originates from excitons. The first one can have the excitonic origin as well as it might have been the interstitial oxygen emission (has maximum at about 2.9 eV) [15]. However, the latter is not probable considering plasma hydrogenation. Hydrogen radicals and ions would rather remove oxygen as discussed above. Therefore, the component E1 is expected to originate from bound excitons and donor-acceptor pairs (DAP).

The E1 band has the tendency to raise as a function of temperature in the range 290-100 K. Then it drops down to the 50-10 K range. The E2 band is increasing in intensity and redshifting to 3.2 eV practically all the time in the 290-10 K range. This indicates the presence of some defects having energy levels at the bottom of the conduction band, most probably, hydrogen. These levels participate in the exciton luminescence processes. Their influence becomes even more pronounced at lower temperatures due to the well-known effect of the bandgap broadening upon cooling. In the case of DAP responsible for the E1 band one could expect electrons to transfer to the excitonic excited states through the hydrogen levels, the effect similar to tunnelling.

4. CONCLUSION

The known observation that the annealing in air at 700 °C significantly suppresses excitonic luminescence (3.24 eV) and at the same time improves defects-related red (1.82 eV) and green (2.4 eV) luminescence in the hydrothermally grown ZnO nano- and microrods has been repeated in the present work. Furthermore, the application of Ar + 5%H2 plasma after the annealing resulted in the strong increase of the excitonic emission (several orders of magnitude), suppression of the red (by about one order of magnitude) and improvement of the green emission (by about one order of magnitude). Er doping level had a strong influence on the luminescence intensity of the mentioned bands improving the defects-related and suppressing the excitonic ones. The observed effects were explained by the participation of the ErxOy QDs serving as nucleation seeds for the ZnO nano- and microrods growth in luminescence properties injecting electrons into the ZnO structures. Plasma hydrogenation cuts off some, probably, bridging, oxygen ions from the ErxOy QDs thus at least partly breaking the bonds between ZnO and ErxOy. This has been proven also by measuring IR PL of the 1.54 µm main Er3+ transition – its intensity has been doubly decreased after the plasma treatment. At the same time plasma dopes the material with hydrogen. This was confirmed by measuring the temperature dependence of the excitonic luminescence. There, the clearly observed charge/energy transfer between excitonic states of different origin through donor levels has been observed.

ACKNOWLEDGEMENTS

The financial support of the Czech Science Foundation project No. 24-12872S and the program "Strategy AV 21" of the Czech Academy of Sciences, specifically work package VP 27 Sustainable Energy (Renewable energy resources and distributed energy systems) are gratefully acknowledged

REFERENCES

[1] AWASTHI, K. Ed. *Nanostructured zinc oxide: synthesis, properties and applications*, 1st ed., Elsevier, Cambridge, 2021.

[2] FEDOTOV, A. K., PASHKEVICH, A. V., FEDOTOVA, J. A., FEDOTOV, A. S., KOŁTUNOWICZ, T. N., ZUKOWSKI, P., RONASSI, A. A., FEDOTOVA, V. V., SVITO, I. A., BUDZYŃSKI, M. Electron transport and thermoelectric properties of ZnO ceramics doped with Fe. *Journal of Alloys and Compounds*. 2021, vol. 854, p. 156169.

[3] PASHKEVICH, A. V., FEDOTOV, A. K., PODDENEZHNY, E. N., BLIZNYUK, L. A., FEDOTOVA, J. A., BASOV, N. A., KHARCHANKA, A. A., ZUKOWSKI, P., KOLTUNOWICZ, T. N., KOROLIK, O. V., FEDOTOVA, V. V. Structure, electric and thermoelectric properties of binary ZnO-based ceramics doped with Fe and Co. *Journal of Alloys and Compounds*. 2022, vol. 895, p. 162621.

[4] SÁNCHEZ-LÓPEZ, A. L., PERFECTO-AVALOS, Y., SANCHEZ-MARTINEZ, A., CEBALLOS-SANCHEZ, O., SEPULVEDA-VILLEGAS, M., RINCÓN-ENRÍQUEZ, G., RODRÍGUEZ-GONZÁLEZ, V., GARCIA-VARELA, R., LOZANO, L. M., ELOYR NAVARRO-LÓPEZ, D., SANCHEZ-ANTE, G., CORONA-ROMERO, K., LÓPEZ-MENA, E. R. Influence of erbium doping on zinc oxide nanoparticles: Structural, optical and antimicrobial aktivity. *Applied Surface Science*. 2022, vol. 575, p. 151764.

[5] ASKAR, A. A., SELIM, M. S., EL-SAFTY, S. A., HASHEM, A. I., SELIM, M. M., SHENASHEN, M. A. Antimicrobial and immunomodulatory potential of nanoscale hierarchical one-dimensional zinc oxide and silicon carbide materials. *Materials Chemistry and Physics*. 2021, vol. 263, p. 124376.

[6] SELIM, M. S., EL-SAFTY, S.A., ABBAS, M. A., SHENASHEN, M. A. Facile design of graphene oxide-ZnO nanorod-based ternary nanocomposite as a superhydrophobic and corrosion-barrier coating. *Colloids and Surfaces A: Physicochemical and Engineering Aspects*. 2021, vol. 611, p. 125793.

[7] SELIM, M. S., SHENASHEN, M. A., ELMARAKBI, A., EL-SAEED, A. M., SELIM, M. M., EL-SAFTY, S. A. Sunflower oil-based hyperbranched alkyd/spherical ZnO nanocomposite modeling for mechanical and anticorrosive applications. *RSC Adv.* 2017, vol. 7, p. 21796.

[8] D'AGOSTINO, D., DI GIORGIO, C., BOBBA, F., DI TROLIO, A., ALIPPI, P., CUCOLO, A. M., AMORE BONAPASTA, A. Effects of cobalt substitution on ZnO surface reactivity and electronic structure. *J. Mater. Chem. C* 2019, vol. 7, p. 8364.

[9] OLEJNIK, M., KERSTING, M., ROSENKRANZ, N., LOZA, K., BREISCH, M., ROSTEK, A., PRYMAK, O., SCHÜRMEYER, L., WESTPHAL, G., KÖLLER, M., BÜNGER, J., EPPLE, M., SENGSTOCK, C. Cell-biological effects of zinc oxide spheres and rods from the nano- to the microscale at sub-toxic levels. *Cell Biol Toxicol.* 2021, vol. 37, p. 573.

[10] KAPAT, K., SHUBHRA, Q. T. H., ZHOU, M., LEEUWENBURGH, S. Piezoelectric Nano-Biomaterials for Biomedicine and Tissue Regeneration. *Adv. Funct. Mater.* 2020, vol. 30, p. 1909045.

[11] SIEBERT, L., LUNA-CERÓN, E., GARCÍA-RIVERA, L. E., OH, J., JANG, J., ROSAS-GÓMEZ, D. A., PÉREZ-GÓMEZ, M. D., MASCHKOWITZ, G., FICKENSCHER, H., OCEGUERA-CUEVAS, D., HOLGUÍN-LEÓN, C. G., BYAMBAA, B., HUSSAIN, M. A., ENCISO-MARTÍNEZ, E., CHO, M., LEE, Y., SOBAHI, N., HASAN, A., ORGILL, D. P., MISHRA, Y. K., ADELUNG, R., LEE, E., SHIN, S. R. Light-Controlled Growth Factors Release on Tetrapodal ZnO-Incorporated 3D-Printed Hydrogels for Developing Smart Wound Scaffold. *Adv. Funct. Mater.* 2021, vol. 31, p. 2007555.

[12] YANG, C., ZHAN, S., LI, Q., WU, Y., JIA, X., LI, C., LIU, K., QU, S., WANG, Z., WANG, Z. Systematic investigation on stability influence factors for organic solar cells. *Nano Energy*. 2022, vol. 98, p. 107299.

[13] PEZHOOLI, N., RAHIMI, J., HASTI, F., MALEKI, A. Synthesis and evaluation of composite TiO_2@ZnO quantum dots on hybrid nanostructure perovskite solar cell. *Sci Rep.* 2022, vol. 12, p. 9885.

[14] LAOKAE, D., PHURUANGRAT, A., WANNAPOP, S., DUMRONGROJTHANATH, P., THONGTEM, T., THONGTEM, S. Preparation, characterization and photocatalytic properties of Er-doped ZnO nanoparticles synthesized by combustion method. *International Journal of Materials Research*. 2023, vol. 114, p. 34.

[15] BURYI, M., REMEŠ, Z., BABIN, V., ARTEMENKO, A., VANĚČEK, V., AUBRECHTOVÁ DRAGOUNOVÁ, K., LANDOVÁ, L., KUČERKOVÁ, R., MIČOVÁ, J. Transformation of free-standing ZnO nanorods upon Er doping. *Applied Surface Science*. 2021, vol. 562, p. 150217.

[16] BURYI, M., REMEŠ, Z., HÁJEK, F., KULDOVÁ, K., BABIN, V., DĚCKÁ, K., HEMATIAN, H., LANDOVÁ, L., NEYKOVA, N., HORYNOVÁ, E., REZEK, B. Peculiarities Related to Er Doping of ZnO Nanorods Simultaneously Grown as Particles and Vertically Arranged Arrays. *J. Phys. Chem. C* 2023, vol. 127, p. 22177.

[17] FRODASON, Y. K., JOHANSEN, K. M., BJØRHEIM, T. S., SVENSSON, B. G., ALKAUSKAS, A. Zn vacancy as a polaronic hole trap in ZnO. *Phys. Rev. B* 2017, vol. 95, p. 094105.

[18] LYONS, J. L., VARLEY, J. B., STEIAUF, D., JANOTTI, A., VAN DE WALLE, C. G. First-principles characterization of native-defect-related optical transitions in ZnO. *Journal of Applied Physics*. 2017, vol. 122, p. 035704.

[19] BURYI, M., REMEŠ, Z., BABIN, V., ARTEMENKO, A., CHERTOPALOV, S., MIČOVÁ, J. Cold plasma treatment of ZnO:Er nano- and microrods: The effect on luminescence and defects creation. *Journal of Alloys and Compounds*. 2022, vol. 895, p. 162671.

SELF-ASSEMBLED MONOLAYERS AS HOLE SELECTIVE CONTACTS FOR PEROVSKITE SOLAR CELLS: EXPERIMENT VERSUS SIMULATIONS

[1,2]Meng-Hsueh KUO, [1]Naini JAIN, [1,2]Neda NEYKOVA, [1]Rupendra Kumar SHARMA, [3]Zhivko VELKOV, [4]Chih-Yu CHANG, [1]Jakub HOLOVSKÝ

[1]Centre for Advanced Photovoltaics, CTU in Prague, Faculty of Electrical Engineering, Prague, Czech Republic, EU

[2]Institute of Physics of the Czech Academy of Sciences, Prague, Czech Republic, EU

[3]Centre Faculty of Mathematics and Natural Sciences, South-West University "Neofit Rilski", Blagoevgrad, Bulgaria, EU

[4]Department of Materials Science and Engineering, National Taiwan University of Science and Technology, Taipei, Taiwan

https://doi.org/10.37904/nanocon.2024.4980

Abstract

Recent advances in the performance of perovskite solar cells (PSCs) have been strongly linked to advancements in interfacial engineering and the development of charge-selective contact. Self-assembled monolayers (SAMs) play an important role in inverted PSCs due to their distinctive and versatile ability to manipulate chemical and physical interface properties. Here we compared the dipole moments and energy levels of (2-(9H-carbazol-9-yl)ethyl)phosphonic acid (2PACz), (4-(3,6-Dimethyl-9H-carbazol-9-yl)butyl)phosphonic acid (Me-4PACz), (4-(3,6-Dimethoxy-9H-carbazol-9-yl)butyl)phosphonic acid (MeO-4PACz), and (2-(3,7-Dibromo-10H-phenothiazin-10-yl)ethyl)phosphonic acid (Br-2EPT) molecules and their impact on the energy conversion of PSC devices. Additionally, we simulated the dipole moment and surface charge effect device behavior to support our hypothesis. Finally, we observed that the fill factor behavior in both devices and simulations exhibited a similar trend, which correlated with the dipole moment of SAMs.

Keywords: Self-assembled monolayers, perovskite solar cells, dipole moment, surface charge, simulations

1. INTRODUCTION

Organic-inorganic hybrid metal halide perovskite solar cells have gained significant attention due to their exceptional optoelectronic properties, with the current power conversion efficiency (PCE) surpassing 26% [1,2]. Furthermore, the wide bandgap single-junction PSCs with a composition of $FA_{0.83}Cs_{0.17}Pb(I_{0.6}Br_{0.4})_3$ (1.72 eV) have achieved an efficiency of 19.4% [3], making them well-suited for application as the top cell in tandem solar cells [4]. In addition to engineering the perovskite compositions and controlling crystallization kinetics [5,6], the precise engineering of the interfaces between the perovskite absorber and contact interlayers is crucial for optimizing performance and stability [7]. Ultrathin self-assembled monolayers (SAMs) have recently been successfully employed as hole transport layers (HTLs) in inverted PSCs due to their ability to optimize interfacial properties [8], namely by improving energy level alignment [9], reducing recombination losses [10], enhancing stability [11], and increasing overall device efficiency [11]. The SAMs offer molecular-level control over interface engineering by introducing a permanent dipole moment, which can improve charge transfer across the interface through energy level alignment [7]. Moreover, fixed charges in SAMs can effectively passivate surface traps, minimizing non-radiative recombination and enhancing charge extraction [12]. The combination of dipole moments and fixed charges significantly reduces the energy barriers in the interface, optimizing the local electric field and enhancing charge transport. This leads to a notable improvement in the PCE and long-term stability of the device [13].

Herein, we report the comparison of four carbazole-based SAMs with various dipole moments, highest occupied molecular orbital (HOMO), and lowest occupied molecular orbital (LUMO) energy levels for application in inverted PSCs. As the absorber for PSCs, we chose wide bandgap material with composition $FA_{0.83}Cs_{0.17}Pb(I_{0.6}Br_{0.4})_3$. Furthermore, we also simulated using various tunneling models, including band-to-band tunneling, quantum tunneling, thermionic tunneling, trap-assisted tunneling, and universal Schottky tunneling, to compare our experimental findings more efficiently. Our combined experimental and simulation results demonstrate that SAMs with higher dipole moments lead to improved fill factors in inverted PSCs, while the correlation of device performance with HOMO and LUMO energies was not confirmed.

2. METHODS

2.1 Perovskite films and solar cell fabrication

$FA_{0.83}Cs_{0.17}Pb(I_{0.6}Br_{0.4})_3$ perovskite films were prepared by dissolving 0.83 mmol of FAI, 0.17 mmol of CsI, 0.4 mmol of PbI_2, and 0.6 mmol of $PbBr_2$ in 1 ml of a DMF and DMSO solvent mixture at a 4:1 ratio. The precursor solution was stirred continuously at 70°C. The solution was then deposited onto the substrate via spin-coating, starting at 1000 rpm for 10 seconds, followed by 5000 rpm for 30 seconds. Five seconds before completing the second spin-coating step, 150 µL of chlorobenzene was applied to the substrate. The samples were subsequently annealed at 100°C for 30 minutes. All procedures were conducted in a nitrogen-filled glovebox.

The ITO-coated glass substrates (10 ohm/sq) were washed by ultrasonic cleaning in a 2% Hellmanex detergent solution, followed by rinsing with deionized water, acetone, and ethanol for 15 minutes each. After drying the substrates using a nitrogen gas gun, they were cleaned with a UV-ozone cleaner for 15 minutes. The single-junction cells structure is ITO/SAMs/$FA_{0.83}Cs_{0.17}Pb(I_{0.6}Br_{0.4})_3$/$PC_{61}BM$/BCP/Ag, where different SAMs materials are used: 2PACz, Me-4PACz: PFN-Br (9.5:0.5), MeO-4PACz, or Br-2EPT. The concentration of 2PACz, MeO-4PACz, and Br-2EPT solution were 1 mmol/l in ethanol. To prepare Me-4PACz:PFN-Br solution, both Me-4PACz and PFN-Br were dissolved in methanol at 0.4 mg/ml each. The SAMs were deposited using spin-coating at 4000 rpm for 30 seconds, followed by annealing at 100 °C for 10 minutes. After cooling down, the SAMs were dynamically washed with ethanol at 4000 rpm for 30 seconds and subsequently annealed again at 100 °C for 3 minutes. Next, the perovskite layer deposited onto SAM fabrication as described above. $PC_{61}BM$ at a concentration of 20 mg/mL in chlorobenzene, was then spin-coated onto perovskite layer at 2000 rpm for 30 seconds, followed by BCP, at a concentration of 0.5 mg/mL in IPA, was spin-coated at 5000 rpm for 30 seconds. Finally, an 80 nm Ag back contact was thermally evaporated onto the BCP layer under high vacuum using a shadow mask, with each electrode having an active area of 0.09 cm^2.

2.2 Device characterizations

Current density-voltage (*J-V*) characterization. The *J-V* measurements were carried out with a Keithley 2400 source meter managed by a computer, under a 100 mW/cm^2 illumination. The illumination was provided by a calibrated solar simulator (Wavelabs) giving AM 1.5 spectrum.

External quantum efficiency (EQE). The EQE measurements were conducted using a Bentham TMC 300 monochromator, a 50 W tungsten halogen lamp, SR570 preamplifier, NI DAQ A-D convertor and a software lock-in analyser.

Photoluminescence (PL). The PL intensity spectra were measured using a fluorescence spectrophotometer (Cary Eclipse) with a 300 nm excitation wavelength.

Scanning electron microscopy (SEM). The SEM measurements were carried out to determine the thickness of the perovskite thin films by examining the sample cross sections using the MAIA 3 scanning electron microscope from TESCAN at a voltage of 5 kV.

All measurements were conducted in an ambient atmosphere at room temperature without encapsulation.

2.3 Simulation

The simulated PSCs include ITO/SAMs/Perovskite/PCBM/BCP/Ag stacks on 400 nm thick $FA_{0.83}Cs_{0.17}Pb(I_{0.6}Br_{0.4})_3$ perovskite absorber, with ambipolar doping (10^{14} cm^{-3}), band gap (1.75 eV), and affinity (4.0 eV). The front and rear contact consists of ITO/SAMs and BCP/Ag interfaces, with a work function of approximately 5.1 eV and 4.0 eV. The DFT calculated parameters for SAM molecules were used in TCAD simulations for analysing the SAMs dipole moment effects on the performance of PSCs.

2.4 Density Functional Theory Calculations

To determine the values of dipole moment and HOMO, LUMO energies Density Functional Theory (DFT) calculations were performed. The calculations of dipole moments were carried out using the DFT [14], as implemented in the Gaussian09 program package. The optimization of the geometry was performed with the Becke 3-parameter hybrid exchange functional combined with the Lee et al. correlation functional (B3LYP) [15,16] with the standard double-ζ 6-31G(d,p) basis set [17]. The optimization was achieved without any geometry constraints. All possible intramolecular interactions were taken into account in the initial geometries. For all structures, the harmonic vibrational frequencies were computed to confirm the true minima on the calculated potential surface.

3. RESULT AND DISSCUSSION

Figure 1 (a) presents the molecular structure of the compounds used in this study: 2PACz, Me-4PACz, MeO-4PACz, and Br-2EPT. We considered 2PACz as a reference due to its extensive study in p-i-n structured PSC using SAMs. To understand the device behaviour with energy level alignment and charge dynamics, we first calculate the dipole moment, E_{HOMO}, and E_{LUMO} parameters for each SAM. As shown in **Table 1**, MeO-4PACz and Br-2EPT exhibit higher dipole moments compared to 2PACz and Me-4PACz, with values exceeding 3.5 debye. Notably, under certain conditions, a higher dipole moment can enhance the efficiency of perovskite solar cells by improving charge separation, reducing recombination, and optimizing energy band alignment. **Figure 1 (b)** presents a schematic of the band edge positions for the SAMs and perovskite. Proper alignment of the band edges, along with a moderate amount of fixed charge at the interface, can facilitate the separation of photo-generated electron-hole pairs. This separation is driven by an electric field, which directs charges in opposite directions, thereby minimizing recombination and enhancing the efficiency of solar cells.

Table 1 Dipole moment, E_{HOMO}, and E_{LUMO} parameters of SAMs

SAMs	Dipole moment (Debye)	E_{HOMO} (eV)	E_{LUMO} (eV)
2PACz	1.94	-5.44	-0.75
Me-4PACz	1.65	-5.15	-0.60
MeO-4PACz	3.61	-4.98	-0.76
Br-2EPT	3.78	-5.32	-1.02

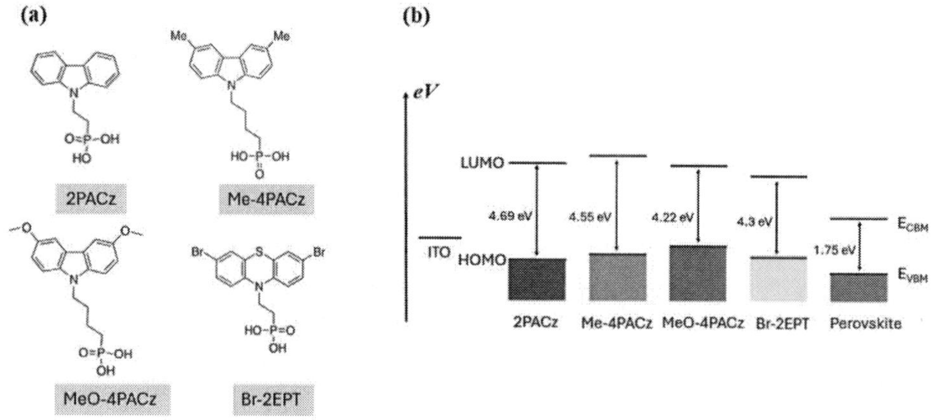

Figure 1 (a) Molecule structure of SAMs and (b) schematic of the band edge positions for the ITO, SAMs, and perovskite

To understand the effects of dipole moments and surface charges in solar devices, we fabricated four p-i-n SAMs PSCs (**Figure 2 (a)**) and analyzed their optical performance. Notably, Me-4PACz poses a hydrophobicity challenge during perovskite formation. To address this issue and enhance the quality of the perovskite film, we mixed Me-4PACz with PFN-Br in a ratio of 9.5:0.5 [18]. In **Figure 2 (b)**, the PL spectra of perovskite on various glass/SAM substrates are presented. The PL peak intensity at approximately 698 nm is slightly lower in the higher dipole moment MeO-4PACz and Br-2EPT samples, demonstrating more efficient charge extraction, compared to the 2PACz and Me-4PACz samples exhibiting lower charge extraction efficiency. This behavior of SMAs refers to the ability to passivate the surface. **Figure 2 (c)** shows the cross-sectional SEM image of the PSC, the SAM is not discernible, potentially due to the tunneling effect. **Table 2** summarizes the Voc, Jsc, FF, and PCE recorded for each of the best-performing cells. In these wide-bandgap PSCs, the Voc exceeded 1.1 V, which is suitable for tandem solar cell applications. Furthermore, the PSC based on MeO-4PACz achieved the highest efficiency of 13%.

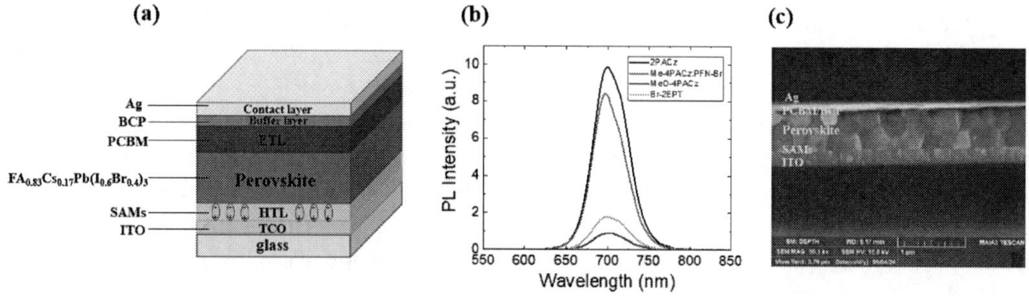

Figure 2 (a) Device architecture of this study; (b) Photoluminescence (PL) spectra; (c) Cross-sectional image for the PSC

Table 2 Summary of photovoltaic parameters extracted from the J−V measurements of best-performing PSCs with different SAMs

SAMs	*Voc* (V)	*Jsc* (mA*cm^{-2})	FF (%)	PCE (%)
2PACz	1.17	14.29	67.0	11.23
Me-4PACz:PFN-Br	1.13	12.42	68.8	9.66
MeO-4PACz	1.16	14.78	77.1	13.16
Br-2EPT	1.12	13.56	75.7	11.48

Figure 3 (a) presents the distribution of photovoltaic parameters for both forward and reverse scans. We compared the device behaviour with the dipole moment and energy levels of the SAMs. Notably, the fill factor graph reveals a trend that corresponds to the dipole moment of the SAMs.

We conducted simulations to investigate the effect of varying interface charges on the dipole moment of SAMs in PSCs. For this investigation, we varied the dipole moment by varying the interface charge density (σ) and separation distance (x) while ensuring the dipole moment remains constant, as determined by DFT simulations.

Figure 3 (b) illustrates the simulation results for the fill factor at the ITO/SAMs interface for the calculated distance and surface charge density (i.e. x = 2 nm, and σ= 2 ~ 5 × 10^{11}), analysed for different SAMs. The results show consistent trends with varying dipole moments and interface charges.

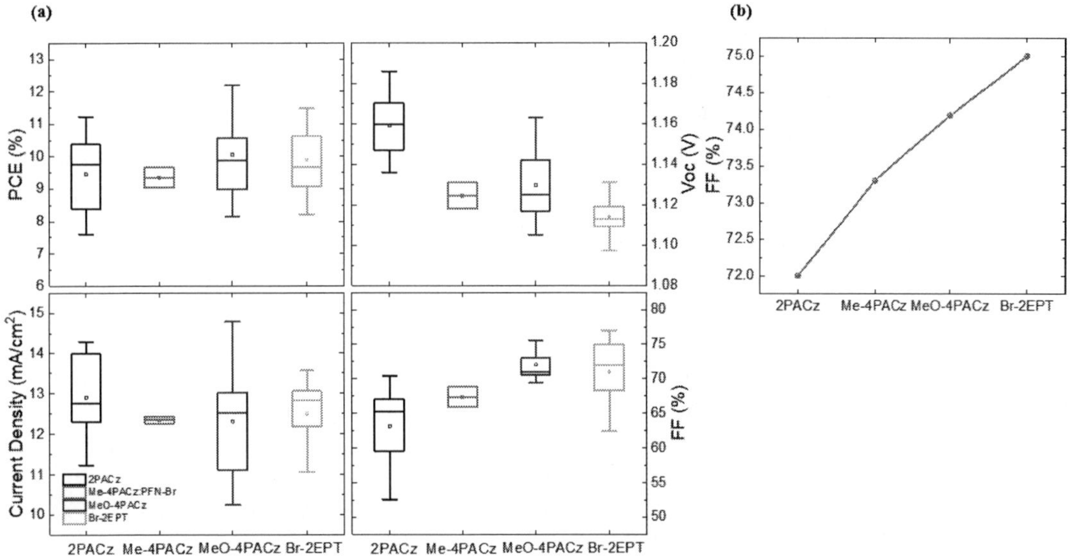

Figure 3 (a) Boxplots of PSCs optical parameters; (b) The simulated fill factor for 2 nm distance in surface charges for the dipole moment.

4. CONCLUSION

This study has analysed the effect of dipole moment and energy level parameters of 2PACz, Me-4PACz, MeO-4PACz, and Br-2EPT, on their electrical performance in inverted PSCs. This was accompanied by DFT and electronic simulations. A clear correlation between dipole moment and fill factor was consistently observed between experimental data and simulation results. These findings underscore the critical role of dipole moment in SAMs for enhancing carrier transport efficiency and suppressing interfacial recombination in PSC devices.

ACKNOWLEDGEMENTS

We acknowledge the support of Czech Science Foundation through grant no 23-06543S and the support of Czech Technical University in Prague and National Taiwan University of Science and Technology through joint Collaborative Research Project CTU-NTUST.

REFERENCES

[1] PARK, J.; KIM, J.; YUN, H.-S.; PAIK, M.J.; NOH, E.; MUN, H.J.; KIM, M.G.; SHIN, T.J.; SEOK, S.I. Controlled Growth of Perovskite Layers with Volatile Alkylammonium Chlorides. *Nature*. 2023, vol. 616, pp. 724–730, Available from: https://doi.org/10.1038/s41586-023-05825-y.

[2] CHEN, H.; LIU, C.; XU, J.; MAXWELL, A.; ZHOU, W.; YANG, Y.; ZHOU, Q.; BATI, A.S.R.; WAN, H.; WANG, Z.; et al. Improved Charge Extraction in Inverted Perovskite Solar Cells with Dual-Site-Binding Ligands. *Science.* 2024, vol. 384, pp. 189–193. Available from: https://doi.org/10.1126/science.adm9474.

[3] GHARIBZADEH, S.; ABDOLLAHI NEJAND, B.; JAKOBY, M.; ABZIEHER, T.; HAUSCHILD, D.; MOGHADAMZADEH, S.; SCHWENZER, J.A.; BRENNER, P.; SCHMAGER, R.; HAGHIGHIRAD, A.A.; et al. Record Open-Circuit Voltage Wide-Bandgap Perovskite Solar Cells Utilizing 2D/3D Perovskite Heterostructure. *Advanced Energy Materials.* 2019, vol. 9, 1803699. Available from: https://doi.org/10.1002/aenm.201803699.

[4] KUMAR, A.; KUMAR, D.; JAIN, N.; KUMAR, M.; GHODAKE, G.; KUMAR, S.; SHARMA, R.K.; HOLOVSKY, J.; SAJI, V.S.; SHARMA, S.K. Enhanced Efficiency and Stability of Electron Transport Layer in Perovskite Tandem Solar Cells: Challenges and Future Perspectives. *Solar Energy.* 2023, vol. 266, 112185. Available from: https://doi.org/10.1016/j.solener.2023.112185.

[5] JEON, N.J.; NOH, J.H.; YANG, W.S.; KIM, Y.C.; RYU, S.; SEO, J.; SEOK, S.I. Compositional Engineering of Perovskite Materials for High-Performance Solar Cells. *Nature.* 2015, vol. 517, pp. 476–480. Available from: https://doi.org/10.1038/nature14133.

[6] CHEN, Y.; LI, N.; WANG, L.; LI, L.; XU, Z.; JIAO, H.; LIU, P.; ZHU, C.; ZAI, H.; SUN, M.; et al. Impacts of Alkaline on the Defects Property and Crystallization Kinetics in Perovskite Solar Cells. *Nat Commun.* 2019, vol. 10, 1112. Available from: https://doi.org/10.1038/s41467-019-09093-1.

[7] WANG, J.; BI, L.; HUANG, X.; FENG, Q.; LIU, M.; CHEN, M.; AN, Y.; JIANG, W.; LIN, F.R.; FU, Q.; et al. Bilayer Interface Engineering through 2D/3D Perovskite and Surface Dipole for Inverted Perovskite Solar Modules. *eScience.* 2024, 100308. Available from: https://doi.org/10.1016/j.esci.2024.100308.

[8] KIM, S.Y.; CHO, S.J.; BYEON, S.E.; HE, X.; YOON, H.J. Self-Assembled Monolayers as Interface Engineering Nanomaterials in Perovskite Solar Cells. *Advanced Energy Materials.* 2020, vol. 10, 2002606. Available from: https://doi.org/10.1002/aenm.202002606.

[9] KAHN, A.; KOCH, N.; GAO, W. Electronic Structure and Electrical Properties of Interfaces between Metals and Π-conjugated Molecular Films. *J Polym Sci B Polym Phys.* 2003, vol. 41, pp. 2529–2548. Available from: https://doi.org/10.1002/polb.10642.

[10] DENG, X.; QI, F.; LI, F.; WU, S.; LIN, F.R.; ZHANG, Z.; GUAN, Z.; YANG, Z.; LEE, C.; JEN, A.K. -Y. Co-assembled Monolayers as Hole-Selective Contact for High-Performance Inverted Perovskite Solar Cells with Optimized Recombination Loss and Long-Term Stability. *Angewandte Chemie.* 2022, vol. 134, e202203088. Available from: https://doi.org/10.1002/ange.202203088.

[11] ALI, F.; ROLDÁN-CARMONA, C.; SOHAIL, M.; NAZEERUDDIN, M.K. Applications of Self-Assembled Monolayers for Perovskite Solar Cells Interface Engineering to Address Efficiency and Stability. *Advanced Energy Materials.* 2020, vol. 10, 2002989. Available from: https://doi.org/10.1002/aenm.202002989.

[12] YANG, Y.; CHENG, S.; ZHU, X.; LI, S.; ZHENG, Z.; ZHAO, K.; JI, L.; LI, R.; LIU, Y.; LIU, C.; et al. Inverted Perovskite Solar Cells with over 2,000 h Operational Stability at 85 °C Using Fixed Charge Passivation. *Nat Energy.* 2023, vol. 9, pp. 37–46. Available from: https://doi.org/10.1038/s41560-023-01377-7.

[13] MA, Y.; GONG, J.; ZENG, P.; LIU, M. Recent Progress in Interfacial Dipole Engineering for Perovskite Solar Cells. *Nano-Micro Lett.* 2023, vol. 15, 173. Available from: https://doi.org/10.1007/s40820-023-01131-4.

[14] KOHN, W.; BECKE, A.D.; PARR, R.G. Density Functional Theory of Electronic Structure. *J. Phys. Chem.* 1996, vol. 100, pp. 12974–12980. Available from: https://doi.org/10.1021/jp960669l.

[15] BECKE, A.D. Density-Functional Exchange-Energy Approximation with Correct Asymptotic Behavior. *Phys. Rev. A.* 1988, vol. 38, pp. 3098–3100. Available from: https://doi.org/10.1103/PhysRevA.38.3098.

[16] LEE, C.; YANG, W.; PARR, R.G. Development of the Colle-Salvetti Correlation-Energy Formula into a Functional of the Electron Density. *Phys. Rev. B.* 1988, vol. 37, pp. 785–789. Available from: https://doi.org/10.1103/PhysRevB.37.785.

[17] PARR, R.G. Density Functional Theory of Atoms and Molecules. In *Horizons of Quantum Chemistry*; Fukui, K., Pullman, B., Eds.; Springer Netherlands: Dordrecht, 1980; pp. 5–15. ISBN 978-94-009-9029-6.

[18] HOSSAIN, K.; KULKARNI, A.; BOTHRA, U.; KLINGEBIEL, B.; KIRCHARTZ, T.; SALIBA, M.; KABRA, D. Resolving the Hydrophobicity of the Me-4PACz Hole Transport Layer for Inverted Perovskite Solar Cells with Efficiency >20%. *ACS Energy Lett.* 2023, vol. 8, pp. 3860–3867. Available from: https://doi.org/10.1021/acsenergylett.3c01385.

COMPARISON OF $LU_3AL_5O_{12}$:CE(ER) BULK AND NANO SINGLE CRYSTALS

[1,2,3]David JOHN, [1]Jakub VOLF, [1]Karol BARTOSIEWICZ, [4]Takahiko HORIAI, [4,5]Kei KAMADA, [4,5,6]Akira YOSHIKAWA, [1]Zdeněk REMEŠ, [1]Lucie LANDOVÁ, [7]Gilles LEDOUX, [7]Julien HOUEL, [1,8]Andrey PROKHOROV, [8]Maksym BURYI

[1]Institute of Physics of the Czech Academy of Sciences, Prague, Czech Republic, EU, john@fzu.cz

[2]Faculty of Nuclear Sciences and Physical Engineering, Czech Technical University, Prague, Czech Republic, EU

[3]Nuclear Physics Institute of the Czech Academy of Sciences, Prague, Czech Republic, EU

[4]New Industry Creation Hatchery Center, Tohoku University, 2-1-1 Katahira Aoba-ku, Sendai, Miyagi, Japan

[5]C&A Corporation, T-Biz, 6-6-10 Aoba, Aramaki, Aoba-ku, Sendai, Miyagi, Japan

[6]Institute for Materials Research, Tohoku University, 2-1-1 Katahira Aoba-ku, Sendai, Miyagi, Japan

[7]Institut Lumière Matière, UMR5306 Université Lyon 1-CNRS, Université de Lyon, Villeurbanne, France, EU

[8]Institute of Plasma Physics of the Czech Academy of Sciences, Prague, Czech Republic, EU

https://doi.org/10.37904/nanocon.2024.4979

Abstract

Lutetium aluminum garnet (LuAG) is a crystal well known for being extraordinarily useful for building highly efficient laser devices or ionizing radiation detectors. To design the most favourable physical properties, different dopants and synthesis methods are constantly developed, primarily focused on the advantages of single crystals. However, the downscaling of the luminescent materials to nanosize also creates additional benefits of increased surface to volume ratio and thus, the efficiency of luminescence or doping can be improved. Therefore, the aim of this work is to compare the single crystals and nanogarnets under the influence of Ce and Er doping. Optical properties of the garnet single crystals are very well known while the nanocrystals should have been studied first. Synergy methods like X-ray diffraction (XRD), photoluminescence (PL), photothermal deflection spectroscopies (PDS) were applied to the LuAG:Ce(Er) nanocrystals whereas the electron paramagnetic resonance (EPR) measurements were carried out in both single and nanocrystals. The main conclusion is that cerium enters nanocrystalline lattice becoming Ce^{4+} compensated by the Cr^{2+}, Fe^{2+} ions (accidental impurities) whereas it is also Ce^{3+} in the single crystal. About one order of magnitude larger Er^{3+} content was observed in the case of nanocrystals as compared to single crystals.

Keywords: Lutetium aluminum garnet, Erbium, Cerium, Single crystal, Nanocrystal, Photoluminescence, EPR

1. INTRODUCTION

Lutetium aluminum garnet (LuAG, $Lu_3Al_5O_{12}$) is a crystalline material known for its exceptional physical properties. LuAG exhibits over the other rare-earth garnets its rich light yield, small lattice parameter, high density and stability combined with fast thermal conductivity, which makes it especially useful for building highly efficient laser devices or ionizing radiation detectors [1]. To design the most favorable physical properties, different dopants and synthesis methods are constantly developed [2], primarily focused on advantages of lanthanoid-doping single-crystals. LuAG has a complex cubic crystal structure based on the presence of lutetium ion. The lutetium is relatively close to other lanthanides which means that even high levels of doping can conserve its favorable physical properties. Most attention is directed to erbium and cerium because their application, e.g. Er wavelength is strongly absorbed by water fluids which makes it promising for medical laser applications, and Ce has fast luminescence emission in the range well suitable for photodiode. The cerium doped isostructural aluminum garnets were studied [3]. Despite the isomorphism between LuAG and YAG, some different and major advantages of LuAG can be expected. For example, the interactions between crystal field and rare earth ions in case of lutetium are higher, which leads to the larger Stark splitting of the lower laser manifold of doped rare earth ions, lower thermal occupation of the lower laser level, and therefore to the improvement of emission efficiency [4]. In this work we focus on optical properties of Er or Ce doped LuAG, we also use the downscaling of the luminescent materials to nanosize which creates additional benefits of increased surface to volume ratio and a substantial increase in the proportion of point defects that can then be more easily investigated. Therefore, the X-ray diffraction (XRD), Raman, photoluminescence (PL) and photothermal deflection spectroscopies (PDS) were applied to the newly synthesized LuAG:Ce(Er) nanopowders, to compare results with classic spectra of their bulk single-crystal counterparts.

2. EXPERIMENTAL CONDITIONS

2.1 Samples preparation

Nanocrystalline garnets of composition $Lu_{3-y}X_yAl_5O_{12}$ (X = Ce, Er) with y = 0.3 were prepared using a sol-gel combustion technique. The starting chemicals were, Lu_2O_3 (4 N, Sigma-Aldrich), $Al(NO_3)_3 \cdot 9H_2O$ (4 N, Sigma-Aldrich), Er_2O_3 (4 N, Sigma-Aldrich) and CeO_2 (4 N, Sigma-Aldrich). These were weighed in a total batch of 1 g of the resulting products. The starting chemicals were then dissolved in HNO_3 (65+ wt%, p.a., PENTA) and the excess HNO_3 was subsequently removed at 90 °C. The dried metal nitrates were then dissolved in 50 mL of deionized water. The metal nitrates in water solution were chelated by addition of citric acid monohydrate (>99.5, Sigma-Aldrich), where the molar ratio of citric acid to all metal ions was 2.5:1. The pH of the stirred solution was then set to pH ~ 8 (at room temperature) by dropwise addition of NH_4OH (24+ wt%, p.a., PENTA). The stirred solution was heated at 90 °C overnight to obtain a homogeneous gel. The dried gel was decomposed in a furnace at 800 °C for 2 h and finally at 1000 ◦C for 1 h.

Single crystals of LuAG were grown using the micro-pulling down (μ-PD) method. Four crystal variants were prepared: pure LuAG and two doped compositions (LuAG:Ce 0.5%, and LuAG:Er 0.2%). The crystal growth was performed in a μ-PD system with a radio frequency (RF) generator at 26 kVA [5]. The growth process utilized an iridium crucible with a 20 mm diameter, fitted with a 3 mm die containing a 0.5 mm diameter capillary channel. The starting materials consisted of high-purity (4N) oxide powders of Lu_2O_3, Er_2O_3, Al_2O_3, and Ce_2O_3, all obtained from Iwatani Corporation. Following the powder mixing in appropriate stoichiometric ratios, the mixture was placed in the iridium crucible. The crystal growth was initiated on a LuAG seed crystal oriented in the <100> direction, with the entire process conducted in an argon atmosphere. A steady crystal pulling rate of 0.05 mm/min was maintained throughout the growth process.

2.2 Experimental techniques

For crystal structure analysis the powder X-ray diffractometer (Empyrean, Malvern Panalytical) with Cu $K_{<\alpha>}$ radiation (at U = 45 kV, I = 30 mA) was used. The X-Ray diffraction patterns were measured in the range of 2θ from 5 till 120 degrees with a step 0.013 degree and time per step 100 s. The characterization of powders was made in the Bragg-Brentano geometry using fixed divergence slit with anti-scatter slit, scanning line detector 1D, and Ni filter. To avoid the reflections from the sample holder powders were on the SiC single crystal wafer with the definite cut. The crystallite sizes were determined using Rietveld refinement method.

Photoluminescence spectroscopy UV-exited photoluminescence (PL) emission spectra were measured in the 375–750 nm spectral range using commercial pulsed UV light emitting diode (LED) #XSL-360-5E (Roither Lasertechnik, GmbH, Vienna, Austria, wavelength 360 nm, output power 1 mW) operating in ac mode, a narrow band pass optical filter BP360 (Thorlabs #FBH360-10 transparent in 350–370 nm spectra range), long pass optical filter LP375 (Edmund Optics #34-302 transparent above 375 nm), a monochromator (Horiba H20VIS, spectral range 300-800 nm, spectral resolution 4 nm), a red-sensitive photomultiplier (Photonis

#XP2203B, spectral range 300-750 nm) and a 50 MHz lock-in amplifier (Zurich Instruments HF2LI). Biased sine modulated power for the UV LED was provided directly by HF2LI via built-in sine-wave generator. PMT photocurrent was converted to voltage signal by current amplifier (Zurich Instruments HF2TA). The HF2LI allowed also to visualize the measured signal using build-in oscilloscope. The setup was spectrally calibrated with a calibrated halogen lamp (Oriel Instruments #63358).

Electron paramagnetic resonance (EPR) spectra of X-band (9.4 GHz) measurements were obtained using a commercial Bruker ELEXSYS E580 spectrometer cooling by liquid Helium.

Custom-made photothermal deflection spectroscope (PDS) consists of previous light source with focusing optics, monochromator and optical filters, mechanical chopper, beamspliter, Si + InGaAs detector, spherical mirror, HeNe probe laser, flat mirror, focusing lenses, position detector, sample immersed in liquid, and anti-vibration table.

The size and morphology of ZnO NRs has been checked by scanning electron microscopy (SEM) using MAIA3, TESCAN electron microscope with the in-beam SE detector placed in objective lens and the electron beam energy 5 keV.

3. RESULTS AND DISCUSSION

3.1 X-ray diffraction and scanning electron microscopy analysis

The microscopy images and XRD patterns of the LuAG nanocrystals samples are shown in **Figure 1**.

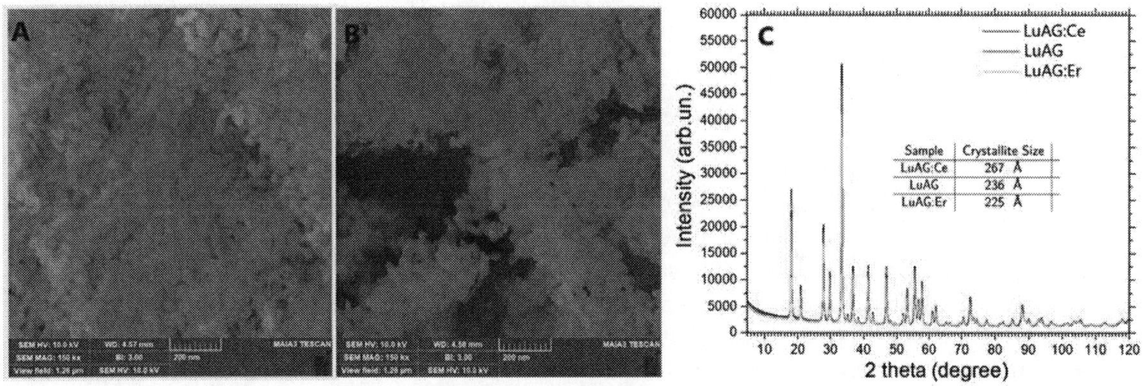

Figure 1 SEM images of the LuAG nanocrystals samples, undoped (A), doped by Ce (B), and XRD patterns of LuAG:Ce(Er) (C). Error of crystallite sizes determination (Rietveld method) was 10%.

From this follows that all samples have the same crystalline phase and very similar pattern, their crystal symmetry is cubic and space group Ia-3d. The crystallite size and micro strain were determined for all samples from exact value 2θ angles higher than 60 degrees. LuAG:Ce, LuAG, LuAG:Er have crystallite size 26.7, 23.6, 22.5 nm and micro strain 5.0197, 5.6643 and 5.949 nm, respectively (see also **Figure 1C**).

To have visual idea of the nanocystallites size, the SEM images of the samples were obtained and shown in **Figure 1A** and **1B**. Indeed, the average size of the nanocrystallites can be expected to be in the range of 20-30 nm. This correlates well with the XRD data.

3.2 Photoluminescence spectroscopy

The luminescence and optical absorption properties of the bulk single-crystal garnets are well known. Therefore, only the luminescence of nanocrystals has been considered in this paper. Two relatively weak peaks at 420 and 440 nm were detected in PL spectrum of all LuAG samples (**Figure 2**) with the same mean decay time around 17 ns (this is evident from **Figure 2A, D** on example of undoped LuAG). These bands originate from surface contamination (the same signals have previously been observed in totally different samples, e.g. in SiO_2) with some organics, probably, the remaining precursors (see Experimental).

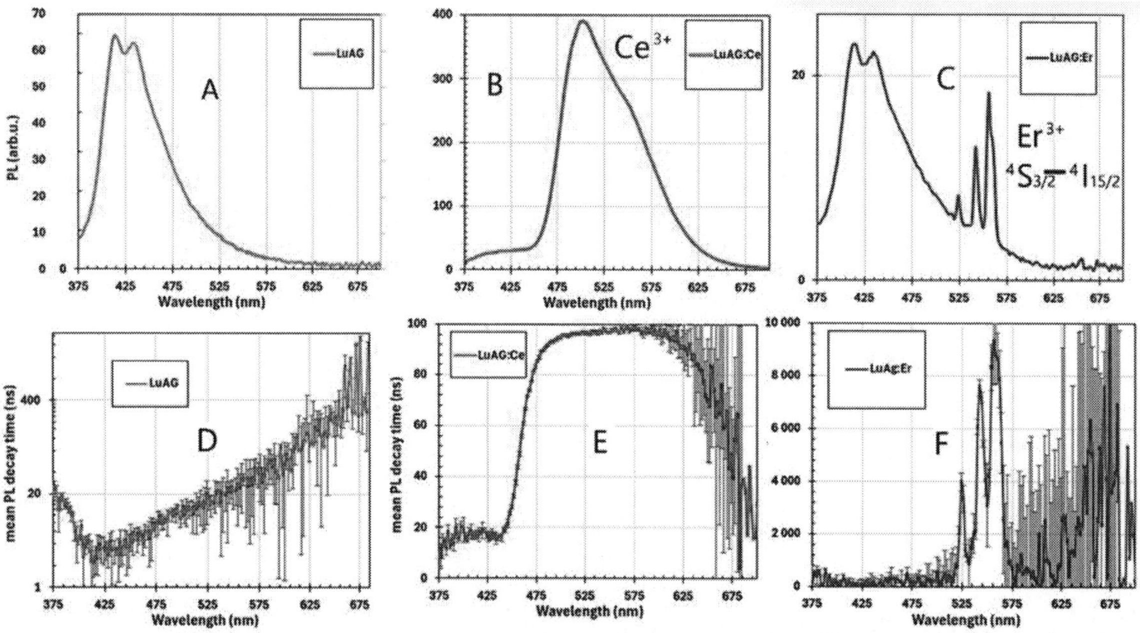

Figure 2 PL in visible range and its mean decay time spectra of the samples A,D) LuAG, B,E) LuAG:Ce and C,F) LuAG:Er, respectively

Otherwise, the pure LuAG generates no other clearly resolved emission bands in visible range (**Figure 2A**). The LuAG:Ce emits at 505 and 552 nm (**Figure 2B**). These are typical 5d - $^2F_{5/2}$ and 5d - $^2F_{7/2}$ transitions of Ce [6]. The energy levels scheme is shown in **Figure 3A**. However, at this point, it is hard to claim for sure which charge state of cerium (Ce^{3+} or Ce^{4+}) is responsible for them. This will be confirmed by EPR and PDS measurements below. The mean decay time measured is constant over the spectral range studied with the approximate value of 100 ns (**Figure 2E**). The LuAG:Er emission is characterized by 3 relatively strong and two very weak peaks in visible range at 524, 542, 556, 656 and 680 nm, respectively (**Figure 2C**). Some of these peaks can be linked to the energy level scheme of Er^{3+} (**Figure 3B**): 524 nm corresponds to $^2H_{11/2}$ -> $^4I_{15/2}$, 556 nm to $^4S_{3/2}$ -> $^4I_{15/2}$ and 654 nm to $^4S_{9/2}$ -> $^4I_{15/2}$ transitions. The mean decay time of the first three peaks is 4, 8, 9 µs, respectively (**Figure 2F**). As for the other two peaks – due to their weak intensity the mean decay time could only be roughly estimated to be in the range 6-10 µs.

The nanoscale and increased surface to volume ratio resulted in no significant increase in photoluminescence range nor intensity as compared to the single crystals or ceramics [7]. To gain better insight into the luminescence properties of the LuAG samples, the IR PL has been measured there as well (**Figure 3C**). The spectra of undoped LuAG and LuAG:Ce were the same - very broad band covering the 1080-1680 nm range, (**Figure 3C** blue/red) overlapped with several small but sharp peaks (like 1623 and 1625 nm) at the ends of spectra.

The LuAG:Er spectra have at least 5 clearly visible peaks in infrared PL spectrum around 1460, 1530, 1580, 1620 and 1660 nm which corresponds to commonly known Er^{3+} transitions between Stark levels ($^4I_{13/2}$ to $^4I_{15/2}$). This indicates the Er incorporation into the LuAG host, especially considering the pure garnet phase confirmed by XRD above (**Figure 1**). The broad IR emission (covering the 1080-1680 nm range) in all the samples should originate from Cr^{2+} [8] as the existence of Cr^{3+} has been proven by EPR below, therefore, one cannot exclude the existence of other charge states. The origin of the 1623 and 1625 nm peaks is unknown.

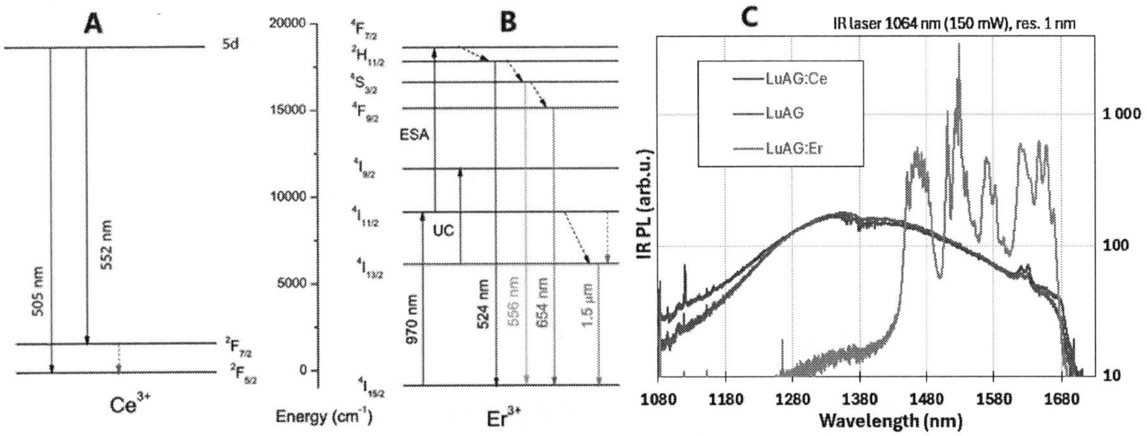

Figure 3 Energy levels scheme of Ce^{3+} (A) and Er^{3+} (B), and IR PL spectra of LuAG:Ce(Er) nanocrystals (C)

3.3 Defect states by photothermal deflection spectroscopy

PDS can be used to reveal more details about deep defect absorption states. The corresponding absorptance spectra are shown in **Figure 4**. All of them have clearly visible backgrounds. They are probably somehow correlated with the origin of the broad IR emission (covering the 1080-1680 nm range) in **Figure 4**. There are new bands starting to appear above 1275 nm in the LuAG:Ce(Er) whereas the absorptance is at the zero level in the case of the undoped LuAG. The origin of these new bands is unknown.

The effect of doping can be seen. Ce enhanced absorption in the region 275-375 nm. This is the typical Ce^{4+} charge transfer (CT) absorption in LuAG:Ce. Er^{3+} transitions are clearly detectable as well in LuAG:Er. The corresponding absorption bands are visible at around 475, 525, 655 and 970 nm, corresponding to the transitions $^4I_{15/2}$ to $^4F_{7/2}$, $^2F_{11/2}$, $^4F_{9/2}$ and $^4I_{11/2}$, respectively.

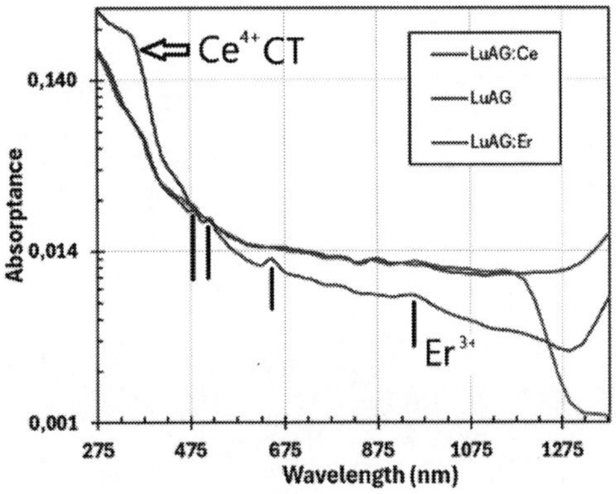

Figure 4 A) LuAG:(Ce,Er) absorptance measured by PDS. Ce^{4+} charge transfer and Er^{3+} specific transitions are indicated

3.4 Charge states and localization of point defects by EPR

To obtain more information about cerium and erbium as well as accidental impurities incorporation, electron paramagnetic resonance (EPR) was measured in all three samples. EPR spectra of the nanoparticles are shown in **Figure 5**.

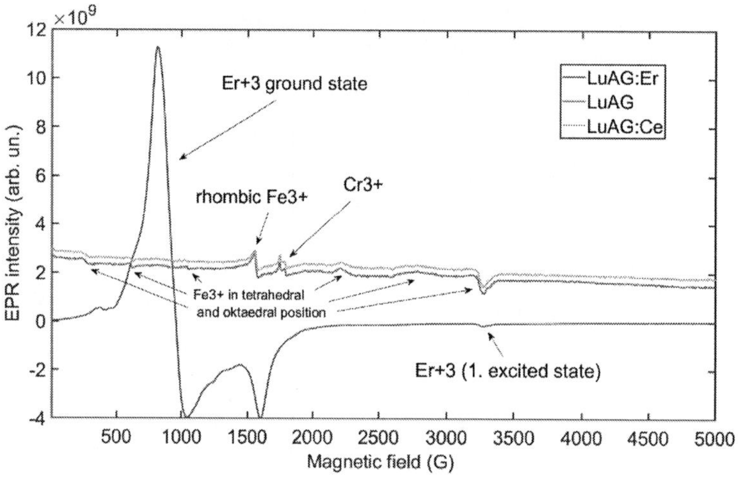

Figure 5 EPR spectra of LuAG:Ce(Er) nanocrystals measured under liquid nitrogen temperature, identified defects are marked with an arrow

EPR signal of Ce^{3+} is known to have very specific temperature dependence. Therefore, to detect its presence in the sample, the spectrum was measured stepwise lowering the temperature from 70 to 30 K. No resonances typical for Ce^{3+} were observed. The LuAG and LuAG:Ce surprisingly generated similar spectra. They were composed of the resonances typical for rhombic Fe^{3+}, Fe^{3+} octahedral and tetrahedral and Cr^{3+} [10]. This indicates only the Ce^{4+} existence (confirmed by PDS) in the LuAG:Ce and, therefore, the observed cerium PL is produced by Ce^{4+}. These results also correlate well with the existence of broad bands in the absorptance and infrared photoluminescence spectra. Er^{3+} signal is very strong and broad. Therefore, one may expect it to be incorporated instead of Lu. Moreover, the dominating is the signal produced by the ground. state, $^4I_{15/2}$, and very weak one produced by the $^4I_{13/2}$, first excited state.

EPR spectra were also measured in the powdered undoped LuAG and LuAG:Ce(Er) single crystals and shown in **Figure 6**. The spectrum of the undoped single crystal LuAG looks very similar to the one measured in the undoped and LuAG:Ce nanocrystalline powder. The typical transitions of the Cr^{3+} were revealed and their positions were indicated with the g factor values.

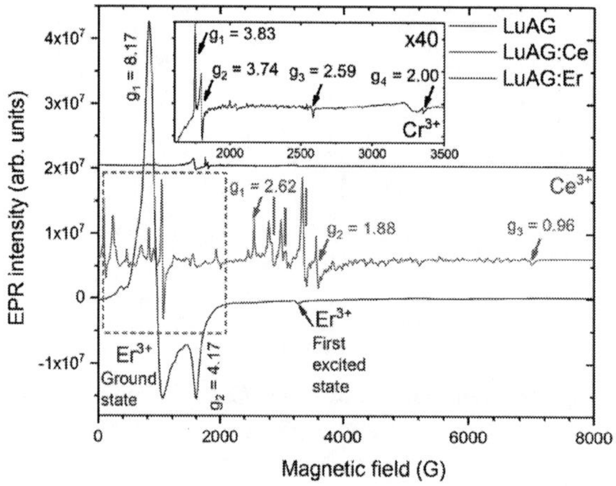

Figure 6 EPR spectra measured at T = 10 K in the powdered single crystals of undoped LuAG and LuAG:Ce(Er). Specific g_i factors indicate the positions of contributions of the specific ions in spectra. Resonances originating from ground and first excited states of Er^{3+} are stressed. Vertical bars and dashed oblong mark the resonances originating from octahedral and tetrahedral Fe^{3+}

The Ce^{3+} incorporation into the LuAG or YAG single crystal is very well known. The corresponding spectrum can be seen in **Figure 6**. Therefore, it is surprising that in the nanocrystals, the whole cerium appears as Ce^{4+}. This must have required additional charge compensation in the nanocrystals, e.g., Fe^{2+}, Cr^{2+} + Ce^{4+} this correlate well to the luminescence measurements. One may conclude that in the case of single crystals the iron (also detected, see **Figure 6**), chromium (mixed up with iron contributions in **Figure 6**) and cerium ions all are in the 3+ charge state and do not appear close to each other enough for the charge transformation to appear. In contrast, considering the size of the LuAG crystallites (about 20 nm, equal to the 3 lattice parameter values) in the case of the nanocrystals – the impurity transition ions appear close to cerium resulting in the charge transformation.

The Er^{3+} spectrum is just the same as in the case of nanocrystalline LuAG (**Figure 6**). The determined *g* factor values are shown in **Figure 6**. The intensity of the Er^{3+} spectrum is about one order of magnitude larger in the nanocrystals as compared to the single crystals (see **Figures 3, 6**). Considering Er doping levels (see Experimenrtal) one may conclude that the segregation coefficient must be larger in the single crystals. Taking into account that Er creates garnet phase as well, one may expect indeed easier tensions relaxation in the case of nanocrystals as compared to the single crystals when Er enters the LuAG lattice.

4. CONCLUSION

According to X-ray diffraction, the nanocrystals are pure phase cubic system of space group Ia-3d, pure phase LuAG:Ce, LuAG or LuAG:Er with homogeneous crystallite size 27, 24, 23 nm and micro strain 0.50197, 0.56643 and 0.5949 Å, respectively. The LuAG:Ce nanocrystals emits at 505 and 552 nm. These peaks are typical for Ce^{4+} (confirmed by the existence of the typical Ce^{4+} charge transfer absorption). LuAG:Er PL spectrum has 5 peaks of about 1460, 1530, 1580, 1620 and 1660 nm which are commonly known Er^{3+} transitions between Stark levels. The pure LuAG emits only negligible, however, the large surface area of nanoparticles traps contamination from the precursors used, which weakly emits at two peaks of 420 and 440 nm. In LuAG and LuAG:Ce infrared spectra very broad (1180 to 1680 nm) band of uncertain origin, probably Cr^{2+} is present. Chrom and iron are present as accidental impurities. They play a significant role in the Ce^{4+} stabilization in the LuAG nanocrystals. Erbium content is about one order of magnitude larger in the nanocrystals as compared to the single crystals. This is because of easier tensions release in the case of nanocrystals whereas in the single crystals this can be more complicated due to large volume resulting in the lowered flexibility of the lattice in contrast to the nanocrystals.

ACKNOWLEDGEMENTS

The financial support of the Leading Agency National Science Centre - Czech Science Foundation project No. 24-14580L is gratefully acknowledged. The work was also supported by Ministry of Education, Youth and Sports of Czech Republic (MEYS) Danube project 8X23025 and the program "Strategy AV 21" of the Czech Academy of Sciences, specifically work package VP 27 Sustainable Energy (Renewable energy resources and distributed energy systems).

REFERENCES

[1] CHEREPY, N. J.; KUNTZ, J. D.; TILLOTSON, T. M.; SPEAKS, D. T.; PAYNE, S. A.; CHAI, B. H. T.; PORTER-CHAPMAN, Y.; DERENZO, S. E. Cerium-doped single crystal and transparent ceramic lutetium aluminum garnet scintillators. *Nuclear Instruments and Methods in Physics Research Section a Accelerators Spectrometers Detectors and Associated Equipment.* 2007, vol. 579, no. 1, pp. 38–41. Available from: https://doi.org/10.1016/j.nima.2007.04.009.

[2] TRETYAK E.V., SHEVCHENKO G.P., KORJIK M.V. Formation of high-density scintillation ceramic from LuAG:Ce + Lu2O3 powders obtained by co-precipitation method. *Optical Materials*. 2015, vol. 46, pp. 596-600, Available from: https://10.1016/j.optmat.2015.05.036

[3] LI, J.; SAHI, S.; GROZA, M.; PAN, Y.; BURGER, A.; KENARANGUI, R.; CHEN, W. Optical and scintillation properties of CE3+-Doped LUAG and YAG Transparent ceramics: a comparative study. *Journal of the American Ceramic Society.* 2016, vol. 100, no. 1, pp. 150–156. Available from: https://doi.org/10.1111/jace.14461.

[4] PEIXIONG Z., ZHENQIANG C., YIN H., ZHEN L., HAO Y., SIQI Z., SHENHE F., ANMING L., Enhanced emission of the 1.50–1.67 μm fluorescence in Er3+, Ce3+–codoped Lu3Al5O12 crystal. *Journal of Alloys and Compounds.* 2017, vol. 696, pp. 795-798. ISSN 0925-8388. Available from: https://doi.org/10.1016/j.jallcom.2016.12.003.

[5] VEBER, P.; BARTOSIEWICZ, K.; DEBRAY, J.; ALOMBERT-GOGET, G.; BENAMARA, O.; MOTTO-ROS, V.; THI, M. P.; BORTA-BOYON, A.; CABANE, H.; LEBBOU, K.; LEVASSORT, F.; KAMADA, K.; YOSHIKAWA, A.; MAGLIONE, M. Lead-free piezoelectric crystals grown by the micro-pulling down technique in the BaTiO3–CaTiO3–BaZrO3 system. *CrystEngComm.* 2019, vol. *21,* no. 25, pp. 3844–3853. Available from: https://doi.org/10.1039/c9ce00405j.

[6] LI, H.-L.; LIU, X.-J.; HUANG, L.-P. Luminescent properties of LuAG:Ce phosphors with different Ce contents prepared by a sol–gel combustion method. *Optical Materials.* 2006, vol. *29,* no. 9, pp. 1138–1142. Available from: https://doi.org/10.1016/j.optmat.2006.05.002.

[7] KHANIN, V.; MEIJERINK, A.; HOUTEPEN, A. J.; JAGT, H. J. B.; DE BOER, D. K. G. Photosaturation in luminescent LUAG:CE Garnet Concentrator Rods. *Advanced Photonics Research.* 2021, vol. 2, no. 8. Available from: https://doi.org/10.1002/adpr.202100055.

[8] FEDOROV, V.; CARLSON, T.; MIROV, S. Energy transfer in iron-chromium co-doped ZnSe middle-infrared laser crystals. *Optical Materials Express.* 2019, vol. 9, no. 5, 2340. Available from: https://doi.org/10.1364/ome.9.002340

DEMONSTRATION OF COPPER:PLA-LIKE NANOCOMPOSITE-BASED DISTRIBUTED BRAGG REFLECTOR GAS SENSOR

[1,2]Zdeněk KRTOUŠ, [1]Pavel SOLAŘ, [3]Oleksandr POLONSKYI, [1,4]Jaroslav KOUSAL

[1]*Charles University, Faculty of Mathematics and Physics, Prague, Czech Republic, EU,* *jaroslav.kousal@mff.cuni.cz*

[2]*Polytechnique Montreal, Department of Engineering Physics, Montreal, CA*

[3]*University of California, Santa Barbara, Department of Chemical Engineering, Santa Barbara (CA), USA*

[4]*Czech Technical University in Prague, Faculty of Mechanical Engineering, Prague, Czech Republic, EU*

https://doi.org/10.37904/nanocon.2024.5003

Abstract

Plasma polymerization, commonly considered a type of plasma-enhanced chemical vapour deposition (PECVD), is a popular method for depositing organic thin films. However, it often produces films with limited molecular complexity due to the necessity to use relatively low molar mass precursors that can be vaporized. To address these limitations, plasma-assisted vapour thermal deposition (PAVTD) was developed. In PAVTD, a solid polymer undergoes thermal degradation (evaporation) in a crucible, producing oligomers with higher molar masses (10^2-10^3 g.mol^{-1}) than typical PECVD precursors. These oligomers are then re-polymerized in RF plasma, allowing PAVTD films to exhibit properties characteristic of classical polymer physics and chemistry, a rarity for plasma polymers. This process enables the precise control of properties such as biodegradability and hydrolyzability, as demonstrated in polylactic acid (PLA)-based films.

PAVTD effectively bridges the gap between classical and plasma polymers. To enhance stability and deposition rates, continuous-PAVTD has been developed using standard FDM 3D printing filaments, achieving deposition rates up to several nanometers per second. This advancement addresses deposition duration and stability issues, making PAVTD a practical tool for studying plasma polymerization. Furthermore, PAVTD can be combined with other vacuum-based thin film deposition techniques like gas aggregation source of nanoparticles (GAS). This capability was demonstrated by fabricating Cu:PLA-like nanocomposite-based distributed Bragg reflector (DBR), where the reproducibility of the deposition rate matters significantly. This reflector was tested as a gas sensor for ethanol vapours, exhibiting strong reflectance peak shifts.

Keywords: Plasma-assisted vapour thermal deposition, gas aggregation source, distributed Bragg reflector, nanocomposite, gas sensor

1. INTRODUCTION

Metal-polymer nanocomposites have diverse applications, like antibacterial films or magnetic films [1-3]. One promising application for metal-dielectric nanocomposites is optical thin films, particularly due to their plasmonic (LSPR) properties [4-8]. Their use ranges from photovoltaic coatings or absorbers to Distributed Bragg Reflectors (DBRs) that can used, e.g. as gas sensors.

DBRs are structures (often fully dielectric) exhibiting band- or peak-like reflectance that are composed of layers with alternating refractive indexes ("low-n" and "high-n"), being thus a kind of so-called 1-D photonic crystals. In the most simple form of DBR the layer thicknesses correspond to quarter-wave interference for the wavelength of the reflection peak. While polymers or oxides are often used for the "low-n" layers, the "high-n" layers can be prepared using nanocomposites. The dielectric:dielectric nanocomposites have a rather limited range of refractive indexes, requiring a large number of layers in a DBR to be effective. Metallic nanoparticle-based nanocomposites can overcome this limitation. DBRs based on nanocomposites can be fabricated using vacuum-based methods, e.g., Au-Teflon, Ag-PTFE or Au-SiO DBRs [9-12].

In this paper, we demonstrate the fabrication of DBR with nanocomposite films based on nanoparticles prepared using a gas aggregation source (GAS) [13]. As the "low-n" layers and as the matrix for the nanocomposite "high-n" layers, PLA-like plasma polymer deposited using plasma-assisted vapour thermal deposition (PAVTD) was used [14]. The use of the PAVTD was possible due to recent improvements in the technique [15].

One application of nanocomposite-based DBRs is gas sensors [16-21]. Typically, the matrix reacts to the presence of sensed gas vapours. Classical polymers exhibit various kinds of "swelling" that change the structural parameters (layer thicknesses) of the DBR and optical properties of its constituents, especially of the nanocomposite. Plasma polymers are typically too crosslinked to exhibit this behaviour efficiently. PAVTD can prepare plasma polymers that exhibit classical polymer-like behaviour [14]. This enables the use of the DBR prepared with this technique as a gas sensor, as demonstrated in this paper.

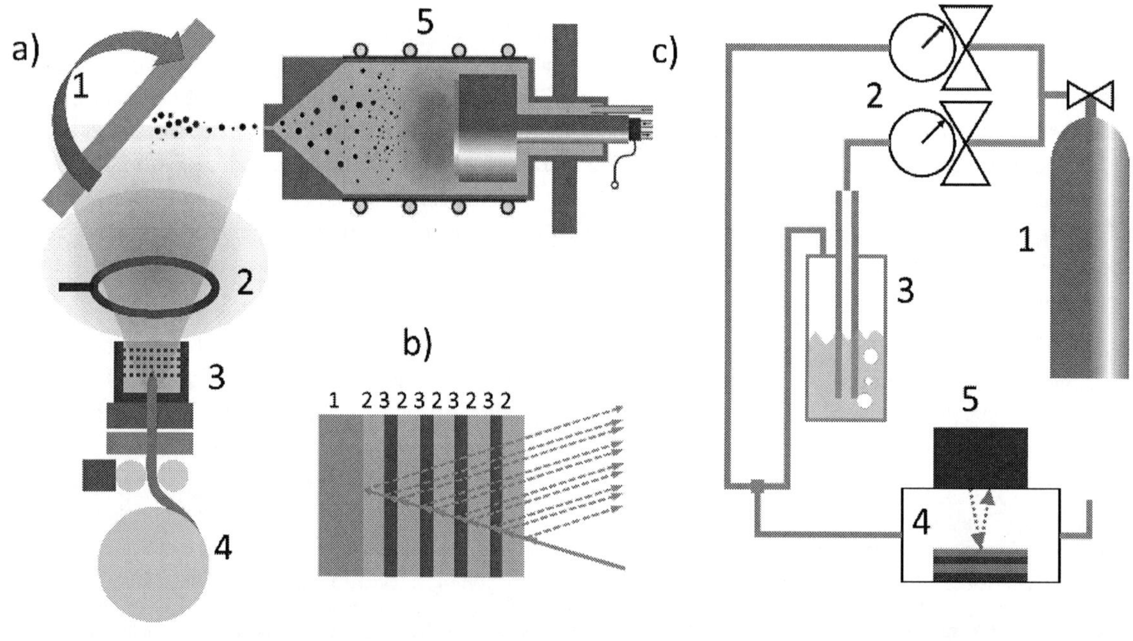

Figure 1 a) Scheme of the deposition setup. 1 – rotatable substrate holder; 2 – RF electrode; 3 – crucible; 4 – polymer (3D printing FDM fiber) storage and feed; 5 – magnetron-based gas aggregation nanoparticle source with copper target b) Scheme of distributed Bragg reflector. 1 – substrate; 2 – low-n film (polymer); 3 – high-n film (metal-polymer nanocomposite) c) Scheme of the gas sensing setup. 1 – gas tank; 2 – flowcontrollers; 3 – bubbler with test chemical; 4 – flow cell with the sample; 5 – UV/VIS spectrophotometer

2. EXPERIMENTAL

First, the distributed Brag reflector (DBR) structure was prepared as a stack of PLA-like layers deposited using plasma-assisted vapour thermal deposition (PAVTD) and nanocomposite films using the same material as a matrix and nanoparticles produced by gas aggregation source (GAS) as a filler.

Second, the DBR structure was checked, and its optical reflectance spectra were measured in an inert and ethanol-containing atmosphere.

2.1 Distributed Bragg reflector preparation

2.1.1 Plasma-assisted vapour thermal deposition

The plasma polymer layers ("low-n" layers of DBR) were deposited using a continuous-PAVTD setup [15] with polylactic acid (PLA) FDM 3D printing fibre ("natural" PLA, Gembird) fed at the rate of 2 g/h. The deposition chamber was fed with argon at the pressure of 0.5 Pa. Above the crucible, plasma discharge at RF (13.56 Mhz) power 5 W was used to induce mild repolymerization in the film (**Figure 1a**).

2.1.2 Nanocomposite layers deposition

The metal-polymer nanocomposite layers ("high-n" layers of DBR) were deposited using the codeposition of PLA-like material (as described in the previous section) with copper nanoparticles produced using GAS with Cu target on DC magnetron, run at the 500 mA current and 70 Pa of argon pressure inside GAS (**Figure 1a**).

2.1.3 Reflector stack deposition

The DBR stack (schematically shown in **Figure 1b**) was deposited on a silicon substrate as 11 alternating layers (6 "low-n", 5 "high-n") by keeping the PAVTD setup running continuously and switching the GAS alternatively. For demonstration purposes, all layers were deposited to a thickness around 100 nm each, the "high-n" nanocomposite layers thinner than the "low-n" polymer layers, as is needed for the DBR structures. The total deposition time of the DBR stack was about 20 minutes.

2.2 DBD stack characterization

The structure of the DBR stack was checked using an SEM microscope (Jeol JSM-7900F) by imaging the crosssection of the DBR on a cut substrate. A UV-VIS-NIR spectrophotometer (Ossila Optical Spectrometer) was used to characterize the DBR reflectance spectra. The DBR was placed into a 3D-printed gas flow cell (internal volume 350 ml) for reflectance measurements (angle of incidence 9°). This flow cell was used to characterize the changes in DBR reflectance spectra with the presence of admixture in the atmosphere. As a model gas, ethanol vapours were chosen. To control the concentration of ethanol in the gas (argon), the gas feed to the flow cell was split into two gas lines: one with pure argon (0.25 $liter_{STP}$/min) and the other going through the bubbler with ethanol (0-1.2 $liter_{STP}$/min), varying the resulting ethanol vapours concentration between 0% and 6% (**Figure 1c**). (Note: The vapour pressure of ethanol at the laboratory temperature was 7.2kPa.)

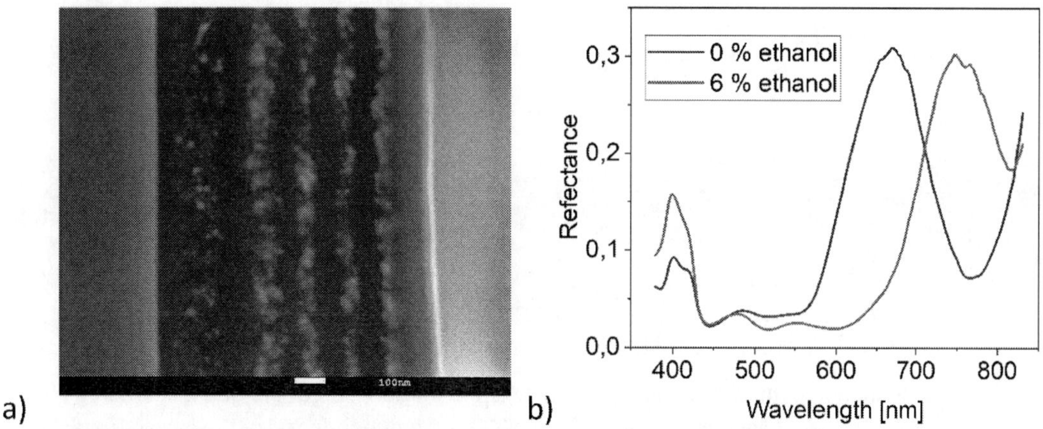

Figure 2 a) SEM image of the crossection of the DBR (substrate side on the left, scale bar 100 nm) (enhanced contrast) b) Reflectance spectra of the DBR in dry argon and Ar+ethanol mixture.

3. RESULTS AND DISCUSSION

3.1 Structure and optical properties

The structure of the deposited DBR, as observed by SEM (**Figure 2a**), was found to be satisfactory, clearly showing the nanocomposite character of the "high-n" layers. The total thickness of the DBR was 880 nm. Due to the visible deviations from purely regular structure, the reflectance was expected to be significantly reduced compared to an ideal DBR. Still, the optical reflectance spectra have clearly shown a Bragg reflectance peak around 660 nm (**Figure 2b**).

3.2 Gas sensing demonstration

With the presence of ethanol vapours in the UV-VIS flow cell, the DBR reflectance peak exhibited a significant spectral shift (**Figure 2b**). For the characterization, the flow cell was repeatedly flushed with pure argon and argon-ethanol mixture. Each ethanol vapour flow took about 100 seconds. The DBR was initially cycled 5 times under an ethanol vapour concentration of 6%. While some relaxation of the DBR reflectance peak shift was observed (the peak shift decreased by about 9 nm), the repeated measurements at the same ethanol concentration (see below) were reproducible within a few nm. Also some relaxation was also observed during each period of ethanol vapour introduction into the cell when the vapour concentration was over ~5.5%.

The sensitivity of the shift of the reflectance peak to the ethanol concentration was then characterized. Time dependences for the peak shift after the introduction of varying concentrations of ethanol into the flow cell are shown in **Figure 3a**. It can be seen that the dynamics of the process are mostly limited by the flow cell volume, especially at the lowest ethanol concentration, where the total gas flow was sufficient to fill the volume of the cell just about 2 times per the filling cycle. In these cases, the real peak shift of the DBR reflectance peak for a corresponding ethanol concentration can be higher than registered. In general, the DBR itself must react under ~30s.

The summary of the sensitivity of the reflectance peak shift is shown in the **Figure 3b**. The peak shift shows nonlinear dependence on the ethanol vapour concentration, reaching a ~100 nm shift for the ethanol concentration of ~6% when the peak moved to about 750 nm. It shall be noted that the DBR stack was neither perfect nor optimized for the maximal physical/chemical interaction with the detected substance.

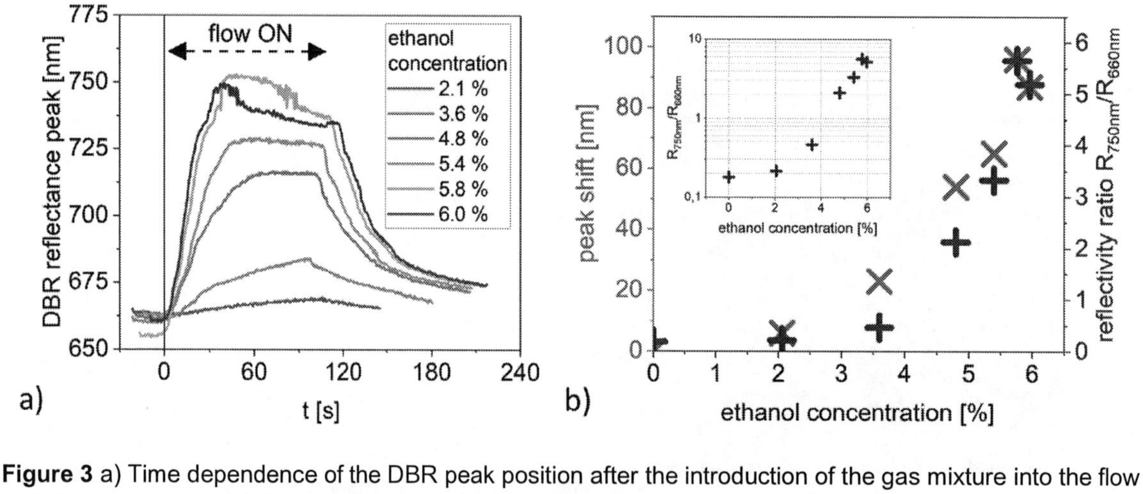

Figure 3 a) Time dependence of the DBR peak position after the introduction of the gas mixture into the flow cell b) Dependence of the peak reflectance on the DBR shift and the dependence of the ratio of reflectances at 660 nm and 750 nm (R_{660nm}/R_{750nm}) on the ethanol concentration (Inset: R_{660nm}/R_{750nm} in semilogarithmic scale).

If some gas sensors are to be based on this kind of DBR stack, instead of full UV-VIS reflectance measurement, two spectral lines reflectance ratio measurement would be much simpler. Selecting the spectral lines of 660 nm and 750 nm (corresponding to measured extremes of reflectance peak position), the R_{750nm}/R_{660nm} reflectance ratio is also shown in **Figure 3b**. Above ~2% ethanol concentration in the atmosphere, the dependence of this ratio on ethanol concentration is almost exponential (**Figure 3b**, inset).

The high peak shift attainable with this DBR as a sensor can be attributed to the reversible swelling of the plasma polymer component of the DBR structure, stretching the thickness of the DBR itself by ~15%. While such significant swelling by solvent adsorption was observed on PAVTD-deposited films [14], it is untypical for common densely crosslinked plasma polymers.

4. CONCLUSIONS

In this paper, the preparation of Cu:PLA-like nanocomposite-based distributed Bragg reflector was demonstrated. The continuous-PAVTD technique is shown to be controllable on the level permitting the preparation of nontrivial optical multilayer stacks and combinable with other vacuum-based deposition techniques, such as GAS, as in this case.

The resulting DBR reflector shows the DBR reflectance peak, limited mostly by the imperfections in the deposition process. Such a DBR stack was then used as a simple gas sensor device, showing clear, high shifts of the reflectance peak that could be easily detected. These high peak shifts are allowed by the ability of the PAVTD technique to produce thin plasma polymer films that exhibit the behaviour of a classical polymer. This property of the films allows expansion of the DBR structure to the levels not attainable with common plasma polymer films.

ACKNOWLEDGEMENTS

This work was supported by the grants GA22-21007S of the Czech Science Foundation and GAUK 411822 of the Charles University Grant Agency.

REFERENCES

[1] ZADEHNAZARI, A. Metal oxide/polymer nanocomposites: A review on recent advances in fabrication and applications. *Polymer-Plastics Technology and Materials.* 2022, vol. 62, no. 5, pp. 655–700, Available from: https://doi.org/10.1080/25740881.2022.2129387

[2] CHOUKOUROV, A. and MANGOLINI, L. Special issue: Plasma synthesis of nanoparticles and nanocomposite coatings. *Plasma Processes and Polymers.* 2020, vol. 17, no. 5, 2090003. Available from: https://doi.org/10.1002/ppap.202090003

[3] SNYDERS, R., HEGEMANN, D., THIRY, D., ZABEIDA, O., KLEMBERG-SAPIEHA, J., MARTINU, L. Foundations of plasma enhanced chemical vapor deposition of functional coatings. *Plasma Sources Science and Technology.* 2023, vol. 32, no. 7, 074001. Available from: https://doi.org/10.1088/1361-6595/acdabc

[4] SUN, L., GRANT, J. T., JONES, J. G., MURPHY, N. R., VERNON, J. P., STEVENSON, P. R. Tailoring the Optical Properties of Nanoscale-Thick Metal–Dielectric Ag–SiO2 Nanocomposite Films for Precision Optical Coating Integration. *ACS Applied Nano Materials.* 2023, vol. 6, no. 9, pp. 7704-7714. Available from: https://doi.org/10.1021/acsanm.3c00853

[5] FAUPEL, F., ZAPOROJTCHENKO, V., STRUNSKUS, T., ELBAHRI, M. Metal-Polymer Nanocomposites for Functional Applications. *Advanced Engineering Materials.* 2010, vol. 12, no. 12, pp. 1177-1190. Available from: https://doi.org/10.1002/adem.201000231

[6] WANG, L., KAFSHGARI, M., MEUNIER, M. Optical Properties and Applications of Plasmonic-Metal Nanoparticles. *Advanced Functional Materials.* 2020, vol. 30, no. 51, 2005400. Available from: https://doi.org/10.1002/adfm.202005400

[7] PRAKASH, J., SUN, S., SWART, H., GUPTA, R. Noble metals-TiO2 nanocomposites: From fundamental mechanisms to photocatalysis, surface enhanced Raman scattering and antibacterial applications. *Applied Materials Today.* 2018, vol. 11, pp. 82-135. Available from: https://doi.org/10.1016/j.apmt.2018.02.002

[8] MISRA, S., LI, L., JIAN, J., HUANG, J., WANG, X., ZEMLYANOV, D., JANG, J., RIBEIRO, F., WANG, H. Tailorable Au Nanoparticles Embedded in Epitaxial TiO2 Thin Films for Tunable Optical Properties. *ACS Applied Materials & Interfaces.* 2018, vol. 10, is. 38, pp. 32895-32902. Available from: https://doi.org/10.1021/acsami.8b12210

[9] DRUFFEL, T., MANDZY, N., SUNKARA, M., GRULKE, E. Polymer nanocomposite thin film mirror for the infrared region. *Small.* 2008, vol. 4, no. 4, pp. 459-461. Available from: https://doi.org/10.1002/smll.200700680

[10] SUN, Y., WANG, G., ZHANG, T., LIU, C., WANG, J. Periodically alternated metallic/dielectric nanocomposites and dielectric films for the fabrication of high-efficiency Bragg reflectors: a case study. *Applied Physics Express.* 2020, vol. 13, no. 7, 072001. Available from: https://doi.org/10.35848/1882-0786/ab93e8

[11] QU, L., ZHAO, J., YANG, J., SUN, Y., LIU, C., Influence of the thermal evolution of Au nanoparticles induced by ion implantation on the reflectivity of multilayer structures. *Optical Materials Express.* 2021, vol. 11, no. 4, pp. 1176-1184. Available from: https://doi.org/10.1364/OME.419456

[12] FAUPEL, F., ZAPOROJTCHENKO, V., GREVE, H., SCHÜRMANN, U., CHAKRAVADHANULA, V.S.K., HANISCH, C., KULKARNI, A., GERBER, A., QUANDT, E. and PODSCHUN, R. Deposition of Nanocomposites by Plasmas. *Contributions to Plasma Physics.* 2007, vol. 47, no. 7, pp. 537-544. Available from: https://doi.org/10.1002/ctpp.200710069

[13] KYLIÁN, O., NIKITIN, D., HANUŠ, J., ALI-OGLY, S., PLESKUNOV, P., BIEDERMAN, H. Plasma-assisted gas-phase aggregation of clusters for functional nanomaterials. *Journal of Vacuum Science & Technology A.* 2023, vol. 41, no. 2, 020802. Available from: https://doi.org/10.1116/6.0002374

[14] KRTOUŠ, Z., KOUSAL, J., SEDLAŘÍKOVÁ, J., KOLÁŘOVÁ RAŠKOVÁ, Z., KUČEROVÁ, L., KRAKOVSKÝ, I., KUČERA, J., ALI-OGLY, S., PLESKUNOV, P., CHOUKOUROV, A. Thin films of crosslinked polylactic acid as tailored platforms for controlled drug release. *Surface and Coatings Technology.* 2021, vol. 421, 127402. Available from: https://doi.org/10.1016/j.surfcoat.2021.127402

[15] KOUSAL, J., KRTOUŠ, Z., SOLAŘ, P., KŘIVKA, I., KRAKOVSKÝ, I., Plasma-assisted vapour thermal deposition with continuous material feed. In: *Proceedings of 14th Nanocon 2022.* Brno: Tanger, 2023, pp. 261-266. ISSN: 2694-930X. Available from: https://doi.org/10.37904/nanocon.2022.4579

[16] CONVERTINO, A., CAPOBIANCHI, A., VALENTINI, A. and CIRILLO, E.N.M. A New Approach to Organic Solvent Detection: High-Reflectivity Bragg Reflectors Based on a Gold Nanoparticle/Teflon-like Composite Material. *Advanced Materials.* 2003, vol. 15, no. 13, pp. 1103-1105, Available from: https://doi.org/10.1002/adma.200304777

[17] JALKANEN, T., TORRES-COSTA, V., SALONEN, J., BJÖRKQVIST, M., MÄKILÄ, E., MARTÍNEZ-DUART, J.M., LEHTO, V.-P. Optical gas sensing properties of thermally hydrocarbonized porous silicon Bragg reflectors. *Optics Express.* 2009, vol. 17, no. 7, pp. 5446-5456. Available from: https://doi.org/10.1364/oe.17.005446

[18] LOVA, P., MANFREDI, G., BOARINO, L., COMITE, A., LAUS, M., PATRINI, M., MARABELLI, F., SOCI, C., COMORETTO, D. Polymer Distributed Bragg Reflectors for Vapor Sensing. *ACS Photonics.* 2015, vol. 2, no.4, pp. 537−543. Available from: https://doi.org/10.1021/ph500461w

[19] KOU, D., ZHANG, Y., ZHANG, S., WU, S., MA, W. High-sensitive and stable photonic crystal sensors for visual detection and discrimination of volatile aromatic hydrocarbon vapors. *Chemical Engineering Journal.* 2019, vol. 375, 121987. Available from: https://doi.org/10.1016/j.cej.2019.121987

[20] PALO, E., DASKALAKIS. K.S., Prospects in Broadening the Application of Planar Solution-Based Distributed Bragg Reflectors. *Advanced Materials Interfaces.* 2023, vol. 10, no. 19, 2202206. Available from: https://doi.org/10.1002/admi.202202206

[21] WEI, J., YI, Z., YANG, L., ZHANG, L., YANG, J., QIN, M., CAO, S. Photonic crystal gas sensors based on metal–organic frameworks and polymers. *Analytical Methods.* 2024, vol 16, no. 29, pp. 4901-4916. Available from: https://doi.org/10.1039/D4AY00764F

INVESTIGATING THE IMPACT OF SYNTHESIS PROCEDURES ON ELECTROCHEMICAL PROPERTIES OF rGO-CeO_2 NANOCOMPOSITE

[1]Xuan Duc NGHIEM, [2]Eliška BIRGUSOVÁ, [1]Jana PEKÁRKOVÁ, [2]Zuzana BYTEŠNÍKOVÁ, [2]Lukáš RICHTERA

[1]Brno University of Technology, Central European Institute of Technology, Brno, Czech Republic, EU, xuan.duc.nghiem@ceitec.vutbr.cz, pekarkova@vutbr.cz

[2]Mendel University in Brno, Department of Chemistry and Biochemistry, Brno, Czech Republic, EU, zuzana.bytesnikova@mendelu.cz, xbirguso@mendelu.cz, richtera@mendelu.cz

https://doi.org/10.37904/nanocon.2024.4994

Abstract

rGO-CeO_2 nanocomposite has been employed and significantly improved electrochemical sensors in terms of facility, sensitivity, and accuracy. Additionally, since synthesis procedures play an important role in the structure and properties of nanomaterials, it is necessary to investigate the effect of synthesis procedures to maximize the electrochemical performance of the sensor using rGO-CeO_2. In this study, two different approaches, namely *in situ* and *ex situ* methods, were used to synthesize GO-CeO_2, then GO-CeO_2 was reduced by hydrazine to form the final product rGO-CeO_2. Subsequently, the morphology and elemental analysis/mapping of these nanocomposites were examined by Scanning Electron Microscopy and Energy Dispersive X-ray Spectroscopy, respectively. Afterwards, the nanocomposites were drop-casted on the surface of glassy carbon electrodes, and the electrochemical properties were evaluated by Cyclic Voltammetry and Electrochemical Impedance Spectroscopy in the presence of ferrocyanide/ferricyanide redox probe. The results indicated that CeO_2 nanoparticles were evenly distributed on the surface of wrinkled rGO sheets; also, in comparison to bare glassy carbon electrode, electrodes modified with the nanocomposites from both approaches reported lower charge transfer resistance and peak-to-peak separation, indicating enhanced electron transfer and electrical conductivity, and the electrode modified with rGO-CeO_2 from *in situ* approach having better improvement than the one with rGO-CeO_2 from *ex situ* approach.

Keywords: Reduced graphene oxide, nanoceria, nanocomposite, electrochemical properties

1. INTRODUCTION

Nanomaterials have been extensively incorporated into electrochemical sensors for multiple applications, for instance, working as labels for biomolecules, enhancing electron transfer, and catalyzing reactions. Consequently, this has led to considerable improvement for this type of sensor regarding facility, sensitivity, and accuracy [1]. Various classes of nanomaterials have been deployed, and among those, carbon-based nanomaterials, particularly reduced graphene oxide (rGO) has attracted considerable interest thanks to its appealing characteristics, consisting of large surface area, outstanding electrical and thermal conductivity, and high possibility of surface modifications. The latest characteristic is significantly important since it enables the incorporation of rGO with different groups of nanomaterials to form rGO-based nanocomposites. In comparison to pristine rGO, rGO-based nanocomposites are superior in terms of (1) overcoming the restacking and aggregation nature of rGO sheet, thus preventing the reduction in active surface area, and (2) combining the outstanding attributes of each pristine material to achieve a synergistic effect [2]. To combine with rGO, metal and metal oxide nanoparticles are frequently employed, and one outstanding example of this class of nanomaterials is nanoceria. CeO_2 is beneficial for electrochemical detection systems since this nanoparticle is affordable, non-toxic, biocompatible, has high surface area, and is capable of improving electron transfer [3,4].

The integration of rGO and CeO_2 results in the formation of rGO-CeO_2 nanocomposite that has been noticed and applied to the electrochemical detection of multiple analytes, for example, neurotransmitter [5] and insecticide [6].

Although nanomaterials can greatly improve the performance of a sensor, the degree of enhancement is dependent on the characteristics, including size and morphology, of such nanomaterials [7]. Furthermore, synthesis procedures play an important role in the control of size and morphology [8], thus controlling the electrochemical properties of the product. Hence, to maximize the performance of an electrochemical determination, it is necessary to investigate the influence of synthesis procedures on the electrochemical properties of rGO-based nanocomposite. Herein, this study aims to evaluate how synthesis procedures impact the morphological characteristics and electrochemical performance of rGO-CeO_2 nanocomposites. More specifically, in terms of synthesis procedures, the loading of CeO_2 into graphene oxide (GO) sheets were implemented by two different approaches: *in situ* and *ex situ*. *In situ* approach refers to the formation and growth of CeO_2 from its precursors in the presence of GO; by contrast, *ex situ* method demonstrates the mechanical combination of as-synthesized CeO_2 and GO sheets [9]. Subsequently, GO-CeO_2 from both approaches were treated with hydrazine for the reduction of GO sheets to form rGO-CeO_2. Furthermore, to fulfill the ultimate goal of this study, three main objectives are proposed as follows: (1) Synthesis of rGO-CeO_2 by *in situ* and *ex situ* approaches; (2) Characterization of rGO-CeO_2 by Scanning Electron Microscopy (SEM) and Energy Dispersive X-ray Spectroscopy (EDX); and (3) Examination of the rGO-CeO_2-modified electrode by Cyclic Voltammetry (CV) and Electrochemical Impedance Spectroscopy (EIS).

2. MATERIALS AND METHODS

2.1 Materials

For the synthesis of nanomaterials, Graphite flakes, Potassium permanganate ($KMnO_4$), Sulfuric acid (H_2SO_4), Hydrogen peroxide (H_2O_2), Cerium(III) nitrate hexahydrate ($Ce(NO_3)_3 \cdot 6H_2O$), Hydrazine monohydrate ($N_2H_4 \cdot H_2O$) were analytical grade and purchased from Merck KGaA (Darmstadt, Germany). Aqueous solution of ammonia (NH_3) with analytical grade quality was purchased from Lachner (Neratovice, Czech Republic). To clean the electrode surface, ethanol absolute (Penta, Prague, Czech Republic), four polishing suspensions (Electron Microscopy Sciences, Hatfield, PA, USA) of 0.5-micron diamond suspension, and alumina slurry with three different sizes of 1 µm, 0.3 µm, and 0.05 µm were employed. To prepare the working solution for electrochemical measurements, Potassium chloride (KCl), Potassium hexacyanoferrate(II) trihydrate ($K_4Fe(CN)_6 \cdot 3H_2O$), and Potassium hexacyanoferrate(III) ($K_3Fe(CN)_6$) were analytical grade and purchased from Merck KGaA (Darmstadt, Germany). Lastly, ultrapure water (18.2 MΩ·cm) was deployed during the course of this study.

2.2 Synthesis of rGO-CeO_2 by *in situ* and *ex situ* approaches

GO was prepared as a precursor for the preparation of rGO-CeO_2 based on the simplified Tour's method [10]. Afterwards, rGO-CeO_2 synthesized by *in situ* and *ex situ* approaches were conducted as follows.

Ex situ approach: This approach was derived from an earlier study [11]. Briefly, $Ce(NO_3)_3 \cdot 6H_2O$ was dissolved in ultrapure water, and the pH was adjusted to 10 by the addition of aqueous solution of NH_3. Subsequently, this solution was mixed with a suspension of GO in ultrapure water and the mixture was ultrasonicated for 20 minutes. Afterwards, the mixture was heated to 90 °C prior to the addition of $N_2H_4 \cdot H_2O$ dropwise and kept under reflux for one hour at the same temperature. Finally, the obtained product was washed 3 times with ultrapure water and denoted as rGO-CeO_2-1.

In situ approach: In comparison to *ex situ* approach, the only difference in this approach was that the solution of $Ce(NO_3)_3$ was mixed with the suspension of GO prior to the addition of aqueous solution of NH_3. Afterwards, the mixture was sonicated and treated with $N_2H_4 \cdot H_2O$, and the final product was washed in the same manner as aforementioned. From this approach, the nanocomposite was denoted as rGO-CeO_2-2.

2.3 Characterization of rGO-CeO_2 by SEM and EDX

SEM and EDX analysis were conducted by TESCAN MIRA3 XMU SEM coupled with Oxford Instruments EDX. The images from the SEM were obtained at resolution mode, accelerating voltage of 5kV, In-Beam secondary electron detector, and working distance of 3.0 mm. In terms of EDX analysis, the conditions of accelerating voltage and working distance were changed to 15kV and 15.0 mm, respectively; and the time for mapping was set to 20 minutes.

2.4 Evaluation of electrochemical properties of rGO-CeO_2 by CV and EIS

Before the modification of glassy carbon electrodes (GCE) by rGO-CeO_2, GCE were polished on polishing pads with 0.5-micron diamond suspension, and alumina slurry with three different sizes of 1 µm, 0.3 µm, and 0.05 µm, respectively, following by sonication in ethanol absolute and ultrapure water. Afterwards, GCE were left to be dried at room temperature. Subsequently, 5 µl of 1mg/ml suspension of either rGO-CeO_2-1 or rGO-CeO_2-2 was drop-casted to the surface of GCE and allowed to be dried at room temperature.

Autolab PGSTAT302N was employed for all electrochemical measurements. A three-electrode system was set up, consisting of a platinum electrode as counter electrode, Ag/AgCl 1.0 mol/l KCl as reference electrode, and GCE as working electrode. The working solution containing 5mM of $[Fe(CN)_6]^{3-/4-}$ redox probe and 0.1M KCl electrolyte was used for both CV and EIS. Afterwards, EIS was implemented to compare the charge transfer resistance of bare GCE to GCE modified with nanocomposites. EIS was conducted at these conditions: frequency range from 0.1 to 10 kHz with an amplitude of 10 mV, and applied voltage of 0.2 V. Also, CV was employed to assess the electrochemical performance of bare and modified GCE, and CV was performed within the potential range of -0.4 V to 0.9 V and scan rate of 0.05 V/s.

3. RESULTS AND DISCUSSION

3.1 Physicochemical properties of rGO-CeO_2 nanocomposites

As aforementioned, rGO-CeO_2 nanocomposites were synthesized by two different approaches, and the difference between these two approaches lies in the anchoring of either precursors of nanoceria or pre-synthesized nanoceria to the surface of GO. In the *ex situ* approach, CeO_2 nanoparticles were formed due to the addition of aqueous solution of NH_3 to precipitate Ce^{3+} ions (precipitation method). It is notable that this synthesis was implemented under ultrasonication. Acoustic cavitation in ultrasound irradiation generated cavitation bubbles and hot spots, with the former operated as nucleation sites and the latter provided sufficient energy for the formation of CeO_2 nuclei. Thanks to the adsorption of cavitation bubbles to the surface of CeO_2 nuclei, the interfacial free energy between the nuclei and the reacting solution was lower, thus avoiding the agglomeration of nanoparticles [11, 12]. Afterwards, CeO_2 nanoparticles were loaded onto GO, and CeO_2 is expected to distribute evenly on the surface of GO as well as rGO. For the *in situ* approach, at the beginning, Ce^{3+} ions were loaded onto the surface of GO, and upon the addition of aqueous solution of NH_3 for the precipitation method under ultrasound, two phenomena occurred. The first one was heterogeneous nucleation in which CeO_2 nuclei was generated on a foreign body as GO, and in liquid phase, these nuclei formed with ease due to the presence of GO as a stable nucleating surface [13]. The second phenomenon was due to the acoustic cavitation as aforementioned; thereby CeO_2 from this approach is expected to be decorated evenly on the surface of GO and then rGO. As can be seen in **Figure 1**, in both approaches, no agglomeration of CeO_2 nanoparticles was observed as expected, and the formation of wrinkled rGO sheet indicated successful reduction of GO to rGO by hydrazine.

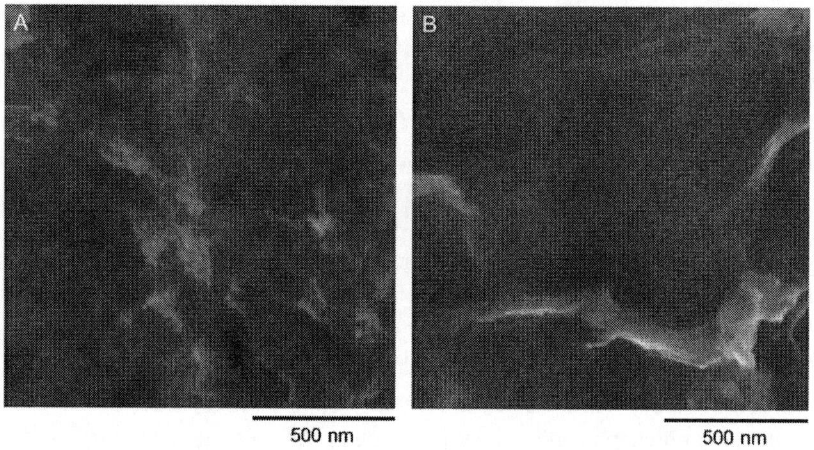

Figure 1 SEM images of rGO-CeO_2-1 (A) and rGO-CeO_2-2 (B) nanocomposites

The findings from EDX can demonstrate the elemental composition and mapping of each element from these nanocomposites. As can be seen from **Figure 2**, in each nanocomposite, three elements were detected, including carbon, oxygen, and cerium. Moreover, the elemental mapping further confirmed the uniform distribution of CeO_2 nanoparticles on rGO sheets, which is in good agreement with the results from SEM.

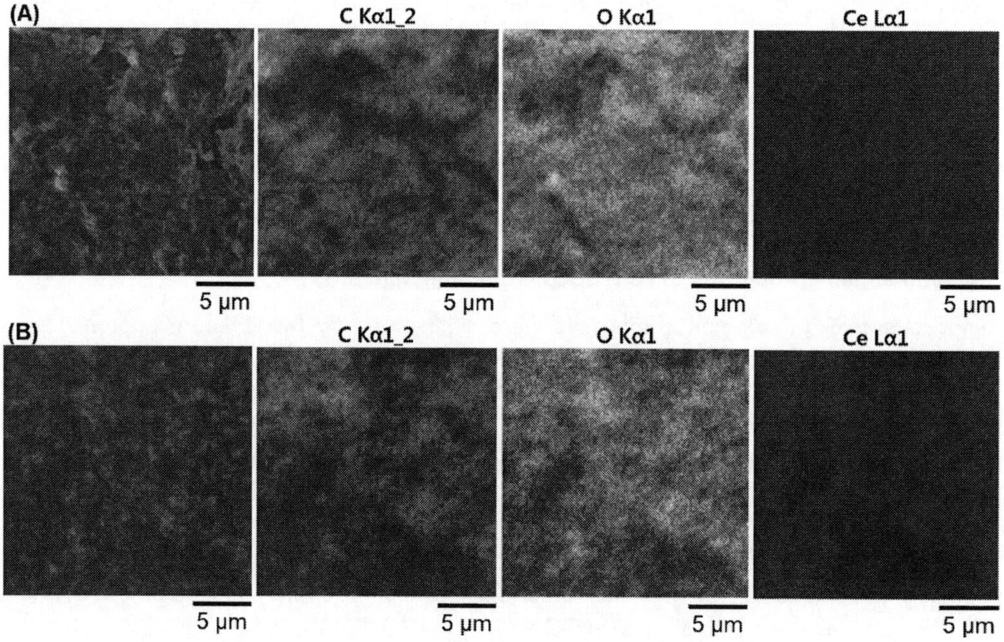

Figure 2 EDX analysis of rGO-CeO_2-1 (A) and rGO-CeO_2-2 (B) (green, yellow, and red maps correspond to carbon, oxygen, and cerium, respectively)

3.2 Electrochemical properties of rGO-CeO_2-modified GCE

EIS is a powerful method to study the charge transfer resistance of electrodes at the electrode-solution interface. Conventionally, EIS spectrum demonstrates a linear part at low frequency, which is attributed to the Warburg impedance for mass transport (diffusion) of redox species to the electrode interface. More importantly, EIS spectrum displays a semi-circle at high frequency, which is correlated with heterogeneous electron transfer process of redox species. The diameter of this semi-circle can be utilized to calculate the value of charge transfer resistance (R_{ct}) [14]. From **Figure 3A**, the diameters of the semi-circle of modified GCE were significantly lower than that of bare GCE. More specifically, R_{ct} values of bare GCE, GCE/rGO-CeO_2-1, and GCE/rGO-CeO_2-2 were estimated as 132.65 Ω, 18.11 Ω, and 8.75 Ω, respectively. This implied an enhanced electron transfer and improved electrical conductivity due to the modification of bare GCE with rGO-CeO_2, and the impact of rGO-CeO_2-2 was slightly better than that of rGO-CeO_2-1.

CV was conducted to assess the electroanalytical performance of bare and modified GCE. As can be seen from **Figure 3B**, larger maximum current values were observed in the case of modified electrodes in both anodic and cathodic directions of potential scan. In addition, the peak-to-peak separation values for bare GCE, GCE/rGO-CeO_2-1, and GCE/rGO-CeO_2-2 were 134.28 mV, 107.42 mV, and 95.22 mV, respectively. Hence, faster electron transfer kinetics were observed thanks to the addition of rGO-CeO_2, with rGO-CeO_2-2 having higher improvement than rGO-CeO_2-1.

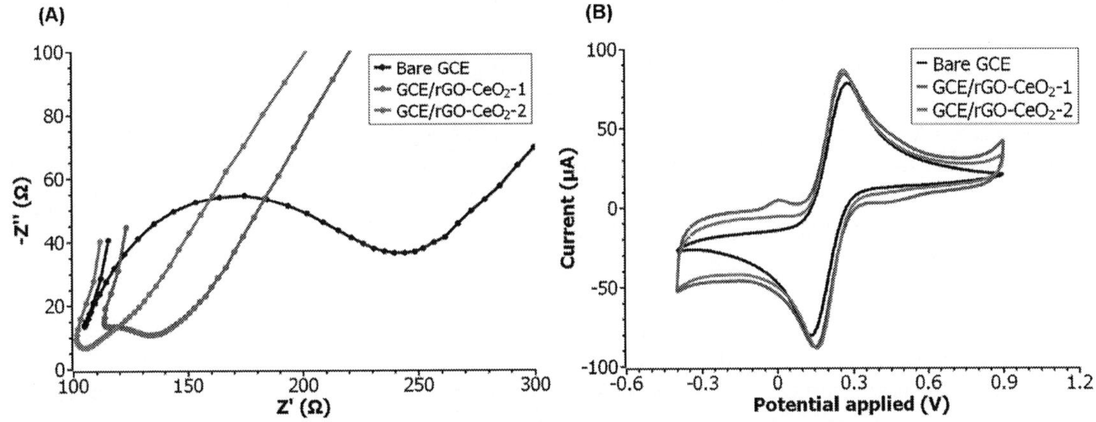

Figure 3 Nyquist plot (A) and cyclic voltammogram (B) of rGO-CeO_2-modified GCE

From both analyses implemented by CV and EIS, the electrochemical properties of modified GCE were superior in comparison to those of bare GCE, indicating the potential of this kind of nanocomposite within the discipline of electrochemical sensors. Moreover, better performance was recorded in the case of rGO-CeO_2 synthesized by *in situ* approach (rGO-CeO_2-2). This can be explained by the degree of uniform distribution of CeO_2 nanoparticles on rGO sheets. Although nanocomposites from both approaches illustrated good anchoring of CeO_2 on the surface of rGO from SEM and EDX results, the *ex situ* approach is more prone to non-uniform coverage of nanoceria by graphene sheets [15], thereby resulting in more agglomeration of nanoceria as well as less active surface area for electrochemical processes.

4. CONCLUSION

In conclusion, this study has demonstrated the impact of synthesis approaches, namely *in situ* and *ex situ* approaches, on the characteristics of rGO-CeO_2 nanocomposites. Firstly, based on the findings of SEM and EDX, rGO-CeO_2 from both approaches exhibited uniform distribution of CeO_2 nanoparticles on the surface of wrinkled rGO sheets. Secondly, from electrochemical measurements by CV and EIS methods, rGO-CeO_2 from both methods proved their capacities to enhance electron transfer and electrical conductivity, and this improvement is important for electrochemical detection of analytes (e.g., vitamin, neurotransmitter, and biogenic amine) in complicated real samples with many interferences. Lastly, the rGO-CeO_2 from *in situ* approach exhibited better electrochemical properties, which can be explained by the nature of this approach to form more uniform decoration of nanoceria on the surface of rGO, leading to less agglomeration of CeO_2 nanoparticles and in turn larger active surface area for electrochemical processes.

ACKNOWLEDGEMENTS

CzechNanoLab project LM2023051 funded by MEYS CR is gratefully acknowledged for the financial support of the measurements/sample fabrication at CEITEC Nano Research Infrastructure.

REFERENCES

[1] HUANG, X.P., ZHU, Y.F., KIANFAR, E. Nano Biosensors: Properties, applications and electrochemical techniques. *Journal of Materials Research and Technology-Jmr&T*. 2021, vol. 12, pp. 1649-1672.

[2] YU, H.T., GUO, W., LU, X., XU, H., YANG, Q., TAN, J.X., ZHANG, W. Reduced graphene oxide nanocomposite based electrochemical biosensors for monitoring foodborne pathogenic bacteria: A review. *Food Control*. 2021, vol. 127.

[3] NIE, X., ZHANG, R., TANG, Z., WANG, H.Y., DENG, P.H., TANG, Y.G. Facile Fabrication of CeO_2/Electrochemically Reduced Graphene Oxide Nanocomposites for Vanillin Detection in Commercial Food Products. *Nanomaterials*. 2020, vol. 10, no. 7.

[4] CHARBGOO, F., RAMEZANI, M., DARROUDI, M. Bio-sensing applications of cerium oxide nanoparticles: Advantages and disadvantages. *Biosensors & Bioelectronics*. 2017, vol. 96, pp. 33-43.

[5] MURALI, A., LAN, Y.P., SARSWAT, P.K., FREE, M.L. Synthesis of CeO_2/reduced graphene oxide nanocomposite for electrochemical determination of ascorbic acid and dopamine and for photocatalytic applications. *Materials Today Chemistry*. 2019, vol. 12, pp. 222-232.

[6] ENSAFI, A.A., NOROOZI, R., ZANDI-ATASHBAR, N., REZAEI, B. Cerium(IV) oxide decorated on reduced graphene oxide, a selective and sensitive electrochemical sensor for fenitrothion determination. *Sensors and Actuators B-Chemical*. 2017, vol. 245, pp. 980-987.

[7] QIAN, J., YANG, X.W., JIANG, L., ZHU, C.D., MAO, H.P., WANG, K. Facile preparation of Fe_3O_4 nanospheres/reduced graphene oxide nanocomposites with high peroxidase-like activity for sensitive and selective colorimetric detection of acetylcholine. *Sensors and Actuators B-Chemical*. 2014, vol. 201, pp. 160-166.

[8] HE, K., ZENG, Z.T., CHEN, A.W., ZENG, G.M., XIAO, R., XU, P., HUANG, Z.Z., SHI, J.B., HU, L., CHEN, G.Q. Advancement of Ag-Graphene Based Nanocomposites: An Overview of Synthesis and Its Applications. *Small*. 2018, vol. 14, no. 32.

[9] VENEGAS, C.J., GUTIERREZ, F.A., REEVES-MCLAREN, N., RIVAS, G.A., RUIZ-LEÓN, D., BOLLO, S. *In situ* or *Ex situ* Synthesis for Electrochemical Detection of Hydrogen Peroxide-An Evaluation of Co_2SnO_4/RGO Nanohybrids. *Micromachines*. 2023, vol. 14, no. 5.

[10] MUKHERJEE, A., ASHRAFI, A.M., BYTESNIKOVA, Z., SVEC, P., RICHTERA, L., ADAM, V. An investigation on the multiple roles of CeO_2 nanoparticle in electrochemical sensing: Biomimetic activity and electron acceptor. *Journal of Electroanalytical Chemistry*. 2023, vol. 935, pp. 11.

[11] DEZFULI, A.S., GANJALI, M.R., NOROUZI, P., FARIDBOD, F. Facile sonochemical synthesis and electrochemical investigation of ceria/graphene nanocomposites. *Journal of Materials Chemistry B*. 2015, vol. 3, no. 11, pp. 2362-2370.

[12] ZHU, S., GUO, J., DONG, J., CUI, Z., LU, T., ZHU, C., ZHANG, D., MA, J. Sonochemical fabrication of Fe_3O_4 nanoparticles on reduced graphene oxide for biosensors. *Ultrasonics Sonochemistry*. 2013, vol. 20, no. 3, pp. 872-880.

[13] THANH, N.T.K., MACLEAN, N., MAHIDDINE, S. Mechanisms of Nucleation and Growth of Nanoparticles in Solution. *Chemical Reviews*. 2014, vol. 114, no. 15, pp. 7610-7630.

[14] LAZANAS, A.C., PRODROMIDIS, M.I. Electrochemical Impedance Spectroscopy—A Tutorial. *ACS Measurement Science Au*. 2023, vol. 3, no. 3, pp. 162-193.

[15] KHAN, M., TAHIR, M.N., ADIL, S.F., KHAN, H.U., SIDDIQUI, M.R.H., AL-WARTHAN, A.A., TREMEL, W. Graphene based metal and metal oxide nanocomposites: synthesis, properties and their applications. *Journal of Materials Chemistry A*. 2015, vol. 3, no. 37, pp. 18753-18808.

METALLIC AND SEMICONDUCTOR NANOPARTICLES INTEGRATED INTO THIN LAYERS OF HYDROGENATED AMORPHOUS SILICON OR SILICONCARBIDE

Jiří STUCHLÍK, Jaroslav KUPČÍK, Jan ČERMÁK, The Ha STUCHLÍKOVÁ, Zdeněk REMEŠ

Institute of Physics, Czech Academy of Sciences, Prague, Czech Republic, EU, stuj@fzu.cz, kupcik@icpf.cas.cz, cermakj@fzu.cz, hotheha@fzu.cz, remes@fzu.cz

https://doi.org/10.37904/nanocon.2024.4995

Abstract

The combination of PECVD - plasma enhanced chemical vapor deposition and VE - vacuum evaporation at one vacuum chamber is in situ method how to integrate semiconductor (Ge) and metallic (Sn) nanoparticles into amorphous hydrogenated thin films and their structures on the base of a-Si:H or a-SiC:H. In this work we focus our effort to deposit a-SiC:H thin films from monomethylsilane (MMS = SiH_3-CH_3) diluted by hydrogen in the ratio 2/100 sccm. The Ge evaporates at vacuum higher than 2E-5 Pa. The TEM - transmission electron microscopy and AFM - atomic force microscopy are used for microscopical characterization of thin films while PDS - photothermal deflection spectroscopy and PL – photoluminescence spectroscopy are used for their spectroscopical characterization.

Keywords: a-SiC:H, MMS, Ge, Sn, nanoparticles

1. INTRODUCTION

This study builds on previous published results achieved on layered structures of amorphous hydrogenated silicon a-Si:H and non-stoichiometric amorphous hydrogenated silicon carbide a-SiC:H with embedded Ge nanoparticles [1]. In previous publications cited in [1], we have already described silane and methane diluted with hydrogen were used as the starting gases for PECVD. We have shown that the presence of methane in the ratio of 1:1 to silane in the discharge increases the carbon content in the deposited layers by only a few percent yet it shifts the transition from amorphous to mixed or even nanocrystalline layers while increasing the deposition temperature [1]. Using CPM, we have shown that these a-SiC:H films deposited in this way show a low amount of defects in band gap compared to a/μc-Si:H films. It has already been confirmed in many past publications [2, 3] that it is not possible to simply increase the carbon content in the films for expanding the band gap of this semiconductor without increasing the methane:silane ratio. Another way is to use monomethylsilane - MMS, in which the carbon is already connected to the silicon by a chemical bond [4,5]. In line with the development trend of using highly temperature-stable SiC-based semiconductor devices, other deposition techniques such as HWCVD [6,7] or non-CVD techniques [8] are being tested. This work focuses on the possible use of conventional PECVD to modify the electrical and/or optical properties of a-SiC:H thin films by integrating Sn or Ge nanoparticles (NPs). Moreover, the formation of Ge NPs in thin films may be interesting not only for tuning the diode structures properties, but also for the study of their magnetic behavior [9,10].

2. EXPERIMENTAL

In previous publications ([1] and relevant references in [1]), we have already described different ways to integrate NPs into thin films and our custom-made system with a vacuum chamber, which allows to alternate two deposition processes - PECVD and VE without interrupting the vacuum - i.e. deposition *in situ*. For the deposition of a-SiC:H thin films we used the following parameters - flow rates of gasses hydrogen 100 sccm and MMS 2 sccm, pressure of gasses during the deposition 30 Pa, RF power 18 W for glow discharge excitation at the industrial frequency 13.56 MHz and deposition temperature about 400°C. The thickness of Ge or Sn evaporated from a Tungsten boat was controlled by the Standard Quartz Crystal Sensor reporting typical thicknesses from 1 to 4 nm. Thin film structures were deposited on fused silica glass substrates or on Corning glass covered by FTO – tin oxide doped by fluor. Always a set of two substrates were used for deposition of a-SiC:H, one for the thin film with and the second without NPs. This second one was used as reference. The samples for TEM analysis were deposited simultaneously under the same conditions, but directly onto standard (commercial) Cu TEM grids covered by carbon foil. The deposited layers were characterized by scanning electron microscope (SEM) Maia3 FE-SEM (Tescan), high resolution transmission electron microscope (HRTEM) Tecnai TF20Xtwin (USA) operated at 200kV equipped with a field-emission gun (FEG) cathode and a high angle annual detector (HAADF) usable in the scanning TEM (STEM) imaging mode and an energy-dispersive X-ray analyzer for elemental analysis and mapping or profiling. Surface topography was scanned also by an atomic force microscope (AFM) NTEGRA Prima (NT-MDT) by a Multi75EG probe (BudgetSensors) in semi contact mode.

The setup for NIR PL spectroscopy uses the 10 mW blue LED (Thorlabs M470F2) triggered by pulse generator (typ. 37 Hz). To increase efficiency of collecting PL signal, the sample is illuminated perpendicularly from thin film side via a long-pass dichroic mirror (Thorlabs DMLP650R, 650 nm cut-on) and the same aspheric fused silica glass lens (focal length 25 mm, N.A. 0.5) that collects PL signal. The excitation and emission are spectrally purified by the OD6 band pass (Thorlabs FBH470-10) and long pass (Thorlabs FELH0550 and FELH0850) optical filters. Emitted PL light is spectrally analyzed using Horiba H20IR monochromator (600 gr/mm concave holographic grating) and focused by CaF_2 lens on an interchangeable Si and InGaAs photodiodes with built-in $2x10^{10}$ V/A ac amplifier (models 2151 and 2152, Newport Corp., USA), connected to a lock-in amplifier (Model SR830, Stanford Research Systems, Sunnyvale, California, USA) referenced to LED frequency. The spectral response was calibrated by a halogen lamp of known output followed by an additional correction [11].

The optical absorption coefficients were measured by PDS - Photothermal Deflection Spectroscopy. Photothermal deflection spectroscopy (PDS) measures directly the optical absorption of thin films on fused silica glass substrates with sensitivity of four orders of magnitude. The PDS spectrophotometer uses a 150 W Xe lamp as a light source and monochromator SpectraPro-150 equipped with two gratings blazed at 300 (1200 gr/mm) and 500 nm (600 gr/mm) operating in a broad spectral range from ultraviolet to infrared region 250−1400 nm. The sample is immersed in a transparent liquid and illuminated by monochromatic light with the probe laser beam passing parallel to the sample surface. The heat absorbed in the sample deflects the probe laser beam detected by the position detector and SR830 lock-in amplifier. Since the amplitude of the probe beam deflection is proportional to the optical absorption, the optical absorptance was evaluated by normalizing the signal on the deflection of the thin black film.

3. RESULTS AND DISCUSSION

Sn and Ge were evaporated on the a-Si:H and a-SiC:H thin films. The first control of deposited thin films on the thin films by AFM and SEM is shown in **Figure 1.** While in the case of Sn evaporation on the a-Si:H surface the creation of Sn NPs is very well visible by both AFM and SEM microscopies the evaporated Ge on the a-SiC:H thin film results in a smooth surface. The rms roughness is 0.83 nm, which is more than three times higher than the control without Ge (0.25 nm). It might indicate that Ge does not form crystalline nanoparticles, but an amorphous thin layer instead. This is supported by the characterization of TEM in **Figure 2.** TEM results obtained on the layered structure of a-SiC:H 9 nm / Ge 3 nm / a-SiC:H 9 nm deposited on a standard Cu grid demonstrated the amorphous character of the deposited films on the surface of a-SiC:H. Therefore, another way to achieve grouping of germanium into nanoparticles in crystalline form should be considered. We conclude that the amorphous substrate and temperature of 400 °C led to higher reactivity, especially germanium with surface silicon.

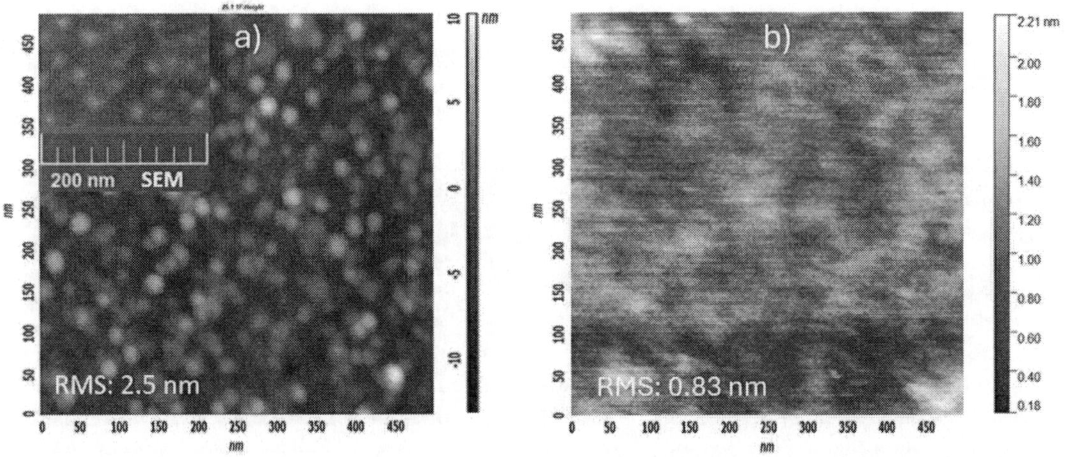

Figure 1 AFM imaging of Sn NPs vaporized on a-Si:H thin film at temperature of approx. 220 °C the insert is SEM imaging (a) and AFM imaging of Ge vaporized on a-SiC:H thin film at temperature of approx. 400 °C (b)

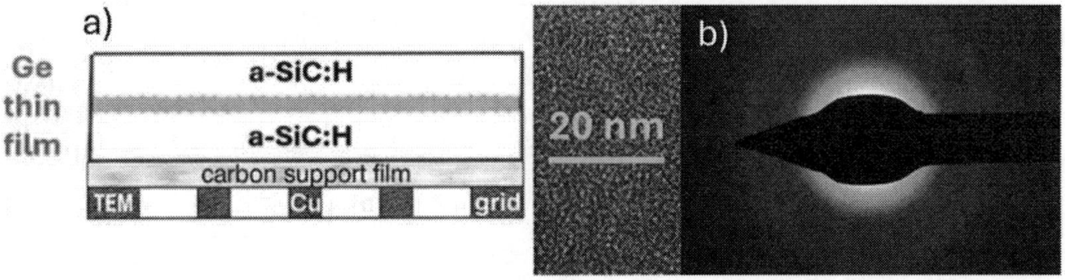

Figure 2 Sample structure prepared for TEM characterization (a) and observation result – bright field (BF) image and corresponding electron diffraction pattern (DP) typical for amorphous material (b)

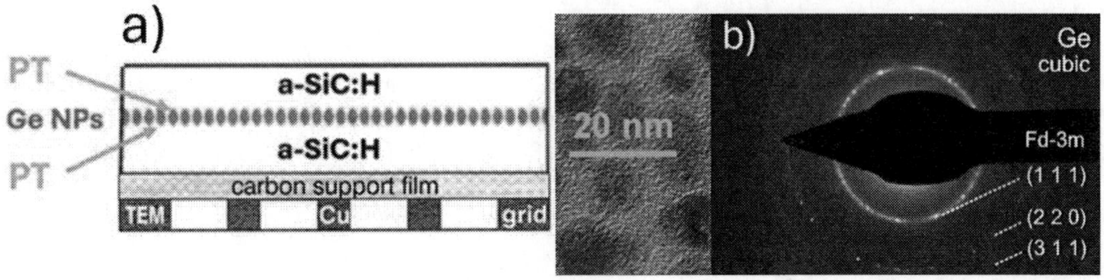

Figure 3 Sample structure prepared for TEM characterization (a) with indicated change in the deposition procedure, which is PT - plasma treatment before and after evaporation of the Ge layer and the result of observation, BF image and corresponding electron DP, assignable to the Ge crystal structure (b)

Because the reactivity of germanium with carbon is lower, we found a solution in the saturation of the surface with carbon using plasma treatment - PT. PT was successful using hydrogen with methane in a ratio of 100:2 [sccm] - see **Figure 3**, which confirms the crystalline structure of the NPs obtained in the manner specified above. Other more precise TEM and EDX characterizations have clearly demonstrated the grouping of germanium into nanoparticles with a crystalline structure. **Figure 4 a)** obtained in BF HRTEM mode shows a typical crystal NP with clearly visible lattice fringes. **Figures 4 b) – 4 f)** show a typical element profile which was obtained during analysis in the STEM mode, which is sensitive also to atomic composition and displays the observed NPs as bright, while the surrounding matter as dark (**Figure 4 b**). The occurrence of Ge (see

Figure 4 f) is higher in the oval bright NPs, while between them drops to zero (see **Figure 4e**) – spectrum A from a dark place A). The Ge-K line intensity corresponds well with the signal of the HAADF detector depicted on the **Figure 4 d**), whereas the Si-K line intensity is higher between NPs and at the particle place is lower but does not drop to zero. O and C do not show significant trends.

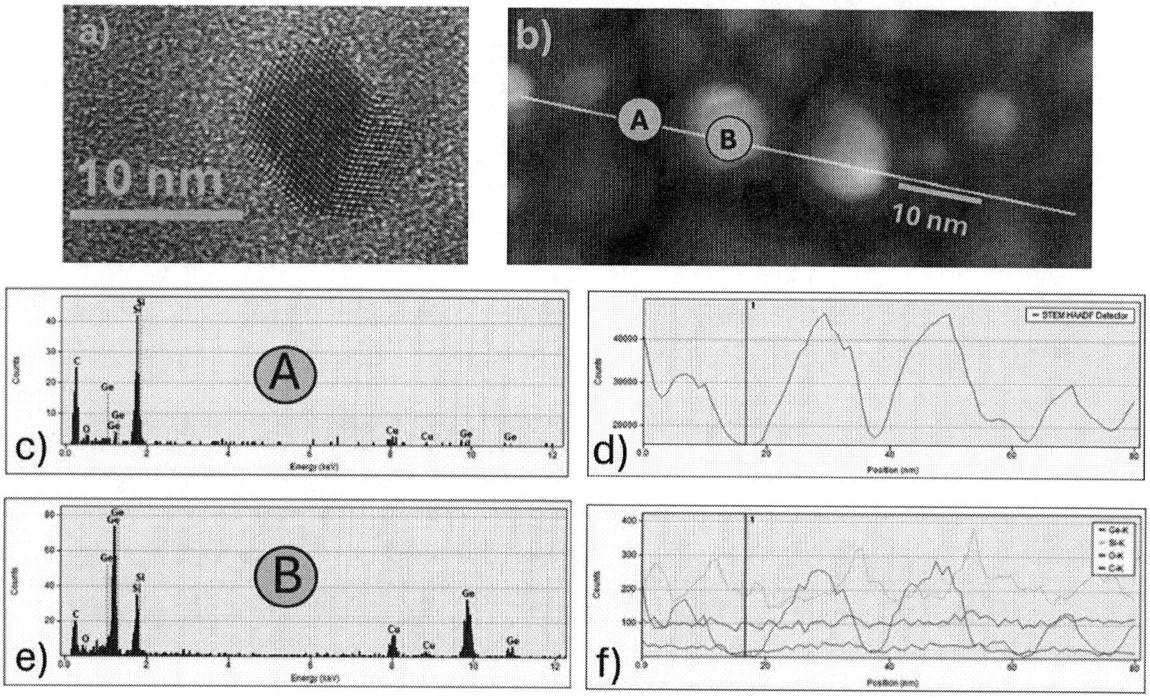

Figure 4 a): HRTEM (BF) image of a single Ge NP, b): STEM-HAADF image of a group of NPs with marking of the EDX profile line and two locations of EDX point spectra acquisition, c): EDX point spectrum from point A, d): HAADF detector intensity corresponded to profile line marked at b), e): EDX point spectrum from point B, and f): occurrence of elements (depicted as intensity of character X-line of the given element) along the profile line: green, yellow, red or orange lines correspond to Ge, Si, C or O, respectively. The elemental profile was obtained on an 80 nm line with a 1 nm step, 1 EDX spectrum from each point, spectral range 0-40 keV, 20 eV/channel, 4000 ms acquisition time.

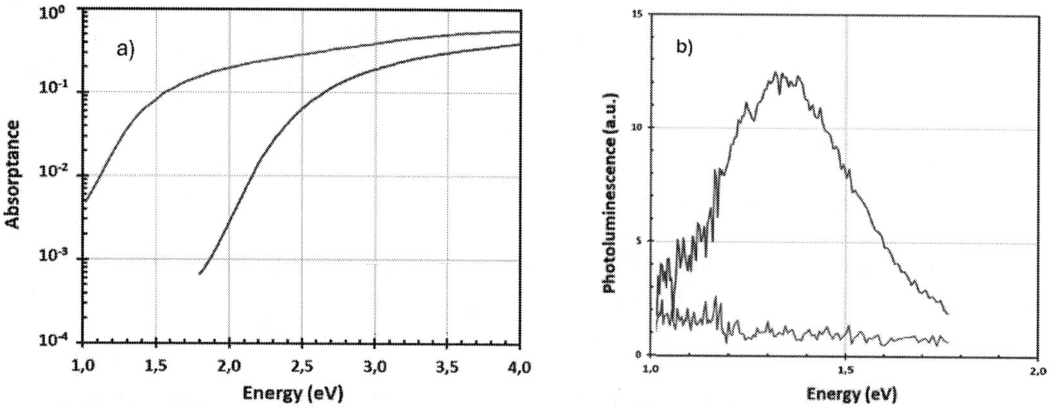

Figure 5 PDS measurements showed an increase in absorption and shifting of the absorption edge to lower energies (a), PL measurements reveal the intrinsic photoluminescence of the a-SiC:H film and the negative effect of deposited Ge NPs (b)

For optical measurements, PDS and PL see **Figure 5.**, the structure of a-SiC:H / Ge / a-SiC:H / Ge / a-SiC:H / Ge / a-SiC:H with thicknesses of 9 / 3 / 9 / 3 / 9 / 3 / 9 nm was deposited on a quartz substrate and for comparison an a-SiC:H layer with a total thickness of 4 x 9 = 36 nm. The intrinsic a-SiC:H thin layers are highly transparent in NIR spectral region and exhibit NIR PL band with the maximum at 1.5 eV. The weak PL in a-SiC:H films is related to radiative recombination on defects in the band gap of a-SiC:H.

We attribute the decrease of PL to the non-radiative recombination of photoexcited free carriers in Ge sub-layers. Thus, we can conclude that the incorporation of Ge in a-SiC:H thin layers extend the absorption spectrum into the near-infrared region, which is beneficial for capturing more IR light and increasing photocurrent sensitivity. However, the decrease in intrinsic PL shows that Ge acts as a non-radiative recombination center. The mechanism of recombination to Ge NPs and the possibility of achieving electroluminescence as in the case of diode structures [9] are the subject of further investigation.

4. CONCLUSION

While increasing the deposition temperature over 300 to 350 °C for hydrogenated silicon alone leads from an amorphous structure to a mixed to microcrystalline structure, this limit exceeds temperatures over 400 to 450°C if the layer is deposited from a mixture of Silane and Methane. The role of hydrogen gas dilution and the applied RF power also have a significant effect on the temperature for the transition from the amorphous to the microcrystalline phase. When using monomethylsilane (MMS - SiH_3-CH_3) the situation is similar, yet it should be also supporting the formation of one-dimensional Ge NPs in the crystalline form within the still amorphous SiC:H layer. However, the increased deposition temperature to 400 °C is also causing the increase in reactivity between the a-SiC:H surface and vaporized Ge. Germanium does not form separate islands on the surface, but a continuous amorphous layer. However, the used plasma treatment using methane diluted with hydrogen and therefore the assumed weak carbonization of the a-SiC:H surface leads back to the grouping of the vaporized germanium and to its crystalline structure.

ACKNOWLEDGEMENTS

The work was supported by the Czech Science Foundation - Project 24-10607J, by the Ministry of Education, Youth and Sports of Czech Republic (Meys) – Danube Project 8X23025 and Czechnanolab Research Infrastructure LM2023051 and by the CAS and SAS - Mobility Plus Project SAV-23-13 (CAS-SAS-2022-08)

REFERENCES

[1] STUCHLÍKOVÁ, T.H., STUCHLÍK, J., REMEŠ, Z., TAYLOR, A., MORTET, V., ASHCHEULOV, P., GREGORA, I., KRIVYAKIN, G., VOLODIN, V. High temperature PIN diodes based on amorphous hydrogenated silicon carbon alloys and boron-doped diamond thin films. *Phys. Status Solidi b*, 2020, vol. 257, no. 6, 1900247(1) - 1900247(6).

[2] STREET, R. A. *Hydrogenated amorphous silicon*. New York : Cambridge University Press, 1991

[3] TORCHYNSKA, T. V. and VOROBIEV, Yu V. (eds.). *Nanocrystals and quantum dots of group IV semiconductors.* Stevenson Ranch, Calif : American Scientific Publ., Nanotechnology book series. ISBN 1-58883-154-X. 2010

[4] HABUKA, H., OHMORI, H., ANDO, Y. Silicon carbide film deposition at low temperatures using monomethylsilane gas. *Surface and Coatings Technology*. 2010, vol. 204, Issues 9–10, pp 1432-1437.

[5] YUNAZ, I.A., HASHIZUME, K., MIYAJIMA, S., YAMADA, A., KONAGAI, M. Fabrication of amorphous silicon carbide films using VHF-PECVD for triple-junction thin-film solar cell applications. *Solar Energy Materials and Solar Cells*. 2009, vol. 93, Issues 6–7, pp 1056-1061

[6] SWAIN, B., DUSANE, R.O. Multiphase structure of hydrogen diluted a-SiC:H deposited by HWCVD. *Materials Chemistry and Physics*. 2006, vol 99, Issues 2–3, pp. 240-246.

[7] CHEN, T., YUELONG HUANG, Y., WANG, H., YANG, D., DASGUPTA, A., CARIUS, R., FINGER, F., Microcrystalline silicon carbide thin films grown by HWCVD at different filament temperatures and their application in n-i-p microcrystalline silicon solar cells. *Thin Solid Films*. 2009, vol. 517, Issue 12, pp 3513-3515.

[8] KALOYEROS. A.E., ARKLES, B. Review—Silicon Carbide Thin Film Technologies: Recent Advances in Processing, Properties, and Applications: Part II. PVD and Alternative (Non-PVD and Non-CVD) Deposition Techniques. *J. Solid State Sci. Technol*. 2024, vol. 13, no. 4, 043001.

[9] LIOU Y., SHEN. Y. L. Magnetic Properties of Germanium Quantum Dots. *Journal Metrics: Advanced Materials*, 2008, pp 639-848.

[10] KATKAR A.S. Chapter 9 - Magnetic properties of doped germanium nanostructures. *Micro and Nano Technologies*. 2021, pp 213-234

[11] MOONEY, J.; KAMBHAMPATI, P. Get the Basics Right: Jacobian Conversion of Wavelength and Energy Scales for Quantitative Analysis of Emission Spectra. *J. Phys. Chem. Lett*. 2013, vol. 4, pp 3316–3318

TAILORABLE DIAMOND CAVITIES: A NOVEL APPROACH TO SELF-STANDING DIAMOND FILM FABRICATION

[1]Tibor IZSÁK, [2]Ondrej SZABÓ, [1]Bohumír ZAŤKO, [2]Alexander KROMKA

[1]*Institute of Electrical Engineering SAV, Bratislava, Slovakia, EU,* tibor.izsak@savba.sk, bohumir.zatko@savba.sk

[2]*Institute of Physics AS CR v.v.i., Prague, Czech Republic, EU,* szabo@fzu.cz, kromka@fzu.cz

https://doi.org/10.37904/nanocon.2024.4997

Abstract

This study presents a novel approach for fabricating self-standing diamond films with tailorable cavities, eliminating the need for reactive ion etching (RIE). We employed selective area deposition on SiO_2/Si substrates in a microwave plasma CVD system at high temperatures (1000°C). The method exploits the thermal expansion mismatch between diamond and SiO_2/Si substrates to induce film delamination, resulting in free-standing structures. Diamond films with thicknesses of approximately 50 μm were produced, featuring cavities whose dimensions could be controlled by lithography mask dimensions and deposition time. Raman spectroscopy confirmed the high quality of the diamond films, showing a dominant peak centered at 1330 cm^{-1}. The presented technological approach offers a promising solution for microstructuring of diamond films with potential applications in high-tech applications.

Keywords: Diamond, chemical vapor deposition, SEM, self-standing, MEMS

1. INTRODUCTION

The diamond film exhibits a multitude of exceptional properties [1,2], including high mechanical hardness, Young's modulus, thermal conductivity, and a wide bandgap, which collectively afford it a vast array of applications in micro-electro-mechanical systems (MEMS), deep ultraviolet photoelectronics, and radiation detectors [3]. In addition, diamond's biocompatibility is highly advantageous as a functional layer in biosensors or artificial substrates for cell culture [4]. In such applications, the surface morphology significantly impacts cell differentiation and proliferation [5], and hermetically coated 3D substrates with diamond are preferred due to minimal toxicity. In particular, a self-standing diamond film with defined holes/cavities could be highly advantageous for these applications.

All these applications necessitate the microstructuring of diamond films. However, diamond's extremely high chemical inertness renders wet chemical etching impractical. An alternative method is dry reactive ion etching (RIE) [6,7], which using lithography masks can produce the desired patterns over large areas. The selective area deposition (SAD) [8] can circumvent the necessity for deep RIE. This technique, along with selective area nucleation (SAN), offers promising alternatives for creating self-standing diamond substrates, eliminating the need for deep RIE of thick diamond films.

In this paper, we discuss technological progress in SAN and SAD of self-standing diamond substrates, focusing on their potential to overcome the limitations of traditional microstructuring methods.

2. EXPERIMENTAL

The experiments employed 10x10 mm^2 Si substrates coated with a 1.2 μm thick SiO_2 layer. Prior to the deposition of the diamond, SAN was conducted using nanodiamond particles. In the SAN process [8], the substrates are initially coated with a photosensitive resist (ma-P1215) and subsequently subjected to a soft baking at 100°C. Subsequently, the substrate is seeded with diamond nanoparticles through ultrasonic agitation in a diamond suspension, followed by a second photoresist layer deposition via spin coating. Optical lithography is employed to pattern the seeding layer. A periodic array of circles with diameters of 100, 200, or 300 μm was employed as a lithography mask (**Figure 1a**). As the final step, the development of the photoresist followed by oxygen-plasma RIE are employed to eliminate the diamond residuals, and the upper photoresist layer (**Figure 1b**). Subsequently, the SAN processed samples were loaded in the MWCVD ellipsoidal cavity reactor (Aixtron P6, [9]) to initialize the diamond growth on desired areas (**Figure 1c**).

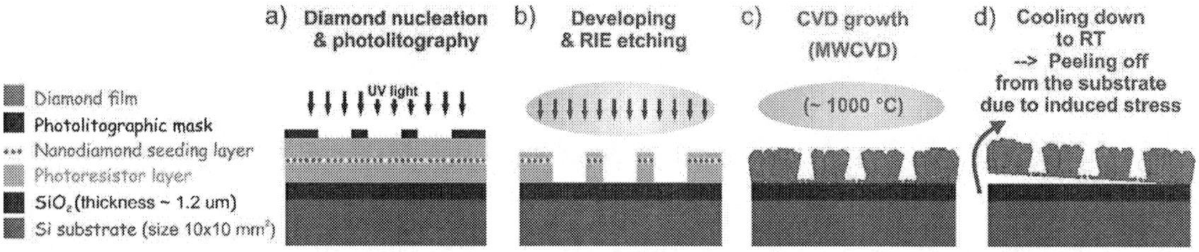

Figure 1 Technological procedure for fabricating self-standing diamond films using high-temperature deposition on SiO_2/Si substrates

The diamond growth conditions were as follows: microwave power 4200 W, pressure 90 mbar, gas mixture 5% CH_4 and 1.5% CO_2 in H_2, process time 40-60 h. The high microwave power and relatively high pressure resulted in a process temperature of 960-1000°C. The elevated deposition temperature and the discrepancy between the thermal expansion coefficients (TEC) of the diamond and SiO_2 layers and the Si substrate resulted in the peeling off of the diamond film from the substrate when the sample was cooled to room temperature (**Figure 1d**).

The morphology and chemical character of diamond films were characterized with scanning electron microscopy (MAIA3, Tescan Ltd., Czechia) and Raman microscopy (Renishaw InVia Reflex Raman spectrometer, UK) with an excitation wavelength of 442 nm.

3. RESULTS AND DISCUSSION

The surface morphology of the diamond film exhibited a replication of the area where diamond nanoparticles had selectively nucleated. **Figure 2a** depicts the final surface morphology of the sample to which a lithography mask with a periodic array of 100 μm circles was applied. Following the growth of the diamond film (with a thickness of approximately 50 μm), the initial 100 μm wide area devoid of diamond was reduced to a 22 μm wide cavity. This observation corroborates the hypothesis that the diamond film grows not only in the vertical direction but also in the lateral direction. It can be expected that increasing the deposition time will result in fully closed diamond cavities. So, the width of the diamond cavities can be controlled by the lithography mask (100, 200, or 300 μm wide circles) and the diamond deposition time.

Figure 2b depicts the peeling off/delamination of the diamond film from the substrate during the cooling process from deposition to room temperature. The high deposition temperature is a critical factor in the fabrication of self-standing diamond films. The disparate thermal expansion coefficients of diamond and SiO_2 (or Si) give rise to considerable internal stress at the interface between these dissimilar materials [10–14], ultimately resulting in the delamination of the diamond film from the substrate.

Figure 2c depicts optical photographs of the 10x10 mm^2 free-standing diamond films from the back and front sides. The optical photograph of the back side exhibits a markedly glossier appearance than the front side, reflecting the smooth surface roughness of the SiO_2/Si substrate. In contrast, the surface roughness of the front (top) side of the sample is considerably higher, resulting in the scattering of light and a less reflective surface.

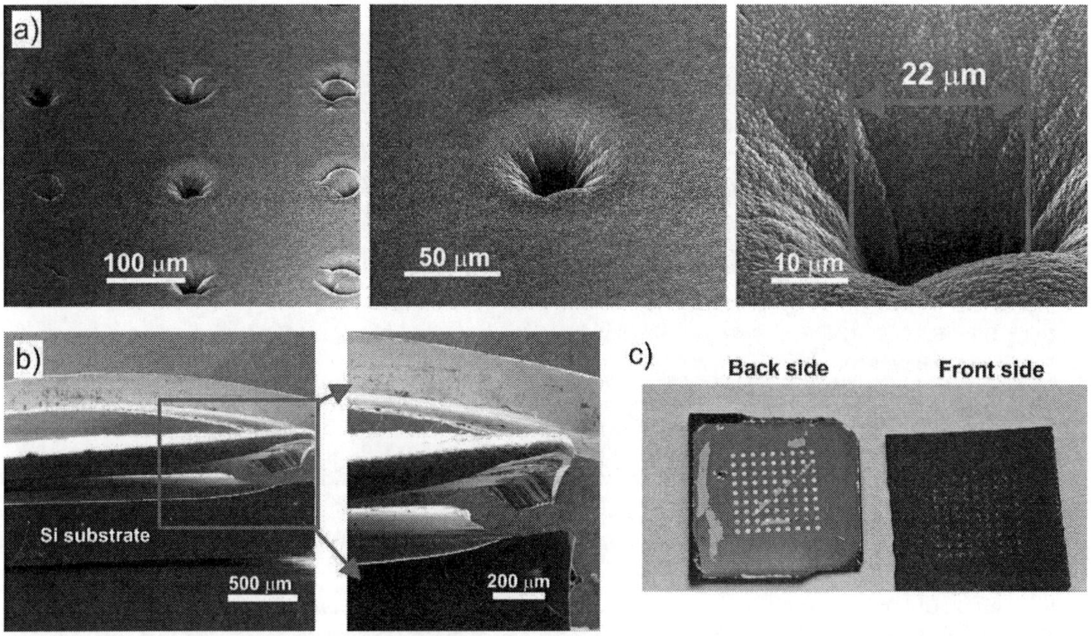

Figure 2 a) Top view SEM images of selective area-grown diamond film with periodically aligned cavities (the initial diameter of the circles in the SAN process was 100 µm). b) Illustration of peeling/delamination of the diamond film from the substrate taken by SEM. c) Optical images of 10x10 mm^2 free-standing diamond films taken from the back and front.

Raman spectroscopy also characterized the diamond film, utilizing an excitation wavelength of 442 nm. The Raman spectra exhibited high similarity across all samples, displaying a prominent diamond peak at 1330 cm^{-1} and a relatively weak G-band at 1532 cm^{-1} [15]. These findings substantiated the presence of a high-quality diamond film with minimal graphitic phases. Moreover, no Si peak was detected on the reverse side of the self-standing diamond film, suggesting that the diamond film is delaminated directly from the SiO_2 layer. In summary, these preliminary experiments have validated the potential of combining SiO_2/Si substrates with high-temperature diamond deposition to fabricate self-standing diamond films without etching. This approach offers a promising alternative to traditional microstructuring methods, potentially simplifying the production process of diamond-based devices for various applications. However, further experiments and simulations are necessary to provide a more comprehensive understanding of this research area. Future studies should focus on optimizing deposition parameters, investigating the effects of substrate properties on film quality, and exploring the scalability of this technique for larger-area diamond films.

4. CONCLUSION

Self-standing diamond films with tailorable cavities offer numerous potential applications, particularly in the fields of radiation detectors, (bio-) sensors, MEMS, and biotechnology. This study presented a straightforward approach for fabricating such films without the need for a RIE process. Tailorable diamond cavities were achieved using SAD in a microwave plasma chemical vapor deposition system.

The diamond growth process was conducted at a high temperature of approximately 1000°C, which was crucial in fabricating self-standing diamond films. Due to the differing thermal expansion coefficients of diamond and SiO_2 (and/or Si), high internal stress is induced at the interface of these dissimilar materials, resulting in the delamination of the diamond film from the substrate. The diameter of the diamond cavities can be controlled by adjusting the lithography mask dimensions and deposition time. Finally, this allows for the customization of diamond film structures to meet specific application requirements.

ACKNOWLEDGEMENTS

This work was supported by the bilateral research project of the Slovak Research and Development Agency grant no. SK-CZ-RD-21-0116 (CZ no. LU-ASK22147), by the Slovak Research and Development Agency, grant no. APVV-22-0382 and the MEYS OP JAC project SENDISO no. CZ.02.01.01/00/22_008/0004596.

REFERENCES

[1] COE, S.E., SUSSMANN, R.S. Optical, thermal and mechanical properties of CVD diamond. *Diamond and Related Materials*. 2000, vol. 9, pp. 1726–9.

[2] ZHANG, C., VISPUTE, R, D., FU, K., NI, C. A review of thermal properties of CVD diamond films. *Journal of Materials Science*. 2023, vol. 58, pp. 3485–507.

[3] LIAO, M. Progress in semiconductor diamond photodetectors and MEMS sensors. *Functional Diamond*. 2021, vol. 1, pp. 29–46.

[4] NISTOR, P.A., MAY, P.W. Diamond thin films: giving biomedical applications a new shine. *Journal of The Royal Society Interface*. 2017, vol. 14, pp. 20170382.

[5] KRESS, S., NEUMANN, A., WEYAND, B., KASPER, C. *Stem Cell Differentiation Depending on Different Surfaces*. In: Kasper, C, Witte, F, Pörtner, R, eds. Tissue Eng. III Cell - Surf. Interact. Tissue Cult., vol. 126. Springer Berlin Heidelberg: Berlin, Heidelberg, 2011:263–83.

[6] IZAK, T., KROMKA, A., BABCHENKO, O., LEDINSKY, M., HRUSKA, K., VERVENIOTIS, E. Comparative study on dry etching of polycrystalline diamond thin films. *Vacuum*. 2012, vol. 86, pp. 799–802.

[7] ZHENG, Y., YE, H., LIU, J., WEI, J., CHEN, L., LI, C. Surface morphology evolution of a polycrystalline diamond by inductively coupled plasma reactive ion etching (ICP-RIE). *Materials Letters*. 2019, vol. 253, pp. 276–80.

[8] IZAK, T., BABCHENKO, O., JIRÁSEK, V., VANKO, G., VALLO, M., VOJS, M., et al. Selective area deposition of diamond films on AlGaN/GaN heterostructures. *Physica Status Solidi (b)*. 2014, vol. 251, pp. 2574–80.

[9] FÜNER, M., WILD, C., KOIDL, P. Novel microwave plasma reactor for diamond synthesis. *Applied Physics Letters*. 1998, vol. 72, pp. 1149–51.

[10] JIRÁSEK, V., IŽÁK, T., VARGA, M., BABCHENKO, O., KROMKA, A. Investigation of residual stress in structured diamond films grown on silicon. *Thin Solid Films*. 2015, vol. 589, pp. 857–63.

[11] MICHLER, J., MERMOUX, M., VON KAENEL, Y., HAOUNI, A., LUCAZEAU, G., BLANK, E. Residual stress in diamond films: origins and modelling. *Thin Solid Films*. 1999, vol. 357, pp. 189–201.

[12] GUNNARS, J., ALAHELISTEN, A. Thermal stresses in diamond coatings and their influence on coating wear and failure. *Surface and Coatings Technology*. 1996, vol. 80, pp. 303–12.

[13] FAN, Q, H., GRÁCIO, J., PEREIRA, E. Residual stresses in chemical vapour deposited diamond films. *Diamond and Related Materials*. 2000, vol. 9, pp. 1739–43.

[14] FERREIRA, N, G., ABRAMOF, E., LEITE, N.F., CORAT, E.J., TRAVA-AIROLDI, V.J. Analysis of residual stress in diamond films by x-ray diffraction and micro-Raman spectroscopy. *Journal of Applied Physics*. 2002, vol. 91, pp. 2466–72.

[15] KLAUSER, F., STEINMÜLLER-NETHL, D., KAINDL, R., BERTEL, E., MEMMEL, N. Raman Studies of Nano- and Ultra-nanocrystalline Diamond Films Grown by Hot-Filament CVD. *Chemical Vapor Deposition*. 2010, vol. 16, pp. 127–35.

IRON-NICKEL NANOCRYSTALLINE PARTICLES – EFFECT OF PREPARATION OF PRECURSORS

[1]Eva ŠVÁBENSKÁ, [1,2]Pavla ROUPCOVÁ, [1,2]Lubomír HAVLÍČEK, [1]Oldřich SCHNEEWEISS

[1]*Institute of Physics of Materials, AS CR, Brno, Czech Republic, EU, svabenska@ipm.cz*

[2]*CEITEC Brno University of Technology, Brno, Czech Republic, EU,*

https://doi.org/10.37904/nanocon.2024.5000

Abstract

Fe-Ni-based nanoparticles are promising materials for the production of high-performance permanent magnets. The research explores the preparation of Fe-Ni alloys using iron-nickel oxalate precursors, which were synthesized via the coprecipitation method and subsequently reduced in a hydrogen atmosphere. One of the aims of the work is to compare the effect of different methods of preparation of precursors on the particle size and composition of the resulting materials. Structural, magnetic, and morphological properties were analyzed using a combination of Mössbauer spectroscopy, X-ray powder diffraction (XRD), and scanning electron microscopy (SEM).

The relative proportions of iron and nickel content in the prepared oxalate precursors were approximately 50% by weight, according to EDS analysis. The precursor samples consisted of clusters of particles with undefined shapes, exhibiting a wide range of particle sizes. After annealing, multiple Fe-Ni phases were present in the sample. Structure investigation and magnetic measurements confirmed the formation of phases suitable for magnetic applications.

Keywords: Magnetic materials, thermal decomposition, Mössbauer spectroscopy, microscopy

1. INTRODUCTION

Magnetic Fe-Ni particles have a wide range of applications, including in the production of inductors, transformers, catalysts, and biomedical devices. [1, 2] The magnetic properties of these materials are heavily influenced by factors such as particle morphology, size, crystallinity, and purity, which are, in turn, dependent on the preparation method used. Various methods for synthesizing Fe-Ni particles have been explored, including mechanical milling [3, 4], coprecipitation [5, 6], chemical reduction [7], etc.

The present work is devoted to preparing Fe-Ni phases with suitable magnetic properties. The first part focuses on synthesizing and characterizing various oxalate precursors. The second part explores the formation of Fe-Ni alloys through thermal decomposition in a reducing atmosphere. The structural characteristics, particle sizes, and magnetic properties of the samples were comprehensively analyzed using a combination of Mössbauer spectroscopy (MS), X-ray diffraction (XRD), magnetization measurements (VSM), scanning electron microscopy (SEM), and energy-dispersive X-ray spectroscopy (EDS).

2. MATERIALS AND METHODS

2.1 Sample preparation

Mixed Fe-Ni oxalate dihydrate crystals were prepared by precipitation. The samples were prepared using 0.1 M solutions of $FeSO_4{\cdot}7H_2O$, $NiSO_4{\cdot}6H_2O$, $FeSO_4{\cdot}(NH_4)_2SO_4{\cdot}6H_2O$, and $NiSO_4{\cdot}(NH_4)_2SO_4{\cdot}6H_2O$, along with a 0.2 M solution of oxalic acid. Oxalate precursor samples were synthesized using the coprecipitation method, with slight variations in the preparation of each sample. For sample ES1, both sulfates were combined and then added to the oxalic acid solution. In the case of sample ES2, sulfate solutions were alternately introduced into the oxalic acid. Sample ES3 followed the same sequential procedure as ES2, but the solutions were additionally treated with ultrasound. For samples ES4 and ES5, one of the sulfates was replaced with ammonium iron sulfate or ammonium nickel sulfate, respectively. The precipitate was isolated by filtration, washed several times with ethanol, and then dried

The oxalate precursor was characterized using microscopy, powder X-ray diffraction, and Mössbauer spectroscopy. After characterization, the samples were annealed in a laboratory furnace under a hydrogen atmosphere to study the effects of thermal treatment on the precursor materials.

2.3 Methods

The chemical composition of the samples was analyzed using a TESCAN LYRA 3XMU FEG/SEM scanning electron microscope equipped with an Oxford Instruments XMax80 detector for energy-dispersive X-ray spectroscopy. Secondary electron images and EDS spectra were obtained at accelerating voltages of 15 kV.

Phase composition studies were conducted using an Empyrean diffractometer (Panalytical) in Bragg-Brentano geometry with CoKα radiation (λ = 0.17902 nm) at room temperature. The resulting patterns were analyzed using the HighScore® software and the ICDD (PanAnalytical) database.

Transmission ^{57}Fe Mössbauer spectra of the studied samples were measured in the 5–293 K temperature range using a standard Mössbauer spectrometer in a constant acceleration mode with a ^{57}Co (Rh) radioactive source. The isomer shift values were referred to as the value of an α-Fe foil sample. The computer processing of the spectra was done using CONFIT program package [8].

Magnetic properties were measured with the EverCool II PPMS (Quantum Design) system using a VSM mode in the temperature range 4-300 K and with the magnetic field range of – 9T to + 9T. Additionally, a Vibrating Sample Magnetometer (VSM) equipped with a furnace capable of reaching up to 800°C and a magnetic field range of -1 T to +1 T was used for magnetic measurements.

3. RESULTS AND DISCUSSION

3.1 Characterization of precursor

The morphology of $Fe_{(1-x)}Ni_xC_2O_4$ particles was investigated using SEM. The samples were made up of a conglomerate of particles of different sizes. The particles displayed a wide range of shapes, from flakes to rectangular structures. According to EDS analysis, the relative proportion of iron and nickel in the prepared oxalate precursor was approximately 48% and 52% by weight, respectively. EDS maps show a homogeneous distribution of iron and nickel in the particles of individual samples.

The XRD patterns of the precursor were identified as Iron Oxalate Dihydrate ($FeC_2O_4{\cdot}2H_2O$). This result might be attributed to the substitution of ions or overlapping diffraction patterns, as iron and nickel oxalates have very similar peak positions, making it challenging to distinguish between them accurately. The XRD analysis estimated an average particle size of around 37 nm. The smallest particle size, approximately 27 nm, was observed in sample ES4, which was prepared using a combination of $FeSO_4{\cdot}(NH_4)_2SO_4{\cdot}6H_2O$ and $NiSO_4{\cdot}6H_2O$.

The Mössbauer spectra of the precursors consist of two quadrupole doublets. The first doublet, with an isomer shift IS = 1.19±0.01 mm/s and quadrupole splitting QS_{pl} = of 1.70±0.02 mm/s, corresponds to Fe-Ni mixed-metal oxalates [9,10]. The second doublet, with IS = 1.29±0.02 mm/s and QS_{pl} = 2.3±0.04 mm/s, is associated with iron sulfate monohydrate. The observed quadrupole splitting values are lower than the reported 2.71 mm/s in the literature [11], which suggests that the increase in QS may be due to the presence of nickel ions in the sample. The phase composition, represented as the atomic fraction of Fe, ranges from 0.89 to 0.96 for the first doublet (D1) and from 0.03 to 0.10 for the second doublet (D2). Detailed phase compositions for each sample are provided in **Table 1**.

Table 1 Hyperfine parameters resulting from the Mössbauer spectra analysis of oxalate precursors (IS - isomer shift, QS_{pl} - quadrupole splitting, A - phase content in atomic fraction of Fe).

Sample	D1			D2		
	IS (mm/s)	QS_{pl} (mm/s)	A	IS (mm/s)	QS_{pl} (mm/s)	A
ES1	1.18±0.01	1.71±0.01	0.96	1.29±0.01	2.39±0.04	0.04
ES2	1.19±0.01	1.71±0.01	0.95	1.29±0.01	2.35±0.01	0.05
ES3	1.18±0.01	1.70±0.01	0.93	1.29±0.01	2.34±0.06	0.07
ES4	1.19±0.01	1.72±0.01	0.93	1.29±0.01	2.35±0.05	0.07
ES5	1.19±0.01	1.70±0.01	0.89	1.29±0.01	2.25±0.10	0.10

Figure 1 SEM images of morphological changes observed on sample ES 3: oxalate precursor as starting material (on the left); FeNi powder after thermal decomposition (on the right).

3.2 Characterization of thermal decomposition of ES3 precursor

Sample ES3 underwent thermal decomposition in a hydrogen atmosphere. The sample was heated to 600 °C and then cooled to room temperature. This process was conducted simultaneously in a laboratory furnace and in a vibrating-sample magnetometer equipped with the furnace, enabling the study of changes in the magnetic properties of the sample during annealing.

Microscopic observations of the sample after thermal decomposition revealed that the ratio of iron to nickel remained unchanged. The annealing time influenced the sample's morphology. Unlike to preceding study, where the sample formed a porous structure after being annealed at 320 °C for 20 hours in a hydrogen atmosphere [12], no such porous structure was observed in the ES3 sample. A comparison of the morphology of sample ES3 before and after thermal decomposition is presented in **Figure 1**.

The results of the magnetic measurements indicate that the decomposition of the precursor in H_2 begins at approximately 360 °C (**Figure 2**) and this first stage completes around 400 °C. Phase composition according to Mössbauer spectroscopy measurement at room temperature is as follows (in iron atomic fraction): 0.64 magnetite Fe_3O_4, 0.25 Fe-Ni alloy, and 0.11 remains of precursor. The spectrum is shown in **Figure 3**. The subspectra can be described by following hyperfine field parameters. Magnetite particles are represented with two sextets. The first one with hyperfine induction B_{hf} = 48.3±0.1 T, isomer shift IS = 0.31±0.01 mm/s, quadrupole shift QS_{sh}= 0.00±0.01 mm/s, and relative spectral area (equal approx. to atomic fraction of iron atoms) A_1=0.31±0.01. The second sextet has B_{hf} = 45.4±0.5 T, IS = 0.65±0.01 mm/s, QS_{sh}= 0.02±0.01 mm/s, and A_2=0.33±0.01. Subspectrum ascribed to FeNi alloy particles are characterized by the sextet with B_{hf} = 28.3±0.1 T, IS = 0.04±0.02 mm/s, QS_{sh}=- 0.01±0.03 mm/s, and A=0.25±0.01. The third phase is represented by singlet with IS = 0.21±0.03 mm/s and A=0.10±0.01 and can be ascribed to precursor remains. The Mössbauer spectrum of the sample ES 3 after reduction in hydrogen atmosphere up to 600 °C was fitted using three subspectra which are ascribed to magnetite and FeNi particles. The first sextet of magnetite has B_{hf} = 48.8±0.6 T, IS = 0.11±0.07 mm/s, QS_{sh}= 0.10±0.09 mm/s, and A_1=0.22±0.02, the second one has B_{hf} = 45.8±0.5 T, IS = 0.53±0.05 mm/s, QS_{sh}= 0.02±0.01 mm/s, and A_2=0.31±0.01. The FeNi phase has subspectrum B_{hf} = 30.4±0.1 T, IS = 0.17±0.02 mm/s, QS_{sh}=- 0. 06±0.03 mm/s, and A=0.46±0.02. The phase composition and Mössbauer spectra parameters are in good agreement with previously published data. [6, 13, 14]. The change in the ratio of A_1/A_2 after annealing up to 600 °C in comparison with the annealing up to 400 °C very probably corresponds with oxygen stoichiometry in the magnetite.

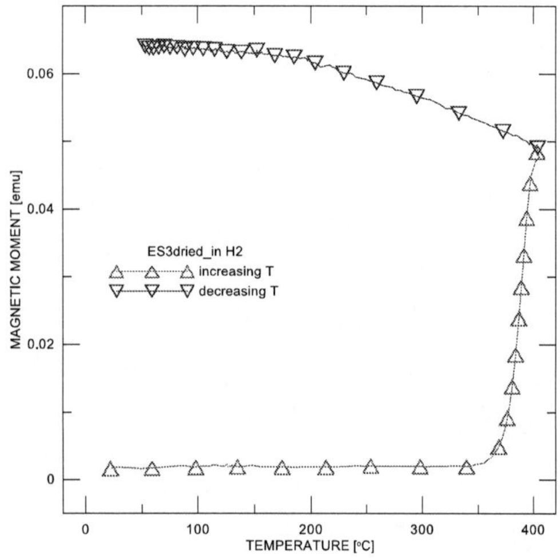

Figure 2 Magnetic measurement in H2 in field 50 Oe

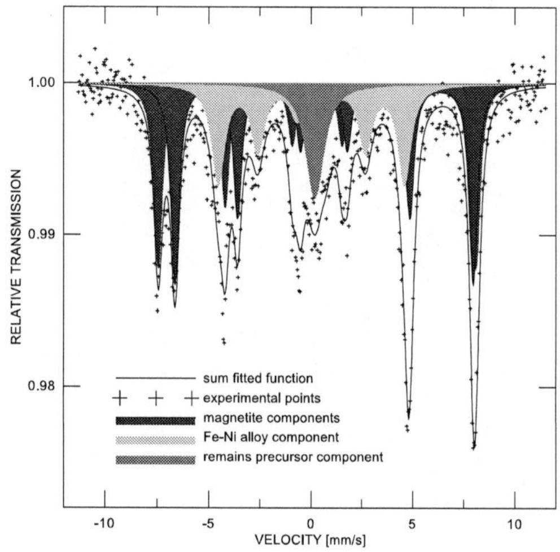

Figure 3 Mössbauer spectrum taken after magnetic measurement shown in Figure 2

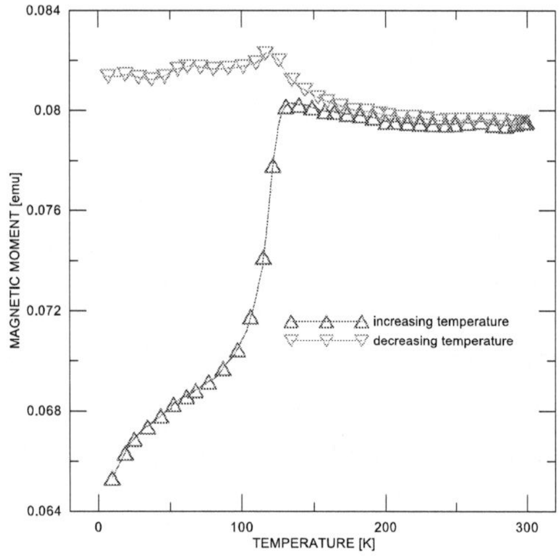

Figure 4 Temperature dependence of magnetic moment ZFC-FC of S3 sample after annealing in H2 at 600 °C.

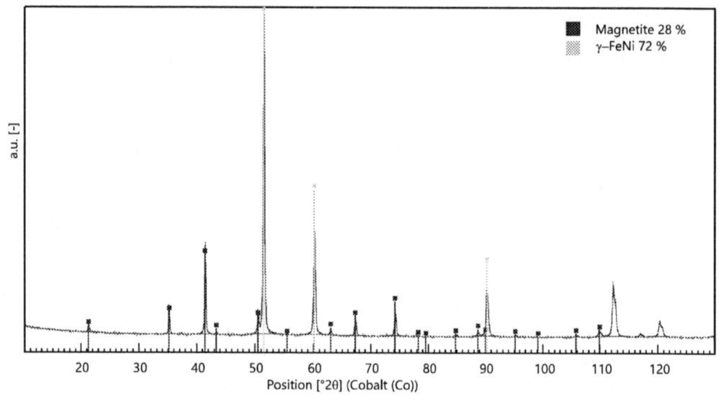

Figure 5 XRD of S3 sample after annealing in H_2 at 600 °C taken at room temperature.

Magnetic measurements of the ES3 precursor sample after annealing in H_2 up to 600 °C indicate a Curie temperature of 460 °C, which can be attributed to the Fe-Ni alloy component. Temperature dependence of magnetic moment ZFC-FC (Zero Field Cooled – Field Cooled) (**Figure 4**) confirms the presence of magnetite with the jumps at ~ 120 K (~ 150°C) corresponding to its Verwey transition.

The result of XRD (**Figure 5**) taken at room temperature confirms the magnetite and fcc FeNi phases in the ES3 sample after annealing in H_2 at 600 °C with mean coherent sizes 62 nm and 58 nm, respectively.

4. CONCLUSION

The morphology of $Fe_{(1-x)}Ni_xC_2O_4$ particles, studied through SEM, revealed a conglomerate of particles with diverse shapes, ranging from flakes to rectangular structures. EDS analysis confirmed a uniform distribution of iron and nickel, with the relative composition at 48% and 52% by weight, respectively. XRD identified the precursor as Iron Oxalate Dihydrate with an average particle size of 37 nm, while Mössbauer spectroscopy indicated the presence of Fe-Ni mixed-metal oxalates and iron sulphate monohydrate phases.

Characterization of the ES3 sample after thermal decomposition in a hydrogen atmosphere revealed the formation of magnetite and FeNi alloy particles. After annealing at 600°C, the mean crystallite sizes were determined to be 62 nm and 58 nm, respectively. These results suggest that the ES3 precursor is suitable for producing a mixture of magnetic nanoparticles with a Fe-Ni phase close to the desired $Fe_{50}Ni_{50}$ composition.

ACKNOWLEDGEMENTS

This work has been created by financial support and using of the research infrastructure of the Institute of Physics of Materials Czech Academy of Sciences, v. v. i.; CzechNanoLab project LM2023051 funded by MEYS CR is gratefully acknowledged for the financial support of the measurements at CEITEC Nano Research Infrastructure.

REFERENCES

[1] COEY, J.M.D. Pernament magnets: Plugging the gap. *Scripta Materialia.* 2012, vol. 67, pp. 524-529. https://doi.org/10.1016/j.scriptamat.2012.04.036

[2] CUI, J., KRAMER, M., ZHOU, L., LIU, F., GABAY, A., HADJIPANAYIS, G., BALASUBRAMANIAN, B., SELLMYER, D. Current progress and future challenges in rare-earth-free permanent Magnets. *Acta Materialia.* 2018, vol. 158, pp. 118-137. https://doi.org/10.1016/j.actamat.2018.07.049

[3] ABDU, Y.A., ERICSSON, T., ANNERSTEN, H. Coexisting antiferromagnetism and ferromagnetism in mechanically alloyed Fe-rich Fe–Ni alloys: implications regarding the Fe–Ni phase diagram below 400°C. *Journal of Magnetism and Magnetic Materials*. 2004, vol. 280, issue 1-2, pp. 395-403. https://doi.org/10.1016/j.jmmm.2004.03.036

[4] LEE, S., EDALATI, K., IWAOKA, H., HORITA, Z., OHTSUKI, T., OHKOCHI, T., KOTSUGI, M., KOJIMA, T., MIZUQUCHI, M., TAKANASHI, K. Formation of FeNi with L10-ordered structure using high-pressure torsion. *Philosophical Magazine Letters*. 2014, vol. 94, pp. 639–646. https://doi.org/10.1080/09500839.2014.955546

[5] YAO, Y., ZHANG, Ch., FAN, Y., ZHAN, J. Preparation and microwave absorbing property of porous FeNi powders. *Advanced Powder Technology*. 2016, vol. 27, issue 5, pp. 2285–2290. https://doi.org/10.1016/j.apt.2016.08.022

[6] ZHU, S., LI, N., ZHANG, D., YAN, T. Metal/oxide heterostructures derived from Prussian blue analogues for efficient photocatalytic CO2 hydrogenation to hydrocarbons. *Journal of CO2 Utilization*. 2022, vol. 64,102177 https://doi.org/10.1016/j.jcou.2022.102177

[7] DHANALAKSHMI, G., RAVICHANDRAN, V. Synthesis of nanocrystalline nickel-iron alloys-A novel chemical reduction method. *Chemical Physics Impact.* 2023. vol. 6,100202. https://doi.org/10.1016/j.chphi.2023.100202

[8] ŽÁK, T., JIRÁSKOVÁ, Y. Confit: Mössbauer spectra fitting program. Surface and Interface Analysis. 2006, vol. 38, pp. 710–714. https://doi.org/10.1002/sia.2285

[9] RAMANI, SATHYAVATHIAMMA, M. P., PUTTASWAMY, N. G., MALLYA R. M. Mössbauer Effect Study of the Thermal Decomposition in Hydrogen of Homogeneously Mixed Ferrous Nickel Oxalates with Different Iron to Nickel Ratios. *Physica Status Solidi (a)*. 1983, vol. 77, issue 1, pp. 87–96. https://doi.org/10.1002/pssa.2210770110

[10] DEVILLERS, M., LADRIÈRE, J., APERS, D. Mössbauer study of 57Fe-doped simple and mixed dihydrated oxalates of bivalent metals (M = Mg, Mn, Fe, Co, Ni, Zn). *Inorganica Chimica Acta.*1987, vol.126, issue 1, pp. 71-77. https://doi.org/10.1016/S0020-1693(00)81242-6

[11] PETKOVA, V., PELOVSKI, Y., PANEVA, D., MITOV, I. Influence of gas media on the thermal decomposition of second valence iron sulphates. J*ournal of Thermal Analysis and Calorimetry*. 2011, vol. 105, pp. 793–803 https://doi.org/10.1007/s10973-010-1242-6

[12] ŠVÁBENSKÁ, E., ROUPCOVÁ, P., HAVLÍČEK, L., SCHNEEWEISS, O. Properties of nonocrystalline Fe-Ni particles prepared by thermal reduction of oxalate precursors. In: NANOCON 2023 Conference Proceedings 2023. Ostrava: TANGER Ltd., 2024, pp 76-82.

[13] LEHLOOH, A.F.D., MAHMOOD, S.H. Mössbauer Spectroscopy Study of Iron Nickel Alloys. *Hyperfine Interactions*. 2002, vol. 139, pp. 387–392. https://doi.org/10.1023/A:1021230926166

[14] VANDENBERGHE, R., BARRERO, C., DA COSTA, G., VAN SAN, E., DE GRAVE, E. Mössbauer characterization of iron oxides and (oxy)hydroxides: the present state of the art. *Hyperfine Interactions.* 2000. vol. 126, pp. 247–259. https://doi.org/10.1023/A:1012603603203

THE EFFECT OF CONDITIONS OF THERMAL TREATMENT OF Ti+Al+C MIXTURE ON THE FORMATION OF MAX PHASES

Taťána BARVÍKOVÁ, Silvie VALLOVÁ, Kryštof FONIOK, Radim ŠKUTA, Vlastimil MATĚJKA

VSB - Technical University of Ostrava, Faculty of Materials Science and Technology, Ostrava, Czech Republic, EU, tatana.barvikova.st@vsb.cz, silvie.vallova@vsb.cz, krystof.foniok@vsb.cz, radim.skuta@vsb.cz, vlastimil.matejka@vsb.cz

https://doi.org/10.37904/nanocon.2024.4999

Abstract

MAX phase materials are nano-layered, hexagonal, metal carbides, and nitrides, where M is a transition metal, A is an element of the A-group (mostly groups 13 and 14), and X is C and/or N. These phases exhibit a unique combination of properties of both metallic and ceramic materials. Physical and technological troubles usually affect the high-temperature synthesis of MAX phases, resulting in decreased yields of the MAX phase and promoting the formation of undesirable phases. For example, in the case of Ti-Al-C based MAX phases, the titanium carbide TiC and the Al_3Ti, TiAl and Ti_3Al intermetallic can also be formed under the synthesis process of Ti_3AlC_2 and Ti_2AlC phases. Additionally, Al can be oxidized, and Al_2O_3 often occurs in resulting products. The successful preparation of MAX phases depends on various conditions such as: heating temperature, holding time, heating rate, or gas flow. In this study, the Ti-Al-C MAX phases were prepared, by high-temperature treatment of mixture of Ti, Al, and C powders. The synthesis was conveniently conducted in the furnace of thermal analyzer, which enables precisely control the heating conditions. The effect of various parameters used during the synthesis on the MAX phases formation was tested, and the resulting products were analyzed using powder X-ray diffraction.

Keywords: MAX phase, Ti_3AlC_2, Ti_2AlC, TG/DTA analysis, phase analysis

1. INTRODUCTION

MAX phases are ternary, hexagonal compounds with formula $M_{n+1}AX_n$ where M is the transition metal, A is the element from A – Groups of the periodic table, and X stands for carbon or nitrogen. These crystalline compounds have a layered structure, in which carbide or nitride blocks are separated by monolayers of atoms of Group A [1]. MAX phases combine the advantages of both metal and ceramic materials. Like metals, they are thermally and electrically conductive, readily machinable at room temperature, resistant to thermal shock, have crack propagation resistance, and deform plastically at elevated temperatures. Similarly to ceramics, they have low density, high elastic modulus, low thermal expansion coefficient, excellent heat resistance, and high temperature strength [2].

Ternary compounds in the Ti-Al-C system, mainly Ti_2AlC and Ti_3AlC_2, are the phases of the largest area of interest in the series of MAX phases [1]. In these compounds, TiC layers are alternately interspersed with Al layers, and weak metallic bonds are formed between Ti and Al atoms, while stronger covalent-ionic bonds are formed between Ti and C. The properties of Ti_3AlC_2 make it a great candidate in applications such as electrical contact coating [3,4], high temperature refractory materials [5], as a material for electrodes of batteries and supercapacitors [6], and in combination with magnesium also for hydrogen storage applications [7]. In 2011, Gogotsi, et. al described the exfolation of Ti_3AlC_2 by etching Al out in hydrofluoric acid [8] and discovered a new class of 2D materials called MXenes.

As reported by Desay et al. [9], the MAX phase Ti_2AlC was firstly synthesized by Jeitschko et al. [10] in 1963 through process of Chemical Vapour Deposition. Similarly, Desay et al. [9] mentioned that successful synthesis of Ti_3AlC_2 was firstly reported in 1994 by Peitzka and Schuster [11] who had used a cold compaction of Ti, C and Al_4C_3 powder followed by its sintering in hydrogen atmosphere at 1300 °C for 2 h. Currently, there are a variety of methods by which these compounds may be synthesized, for example hot isostatic pressure (HIP) [12], self-propagating high temperature synthesis (SHS) [13], or spark plasma sintering (SPS) [14]. These synthesis methods are based on high temperature treatment of the mixture of powdered titanium, carbon, and aluminum. The selected process of Ti+Al+C mechanical mixture heating influence the quality of the final product. For example, Zhou et al. [15] observed that the heating conditions used during the synthesis determine the amount of residual phases, especially TiC and high rates of temperature increase is recommended to suppress TiC formation. Ovodok et al. [16] studied the effect of the temperature on Ti_3AlC_2 formation during the heating of Ti+Al+C mechanical mixture under vacuum and observed the formation of the Ti_3AlC_2 already at the temperature 1100 °C. The authors also indicated the lowest amount of TiC and Ti_2AlC impurities for the sample prepared at 1300 °C.

As evident, the heating conditions during the preparation of Ti-Al-C MAX phases play an important role in the formation and purity of the final product. In our research, we focused on the indication of how the heating rate, temperature, and holding time at this temperature influence the phase composition of the product obtained by heating of the Ti+Al+C mechanical mixture.

2. MATERIALS AND METHODS

As a raw materials, the titanium powder from AP&C company containing spherical particles of size 15-45 µm, the aluminium powder (99,5% purity, particle size > 120 mesh) from International Enzymes Ltd. Company (United Kingdom) and graphite in its powder form (99,8% purity, <0.1 mm granulometry) from Fluka Chemie AG (Germany) were used. All reagents were used as received without any further pre-treatment. Individual components (Ti, Al and C) were mixed in molar ratio 3 : 1.1 : 2 and homogenized 30 min in in planetary ball mill.

Thermogravimetry analysis/differential scanning calorimetry (TGA/DSC) was carried out on a simultaneous thermal analyzer SDT 650 TA Instruments, USA which has a DSC/TGA system with horizontal dual-beam design for heat flow and weight measurements. Each sample (~ 20 mg) was inserted into α-Al_2O_3 crucible and heated up according to the conditions specified in **Table 1**, argon atmosphere (6.0) of flow rate 200 $cm^3\ min^{-1}$ was used for all the experiments. The crucibles were covered with α-Al_2O_3 lid during the experiments. All the tested parameters during the thermal treatment of Ti+Al+C mechanical mixtures are listed in **Table 1**.

Table 1 Conditions during the high temperature treatment of Ti+Al+C mechanical mixtures

Sample	heating rate (°C min^{-1})	holding time (min)	gas flow ($cm^3\ min^{-1}$)	temperature (°C)
P1_A	10	60	200	1300
P1_B	50	60	200	1300
P1_C	50	0	200	1300
P1_D	10	60	200	1400
P1_E	10	60	200	1500

The phase composition of the resulting products was characterized by X-ray powder diffraction method using diffractometer MiniFlex600 (Rigaku, Japan). Registered diffraction patterns were evaluated using SmartLab Studio II software (Rigaku, Japan), the database PDF 2 release 2019 (International Centre for Diffraction Data, USA) was used for the identification of the crystalline phases.

3. RESULTS AND DISCUSSION

As already mentioned in section Introduction, Zhou et al. [15] reported that higher heating rates are preferable for the elimination of the impurities, mainly TiC during the high temperature synthesis of Ti-Al-C MAX phases. Based on this recommendation, we tested heating rates 10 and 50 °C·min^{-1} respectively and obtained samples P1_A and P1_B (all the parameters are listed in **Table 1**). Registered diffraction patterns of these samples are compared in **Figure 1**.

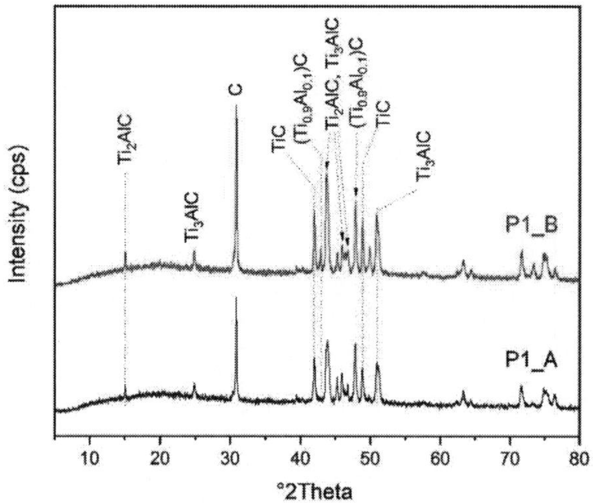

Figure 1 X-ray diffraction patterns documenting the effect of heating rate on the phase composition of final product (P1_A (10°C min^{-1}) and P1_B (50°C min^{-1}))

The phase composition documented by X-ray diffraction patterns of samples prepared at different heating rates (**Figure 1**) did not reveal significant difference in phase composition of the samples. With respect to this observation, the heating rate 50°C min^{-1} was selected for next experiment with holding time.

The effect of the time at which the sample is kept at given temperature (holding time 0 and 60 min) on the phase composition of the resulting products is documented with diffraction patterns pictured in **Figure 2**.

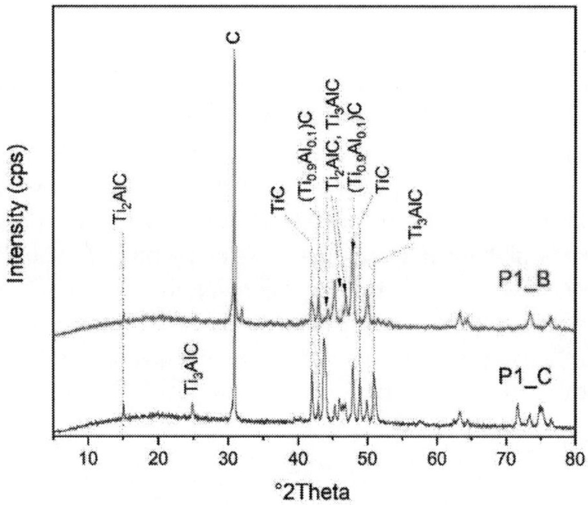

Figure 2 X-ray diffraction patterns documenting the effect of holding time on the phase composition of the final products (P1_C (0 min) and P1_B (60 min))

Comparing the phase composition of the samples, it is noticed, that with prolonged time of heating, the product contains lower amount of TiC, and does not contain Ti_3AlC phase, on the other hand the content of $(Ti_{0.9}Al_{0.1})C$ phase was increasing. The MAX phase Ti_2AlC originated for both tested holding times (0 and 60 min). Higher intensities of diffraction peaks belonging to Ti_2AlC was observed for holding time 60 min, and this value was used for next experiments. The positive effect of prolonged holding time led also to decision to conduct the experiments with lower value of heating rate 10 °C·min^{-1}, which in fact also affect the total time at which the sample stays in higher temperature range.

The effect of the final temperature (1300, 1400 and 1500 °C) on the phase composition of the resulting products is documented with diffraction patterns pictured in **Figure 3**. Only at the temperature 1500 °C the phase Ti_3AlC_2 was formed as evident from diffraction pattern of sample P1_E showed in **Figure 3**. The diffraction pattern of this sample proved also the presence of Ti_2AlC, low intensive diffraction peak centred at approximately 41 °2Theta (CoKα) demonstrate also the presence of Al_2O_3 phase.

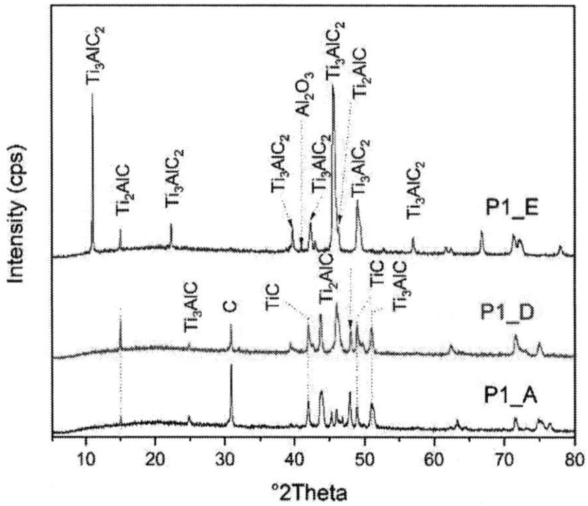

Figure 3 X-ray diffraction patterns documenting the effect of the final temperature on the phase composition of final products (P1_C (1300 °C), P1_D (1400 °C) and P1_E (1500 °C))

4. CONCLUSIONS

The aim of this study was to investigate the effect of different parameters of heating on synthesis of MAX phases based on the system Ti-Al-C in the furnace of thermal analyzer. The X-ray diffraction patterns showed that the heating rate had no significant effect on the composition. In the case of duration of the sample retention at final temperature, it was observed, that longer holding time resulted in the lower amounts of TiC and higher amount of one of the desired MAX phase. Lastly the effect of temperature on MAX phases was investigated. The diffraction patterns showed that the sample prepared at temperature of 1500 °C contained both, Ti_3AlC_2 and Ti_2AlC, while the treatment of the sample at lower temperatures caused the formation of only Ti_2AlC as the target MAX phase. Performed study provided us with the conditions at which the MAX phase Ti_3AlC_2 and Ti_2AlC can originate and will be used for the treatment of the Ti+Al+C mixtures on a larger scale in a tube furnace.

DATA AVAILABILITY

10.5281/zenodo.13918242

ACKNOWLEDGEMENTS

This research was supported by project No. CZ.02.01.01/00/22_008/0004631 Materials and technologies for sustainable development within the Jan Amos Komensky Operational Program financed by the European Union and from the state budget of the Czech Republic. Authors also thank the VSB-Technical University of Ostrava project SP2024/025) for the financial support.

REFERENCES

[1] VERSHINNIKOV, V.I., KOVALEV, D.Y. Preparation of Ti2AlC and Ti3AlC2 MAX Phases by Self-Propagating High-Temperature Synthesis with the Reduction Stage. *Russian Journal of Non-Ferrous Metals.* 2020, vol. 61, pp. 554–558.

[2] POTANIN, A., LOGINOV, P., LEVASHOV, E., POGOZHEV, Y., PATSERA, E., KOCHETOV, N. Effect of Mechanical Activation on Ti3AlC2 Max Phase Formation under Self-Propagating High-Temperature Synthesis. *Eurasian Chemico-Technological Journal.* 2015, vol. 17, pp. 233-242.

[3] FENG, L., MEIQIAN, L.V., QIAN, Q., LUO, R., HUANG, B. The synthesis of high purity Ti3AlC2 MAX phase via molten salt method. *Advanced Powder Technology.* 2023, vol. 34, pp. 103920.

[4] ZHU, Y., TIAN, W., WANG, D. et al. Fabrication and properties of silver-based composites reinforced with carbon-coated Ti3AlC2. *International Journal of Minerals, Metallurgy and Materials.* 2021, vol. 28, pp. 1836–1843.

[5] CHEN, J., LI, N., YAN, W. Influence of Ti_3AlC_2 on corrosion resistance and microstructure of Al_2O_3–Ti_3AlC_2–C refractories in contact with ladle slag. *Journal of the European Ceramic Society.* 2016, vol. 36, no. 6, pp. 1505-1511.

[6] KEWATE, O. J., PUNNIYAKOTI, S. Ti_3AlC_2 MAX phase and $Ti_3C_2T_X$ MXene-based composites towards supercapacitor applications: A comprehensive review of synthesis, recent progress, and challenges. *Journal of Energy Storage.* 2023, vol. 72, pp. 108501.

[7] WANG, K., DU, H., WANG, Z., GAO, M., PAN, H., LIU, Y. Novel MAX-phase Ti_3AlC_2 catalyst for improving the reversible hydrogen storage properties of MgH2. *International Journal of Hydrogen Energy.* 2017, vol. 42, no. 7, pp. 4244-4251.

[8] NAGUIB, M., KURTOGLU, M., PRESSER, V., LU, J., NIU, J., HEON, M., HULTMAN, L., GOGOTSI, Y., BARSOUM, M.W. Two-Dimensional Nanocrystals: Two-Dimensional Nanocrystals Produced by Exfoliation of Ti_3AlC_2. *Advanced Materials.* 2011, vol. 23, pp. 4207-4207.

[9] DESAI, W., SHRIVASTAVA, A., ZALA, A., PAREKH, T., SUROJIT, G., JAMNAPARA, N.I., Pressureless manufacturing of high purity Ti_3AlC_2 MAX phase material: Synthesis and characterisation. *Vacuum.* 2023, vol. 214, pp. 112221.

[10] JEITSCHKO, W., NNOWOTNY, H., BENESOVSKY, F. Kohlenstoffhaltige ternäre Verbindungen (H-Phase). *Monatshefte für Chemie.* 1963, vol. 94, pp. 672–676.

[11] PIETZKA, M.A., SCHUSTER, J.C. Summary of constitutional data on the Aluminum-Carbon-Titanium system. *Journal of Phase Equilibria.* 1994, vol. 15, pp. 392–400.

[12] BAI, Y., HE, X., ZHU, C., CHEN, G. Microstructures, Electrical, Thermal, and Mechanical Properties of Bulk Ti_2AlC Synthesized by Self-Propagating High-Temperature Combustion Synthesis with Pseudo Hot Isostatic Pressing. *Journal of the American Ceramic Society.* 2012, vol 95, no. 1, pp. 358-364.

[13] PAZNIAK, A., BAZHIN, P., SHCHETININ, I., KOLESNIKOV, E., PROKOPETS A., SHPLIS, N., STOLIN, A., KUZNETSOV, D., Dense Ti_3AlC_2 based materials obtained by SHS-extrusion and compression methods. *Ceramics International.* 2019, vol 45, no. 2, pp. 2020-2027.

[14] YUNUS, M., KUMAR, R., MAJI, B.C., KRISHAN, M. An optimized method for synthesizing phase-pure Ti_3AlC_2 MAX-phase through spark plasma sintering. *Journal of the European Ceramic Society.* 2022, vol. 42, no. 2, pp. 354-363.

[15] ZHOU, A., WANG, CA., GE, Z. *et al.* Preparation of Ti_3AlC_2 and Ti_2AlC by self-propagating high-temperature synthesis. *Journal of Materials Science Letters.* 2001, vol. 20, pp. 1971–1973.

[16] OVODOK, E. A., IVANOVSKAYA, M. I., POZNYAK, S. K., MALTANOVA, A. M., AZARKO I. I., MICUSIK, M., OMASTAVA, M., ANISKEVICH, A. Synthesis of Ti_3AlC_2 max phase under vacuum, its structural characterization and using for $Ti_3C_2T_x$ MXene preparation. *Thin Solid Films.* 2023, vol. 771, pp. 139759.

THEORETICAL STUDY OF STRUCTURAL, ELECTRONICS AND THERMOELECTRIC PROPERTIES OF H-BN NANOTUBE ACTIVATED BY FE IONS

[1]F. BOUAMRA, [1]R. BOUZIANE, [1]I. BOUDJEMLINE, [2]M. RÉRAT

[1]LPCMIA Laboratory, Blida1 University, Faculty of Science, Department of Physics, Blida, Algeria, bouamra_faiza@univ-blida.dz, boudjemline_ihssen@univ-blida.dz

[2]IPREM UMR5254, Université de Pau et des Pays de l'Adour, Pau, France, michel.rerat@univ-pau.fr,

https://doi.org/10.37904/nanocon.2024.4998

Abstract

Actually, Boron Nitride nanotubes (BNNT) have received great interest as material dedicated to many applications such as electronic and thermoelectrical devices. This work is a theoretical study of the structural, electronic and thermic transport properties of pure and iron (Fe) doped Boron Nitride nanotubes (BNNT) in its hexagonal phase (h-BN). This theoretical study was performed with the CRYSTAL program (www.crystal.unito.it) providing crystalline orbitals built on localized orbitals at the DFT (B3LYP hybrid hamiltonian) level of approximation. We have shown that Fe-doping nanotubes could behave as narrow bandgap semiconductors, the presence of Fe atoms modifying the gap value because of the presence of states participating to the thermoelectrical transport.

Our calculations show that the undoped nanotubes are ceramic materials with a large band gap around 6 eV while this latter value can be divided by 2 when the nanotube is doped by Fe. The related Seebeck coefficient, and electrical and thermal conductivity properties allow to deduce the values of the figure of merit ZT. Obtained results indicate that BNNT can be considered promising as thermoelectrical materials.

Keywords: BN-Fe nanotubes, DFT, electronic, thermoelectric

1. INTRODUCTION

Based on the similarities between carbon-based materials and boron nitride (BN)-based materials, the existence of boron nitride nanotubes (BNNTs) has been proposed [1,2]. These novel nanomaterials have shown great promise across various fields, including medicine, biology, and the nanotechnology industry. Electronic structure calculations reveal that, unlike carbon nanotubes, BNNTs are consistent bandgap materials, opening intriguing possibilities for potential device applications. Their unique properties make BNNTs appealing for a range of applications, such as optoelectronic nanodevices, functional composites, hydrogen storage, and electrically insulating substrates, including nanotube and nanosheet forms [3, 4].

Additionally, BNNTs exhibit exceptional thermal and mechanical stability, making them ideal candidates for use in nanoelectromechanical systems (NEMS) actuators. Theoretical and experimental studies indicate that BNNTs possess a high crystalline structure with minimal defects, enhancing their utility in ceramics and alloys to reduce friction on solid surfaces. The measured axial Young's modulus is approximately (1.22 ± 0.24), [5]. Consequently, BNNTs find applications in corrosion-resistant materials, transistor heat sinks, nuclear reactor control rods, crucibles for metal evaporation, and neutron absorbers [4]. Furthermore, they show potential for solar cell device applications and are recognized for their superior thermal shock resistance due to their high electrical resistance and low dielectric constant.

Various studies have demonstrated that defects in boron nitride nanotubes (BNNTs) can significantly impact their electronic, optical, mechanical, and thermoelectrical properties [6,7]. These defects are crucial for introducing new properties and applications to BNNT materials. Among these defects are vacancies and the attachment of metal nanoparticles (such as Ag, Au, Pt, and Fe) onto BNNTs [4,13]. Additionally, edge-functionalization through hydroxylation of BNNTs presents a significant method for modifying band gaps. Through ab-initio calculations, it has shown that the energy band gap is dependent on the diameter of the nanotubes, with band gaps decreasing as tube size diminishes [8].

Currently, the pursuit of suitable thermoelectric materials is one of the most active areas of research. These materials can generate voltage from temperature differences or create temperature gradients from applied potential differences, making them applicable in two advanced industries: power generation and cooling. Modern solid-state thermoelectric devices offer effective solutions for managing heat dissipation on both large and small scales. In these devices, solid-state materials are employed for heat management, where electrons and holes behave similarly to liquids in semiconductor materials. The engineering goal for these materials in energy conversion applications is to maximize electrical power output [9-12].

In this context, our study aims to investigate the effects of iron (Fe) doping on the properties of boron nitride nanotubes (BNNTs), specifically examining how the introduction of Fe atoms influences their electronic and thermoelectric properties. A key objective is to analyze how Fe doping can modify the energy band gaps of BNNTs, potentially enhancing their performance in various applications. Additionally, the research seeks to establish the relationship between the concentration of Fe dopants and the resulting changes in the structural and functional properties of BNNTs. By optimizing Fe doping strategies, the study aims to develop advanced thermoelectric materials that leverage the unique properties of doped BNNTs for improved efficiency in power generation and cooling applications. Ultimately, the goal is to engineer these materials for maximum electrical power conversion in thermoelectric devices through effective iron doping techniques.

2. THEORETICAL AND COMPUTATIONAL DETAILS

Our study is began from a perfect one-dimensional (1D) boron nitride (h-BN) nanotubes obtained by rolling a (2D) hexagonal sheet of the BN (modelized by SLAB model), (n,m)=(10,0) giving a BNNT of radius of 3.98Å. To study the effect of iron on BNNTs, two cases are considered, the substitution of one and two atoms of Fe with one and two atoms of N, respectively **Figure 1**. Positions of Fe atoms are selected with respecting of symmetry and to ensure to the conditions of convergence of calculations.

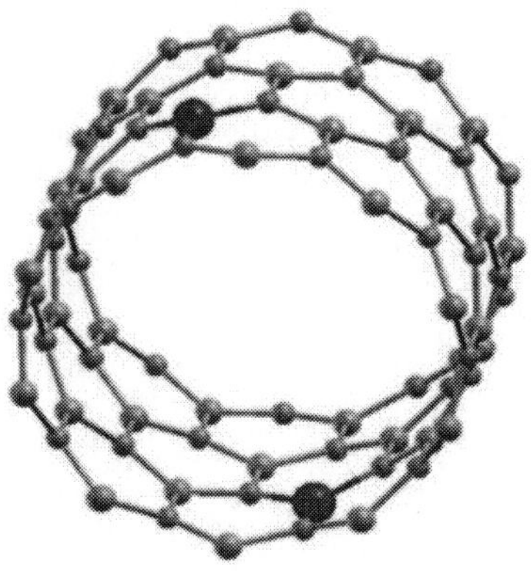

Figure1 Non relaxed BNNT 3x1x1 supercell doped Fe atoms. Blue, purple and red circles represent, respectively; B, N and Fe atoms

First-principles calculations were all carried out using CRYSTAL 17 software package [14] within the framework of the density functional theory (DFT) at the level of B3LYP hybrid hamiltonian [15]. Consistent Gaussian basis-sets of triple-zeta (TZVP) and double-zeta (DZVP_rev2) valence with polarization quality are used for B and N atoms, respectively, and all electrons (86-411d41G) for Fe atom [14] are used. To define a reasonable orbital occupancy in presence of Fe atoms, a spinlock option is used and optimized. These forces a given (n_α-n_β) electrons value to obtain a correct atomic spin density to start SCF process. Calculations are released for BNNT (10, 0) doped with one and two Fe atoms, for this a supercell of 3 times the primary cell along the principle tube axis is used (3x1x1)p. undoped BNNT structures are studied and used as references. A set of 12 × 1 × 1 Monkhorst-Pack corresponding to 7 special *k* points is used to sample the Brillouin zone of the system. Before starting the actual calculations and to obtain minimal value of energy, the geometry of the structures was optimized by consideration of all atoms of our structure including Fe atoms. All atoms were free to move in all directions. As default, the model of optimization adopted in the Crystal code is of Hessian [14].

3. RESULTS AND DISCUSSION

3.1. Structural and Electonic properties

Relaxed structure obtained after optimization of the supercell (3x1x1)p containing 120 atoms of (10,0) is used to simulate BNNT doped Fe materials. Our calculations show that non doped structure is not relaxed.

The substitution on a single Fe atom by N atoms has an important effect on the relaxation of the structure. For this, full structural relaxation is performed and obtained structure is represented in the **Figure 2**.

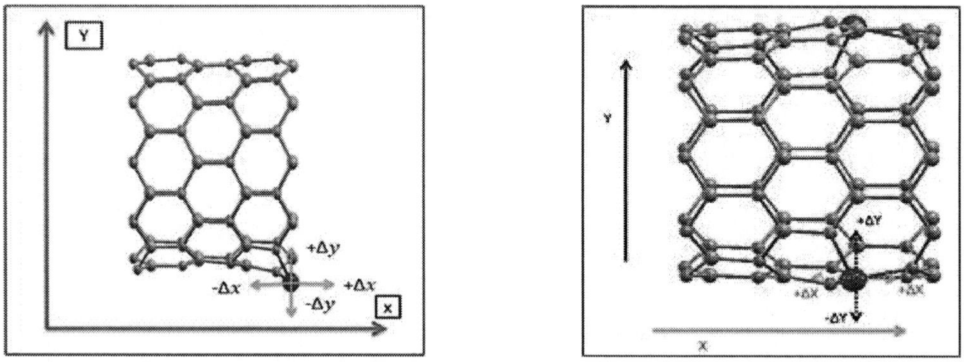

Figure2 relaxation parameters Δx and Δy for BNNT (a) doped with single Fe atoms and (b) doped with two Fe atoms

An important displacement of Fe atoms and her neighboring is presented by the structures. How inner and outer variations of relaxation parameters (Δx and Δy) varied between 0.3Å to 1Å for the BNNT:1Fe and between 0.6Å to 1Å to BNNT:2Fe are measured as mentioned at the **Figure 2**. We notice that there is no reconstruction of all structures.

The nanotube deformation confirms the formation of a strong chemical bonding between Fe atoms and B ones.

a) b)

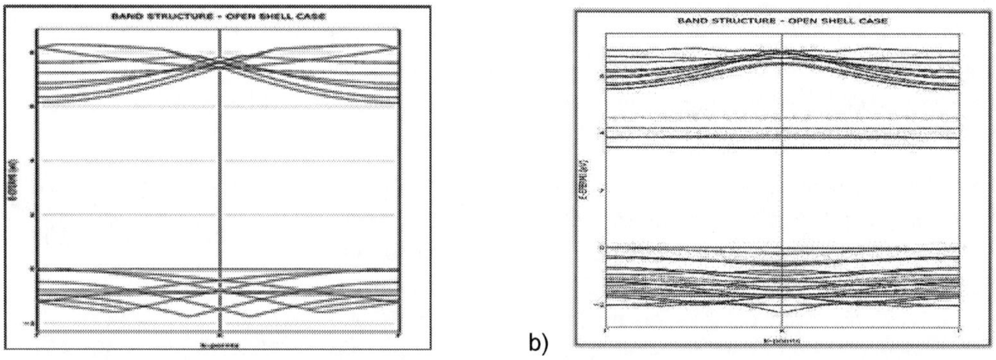

c)

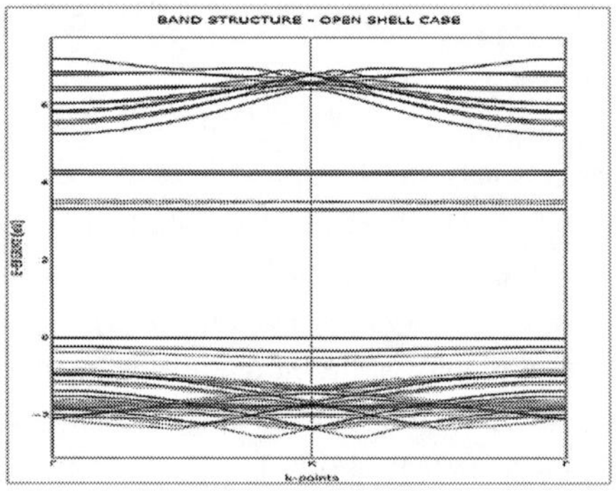

Figure 3 band structures of BNNT structures (a) undoped, (b) doped with 1Fe and (c) 2Fe

Band structures and density of states (DOS) from Γ (0, 0, 0) to K (1, 1,0) points corresponding to the direction Σ of the Brillouin zone are computed for non-doped and doped BNNTs **Figure 3**.

The band gap calculated value of non-doped BNNTs is 6.16 eV. The presence of Fe atoms caused a diminution of this value to 3.5 eV for BNNT:1Fe and to 3.31 eV for BNNT:2Fe. From **Figure 3**, we find that this diminution is caused by presence of impurities states. Knowledge of the nature of these contributions is release by calculating of the projected density of state (PDOS) represented in the **Figure 4**. It is apparent that this is the result of participation of (3d) AOs of Fe atoms in the band gap. In addition, the asymmetry of the PDOS in the range of 0 to -0.4 eV and from 3 to 4 eV caused by the spin state difference caused by the presence of the Fe atoms. We can conclude the increases of the magnetic characteristics in this compound.

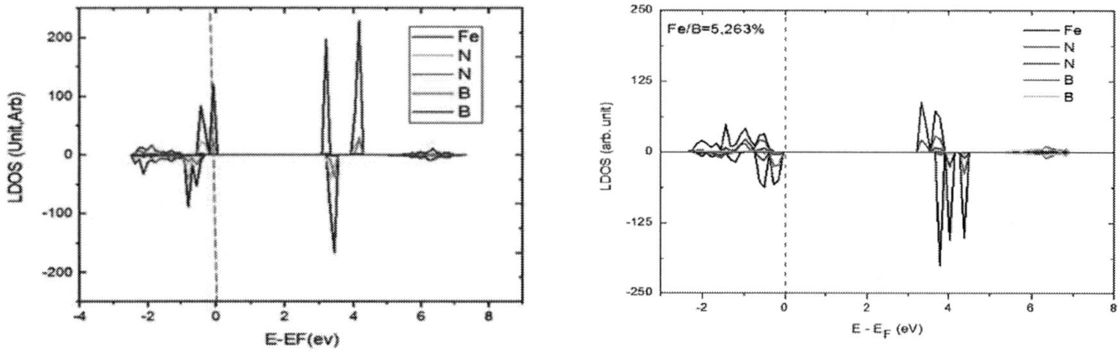

Figure4 projected density of state (PDOS) on the AOs of the Fe atoms for BNNT:1Fe and BNNT:2Fe.

The comparison with another works (**table 1**), we see that our results are slightly different from those obtained in the literature.

Table 1 energy band gap values comparison

	BNNT	**BNNT:Fe**
Our results	6.16	3.31 / 3.50
References	5.5 [16] - 5.8 [6]	3.16 [7]

3.2. TRANSPORT PROPERTIES

The Seebeck coefficient, Electrical and thermal conductivity, can be estimated by calculating the three transport coefficients. These expressions are obtained by solving Boltzmann's transport equation within the relaxation time approximation [9]. We note that, using Crystal code, the thermal conductivity is derived from only the electronic thermal conductivity k_{el}.

The effect of doping with Fe atoms on the thermoelectric properties of BNNT is represented here. In **Figure 5**, the variation of Seebeck coefficient (S) of the different BN nanotubes are shown as a function of the, where μ is the chemical potential at different values of temperature: 900 K, 1200 K and 1500 K. Here, (μ-E_F) is positive for n-type doping and negative for p-type doping.

The Seebeck coefficient displays a symmetric behavior with respect to the charge neutrality point (μ = 0.0 eV) and its maximum value varies from 0.003V/K for 900K to 0.002V/K for 1500V/K for undoped BNNT. The positive and negative values at Fermi energy level confirm the existence of band gap and our materials are a semiconductor. Now, introducing Fe atoms has decreased the value of Seebeck coefficient to 0.002V/K at the temperatures 900 K and to 0.001 V/K at 500 K for different concentrations of Fe atoms **Figures 5(b-c)**. which is comparable to the other suitable thermoelectric compounds, as 110 μ V/K for BN-Be (8.0) [18] and 0.06 x

10^{-3} eV/K for (pristine (6,6)-CNT) up to 4.0 x 10^{-3} eV/K (pristine (6,6)-BNNT) upon increasing the BN concentration [17].

The corresponding μ values of S_{max} for doped BNNTs move far from the charge neutrality point more and more according to the doping rates. This displacement is due because of decreasing of the electronic bandgap caused by presence of "3d" states of Fe atoms. In addition, we notice in the interval of the 1 to 3 eV, appearance of new states participating to the thermal transport. The significant Seebeck coefficient demonstrates that a small temperature difference creates a large voltage on both sides of the material, so a large Seebeck coefficient is essential for a thermoelectric material.

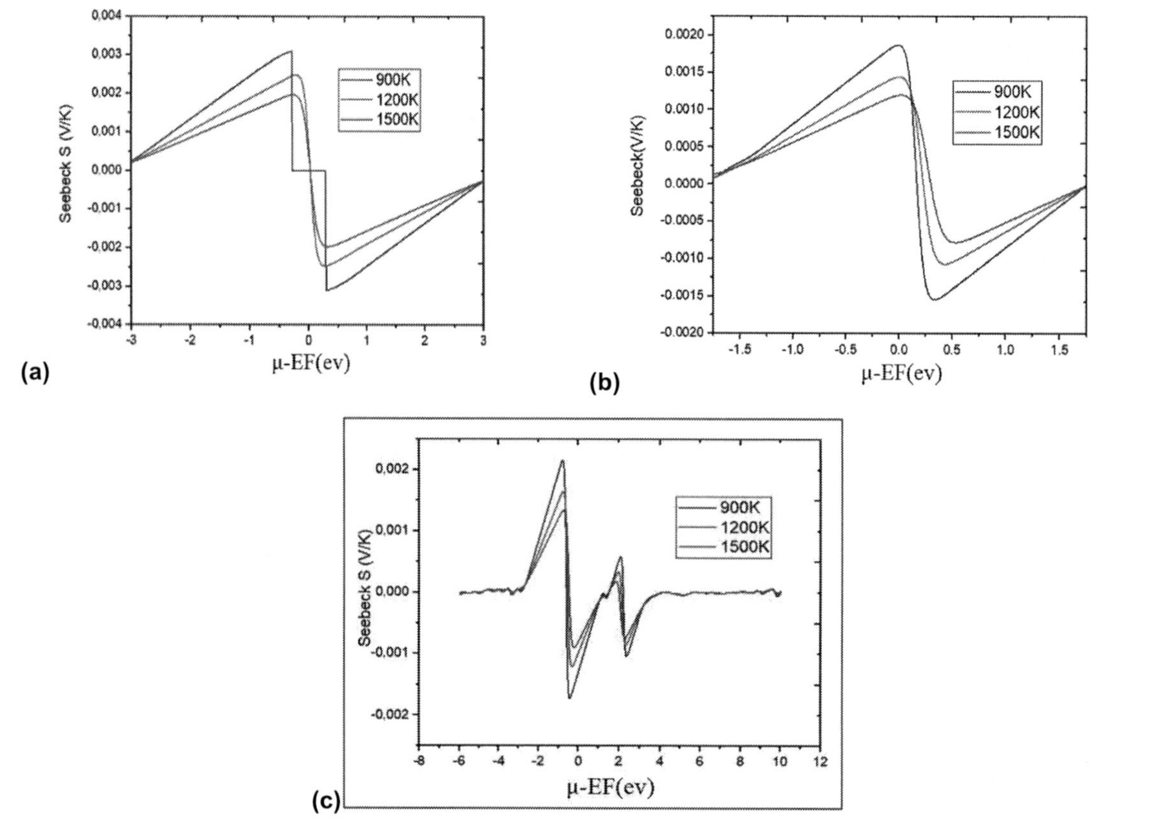

Figure 5 Seebeck coefficient at various temperatures; (a) pure BNNTs, (b) BNNT:1Fe and (c) BNNT:2Fe

It is well known that good thermoelectric materials must have high electrical conductivity. This is the case of our material. As mentioned in the **Figure 6-a**, electrical conductivity (σ) and electronic thermal conductivity (K) presents an important value for different structures. Thus, values of electrical conductivity varied for 900 K from $40x10^{+6}$ $\Omega^{-1}.m^{-1}$ for undoped BNNT to $6x10^{+8}$ $\Omega^{-1}.m^{-1}$ for BNNT:2Fe. In addition, the electronic thermal conductivity oscillates 2500 W/mK for undoped BNNT to 11000 W/mK for BNNT:2Fe. Comparing to the CNT, how it's characterized by a thermal conductivity superior as 3000 W/mK. Our results have verified the high thermal conductivity how reflect a good choice of conditions of calculation as compared with other works [11,17,18].

The thermoelectric efficiency of materials, important to know if the materials are good thermoelectrically or not, can be measured by calculating the dimensionless figure of merit, ZT, and the power factor, which can be calculated using the following formula. $ZT = S^2 \sigma T K$, where T is the absolute temperature, S is the Seebeck coefficient or thermopower, σ is the electrical conductivity, and K is the thermal conductivity, including the electron contribution only (K=K_{el}) see **Figures 6-b**. For all cases we observe that the (ZT) is at his maximum values (~0.98) near to the value "1" and we noticed the appearance of the of "d" states intensively in the case of 2Fe doping **Figures 6-c**. Moreover, ZT values increase strongly with chemical potential and its maximum value is observed for a range of -3.3 to 3 eV for both cases of doping. These results proves that BNNTs materials are promising to thermoelectrical applications.

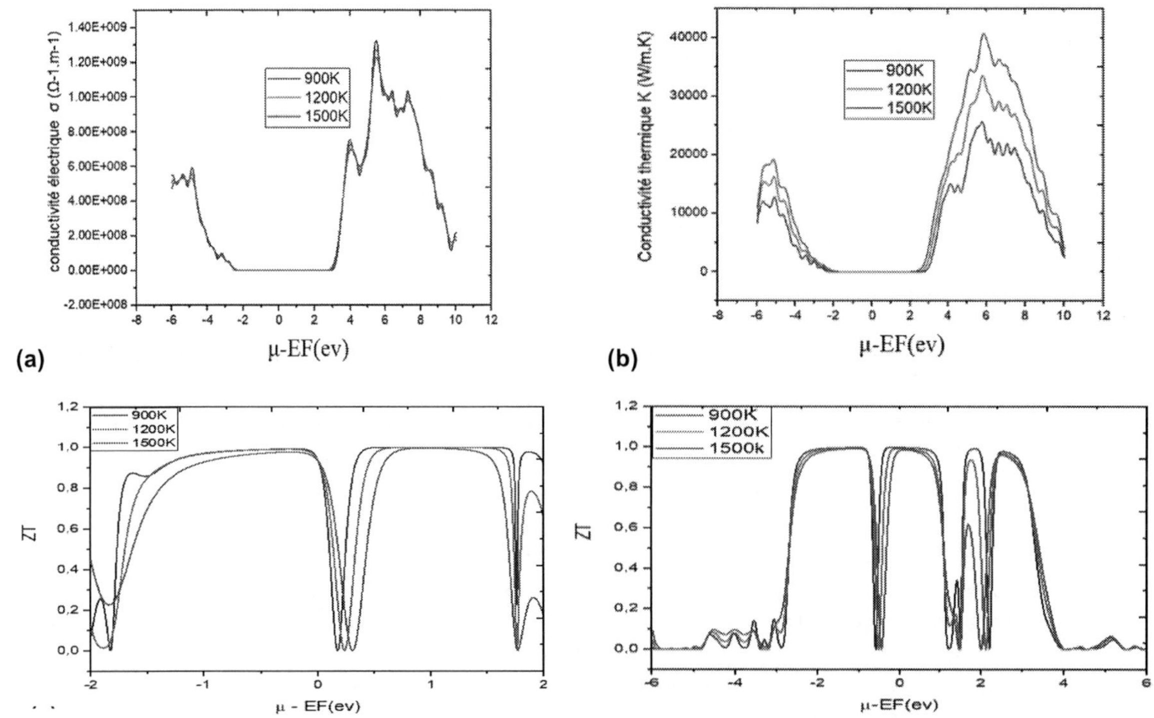

Figure 6 (a) electrical conductivity, (b) electronic thermal conductivity (Ke) for BNNT:2Fe and (c) figure of merit for BNNT:1Fe and BNNT:2Fe for different temperatures

4. CONCLUSION

This work presents an *ab-initio* calculation of the structural, electronic, and thermoelectric properties of BNNT activated by Fe atoms. Doping with metallic species significantly contributes to thermal transport due to the appearance of new impurity states. The substantial Seebeck coefficient indicates that a small temperature difference generates a large voltage across the material, making a high Seebeck coefficient crucial for thermoelectric materials. The results show that the presence of Fe atoms reduces the band gap due to the involvement of "d" states, which contribute to the thermal transport in BNNT:Fe.

ACKNOWLEDGEMENTS

This work was supported by "The General Directorate of Scientific Research and Technological Development (DGRSDT)-Algeria". The authors gratefully acknowledge the Chemical-physics group of IPREM of Pau University for the performed calculation.

REFERENCES

[1] SOTUDEH, M., BOOCHANI, A., PARHIZGAR, S. S., & MASHARIAN, S. R. Electronic, optical and thermoelectric properties of BN-Be (8, 0) nanotube: DFT study. *Solid State Communications*, 2022, vol. 353, 114822.

[2] GOU, G. Y., PAN, B. C., et SHI, L. Interaction of iron atoms with pristine and defective (8, 0) boron nitride nanotubes. *The Journal of Physical Chemistry C*. 2008, vol. 112, no 35, p. 13571-13578.

[3] GOLBERG, DMITRI, BANDO, YOSHIO, HUANG, YANG, *et al.* Boron nitride nanotubes and nanosheets. *ACS nano*. 2010, vol. 4, no. 6, pp. 2979-2993.

[4] WENG Qunhong, WANG Xuebin, WANG Xi, *et al.* Functionalized hexagonal boron nitride nanomaterials: emerging properties and applications. *Chemical Society Reviews*. 2016, vol. 45, no 14, p. 3989-4012.

[5] CHOPRA, Nasreen G. et ZETTL, A. Measurement of the elastic modulus of a multi-wall boron nitride nanotube. *Solid State Communications*. 1998, vol. 105, no 5, p. 297-300.

[6] ZUNGER Alex, KATZIR Abraham, et HALPERIN A. Optical properties of hexagonal boron nitride. *Physical Review B*. 1976, vol. 13, no 12, p. 5560.

[7] JALALINEJAD, A., YEGANEGY, M., FAROKHZAD, M., *et al.* Structural, magnetic, and electronic properties and stability of 3d-TM substituted single-walled zigzag BNNTs: A density functional theory study. *Computational Condensed Matter*. 2021, vol. 28, p. e00575.

[8] ABDULLAH, B. J. Diameter dependence of band gap of single-walled boron nitride nanotubes. *Physics of the Solid State*. 2021, pp. 1-5.

[9] BOUCHELAREM, N., BOUAMRA, F., DERBAL, M., *et al.* Structural, electronic, and thermoelectric properties of hydroxyl groups adsorption on SnO_2 (110) surface: A first-principles study. *Surface and Interface Analysis*. 2023, vol. 55, no 12, pp. 925-934.

[10] IVANOV Yuri V., BURKOV Alexander T., et PSHENAY-SEVERIN Dmitry A. Thermoelectric properties of topological insulators. *physica status solidi (b)*. 2018, vol. 255, no 7, p. 1800020.

[11] BENTOUAF, Ali. Elastic, Half-Metallicity, Thermodynamic, and Transport Properties of Ru2VGe and Ru2VSb Full-Heusler Compounds: a First-Principle Study. *Journal of Superconductivity and Novel Magnetism*. 2021, vol. 34, no. 1, p. 157-167.

[12] SAKR Mahmoud AS, ABDELSALAM Hazem, TELEB Nahed H., *et al.* Exploring the structural, electronic, and hydrogen storage properties of hexagonal boron nitride and carbon nanotubes: insights from single-walled to doped double-walled configurations. *Scientific Reports*. 2024, vol. 14, no 1, p. 4970.

[13] KHAN Muhammad Isa, AMIN Muhammad Usman, BUZDAR Saeed Ahmad, *et al.* Computational study of Fe-and Mn-decocted hexagonal boron nitride for hydrogen storage applications. *International Journal of Hydrogen Energy*. 2024, vol. 65, p. 727-739.

[14] R. DOVESI, V. R. SAUNDERS, C. ROETTI, R. ORLANDO, C. M. ZICOVICH-WILSON, F. PASCALE, B. CIVALLERI, K. DOLL, N. M. HARRISON, I. J. BUSH, P. D'ARCO, M. LLUNELL, M. CAUSÀ, Y. NOËL, L. MASCHIO, A. ERBA, M. Rerat and S. Casassa. *CRYSTAL17 User's Manual*. University of Torino, Torino, 2017.

[15] LEE Chengteh, YANG Weitao, et PARR Robert G. Development of the Colle-Salvetti correlation-energy formula into a functional of the electron density. *Physical review B*. 1988, vol. 37, no 2, p. 785.

[16] BLASE, X., RUBIO, ANGEL, LOUIE, STEVEN G., *et al.* Stability and band gap constancy of boron nitride nanotubes. *Europhysics Letters*. 1994, vol. 28, no 5, p. 335.

[17] SANDONAS Leonardo Medrano, CUBA-SUPANTA Gustavo, GUTIERREZ Rafael, *et al.* Doping engineering of thermoelectric transport in BNC heteronanotubes. *Physical Chemistry Chemical Physics*. 2019, vol. 21, no 4, pp. 1904-1911.

[18] MÜCHLER Lukas, CASPER Frederick, YAN Binghai, *et al.* Topological insulators and thermoelectric materials. *Physica status solidi (RRL)–Rapid Research Letters*. 2013, vol. 7, no 1-2, pp. 91-100.

A LOW-TEMPERATURE SOL-MICROWAVE METHOD FOR ULTRA-SMALL RUTILE TITANIUM DIOXIDE NANOPARTICLES

[1,2]Aiswarya Vijayakumar THELAPPURATH, [1]Naděžda PIZÚROVÁ

[1]*Institute of Physics of Materials, v.v.i., Czech Academy of Sciences, Brno, Czech Republic, EU, pizurova@ipm.cz*

[2]*Central European Institute of Technology, Brno University of Technology (CEITEC), Brno, Czech Republic, EU*

https://doi.org/10.37904/nanocon.2024.5001

Abstract

The rutile phase of TiO_2 can generally be obtained by high-temperature calcination of anatase nanoparticles. However, calcination undoubtedly leads to agglomeration and growth, hence increasing particle size. So, fabrication of rutile titania at low temperatures is of great importance since the formation of bigger particles can be avoided. Here, ultra-small rutile TiO_2 nanoparticles (NPs) of ~ 5 nm have been synthesized using a sol-microwave method at low temperature (150 °C). These rutile TiO_2 NPs have shown no surface defects, such as trivalent titanium (Ti^{3+}) ions and oxygen vacancies. The resulting samples were analyzed by the following methods: X-ray diffraction (XRD) to determine the crystallinity and phase composition, Raman spectroscopy to refine the phase composition, X-ray Photoelectron Spectroscopy (XPS) to identify oxidation states and phase composition in surface layers, and High-Resolution Transmission Electron Microscopy (HRTEM) to study morphology, phase analysis, and surface quality assessment.

Keywords: TiO_2 nanoparticles, rutile, low temperature, microwave synthesis, defect free

1. INTRODUCTION

Nanoscale TiO_2 is a very well-known and well-researched material due to its unique physical, [1] chemical, [2] optical [3] and electrical properties,[4] and chemical and thermal stability, [5,6] which found application in various fields, including photocatalysis,[7] electrochemistry, [8,9] and production of cosmetics [10] pigments, [11] and biomedicine [12] Furthermore, titanium oxide nanoparticles also exhibit unique surface chemistry and morphologies [13]. The specific surface area and surface-to-volume ratio increase dramatically as the size of a material decreases [14]. The high surface area accomplished by small particle size is favorable to various applications [15]. Thus, the performance of TiO_2 in its applications is predominantly influenced by the sizes of the building units, apparently at the nanometer scale. TiO_2 exists in several polymorphs [16]. Due to the novel characteristics and promising properties, naturally occurring polymorphs such as rutile, anatase and brookite have been widely synthesized [17]. From a thermodynamic point of view, rutile is the more stable phase, whereas anatase and brookite are metastable at all temperatures and transform to rutile when heated [18]. Rutile phase is the only high refractive index phase of titania [19]. Moreover, it has higher relative permittivity and ultraviolet ray absorption rate [20,21]. Accordingly, it is important to obtain and investigate rutile TiO_2 nanoparticles. The widely known method for the preparation of nano-sized rutile TiO_2 is the phase transformation from anatase at higher temperatures [22]. This method leads to the formation of large particle size, agglomeration, and low surface area [23]. Therefore, the fabrication of rutile titania at low temperature is significant.

There are a few methods reported to synthesize rutile TiO_2 nanoparticles at low temperature including hydrothermal method, [24] sol-gel synthesis, [25] chemical precipitation route, [26] and low-temperature precipitation method [27]. We hereby report a facile and easily reproducible microwave-assisted synthesis of rutile TiO_2 NPs with a size less than 5 nm at low temperature. The microwave (MW) synthesis method dramatically reduces reaction times, increases product yields, and enhances product purities compared to conventionally heated experiments [28]. In addition, the method is economical, rapid, less time-consuming, and atom-economical for preparing nanostructures [29]. Furthermore, this work investigates the structural characteristics and morphological properties of the synthesized ultra-fine rutile TiO_2 NPs.

2. EXPERIMENTAL

2.1 Materials

Titanium (IV) acetylacetonate (ca. 63% in Isopropyl alcohol, TCI), Isopropanol (99.5%, anhydrous, Merck), Ammonium hydroxide (NH_4OH) (25%, Lach:ner), Ethanol (99.8%, Penta Chemicals) and DI water were used for the experiment. All the chemicals were used as received.

2.2 Synthesis of Titanium oxide nanoparticles

Titanium (IV) acetylacetonate was mixed with 100 mL of isopropanol, 14.4 mL of distilled water and 25 mL of NH_4OH solution in a 250 mL beaker and stirred at 500 rpm for 5 minutes to create a homogenous reaction mixture. Then, the resulting sol was introduced in a 30 mL glass vial sealed with a snap cap comprising silicone septum. The filled vial was enclosed into a synthesis reactor, and the reactor was heated to the temperature of 150 °C, followed by microwave heating at the same temperature for 10 min at 900 rpm. The obtained product was dried in an oven at 100 °C for 2 hrs. After that, the dried sample was washed using DI water and ethanol to remove organic or inorganic impurities and dried for 2 h at 100 °C.

2.3 Characterization

The TiO_2 NPs have been synthesized in an Anton Paar Monowave–400 microwave reactor. All the NPs were analyzed using the X-ray diffractometry (XRD; Smart Lab 3 kW from Rigaku, Japan), which was set in Bragg-Brentano geometry with Cu-Kα radiation (λ = 0.154 nm) and fitted with the Dtex-Ultra 1D detector. The Cu radiation was powered by a 30-mA current and a 40-kV voltage. The diffraction patterns were acquired using a step size of 0.02° and a scanning speed of 4°/min from 20° to 80°. Phase analysis and chemical composition were performed using the PDF2 database. A Witec alpha-300R instrument (WITec, Ulm, Germany) was used for Raman spectroscopy. Raman spectra were recorded in continuous scanning mode at a laser excitation wavelength of 532 nm. A 100x objective lens focused the laser beam, producing a spot about 1 µm in diameter. The measurement signal was reconstructed using 50 accumulations and a 20 s integration period. The chemical composition investigation was conducted using X-ray photoelectron spectroscopy (XPS, Kratos Analytical Axis Supra), and the spectra were calibrated against the C 1s peak (284.8 eV) using the CasaXPS software. Transmission electron microscopy (TEM) was used to confirm the NP size and size distribution and impart detailed structural characterization. TEM images were carried out on a TALOS F200i (ThermoFisher Scientific) instrument, operating at an acceleration voltage of 200 kV. The TEM samples were prepared by putting a drop of sample solution on a commercially prepared Cu grid coated with an amorphous carbon layer. The corresponding filtered Fast Fourier transform patterns were evaluated with JEMS software.

3. RESULTS AND DISCUSSION

To accomplish low-temperature synthesis for nano-sized rutile TiO_2, a sol- microwave-assisted method of synthesis has been designed and executed at the temperature of 150 °C, using titanium (IV) acetylacetonate as the precursor and ammonium hydroxide as precipitating agent, followed by heating at 100 °C. The sample was yellow in colour.

3.1 X-ray diffraction

Figure 1A shows the XRD patterns of the synthesized TiO_2 samples. The broad diffraction peaks can be attributed to the particles' low crystallinity or relatively small size. **Figure 1A** revealed that the sample possessed an intense peak at 48.5° corresponds to rutile [30] and another broad rutile peak at 62° [31]. The phase analysis conducted with an X-ray diffraction method confirmed a small crystallite size (less than 5 nm) in the entire sample. The significant peak broadening does not allow us to confirm the purity of the rutile phase. Hence, the Raman spectra were recorded to confirm the presence of any additional phase.

3.2 Raman spectroscopy

Raman spectroscopy was also carried out further to understand the phase of TiO_2 (**Figure 1B**). The Raman spectra show broad peaks at around ~ 260 cm^{-1} (second-order Raman scattering of rutile), ~ 450 cm^{-1} (E_g rutile), ~ 620 nm (A_{1g} rutile) [32,33] and a broad band in the 600-750 cm^{-1} originated from to the silica substrate used for Raman analysis [34]. According to the previous report, the second-order Raman scattering peak of rutile is slightly shifted to a higher wavenumber, possibly because of the influence of the precipitating agent ammonium hydroxide.

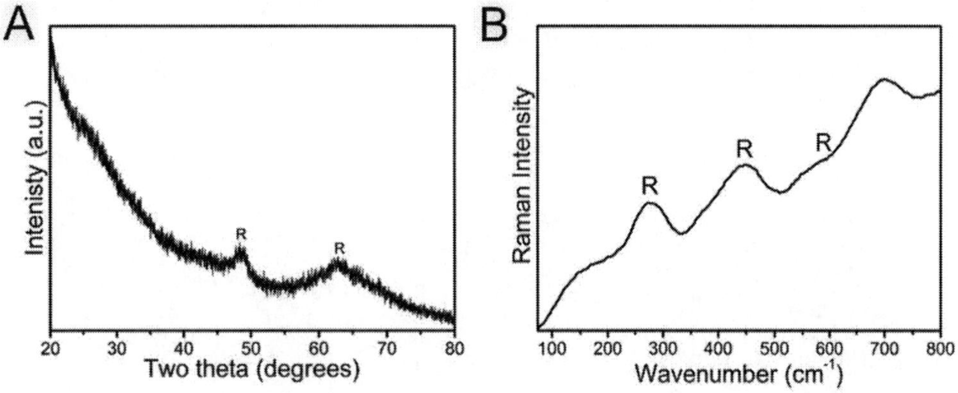

Figure 1 A) X ray diffraction and B) Raman spectrum of rutile (R) TiO_2

3.3 X-ray photoelectron spectroscopy

XPS spectra have been recorded for the synthesized rutile TiO_2 NPs. **Figure 2** represents the wide, Ti2p and O1s XPS of the sample. The rutile TiO_2 possessed Ti2p characteristic peaks corresponding to Ti^{4+}. The $Ti2p_{3/2}$ and $Ti2p_{1/2}$ were observed at 458.78 eV and 464.53 eV [35]. No indication of Ti^{3+} was seen. In O1s, peaks ascribed to Ti-O-Ti, and Ti-OH were found at 530.24 eV, and 531.28 eV, respectively [36].

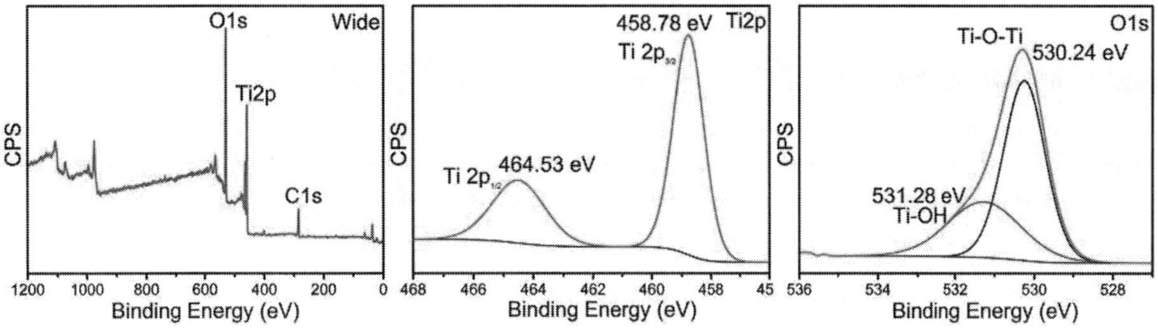

Figure 2 Wide, Ti2p and O1s XPS of rutile TiO_2

3.4 Transmission electron microscopy

Transmission electron microscopy measurements were used to confirm the morphology, size, and phase of the TiO_2 nanoparticles. **Figure 3A** shows the clusters of ultrasmall nanoparticles, where heterogeneity in size and shape can be observed. The obtained nanoparticles were nearly spherical morphology with an average diameter of 5 nm. **Figure 3B** is a high-resolution TEM (HRTEM) image of a single particle. Additionally, in **Figures 3C – 3E,** the Fast Fourier transform (FFT) over the crystalline areas was carried out to validate the presence of the TiO_2 rutile phase.

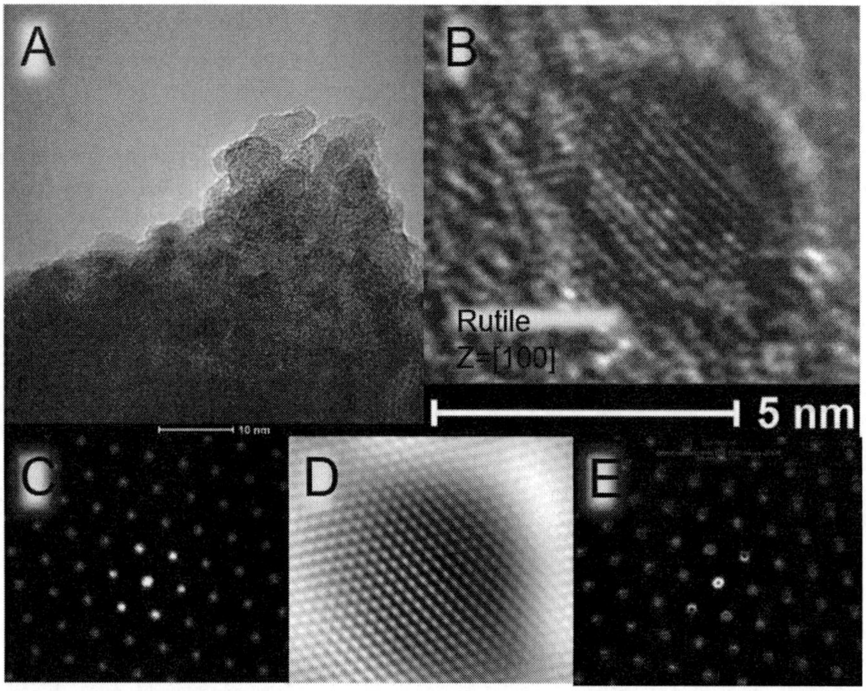

Figure 3 A) TEM, B) HRTEM of selected nanoparticle, C) filtered FFT pattern, D) inverse filtered FFT pattern, and E) filtered FFT pattern evaluated as the rutile TiO_2 phase in the [100] crystallographic direction.

4. CONCLUSIONS

A facile reproducible sol-microwave method has been developed to synthesize rutile TiO_2 NPs at low temperature. These rutile NPs were nearly spherical in shape with ~ 5 nm in size. Furthermore, these rutile TiO_2 NPs possessed only Ti^{4+} and stoichiometric oxygen. In other words, these rutile NPs were observed free from defects such as trivalent titanium ions (Ti^{3+}) and oxygen vacancies.

ACKNOWLEDGEMENTS

The work was supported with the Czech Science Foundation, project No. 21-31852J and CzechNanoLab project LM2018110 funded by MEYS CR is gratefully acknowledged for the financial support of the measurements/sample fabrication at CEITEC Nano Research Infrastructure.

REFERENCES

[1] GUERIDI, B., BOUFERRACHE, K., GHEBOULI, M. A., ROUABAH, F., SLIMANI, Y., CHIHI, T., ... & BENALI, A. Physical properties of rutile-TiO_2 Nanoparticles and effect on PVA/SiO_2 Hybrid films synthesized by sol-gel method. *High Energy Density Physics*. 2024, 101122.

[2] HSU, C. Y., MAHMOUD, Z. H., ABDULLAEV, S., ALI, F. K., NAEEM, Y. A., MIZHER, R. M., ... & HABIBZADEH, S. Nano titanium oxide (nano-TiO_2): a review of synthesis methods, properties, and applications. *Case Studies in Chemical and Environmental Engineering*. 2024, 100626.

[3] TRIPATHI, A. K., SINGH, M. K., MATHPAL, M. C., MISHRA, S. K., & AGARWAL, A. Study of structural transformation in TiO_2 nanoparticles and its optical properties. *Journal of Alloys and Compounds*. 2013, vol. 549, pp. 114-120.

[4] REDDY, K. M., MANORAMA, S. V., & REDDY, A. R. Bandgap studies on anatase titanium dioxide nanoparticles. *Materials Chemistry and Physics*. 2003, vol. 78, no. 1, pp. 239-245.

[5] THAKUR, N., THAKUR, N., KUMAR, A., THAKUR, V. K., KALIA, S., ARYA, V., ... & KYZAS, G. Z. A critical review on the recent trends of photocatalytic, antibacterial, antioxidant and nanohybrid applications of anatase and rutile TiO_2 nanoparticles. *Science of The Total Environment*. 2024, 169815.

[6] FAROOQ, N., KALLEM, P., UR REHMAN, Z., KHAN, M. I., GUPTA, R. K., TAHSEEN, T., ... & SHANABLEH, A. Recent trends of titania (TiO_2) based materials: A review on synthetic approaches and potential applications. *Journal of King Saud University-Science*. 2024, 103210.

[7] TUNTITHAVORNWAT, S., SAISAWANG, C., RATVIJITVECH, T., WATTHANAPHANIT, A., HUNSOM, M., & KANNAN, A. M. Recent development of black TiO_2 nanoparticles for photocatalytic H2 production: An extensive review. *International Journal of Hydrogen Energy*. 2024, vol. 55, pp. 1559-1593.

[8] PUTJUSO, T., PUTJUSO, S., KARAPHUN, A., & SWATSITANG, E. Influence of Li concentration on structural, morphological and electrochemical properties of anatase-TiO_2 nanoparticles. *Scientific Reports*. 2024, vol. 14, no. 1, 11200.

[9] ZHU, Y., LIU, T. H., ZHOU, W., SHI, M., WU, M., SHI, P., ... & LIN, C. T. An “On-Site Transformation” Strategy for Electrochemical Formation of TiO_2 Nanoparticles/$Ti_3C_2T_x$ MXene/Reduced Graphene Oxide Heterojunction Electrode Controllably toward Ultrasensitive Detection of Uric Acid. *Small Structures*. 2024, 2400034.

[10] BERARDINELLI, A., & PARISI, F. TiO_2 in the food industry and cosmetics. In Titanium Dioxide (TiO_2) and Its Applications. *Elsevier*. 2021, pp. 353-371.

[11] GÁZQUEZ, M. J., BOLÍVAR, J. P., GARCIA-TENORIO, R., & VACA, F. A review of the production cycle of titanium dioxide pigment. *Materials Sciences and Applications*. 2014.

[12] SAGADEVAN, S., IMTEYAZ, S., MURUGAN, B., ANITA LETT, J., SRIDEWI, N., WELDEGEBRIEAL, G. K., ... & OH, W. C. A comprehensive review on green synthesis of titanium dioxide nanoparticles and their diverse biomedical applications. *Green Processing and Synthesis*. 2022, vol. 11, no. 1, pp. 44-63.

[13] MINO, L., NEGRI, C., SANTALUCIA, R., CERRATO, G., SPOTO, G., & MARTRA, G. Morphology, surface structure and water adsorption properties of TiO_2 nanoparticles: A comparison of different commercial samples. *Molecules*. 2020, vol. 25, no. 20, 4605.

[14] ALIVISATOS, A. P. Perspectives on the physical chemistry of semiconductor nanocrystals. *The Journal of Physical Chemistry*. 1996, vol. 100, no. 31, pp. 13226-13239.

[15] WU, L., FU, C., & HUANG, W. Surface chemistry of TiO_2 connecting thermal catalysis and photocatalysis. *Physical Chemistry Chemical Physics*. 2020, vol. 22, no. 18, pp. 9875-9909.

[16] SONG, M., LU, Z., & LI, D. Phase transformations among TiO_2 polymorphs. *Nanoscale*. 2020, vol. 12, no. 45, pp. 23183-23190.

[17] MOHAMAD, M., HAQ, B. U., AHMED, R., SHAARI, A., ALI, N., & HUSSAIN, R. A density functional study of structural, electronic and optical properties of titanium dioxide: Characterization of rutile, anatase and brookite polymorphs. *Materials Science in Semiconductor Processing*. 2015, vol. 31, pp. 405-414.

[18] MAHDJOUB, N., ALLEN, N., KELLY, P., & VISHNYAKOV, V. Thermally induced phase and photocatalytic activity evolution of polymorphous titania. *Journal of Photochemistry and Photobiology A: Chemistry*. 2010, vol. 210, no. 2-3, pp. 125-129.

[19] LANCE, R. A. Optical analysis of titania: band gaps of brookite, rutile and anatase. 2018.

[20] BONKERUD, J., ZIMMERMANN, C., WEISER, P. M., VINES, L., & MONAKHOV, E. V. On the permittivity of titanium dioxide. *Scientific reports*. 2021, vol. 11, no. 1, 12443.

[21] CHEN, X. D., WANG, Z., LIAO, Z. F., MAI, Y. L., & ZHANG, M. Q. Roles of anatase and rutile TiO_2 nanoparticles in photooxidation of polyurethane. *Polymer testing*. 2007, vol. 26, no. 2, pp. 202-208.

[22] HANAOR, D. A., & SORRELL, C. C. Review of the anatase to rutile phase transformation. *Journal of Materials Science*. 2011, vol. 46, pp. 855-874.

[23] ZHANG, H., & BANFIELD, J. F. Understanding polymorphic phase transformation behavior during growth of nanocrystalline aggregates: insights from TiO_2. *The Journal of Physical Chemistry B*. 2000, vol. 104, no. 15, pp. 3481-3487.

[24] WANG, H. E., CHEN, Z., LEUNG, Y. H., LUAN, C., LIU, C., TANG, Y., ... & LEE, S. T. Hydrothermal synthesis of ordered single-crystalline rutile TiO_2 nanorod arrays on different substrates. *Applied Physics Letters*. 2010, vol. 96, no. 26.

[25] ANWAR, M. S., KUMAR, S., AHMED, F., ARSHI, N., LEE, C. G., & KOO, B. H. One step synthesis of rutile TiO_2 nanoparticles at low temperature. *Journal of Nanoscience and Nanotechnology*. 2012, vol. 12, no. 2, pp. 1555-1558.

[26] BORSE, P. H., KANKATE, L. S., DASSENOY, F., VOGEL, W., URBAN, J., & KULKARNI, S. K. Synthesis and investigations of rutile phase nanoparticles of TiO_2. *Journal of Materials Science: Materials in Electronics*. 2002, vol. 13, pp. 553-559.

[27] PARK, N. G., CHANG, S. H., Van de LAGEMAAT, J., KIM, K. J., & FRANK, A. J. Effect of Cations on the Open-Circuit Photovoltage and the Charge-Injection Efficiency of Dye-Sensitized Nanocrystalline Rutile TiO_2 Films. *Bulletin-Korean Chemical Society*. 2000, vol. 21, no. 10, pp. 985-988.

[28] DALLINGER, D., LEHMANN, H., MOSELEY, J. D., STADLER, A., & KAPPE, C. O. Scale-up of microwave-assisted reactions in a multimode bench-top reactor. *Organic Process Research & Development*. 2011, vol. 15, no. 4, pp. 841-854.

[29] ULLATTIL, S. G., & PERIYAT, P. (2015). Green microwave switching from oxygen rich yellow anatase to oxygen vacancy rich black anatase TiO_2 solar photocatalyst using Mn (II) as 'anatase phase purifier'. *Nanoscale*. 2015, vol. 7, no. 45, pp. 19184-19192.

[30] ZHANG, H., CHEN, B., BANFIELD, J. F., & WAYCHUNAS, G. A. Atomic structure of nanometer-sized amorphous TiO_2. *Physical Review B—Condensed Matter and Materials Physics*. 2008, vol. 78, no. 21, 214106.

[31] NAGARAJ, G., MOHAMMED, M. K., ABDULZAHRAA, H. G., SASIKUMAR, P., KARTHIKEYAN, S., & TAMILARASU, S. Effects of the surface of solar-light photocatalytic activity of Ag-doped TiO_2 nanohybrid material prepared with a novel approach. *Applied Physics A*. 2021, vol. 127, no. 4, 269.

[32] MATOSSI, F. The vibration spectrum of rutile. *The Journal of Chemical Physics*. 1951, vol. 19, no. 12, pp. 1543-1546.

[33] ZANATTA, A. R. A fast-reliable methodology to estimate the concentration of rutile or anatase phases of TiO_2. *AIP advances*. 2017, vol. 7, no. 7.

[34] BOROWICZ, P., LATEK, M., RZODKIEWICZ, W., ŁASZCZ, A., CZERWINSKI, A., & RATAJCZAK, J. Deep-ultraviolet Raman investigation of silicon oxide: thin film on silicon substrate versus bulk material. *Advances in Natural Sciences: Nanoscience and Nanotechnology*. 2012, vol. 3, no. 4, 045003.

[35] SAHU, D. P., & JAMMALAMADAKA, S. N. Detection of bovine serum albumin using hybrid TiO2 + graphene oxide based Bio–resistive random access memory device. *Scientific reports*. 2019, vol. 9, no. 1, 16141.

[36] KUANG, J., XING, Z., YIN, J., LI, Z., TAN, S., LI, M., ... & ZHOU, W. Ti^{3+} self-doped rutile/anatase/TiO_2 (B) mixed-crystal tri-phase heterojunctions as effective visible-light-driven photocatalysts. *Arabian Journal of Chemistry*. 2020, vol. 13, no. 1, pp. 2568-2578.

SESSION B
INDUSTRIAL AND ENVIRONMENTAL APPLICATIONS OF NANOMATERIALS

Chairmen

Prof. Dr. Radek ZBORIL	Palacky University in Olomouc, CATRIN-RCPTM, Czech Republic, EU
Prof. Dr. Miroslav CERNIK	Technical University of Liberec, Czech Republic, EU
Prof. Dr. Miloslav PEKAR	Brno University of Technology, Czech Republic, EU
Dr. Jan MACAK	CEITEC BUT and University of Pardubice, Czech Republic, EU

PHOTOCATALYTIC HYDROGEN EVOLUTION ACTIVITY OF ANATASE-BROOKITE NANOPARTICLES MODIFIED WITH RARE EARTH METALS (Gd, Lu, Tm, Tb, Pr)

[1, 2]Tetiana KHALIAVKA, [3]Nataliya SHCHERBAN, [3]Ganna KORZHAK, [3]Pavel YAREMOV, [3]Ivan KOPA, [1]Regina BURVE, [1]Elif COŞKUN, [1]Jean-Claude GRIVEL

[1]*Technical University of Denmark, Department of Energy Conversion and Storage, Lyngby, Denmark, EU,* tekha@dtu.dk

[2]*Institute for Sorption and Problems of Endoecology NASU, Kyiv, Ukraine, EU*

[3]*L.V. Pisarzhevskii Institute of Physical Chemistry NASU, Kyiv, Ukraine, EU*

https://doi.org/10.37904/nanocon.2024.4988

Abstract

Rare earth elements (Gd, Lu, Tm, Tb, Pr) modification effects on two-phased TiO_2 (anatase-brookite) were analyzed. The powders were synthesized by a one-step hydrothermal method. The phase composition, crystalline size, and shape of the samples were studied by X-ray diffractometry, transmission and scanning electron microscopies, and energy-dispersive X-ray spectroscopy. The decrease of the crystallite size and increase of the lattice strain due to the modification were confirmed by the Scherrer, modified Scherrer, Williamson-Hall, Size-strain plot (SSP), and Halder-Wagner methods. The textural characteristics of the samples were determined from N_2 adsorption-desorption isotherms using the methods of Brunauer–Emmet–Teller and Barret–Joiner–Halenda and revealed their mesoporous structure. UV-Vis diffuse reflectance spectra of the modified samples showed a blue shift compared to the unmodified sample. The values of band gap energy of the modified powders increased due to a phenomenon of quantum confinement or size quantization. The modification with rare earth elements improved the photocatalytic activity of the anatase-brookite composite in the hydrogen evolution reaction, which was ascribed to physico-chemical properties.

Keywords: Rare earth elements, anatase-brookite nanoparticles, photocatalysis, hydrogen

1. INTRODUCTION

Titanium dioxide (TiO_2) is the most widely used photocatalyst, which exists in three main forms: anatase, rutile and brookite. Anatase is considered the most active form, but it has some disadvantages, such as a wide band gap, a high recombination rate of the photogenerated electron-hole pairs, etc.

To improve the photocatalytic properties of TiO_2, researchers have tried two- or three-phased compositions based on anatase, rutile, or brookite as photocatalysts. It was established that two-phased TiO_2 (P25 Degussa), which contains anatase and rutile, shows high activity because it can effectively stimulate electron transfer from one phase to another thereby improving photocatalytic properties [1].

Recent reports have indicated that anatase-brookite composites demonstrate better photocatalytic performance due to the synergistic effect between these two phases [2]. Also, during the photocatalytic reaction electrons transfer from the more cathodic conduction band of brookite to the anatase conduction band favouring an effective electron–hole separation [3]. Besides, brookite has an appropriate electron trap depth for prolongation of the lifetime of photogenerated electrons and holes and special connection of the octahedral chain leads to exposed O atoms, which can be used as catalytic active atoms and enhance the catalytic activity [1]. Therefore, two-phased TiO_2 containing brookite is a promising material for photocatalysis.

Another way to improve photocatalytic activity is modification, where rare earth elements deserve special attention as dopants. Using these dopants allows to tune structural, optical, textural, morphological, etc. characteristics of TiO_2 [4-10]. Rare earth elements as modifiers for TiO_2 can create defects and oxygen vacancies [11], tuning the bandgap, extending the adsorption spectrum into the visible region, and enhancing the charge separation and transfer processes that favour photocatalytic processes [4-10]. Such materials can be active in the reaction of photocatalytic hydrogen (H_2) evolution. It is known that hydrogen has been suggested as an alternative fuel because it can be eco-friendly and clean depending on its production method. The use of hydrogen allows to avoid greenhouse gas emissions.

In this study the influence of modification of two-phased anatase-brookite TiO_2 (*AB*) with rare earth elements (*RE*) (Gd, Lu, Tm, Tb, Pr) on structural, optical, textural, and photocatalytic properties was analysed. The photocatalytic activity was evaluated in the reaction of hydrogen evolution from water-ethanol solutions under UV light.

2. EXPERIMENTAL

2.1 Samples synthesis

The unmodified and modified with *RE* (Gd, Lu, Tm, Tb, Pr) *AB* nanopowders were synthesized by a one-step hydrothermal method. In a typical synthesis, 40 ml of distilled water or 40 ml solution of rare earth metal acetate (Alfa Aesar) (~2% molar amount per titanium atoms) were applied for the synthesis of *AB* and *RE/AB*, respectively. Then titanium butoxide ($Ti(OBu)_4$) (Aldrich) was slowly added to this solution under continuous stirring. Afterward, the solution was transferred into a Teflon-lined autoclave with a capacity of 100 ml for 24 h at 150°C. The heating rate was 5°C/min. Synthesis was carried out at an initial pH of 10.5. After the cooling of the autoclave to room temperature, the samples were washed with distilled water and dried in the oven at 60 °C for 24 h.

2.2 Characterization

The phase composition of the samples was determined by X-ray diffraction (XRD) method using an Aeris diffractometer (PANalytical) with Cu Kα radiation (λ = 1.5418 Å). The crystallite size was evaluated by the Scherrer (D_{Sch} and $D_{Sch(Aver)}$) and modified Scherrer equation (D_{MSch}), the Williamson-Hall method based on uniform deformation model ($D_{W\text{-}H\text{-}UDM}$), uniform stress deformation model ($D_{W\text{-}H\text{-}USDM}$), and Uniform deformation energy density model ($D_{W\text{-}H\text{-}UDEDM}$) approaches, Size-strain plot (D_{SSP}) and the Halder-Wagner method ($D_{H\text{-}W}$).

Raman spectra of the prepared samples were recorded using an InVia Raman Spectrometer (Renishaw), UK with laser emitting at the excitation wavelengths of 532 nm (100 mW max. power output. To record the spectrum, the laser was set to 10 mW power, total excitation time was 15 seconds (3s x 5).

The morphology, crystallites size and element composition of the samples were studied by scanning electron microscopy (SEM) using a Zeiss Merlin FEGSEM microscope, and a Tabletop TM3000 Hitachi SEM microscope with EDS as well a transmission electron microscopy (TEM) using a Tecnai T20 G2 microscope with a 200 kV accelerating voltage (DTU National Centre for Nano Fabrication and Characterization).

The room-temperature FT-IR spectra of the samples were recorded using a Bruker Tensor 27 FT-IR Spectrometer. UV-Vis diffuse reflectance spectra (DRS) of the powders were measured using a Perkin-Elmer Lambda Bio 35 spectrophotometer. The band gap values (E_g) was estimated using the Wood and Tauc method.

The porous structure of the samples was studied by nitrogen adsorption using a Micromeritics 3Flex instrument. Textural characteristics were determined from N_2 adsorption–desorption isotherms using the methods of Brunauer–Emmet–Teller (BET) and Barret–Joiner–Halenda (BJH).

2.3 Photocatalytic experiments

In a typical experiment of photocatalytic hydrogen evolution, a sample of photocatalyst (0.03 g) was dispersed in 10.0 ml of aqueous solution containing 2 mol/l ethanol (electron donor) in cylindrical glass thermostatic reactors. Co-catalyst Pd/SiO_2 (0.01 g,) was obtained employing $PdCl_2$ reduction with hydrogen on the surface of commercial silica in an alcohol solution (1 wt. % palladium on SiO_2) and was added into the mixture. The reaction mixture was degassed before irradiation using a vacuum pump. The suspension was stirred and irradiated with a high-pressure mercury lamp (DRSH 1000; 1000 W; λ_{max} = 365nm; 50 mW/cm^2), T = 40 ^{0}C. The composition of the gas phase above the solution was determined every 30 min chromatographically. The experiment with each sample was repeated three times.

3. RESULTS AND DISCUSSIONS

The synthesized powders consist of agglomerates containing roundly elongated crystallites (**Figure 1a**). **Figure 1b** shows XRD patterns for *AB* and *RE/AB*. The diffraction peaks at the 2θ ~ 25.3°, 37.8°, 48.0°, 54.0°, 62.7°, 75.1° were indexed to anatase phase according to standard JCPDS Card no. 96-900-9087 and assigned to (101), (004), (200, (105), (204), (215) planes, respectively. The samples' unit cell parameters, unit cell volumes, and axial ratios indicated the tetragonal crystal structure of the samples which is characteristic for anatase. Besides these peaks, the peak at 2θ ~ 30.8° was fixed and related to the (121) crystal plane of brookite according to JCPDS Card no. 96-900-4141. No characteristic peaks were detected for rutile phase.

The Raman spectra (**Figure 1c**) of the samples also showed anatase bands at 144, 199, 399, 517 and 639 cm^{-1}. The peaks located at 144 (E_g), 199 (E_g) and 399 cm^{-1} (B_{1g}) are assigned to the Ti–O bending vibrations of TiO_2, and the peaks at 639 cm^{-1} (E_g) and 517 cm^{-1} (A_{1g} + B_{1g}) refer to Ti-O stretching vibrations [12, 13].

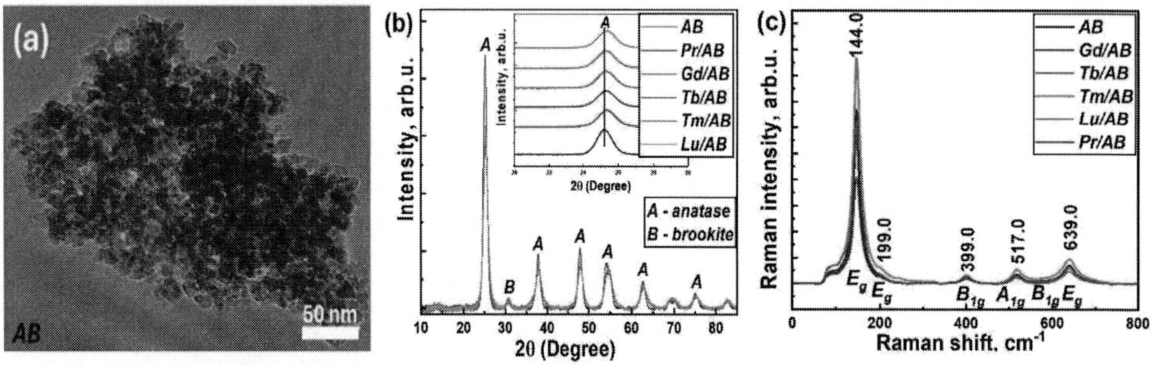

Figure 1 TEM-image (a), XRD patterns (b) and Raman spectra (c) of the samples

According to XRD studies, the phase content of brookite in unmodified *AB* was ~6%. Modification with Gd reduced the brookite content to ~4%, on the contrary modification with Pr increased it to ~9.4%. There were no peaks related to rare earth elements in the spectra of *RE/AB* (**Figure 1b**) due to their low concentration in the materials. The EDX spectra confirmed the presence of *RE* with the atomic ratio Ti:*RE* (Gd, Lu, Tm, Tb, Pr) ~1:0.013; 0.016; 0.016; 0.014; 0.012, respectively and Ti, O, C elements in *RE/AB* powders. The modified samples had an O/Ti ratio higher (2.52; 2.31; 2.72; 2.69; 2.96) than the stoichiometric value (2.0) because of water adsorption on the surface. The Ti-O and Ti-OH bonds in the samples were confirmed by FTIR analysis.

RE have larger ionic radii (0.0745–0.1045 nm) than Ti^{4+} (0.064 nm) therefore they can be distributed, create metal oxide clusters on the surface of TiO_2, occupy interstitial sites in the lattice of TiO_2 or form *RE*-O-Ti bonds [5, 6, 8, 9]. Such TiO_2 modification with *RE* can distort the crystal lattice, produce strain [14] and lead to crystallite size reduction. Indeed, the unit cell parameters and the unit cell volume were slightly decreased for *RE/AB* samples compared to *AB*. The broadening of the full width at half maximum (FWHM) and a slight shift of all peaks towards higher 2θ were observed for modified samples in XRD patterns (**Figure 1b**) and can also indicate the appearance of the strain and crystallite size reduction. To obtain reliable information different methods were used for the calculation of these values (**Figure 2a, Table 1**).

All the methods confirmed the decrease of the crystallite size and increase of strain, stress, and energy density for *RE/AB* compared to *AB*. The samples *Pr/A-B* and *Tm/A-B* had the smallest crystallite size and the highest values of strain, stress, and energy density (**Figure 2a, Table 1**). Therefore, the modification of TiO_2 with rare earths induced stress in the lattice and inhibited the crystallites growth. Such decrease of the crystallite size resulted in an increase of the samples dislocation density meaning that TiO_2 modification with *RE* improved its strength and hardness [15].

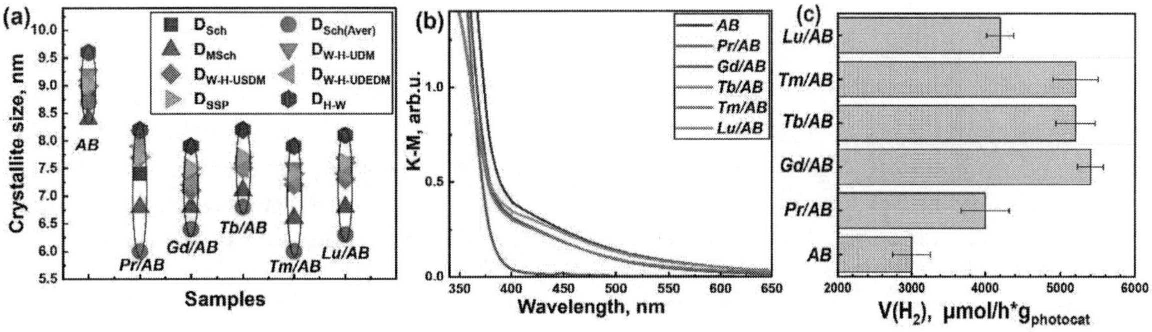

Figure 2 Crystallite size of the samples calculated by different methods(a), UV–Vis spectra (b) and photocatalytic hydrogen evolution from aqueous/ethanol solution in the presence of the obtained samples

Table 1 Strain (ε), stress (σ, MPa) and energy density (U, KJm^{-3}) of the samples calculated by different methods

Sample	W-H						SSP	H-W
	UDM	USDM		UDEDM				
	$\varepsilon \times 10^{-3}$	$\varepsilon \times 10^{-3}$	$\sigma \times 10^{-2}$	$\varepsilon \times 10^{-3}$	$\sigma \times 10^{-2}$	$U \times 10^{-3}$	$\varepsilon \times 10^{-2}$	$\varepsilon \times 10^{-2}$
AB	0.80	0.66	8.37	0.73	9.30	0.034	1.56	1.01
Pr/A-B	3.72	3.05	38.69	3.36	42.65	0.716	4.10	2.66
Gd/A-B	1.50	1.20	15.01	1.33	16.85	0.112	3.00	1.95
Tb/A-B	1.08	0.84	10.63	0.95	12.03	0.057	2.57	1.67
Tm/A-B	2.68	2.10	27.29	2.30	30.39	0.364	3.68	2.39
Lu/A-B	2.21	1.70	22.09	1.95	24.78	0.242	3.38	2.19

Table 2 Textural characteristics of the samples

Sample	*External surface area*, m^2/g	*Mesopore surface area*, m^2/g	*Specific surface area*, m^2/g	*Micropore volume*, cm^3/g	*Mesopore volume*, cm^3/g	*Total pore volume*, cm^3/g	*Pore diameter*, nm
AB	28	140	168	0.0074	0.211	0.273	6.6
Pr/AB	32	153	185	0.0051	0.248	0.309	6.5
Gd/AB	27	165	192	0.0046	0.246	0.298	6.2
Tb/AB	36	154	190	0.0041	0.233	0.302	6.4
Tm/AB	33	153	186	0.0033	0.220	0.286	6.1
Lu/AB	38	150	188	0.0038	0.220	0.299	6.3

Decreasing the crystallite size can also favour increasing the surface area of the samples. Indeed, the specific surface area for undoped *AB* was 168 m^2/g and higher for *RE/AB* (**Table 2**). The isotherms of the samples belong to type IV in accordance with IUPAC classification, namely type IVa with H2b type of hysteresis loop which is characteristic for mesoporous materials. In addition to mesopores the samples contained a negligible volume of micropores that decreased twice for *RE/AB* compared to *AB*.

Changes in the structural and textural characteristics of the *RE/AB* samples affected their optical band gap properties. The UV-Vis absorption spectra of the samples are shown in **Figure 2b**. The absorption edge of *RE/AB* samples was shifted to a shorter wavelength compared to *AB* that indicates an increase in the optical bandgap E_g of the composites. The bandgap for the direct and indirect transitions for *AB* were 3.33 and 3.11 eV, for *RE* (Gd, Lu, Tm, Tb, Pr)/*AB* – 3.37; 3.31; 3.34; 3.33; 3.41 eV and 3.13; 3.12; 3.13; 3.10; 3.16 eV, respectively. Such band gap widening of the nanomaterials is attributed to the dimensional limitation of photogenerated charge carriers in the nanoparticles. This phenomenon is known as quantum confinement or size quantization. The powders *Pr/AB* and *Tm/AB* had the smallest crystallites and the largest E_g.

The samples *RE/AB* demonstrated better photocatalytic performance in the photoreduction reaction of H_2 formation compared to *AB* (**Figure 2c**). The photocatalytic activity of *AB* modified with Gd, Tb and Tm is approximately twice as high than for unmodified and does not change for 5 consecutive cycles. The increase of photocatalytic activity of *RE/AB* samples seemed to be associated with the decrease of the crystallites size and increasing the specific surface area, i.e. the number of active reaction sites on the surface of photocatalysts increases. Rare earth elements can act as electron scavengers and consequently generate more electrons (e^-) available for the reaction, prevent their recombination and increase their lifetime [2]. Besides, brookite has a higher cathodic potential than anatase, which energetically favours the reduction of protons to produce H_2 [16]. Moreover, the presence of anatase–brookite heterojunctions has a positive effect on the activity of the samples mostly due to the increase of the number of charge carrier species available to participate in photoreactions at the photocatalyst surface [17].

4. CONCLUSION

In this work two-phase TiO_2 (anatase-brookite) modified with (Gd, Lu, Tm, Tb, Pr) was obtained by a one-step hydrothermal method. The influence of rare earth elements on morphology, crystal phase, optical, textural, and photocatalytic properties of *AB* was analysed. It was established that the modification inhibited the crystallite growth and induced stress in the TiO_2 lattice. The decreasing of crystallite size and increasing of strain, stress, and energy density values for *RE/AB* were confirmed by the Scherrer equation, modified Scherrer equation, the Williamson-Hall method based on a uniform deformation model, uniform stress deformation model, and uniform deformation energy density model approaches, size-strain plot (SSP) and the Halder-Wagner method. The IVa type of N_2 adsorption-desorption isotherms and H2b type of a hysteresis loop confirmed mesoporous structure of the prepared samples. Modification of TiO_2 with *RE* led to the increase of specific, external and mesopore surface area, total pore volume and mesopore volume. UV-Vis diffuse reflectance spectra of the modified samples showed a blue shift compared to *AB*. The band gap of the modified samples was increased due to a phenomenon of quantum confinement or size quantization. The modified samples showed higher photocatalytic activity in the reaction of photoreduction compared to *AB*.

ACKNOWLEDGEMENTS

This work was supported by the SARU-fellowship program for researchers at risk from Ukrainian universities and by the Technical University of Denmark.

REFERENCES

[1] EDDY, D.R., PERMANA, M.D., SAKTI, L.K., SHEHA, G.A.N., SOLIHUDIN, N., HIDAYAT, S., TAKEI, T., KUMADA, N., RAHAYU, I. Heterophase polymorph of TiO_2 (Anatase, rutile, brookite, TiO_2 (B)) for efficient photocatalyst: fabrication and activity. *Nanomaterials*, 2023, vol. 13, p. 704. https://doi.org/10.3390/nano13040704

[2] KHEDR, T.M., EL-SHEIKH, S.M., KOWALSKA, E., ABDELDAYEM, H.M. The synergistic effect of anatase and brookite for photocatalytic generation of hydrogen and diclofenac degradation. *Journal of Environmental Chemical Engineering*. 2021, vol. 9, p. 106566. https://doi.org/10.1016/j.jece.2021.106566

[3] NACHIT, W., AHSAINE, H.A., RAMZI, Z., TOUHTOUH, S., GONCHAROVA, I., BENKHOUJA, K. Photocatalytic activity of anatase-brookite TiO_2 nanoparticles synthesized by sol gel method at low temperature. *Optical Materials*. 2022, vol. 129, p. 112256. https://doi.org/10.1016/j.optmat.2022.112256

[4] SMIRNOVA, O.V., GREBENYUK, A.G., LOBANOV, V.V., KHALYAVKA T.A., SHCHERBAN N.D., PERMYAKOV V.V., SCHERBAKOV S.N. Experimental and quantum-chemical studies of electronic and spectral properties of titanium dioxide, modified with tin and lanthanum. *Applied Nanoscience*. 2023, vol. 13, pp. 5345–5355. https://doi.org/10.1007/s13204-023-02797-3

[5] KUTUZOVA, A., MORITZ, J.-O., MOUSTAKAS, N.G., DONTSOVA, T., PEPPEL, T., STRUNK, J. Performance of Sm-doped TiO_2 in photocatalytic antibiotic degradation and photocatalytic CO_2 reduction. *Applied Nanoscience*. 2023, vol. 13, pp. 6951–6966. https://doi.org/10.1007/s13204-023-02832-3

[6] MANDARI, K.K., POLICE, A.K.R., YEON, J., DO, KANG, M., BYON, C. Rare earth metal Gd influenced defect sites in N doped TiO_2: Defect mediated improved charge transfer for enhanced photocatalytic hydrogen production. *International Journal of Hydrogen Energy*. 2018, vol. 43, pp. 2073-2082. https://doi.org/10.1016/j.ijhydene.2017.12.050

[7] KHALYAVKA, T.A., SHCHERBAN, N.D., SHYMANOVSKA, V.V., MANUILOV E.V., PERMYAKOV V.V., SHCHERBAKOV S.N. Cerium-doped mesoporous $BaTiO_3/TiO_2$ nanocomposites: structural, optical and photocatalytic properties. *Res Chem Intermed*. 2019, vol. 45, pp. 4029–4042. https://doi.org/10.1007/s11164-019-03888-z

[8] MIKOLAJCZYK, A., WYRZYKOWSKA, E., MAZIERSKI, P., GRZYB, T., WEI, Z., KOWALSKA, E., CAICEDO, P.N.A., ZALESKA-MEDYNSKA, A., PUZYN, T., NADOLNA, J. Visible-light photocatalytic activity of rare-earth-metal-doped TiO_2: Experimental analysis and machine learning for virtual design. *Applied Catalysis. B, Environmental*. 2024, vol. 346, p. 123744. https://doi.org/10.1016/j.apcatb.2024.123744

[9] LI, H., ZHENG, K., SHENG, Y., SONG, Y., ZHANG, H., HUANG, J., HUO, Q., ZOU, H. Facile synthesis and luminescence properties of $TiO_2:Eu^{3+}$ nanobelts. *Optics & Laser Technology/Optics and Laser Technology*, 2013, vol. 49, pp. 33–37. https://doi.org/10.1016/j.optlastec.2012.12.007

[10] KHALYAVKA, T.A., SHYMANOVSKA, V.V., MANUILOV, E.V., SHCHERBAN, N.D., KHYZHUN, O.Y., KORZHAK, G.V., PERMYAKOV, V.V. The influence of La Doping on structural, optical, and photocatalytic properties of TiO_2 in dyes destruction and hydrogen evolution. In: *Fesenko, O., Yatsenko, L. (eds) Nanomaterials and Nanocomposites, Nanostructure Surfaces, and Their Applications. Springer Proceedings in Physics*, Springer, Cham. 2020, vol. 246, pp. 361–380. https://doi.org/10.1007/978-3-030-51905-6_27

[11] ELAHIFARD, M., SADRIAN, M.R., MIRZANEJAD, A., BEHJATMANESH-ARDAKANI, R., AHMADVAND, S. Dispersion of Defects in TiO_2 Semiconductor: Oxygen Vacancies in the Bulk and Surface of Rutile and Anatase. *Catalysts*, 2020, vol. 10, p. 397. https://doi.org/10.3390/catal10040397

[12] RADHA, E., KOMARAIAH, D., SAYANNA, R., SIVAKUMAR, J. Photoluminescence and photocatalytic activity of rare earth ions doped anatase TiO_2 thin films. *Journal of Luminescence*. 2022, vol. 244, p. 118727. https://doi.org/10.1016/j.jlumin.2022.118727

[13] KOMARAIAH, D., RADHA, E., JAMES J., KALARIKKAL, N., SIVAKUMAR, J., REDDY, M.V.R., SAYANNA, R. Effect of particle size and dopant concentration on the Raman and the photoluminescence spectra of $TiO_2:Eu^{3+}$ nanophosphor thin films. *Journal of Luminescence*. 2019, vol. 211, pp. 320-333. https://doi.org/10.1016/j.jlumin.2019.03.050

[14] PAUL, S., CHETRI, P., CHOUDHURY, B., AHMED, G.A., CHOUDHURY, A. Enhanced visible light photocatalytic activity of Gadolinium doped nanocrystalline titania: An experimental and theoretical study. *Journal of Colloid and Interface Science*. 2015, vol. 439, pp. 54–61. https://doi.org/10.1016/j.jcis.2014.09.083

[15] THEIVASANTHI, T., ALAGAR, M. Titanium dioxide (TiO_2) Nanoparticles XRD Analyses: An Insight. *arXiv (Cornell University)*. 2013. https://doi.org/10.48550/arxiv.1307.1091

[16] MANZOLI, M., FREYRIA, F.S., BLANGETTI, N., BONELLI, B. Brookite, a sometimes under evaluated TiO_2 polymorph. *RSC Advances*. 2022, vol. 12, pp. 3322–3334. https://doi.org/10.1039/d1ra09057g

[17] ŽERJAV, G., ŽIŽEK, K., ZAVAŠNIK, J., PINTAR, A. Brookite vs. rutile vs. anatase: What`s behind their various photocatalytic activities? *Journal of Environmental Chemical Engineering*. 2022, vol. 10, p. 107722. https://doi.org/10.1016/j.jece.2022.107722

SILVER NANOWIRE COATED POPLAR-PET CONDUCTIVE NONWOVEN FABRICS FOR PRESSURE SENSING APPLICATIONS

[1]Canan USTA, [1]Aybeniz SEYHAN, [1]Alper GURARSLAN

[1]*Istanbul Technical University, Textile Engineering Department, Istanbul, Türkiye,* cananusta@itu.edu.tr

https://doi.org/10.37904/nanocon.2024.4990

Abstract

This study demonstrates the use of needle-punched nonwoven fabrics of poplar seed fiber and polyester fiber blends in sensor applications. For this purpose, the surfaces of the produced fabrics were functionalized by coating them with silver nanowires (AgNWs). AgNWs were synthesized according to the polyol method and then coated onto the nonwoven fabrics through dip coating process. Surface morphology is characterized by scanning electron microscopy (SEM), and the analysis revealed a uniform coating on the fabric surface. Measurement of electrical surface resistance of treated fabrics confirmed that the presence of silver nanowires significantly improved the electrical conductivity. The capacitor structure was fabricated as a sandwich via placing a stretchable sealing film (Parafilm®) between two conductive nonwoven samples. Variation in the relative capacitance as a result of loading and unloading was observed. Poplar blended nonwoven fabrics exhibited great potential to be used as capacitive sensors to measure variable weight pressures. By virtue of its biodegradability, availability and light-weight solution, bio-waste poplar fibers offer a novel application for developing wearable and flexible tactile sensors.

Keywords: Poplar, pressure sensing, conductive, silver nanowire, coating, wearable sensors

1. INTRODUCTION

The development of wearable pressure sensors utilising eco-friendly materials is of great interest in the application of flexible and stretchable physical sensing devices. These devices facilitate a wide range of applications, including electronic skin (e-skin), health monitoring, medical diagnostics, activity tracking, environmental monitoring, and more [1]. To date, four primary types of sensory models including capacitive, piezo-resistive, piezoelectric and optical type have been successfully developed for the production of flexible pressure sensors [2]. Among these, capacitive sensors are preferred due to their high sensitivity, low power consumption and simple fabrication process [3].

In general, a capacitive sensor has a sandwich structure in which an elastic dielectric layer is placed between two parallel plate electrodes. When external pressure is applied onto a sensor, it can deform the flexible electrodes and the elastomeric dielectric layer, altering the effective electrode area, thickness and permittivity of the dielectric layer. These changes result in variations in measured capacitance value, which can be monitored in real-time to record the applied pressure accurately. Typically, two flexible conductive fabrics serve as the electrode plates, separated by dielectric materials such as foams or soft polymers [1]. Various materials, including conductive polymers, carbon-based substances and metallic materials have been widely investigated in the literature for the fabrication of conductive textiles [4]. In recent years, silver nanowires (AgNWs) have gained significant attention for creating conductive networks in flexible capacitive materials due to their high electrical conductivity, ease of synthesis, antimicrobial activity and excellent stability against oxidation [5-7].

In the development of flexible electronics, the type of substrate, as well as conductive materials and fabrication approaches used in the design of flexible electronics, also has a significant impact. Hence, bio-based resources have gained considerable attention in recent years. As opposed to using non-biodegradable and non-renewable materials, many natural fibers offer promising potential to meet the demands of current consumer electronics, such as wool [6], cotton [2], kapok [8], cattail [9], poplar fibers [10]. Among them, poplar fiber, which is the seed hair of the populus genus of trees in the willow family (Salicaceae), is generally treated as waste material. Poplar fiber, a typical cellulose fiber, has a large hollow lumen and a very thin cell wall (average thickness of 1.2 μm) [11]. These air-filled lumens provide poplar fiber a fluffy texture, softness, extreme lightness and excellent thermal and sound insulation. Poplar, a renewable plant fiber, is abundant, biodegradable and has attracted growing attention by researchers [10]. Traditionally, poplar has been used as thermal and acoustic insulator due to its intrinsic hollow structure [12-13]. Furthermore, poplar fibers have been studied as oil sorbents for oily water cleanup [13-15] and to prepare aerogel materials [16]. Additionally, studies have explored poplar fiber as natural supports for wearable electronic applications [10].

In this study, pressure sensing was performed based on conductive and superhydrophobic pristine PET and poplar/ PET blend nonwoven fabrics. Three different sustainable nonwoven fabrics were produced by needle punching technique using biodegradable poplar fibers and hollow PET fibers at different mass ratios (0:100, 30:100, 60:100, w/w for poplar and PET fiber) and then coated with silver nanowires (AgNWs). The sensor exhibits notable attributes such as high sensitivity and rapid response time. These qualities enable thenonwoven fabric samples to effectively detect varying pressures from standard weights, making them highly suitable for applications in wearable devices such as flexible pressure-sensitive sensors. Furthermore, the use of eco-friendly materials in the production of sensors coincides with the increasing demand for sustainable and environmentally friendly technologies. Poplar fiber-based nonwovens have the potential to be an innovative product in the field of wearable electronics.

2. EXPERIMENTAL

2.1 Materials

Poplar fibers were harvested during the blooming period in mid-May in Türkiye. Hollow polyethylene terephthalate fiber (PET, length: ~64 mm) was kindly supported by SASA Polyester Sanayi A.Ş., Türkiye. Silver nitrate ($AgNO_3$, ≥99.8%, Sigma-Aldrich), polyvinylpyrrolidone (PVP40, average mol. wt. 40.000) (Sigma-Aldrich), ethylene glycol (EG) (Merck), ethanol (Merck), acetone (99.5% pur.) (TEKKİM) were procured from Labor-Teknik Laboratuvar Malzemeleri Sanayi Ve Ticaret A.Ş., Türkiye. All chemical reagents were used as received.

2.2 Methods

Mechanical cleaning of poplar fibers was carried out with steel tweezers to remove contamination such as dust, seeds, leaves etc. on the surface. Subsequently, the cleaned poplar fibers were blended with PET fibers in three different ratios (0:100, 30:100 and 60:100, w/w) without the need for chemical pre-treatment processes. The produced nonwoven fabrics were labeled as PET, P-Po30 and P-Po60 with blend ratios of 0/100, 30/100 and 60/100 (w/w) for the amount of poplar and PET fibers, respectively (as shown in **Table 1**).

In this study, nonwoven fabrics were produced in two stages, the first is carding, where webs were formed by blending fibers in the specified ratios, and the other is needle-punching, where the carded webs were bonded to produce nonwoven fabrics. The thickness and density of nonwoven fabrics were given in **Table 1**.

Table 1 Properties of produced nonwovens

Sample	Blend Ratios (Poplar/PET) (in g mass, w/w)	Thickness (mm)	Mass per Unit Area (g/m^2)
PET	0/100	2.9 ± 0.008	243 ± 2.59
P-Po30	30/100	3.0 ± 0.010	263 ± 2.95
P-Po60	60/100	3.1 ± 0.011	279 ± 3.01

Preparation of Silver Nanowire (AgNW) Solution

Initially, a silver nitrate solution was prepared by dissolving silver nitrate in ethylene glycol in a glass flask for 45 minutes. Simultaneously, a PVP solution was prepared by dissolving PVP40 in ethylene glycol in a three-necked glass flask, placed in an oil bath, and stirred on a hot plate at 180°C for 45 minutes. Following this, a 1 M NaCl- EG solution was added to the PVP solution, and the mixture was allowed to react for an additional 10 minutes. The silver nitrate solution was gradually introduced into the PVP solution at a rate of 1 mL/min using a syringe pump. The combined solution was stirred at 180 ℃ for another 20 minutes before being cooled to room temperature. Once cooled, acetone was added to dilute the mixture, and the silver nanowires (AgNWs) were precipitated via centrifugation. Finally, the precipitated AgNWs were diluted in ethanol (EtOH) at a concentration ratio of 1 g/ 20 ml for application onto the nonwoven fabric.

Fabrication of nonwoven fabrics coated with AgNWs

The nonwoven fabrics were submerged in 15 mL of AgNW suspension for 10 s as a cycle of dip-coating (**Figure 1**). After drying at 60 ℃ for 30 minutes, silver nanowire-functionalized nonwoven fabrics were obtained. The number of dip-coating cycles was increased by 5 to decrease the electrical resistance of the fabrics to about 15 Ohm.

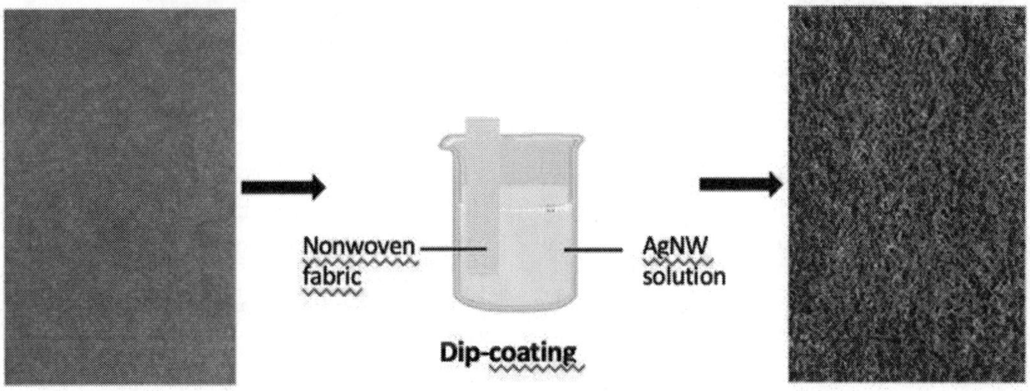

Figure 1 The preparation process of AgNW-coated nonwoven fabric

Fabrication of capacitance using the conductive nonwoven fabrics

A capacitive type fiber-based pressure sensor was constructed by placing a 0.01-mm thick Parafilm® between two AgNW coated same nonwoven fabrics. **Figure 1** illustrates the resulting sandwich structure. All three fabrications were performed on conductive nonwoven fabrics consisting of pure PET and poplar/PET blend fibers under the same environment. This capacitance is used as a pressure sensor. The relationship between capacitance, material properties, and dimensions can be expressed as capacitance (C) being directly proportional to the product of emissivity (ε) and surface area (A), and inversely proportional to the material thickness (d).

Characterization and measurements

The surface morphology of the conductive nonwovens were examined using scanning electron microscopy (SEM, TESCAN VEGA3). The surface electrical resistance of the conductive nonwovens was measured using a Digital Multimeter (FLUKE 17B+) which has ± 0.5% accuracy with wide measurement range. Besides, measurements to detect the capacitance change have been performed using the same multimeter. Pressure is generated by placing standard weights (10- 20- 50- 100- 200-500 g) on the fabric-based capacitor and the response of the sensor to loadings at different weights was recorded.

3. RESULTS AND DISCUSSION

3.1 Surface Morphology of PET and Poplar/PET Blend Nonwovens Decorated with AgNWs

In this work, a simple method to fabricate highly flexible, sensitive and environmentally friendly textile-based pressure sensor was presented. Pristine PET and poplar/PET blend nonwoven fabrics, P-Po30 and P-Po60, were utilized as substrates, while AgNW conductive networks -synthesized using the polyol method- were deposited onto the fabric surfaces via a dip-coating process. This scalable dip-coating technique facilitates practical applications for flexible pressure sensors [2]. Upon complete impregnation of the nonwoven fabrics with an ethanol suspension of AgNWs, the color of the fabrics changed from the original white (for pristine PET fabric) and yellowish (for poplar blends) to grey, indicating successful AgNW coating, as shown in **Figure 1**.

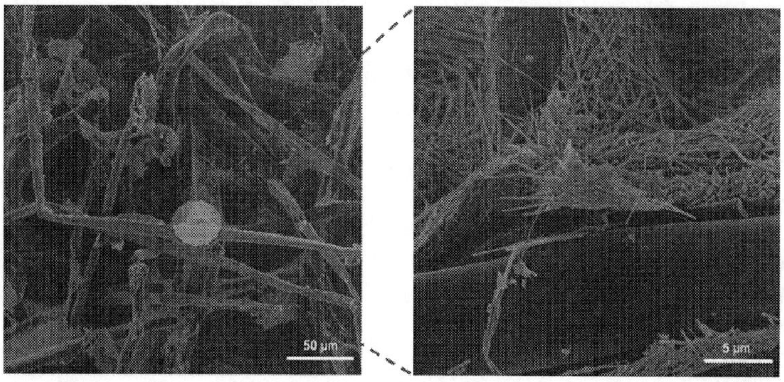

Figure 2 SEM image of the surface morphology of blend nonwoven after AgNW coating. AgNWs formed on the fiber surface ensure an effective interconnection that facilitates electrical conductivity. The inset provides a magnified view of the coating, highlighting AgNW clusters.

The SEM images of the nonwovens clearly illustrate that the surface was completely covered by AgNW conductive networks after 5 dip-coating cycles (**Figure 2**).

3.2 Electrical Resistance

Most textiles are inherently non-conductive, serving as an insulator with high electrical resistance. To utilize textiles in wearable e-textile components, it is essential to convert the substrate into a conductive platform for specific functional processes. In this study, AgNW coating effectively reduced the surface electrical resistance of non-conductive PET, P-Po30 and P-Po60 nonwovens. Increasing the number of dip-coating cycles to five led to a notable reduction in electrical resistance, reaching approximately 15 Ω/cm, as each cycle deposited more metal particles onto the fabric surface. However, addition of more cycles increases the cost and may lead to excessive release during washing, which negatively impacts the environment. In addition, the electrical resistance varied based on the distance between the two multimeter probes; shorter distances resulted in lower resistance, with 15 Ω measured at 1 cm [10]. Hence, five dip-coating cycles and measurement distance of 1 cm were selected for further analysis.

3.3 Pressure Sensing

The capacitance formed by the sandwiched structure of two conductive nonwovens is effectively changed when increasing forces are applied, leading to deformation in thickness [4]. As external pressure compresses the fabric layers, the distance between the electrode fabrics decreases, the contact area between the electrodes increases, which results in increase in capacitance according to Eq.(1). Study of the working mechanism revealed that the layers of conductive nonwoven fabrics have a great influence on the performance of pressure sensors.

The pressure sensitivity of PET and poplar/PET blend nonwoven fabrics was assessed using a setup where a 1 cm long section of the material surface was connected to a digital multimeter (**Figure 3a**) to measure real-time responses to different applied pressures (**Figure 3b**). When a light force was applied, the electrical resistance of the material increased slightly and returned to its original state when the force was removed. When pressed with varying weights (ranging from 10 g to 500 g), the capacitance showed a periodic change following a consistent and repetitive pattern (**Figure 3c**). The observed capacitance changes were more noticeable with heavier weights due to the increased pressure applied to the substrate. However, when the weights were removed, the resistance consistently returned to its baseline value, demonstrating the potential of the substrate to sense a wide range of pressures.

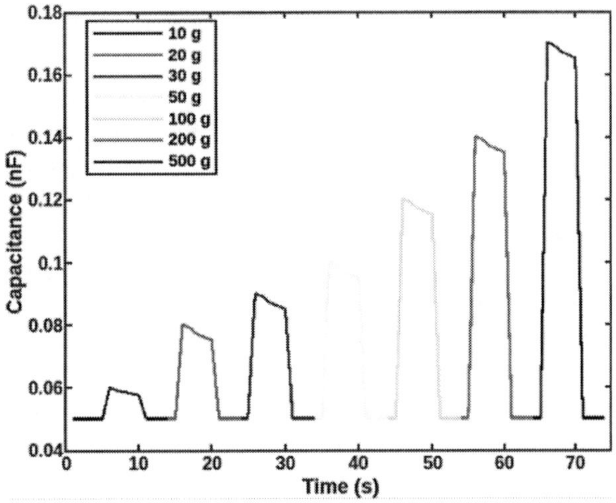

Figure 3 Record the relative capacitance over time while weights were loaded and unloaded

Pressures from different standard weights caused a notable increase in the capacitance value of the nonwoven fabrics and also the increase in the capacitance value was quite similar for all three nonwoven fabrics, so only one graph was presented. As illustrated in **Figure 3**, the capacitance increased from 0.05 nF to 0.06 nF at the lowest weight of 10 g, while it increased much higher to 0.17 nF at the highest weight of 500 g. This is related to the decrease in thickness with the increase in the weight pressure. This result suggested that the needle-punched poplar blended nonwoven fabric–based capacitor can be used as a pressure-sensitive sensor.

4. CONCLUSION

In summary, pure PET and poplar/ PET blend nonwoven fabrics were fabricated using needle-punching technique and subsequently coated with AgNWs. Owing to the high affinity for AgNWs, the nonwoven fabrics achieved high conductivity of about 15 Ohm/cm after five dip-coating cycles. The conductive nonwoven fabrics were then transformed into flexible and wearable capacitors to serve as a pressure-sensitive sensor. The fabrics revealed excellent potential in sensing a wide range of pressures under counterweight pressing, indicating that the capacitor made of poplar/ PET blend fabric can detect weights ranging from 10 g to 500 g. Based on the outstanding performance and cost-effective production process, the poplar fiber-based tactile sensor paves the way to promote new opportunities in the development of next-generation wearable electronics.

ACKNOWLEDGEMENTS

This work was supported by the Scientific and Technological Research Council of Türkiye (TÜBİTAK) under Grants No. 121M308 and Grants No. 124N407.

REFERENCES

[1] ZHANG, J., ZHANG, Y., LI, Y., WANG, P. Textile-Based Flexible Pressure Sensors: A Review. *Polymer Reviews*. 2022, vol. 62, no. 1, pp. 65–94.

[2] WEI, Y., CHEN, S., LIN, Y., YUAN, X., LIU, L. Silver nanowires coated on cotton for flexible pressure sensors. *J Mater Chem C*. 2016, vol. 4, no. 5, pp. 935–43.

[3] KANG, S., LEE, J., LEE, S., KIM, S., KIM, J.K., ALGADI, H. et al. Highly Sensitive Pressure Sensor Based on Bioinspired Porous Structure for Real-Time Tactile Sensing. *Adv Electron Mater*. 2016, vol. 2, no. 12, 1600356.

[4] CHHETRY, A., YOON, H., PARK, J.Y. A flexible and highly sensitive capacitive pressure sensor based on conductive fibers with a microporous dielectric for wearable electronics. *J Mater Chem C*. 2017, vol. 5, no. 38, pp. 10068–76.

[5] LIAN, Y., YU, H., WANG, M., YANG, X., ZHANG, H. Ultrasensitive Wearable Pressure Sensors Based on Silver Nanowire-Coated Fabrics. *Nanoscale Res Lett*. 2020, vol. 15, no. 1, 70.

[6] GURARSLAN, A., ÖZDEMIR, B., BAYAT, İ.H., YELTEN, M.B., KARABULUT Kurt, G. Silver nanowire coated knitted wool fabrics for wearable electronic applications. *J Eng Fibers Fabr*. 2019, 14, 155892501985622.

[7] SHUAI, X., ZHU, P., ZENG, W., HU, Y., LIANG, X., ZHANG, Y. et al. Highly Sensitive Flexible Pressure Sensor Based on Silver Nanowires-Embedded Polydimethylsiloxane Electrode with Microarray Structure. *ACS Appl Mater Interfaces*. 2017, vol. 9, no. 31, pp. 26314–24.

[8] TANG, B., CHEN, X., HE, Y., ZHOU, J., ZHAO, H., CHEN, W. et al. Fabrication of kapok fibers and natural rubber composites for pressure sensor applications. *Cellulose*. 2021, vol. 28, no. 4, pp. 2287–301.

[9] BAI, J., LI, D., TANG, W., WANG, J., SHI, C., CUI, Y. et al. Carbonized biomass cattail for flexible pressure sensor. *Mater Today Commun*. 2023, vol. 35, 105561.

[10] SADI, M.S., NAHAR, N., KUMPIKAITĖ, E. Multifunctional Cu/poplar-PLA conductive nonwoven fabric for electromyography (EMG) sensing, pressure sensing, and thermal heating applications. *J Mater Sci*. 2024, vol. 59, no. 16, pp. 6861–78.

[11] USTA, C., SEYHAN, A., GURARSLAN, A. Utilization of Poplar Fibers in Needle Punched Nonwovens. *Cellulose*. 2024, vol 31, pp. 8351-8367.

[12] CHEN, H.L, CLUVER, B. Assessment of Poplar Seed Hair Fibers as a Potential Bulk Textile Thermal Insulation Material. *Cloth Text Res J*. 2010, vol. 28, no. 4, pp. 255–62.

[13] LIU, Y., LYU, L., XIONG, X., QIAN, Y., LI, H., GUO, J. Structural characteristics and sound absorption properties of poplar seed fibers. 2020, vol 90, pp. 2467-2477.

[14] LIKON, M., REMŠKAR, M., DUCMAN, V., ŠVEGL, F. Populus seed fibers as a natural source for production of oil super absorbents. *Journal of Environmental Management*. 2013, vol 114, pp. 158-167.

[15] XU, Y., SU, Q., SHEN, H., XU, G. Physicochemical and sorption characteristics of poplar seed fiber as a natural oil sorbent. *Text Res J*. 2019, vol. 89, no. 19–20, pp. 4186–94.

[16] WU, Y., SUN, M., WU, X., SHI, T., CHEN, H., WANG, H. Preparation of nanocellulose aerogel from the poplar (Populus tomentosa) catkin fiber. *Forests*. 2019, vol. 10, no. 9, 749.

NANO TWIST AND SHOUT: CHIRAL NANOANTENNAS MAKING WATER SPLITTING COOL

[1]Jana ROSENKRAZOVA, [1]Elena MILIUTINA, [1]Vasilii BURTSEV, [1]Vaclav SVORCIK, [1]Oleksiy LYUTAKOV

[1]*University of Chemistry and Technology, Prague, Czech Republic, EU, rosenkrj@vscht.cz*

https://doi.org/10.37904/nanocon.2024.4989

Abstract

In the last several years, spin polarization of electrons has been reported as a promising method of improving oxygen evolution reaction (OER). Here, we report the development of a system of hybrid multilayer materials with a chiral property for light-assisted water splitting. The chiral electrodes were prepared via a thin patterned metal film, with a dielectric layer. A layer of redox-active was deposited on the top of prepared structure. The resulting samples were characterized at each step of their preparation using SEM-EDX methods. It is hypothesized that electron spin polarization suppresses the formation of hydrogen peroxide and promote the formation of triplet state of O_2 during OER. This assumption was confirmed via electrochemical measurements, particularly using linear sweep voltammetry and chronoamperometry. During the experiments, we found a correlation between the OER efficiency and the electrode chirality.

Keywords: Chiral electrode, hybrid structure, OER, water splitting

1. INTRODUCTION

In recent years, plasmonic nanoantennas have emerged as a promising technology in the field of plasmon-assisted water splitting [1 ,2], possibly paving the way for the future of efficient green hydrogen production [2–4]. Capable of efficiently converting free-propagating optical radiation to localized energy [5] beyond the diffraction limit [6–8] and vice versa [5,6], plasmonic nanoantennas operate on the principle of absorbing the incident electromagnetic radiation at a resonant frequency, which causes movement of free electrons in the plasmonic material and thus generating an alternating electric current [5].

While geometry [5–7], used materials [5,7] and the dielectric environment [6,7] of the nanoantenna significantly affect their properties [1,5–7], the addition of a chiral element to the structure allows for more subtle optical effects, with light propagation and absorption being different for left and right circularly polarized light [9].Chirality can be introduced through meticulous design [9,10] or the addition of chiral substances to the otherwise non-chiral structures [10,11]. In both cases, chirality of the structure imposes a chiral character on the oscillation of free electrons on the electrode surface and bulk [12]. Another important property of chiral structures is the chiral-induced spin selectivity (CISS) effect, which causes spin the polarisation of an electron [13]. In other words, the CISS effect implies the chiral substance preferentially transmits electrons with one spin direction [14].

For water splitting in particular, the sluggish kinetics of oxygen evolution reaction (OER) [13–16] prevent it from being used on a large scale for green hydrogen production [13,14]. During the water splitting process, the interaction of OH^* radicals intermediates determines the formation of either an oxygen molecule or hydrogen peroxide [14]. By allowing the transfer of parallel spin-oriented electrons only, the formation of H_2O_2 is significatnly reduced, thus enhancing the creation of O_2 [14,16]. The differences between chiral and achiral structures for OER, the utilization of spin-control for catalysis and reaction products are represented in **Figure 1** [15]. In this work, we propose a hybrid multilayer structure, which acts as a chiral lock, enhancing the efficiency of a previously proposed design [1].

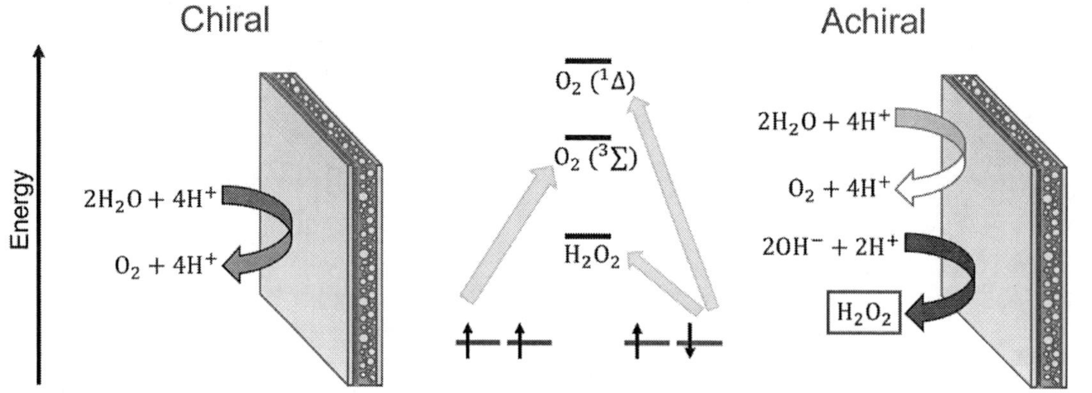

Figure 1 Scheme of chiral and achiral nanoantennas and their utilization for OER, electron spins participating in the reaction, energetic states of potential reaction products and preferential reaction paths

2. EXPERIMENTAL SECTION

2.1 Materials

All chemical substances used for the preparation of the samples are provided in **Table 1**. Platinum target with 4N purity from SAFINA a.s. was used for thin-film preparation. FTO coated glass (TEC 7 FTO) was purchased from MSE Supplies™. Other chemicals used were demineralized water (demi-H_2O) and potassium hydroxide (KOH, 1 mol L^{-1}) from Sigma-Aldrich.

Table 1 Chemical substances used for sample preparation

Substance	Full name	Purity (%)	Manufacturer
PS-*b*-PEO	Poly(styrene-*b*-ethylene oxide)	100	Polymer Source, Inc.
$H_7AuCl_4O_3$	Hydrogen tetrachloroaurate (III) trihydrate	(Au) 49.5	Abcr GmbH
THF	Tetrahydrofuran	99.95	Lach-Ner, s.r.o.
EtOH	Ethanol for analysis	100	Merck
PS	Polystyrene	100	SigmaAldrich
Toluene	Toluene	≥99.9	Lach-Ner, s.r.o.
$C_4H_6NiO_4 \cdot 4H_2O$	Nickel(II) acetate tetrahydrate	≥98	SigmaAldrich
-	Hydrogen peroxide test	-	SigmaAldrich

2.2 Sample preparation

First, a micelle solution was prepared for the preparation of mesoporous gold (porAu). 10 mg of PS-*b*-PEO was weighed and dissolved in 3 ml of THF and left in a water bath (40 °C, 30 min, 275 rpm). Then, 1.5 ml of EtOH was slowly added, followed by 1 ml of 40 mM $HAuCl_4$ and 2.5 ml of demi-H_2O, and the solution was kept at 40 °C for 1 hour. For the preparation of the patterned gold layer, chronoamperometry ($E_{dc} = -0.5$ V, $t_{\text{run}} = 1200$ s) was used in a 5 ml beaker in a three-electrode system with an FTO glass as the working electrode, Ag/AgCl (3 M KCl) as the reference electrode (BVT Technologies, CZ), and a platinum wire as the counter electrode (BASi, USA). PalmSens 4 potentiostat (Palm Instruments, Netherlands) with PSTrace 5.8 software was used for all electrochemical measurements. After chronoamperometric deposition, the FTO glasses with the gold layer were carefully washed alternately in THF and demi-H_2O seven times using a pipette. This preparation was adapted from Li et al. [17].

Subsequently, a dielectric chiral spacer was deposited. Finally, a thin layer of Pt (Quorum Q300T, 20 mA, 110 s, Ar atmosphere) was sputtered. Scheme of the samples' preparation is shown in **Figure 2**. Finally, NiO_x was grown by electrodeposition at -1 V vs. Ag/AgCl (3M KCl) for 200 s in an aqueous electrolyte containing 20 mM $C_4H_6NiO_4 \cdot 4H_2O$.

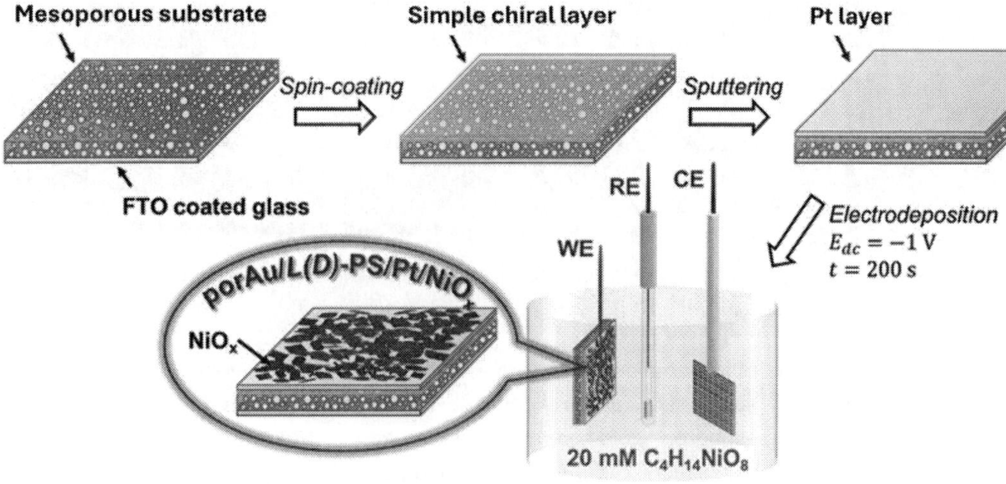

Figure 2 Nanoantenna samples' preparation

2.3 Measurement techniques

To study the surface of the samples at multiple stages of preparation, scanning electron microscopy (SEM) and energy-dispersive X-ray spectroscopy (EDX) (LYRA3 GMU, Tescan, CR) were collected at the operating voltage of 10 kV and a beam current of 600 pA. To better understand the absorption of light of the samples, UV-Vis spectra was obtained using UV/VIS spectrometer HR2000 (Ocean Optics) with AvaLight-DHS light source (Avantes).

2.4 Electrochemical measurements

LSV and chronoamperometry were used for electrochemical analysis in a classic three-electrode system in a 5 ml beaker. The samples were attached as the working electrode, Ag/AgCl (sat. with 3 M KCl) electrode (BVT Technologies, CZ) as the reference electrode and a Platinum Wire electrode (BASi, USA) as the counter electrode, with 0.1 M KOH as the electrolyte.

2.5 H_2O_2 determination

Tests for the presence of H_2O_2 were performed by chronoamperometry in 0.1 M KOH saturated with CO_2 to reduce H_2O_2 decomposition. Each chronoamperometry measurement lasted 20 min at 2 V against Ag/AgCl to sufficiently produce H_2O_2. The electrolytes were then transferred to cleaned polypropylene centrifuge tubes and the procedure indicated in the test instructions was carried out.

3. RESULTS AND DISCUSSION

To better organise the results, the following labels are used for the sample names: porAu/PS/Pt for achiral systems and porAu/L- or D-PS/Pt or for chiral systems. Systems without the dielectric layer are labelled porAu/Pt.

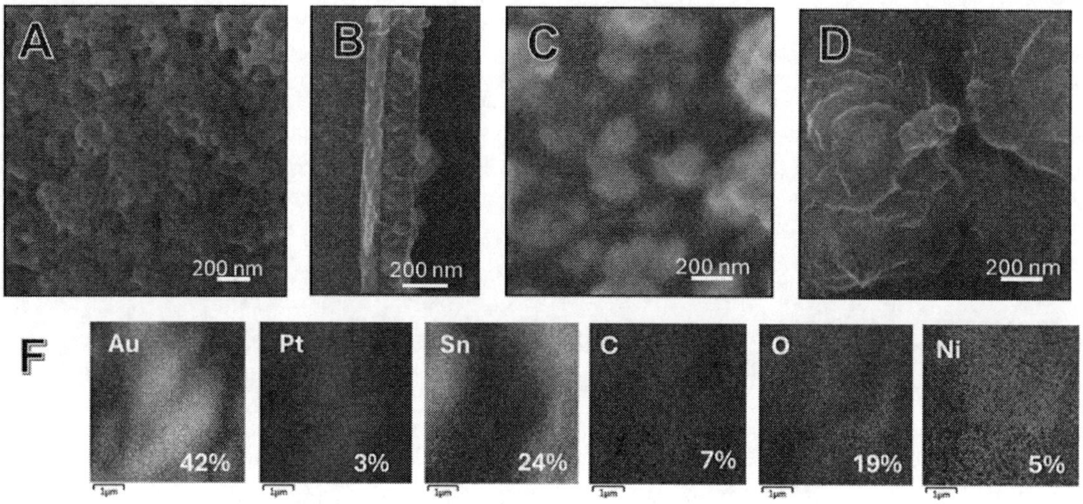

Figure 3 SEM images of nanoantennas at each preparation step: porAu (A), porAu cross-section (B), porAu/L-PS/Pt (C) and porAu/L-PS/Pt/NiO_x (D)

Figure 3 shows the SEM images of all samples containing a chiral or non-chiral dielectric layer with a layer of Pt and NiO_x on top. As expected, sample porAu has pores in the order of 20 - 50 nm. From a cross-section of the sample, the thickness of the mesoporous gold layer was determined as ~180 nm (**Figure 3B**). As expected, after the dielectric and Pt layers deposition, the surface is smoother, and no pores are observed. Surface morphology after NiO_x deposition is presented in **Figure 3D**. In addition, **Figure 3E** shows EDX analysis data, which confirms the presence of Au, Pt and Ni elements on the samples' surface.

Figure 4A shows LSV measurements of chiral and achiral nanoantennas. At first, the LSV was left to run in the dark. After, the light was turned on, as indicated in the figure. The results show that the activity of chiral hybrid structures is much higher than their non-chiral counterparts. This is in agreement with spin-control reaction theory, as explained above [14–16].

Figure 4B shows UV-Vis spectra of the solution analysed for the presence of H_2O_2, as described above. The presence of hydrogen peroxide was clearly detected for both analysed samples with the light off. However, the addition of a chiral substance to the nanoantenna caused the detected amount to decrease significantly while the reaction was left to run with the light on compared to the achiral. This behaviour is, once again, in agreement with spin-control reaction theory, as explained above [14–16].

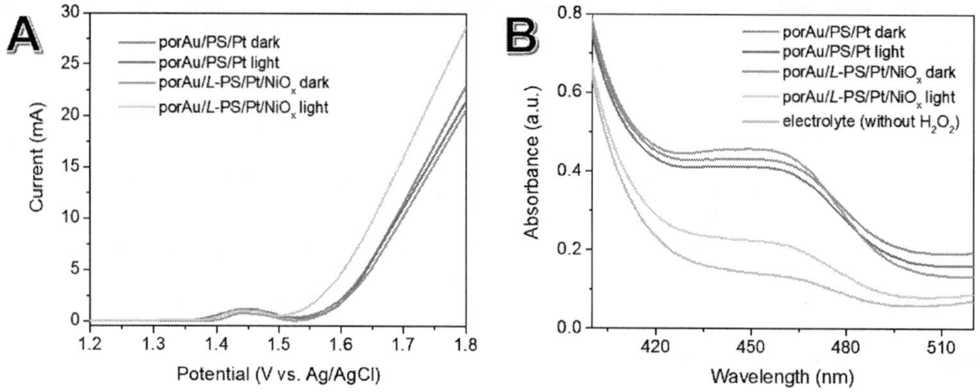

Figure 4 OER activity of chiral and non-chiral nanoantennas in 0.1 M KOH (A) and UV-Vis results of H_2O_2 formation from chronoamperometry at 2 V vs. Ag/AgCl obtained on chiral and achiral structures under and without light (B). The absorption spectra were normalized by the total charge transferred during the reaction.

4. CONCLUSION

In this work, a new way to improve OER was found – using chiral multilayer materials. The incorporation of chiral substances into an already sufficiently efficient system, the rate of the OER exhibited a sharp increase due to the CISS effect, allowing only electrons of a certain spin to participate in the reaction and thus reducing the H_2O_2 production. Therefore, chiral nanoantennas represent a largely unexplored system with significant potential for improving OER.

ACKNOWLEDGEMENTS

This work was supported by the GACR under project 23-08509S.

REFERENCES

[1] ROSENKRANZOVA, J.; BURTSEV, V.; MILIUTINA, E.; SVORCIK, V.; LYUTAKOV, O. Creation of Plasmon-Based Nanoantenna for Hydrogen Production. In: *NANOCON 2023*. Ostrava: Tanger, 2023, pp. 94–99. Available from: https://doi.org/10.37904/nanocon.2023.4754.

[2] GHOBADI, T. G. U.; GHOBADI, A.; SOYDAN, M. C.; VISHLAGHI, M. B.; KAYA, S.; KARADAS, F.; OZBAY, E. Strong Light–Matter Interactions in Au Plasmonic Nanoantennas Coupled with Prussian Blue Catalyst on BiVO4 for Photoelectrochemical Water Splitting. *ChemSusChem*. 2020, vol. 13, no. 10, pp. 2577–2588. Available from: https://doi.org/10.1002/cssc.202000294.

[3] HASSAN, N. S.; JALIL, A. A.; RAJENDRAN, S.; KHUSNUN, N. F.; BAHARI, M. B.; JOHARI, A.; KAMARUDDIN, M. J.; ISMAIL, M. Recent Review and Evaluation of Green Hydrogen Production via Water Electrolysis for a Sustainable and Clean Energy Society. *International Journal of Hydrogen Energy*. 2024, vol. 52, pp. 420–441. Available from: https://doi.org/10.1016/j.ijhydene.2023.09.068.

[4] Di GARBO, C.; LIVRERI, P.; VITALE, G. Solar Nanoantennas Energy Based Characterization. *Renewable Energy and Power Quality Journal*. 2016, vol. 1, pp. 862–867. Available from: https://doi.org/10.24084/repqj14.490.

[5] LI, N.; LAI, Y.; LAM, S. H.; BAI, H.; SHAO, L.; WANG, J. Directional Control of Light with Nanoantennas. *Advanced Optical Materials*. 2021, vol. 9, no. 1, 2001081. Available from: https://doi.org/10.1002/adom.202001081.

[6] KRASNOK, A. E.; MAKSYMOV, I. S.; DENISYUK, A. I.; BELOV, P. A.; MIROSHNICHENKO, A. E.; SIMOVSKI, C. R.; KIVSHAR, Y. S. Optical Nanoantennas. *Phys.-Usp.* 2013, vol. 56, no. 6, 539. Available from: https://doi.org/10.3367/UFNe.0183.201306a.0561.

[7] ZABELINA, A.; MILIUTINA, E.; ZABELIN, D.; BURTSEV, V.; BURAVETS, V.; ELASHNIKOV, R.; NEUBERTOVA, V.; ŠŤASTNÝ, M.; POPELKOVÁ, D.; LANCOK, J.; CHERTOPALOV, S.; PAIDAR, M.; TRELIN, A.; MICHALCOVÁ, A.; ŠVORČÍK, V.; LYUTAKOV, O. Plasmon Coupling inside 2D-like TiB2 Flakes for Water Splitting Half Reactions Enhancement in Acidic and Alkaline Conditions. *Chemical Engineering Journal*. 2023, vol. 454, 140441. Available from: https://doi.org/10.1016/j.cej.2022.140441.

[8] KRAVETS, V. G.; BARNES, W. L.; KABASHIN, A. V.; GRIGORENKO, A. N. Plasmonic Surface Lattice Resonances: A Review of Properties and Applications. *Chemical Reviews*. 2018, vol. 118, no. 12, pp. 5912–5951. Available from: https://doi.org/10.1021/acs.chemrev.8b00243.

[9] KIM, M.; BAZAID, M. O.; LEE, H. J.; JANG, S. S.; JANG, Y. J.; KIM, D. H. Chiral Inorganic Nanostructures from Achiral Platforms: A Universal Synthesis Route via Supramolecular Self-Assembly. *Chem. Mater.* 2023, vol. 35, no. 8, pp. 3073–3082. Available from: https://doi.org/10.1021/acs.chemmater.2c02833.

[10] KALACHYOVA, Y.; GUSELNIKOVA, O.; ELASHNIKOV, R.; PANOV, I.; ŽÁDNÝ, J.; CÍRKVA, V.; STORCH, J.; SYKORA, J.; ZARUBA, K.; ŠVORČÍK, V.; LYUTAKOV, O. Helicene-SPP-Based Chiral Plasmonic Hybrid Structure: Toward Direct Enantiomers SERS Discrimination. *ACS Appl. Mater. Interfaces*. 2019, vol. 11, no. 1, pp. 1555–1562. Available from: https://doi.org/10.1021/acsami.8b15520.

[11] GOERLITZER, E. S. A.; PURI, A. S.; MOSES, J. J.; POULIKAKOS, L. V.; VOGEL, N. The Beginner's Guide to Chiral Plasmonics: Mostly Harmless Theory and the Design of Large-Area Substrates. *Advanced Optical Materials*. 2021, vol. 9, no. 16, 2100378. Available from: https://doi.org/10.1002/adom.202100378.

[12] LIANG, Y.; BANJAC, K.; MARTIN, K.; ZIGON, N.; LEE, S.; VANTHUYNE, N.; GARCÉS-PINEDA, F. A.; GALÁN-MASCARÓS, J. R.; HU, X.; AVARVARI, N.; LINGENFELDER, M. Enhancement of Electrocatalytic Oxygen Evolution by Chiral Molecular Functionalization of Hybrid 2D Electrodes. *Nat Commun.* 2022, vol. 13, no. 1, 3356. Available from: https://doi.org/10.1038/s41467-022-31096-8.

[13] MINGOES, C. J.; SCHROEDER, B. C.; JORGE SOBRIDO, A. B. Electron Spin Selective Iridium Electrocatalysts for the Oxygen Evolution Reaction. *ACS Mater. Au.* 2024, vol. 4, no. 2, pp. 204–213. Available from: https://doi.org/10.1021/acsmaterialsau.3c00084.

[14] GHOSH, K. B.; ZHANG, W.; TASSINARI, F.; MASTAI, Y.; LIDOR-SHALEV, O.; NAAMAN, R.; MÖLLERS, P.; NÜRENBERG, D.; ZACHARIAS, H.; WEI, J.; WIERZBINSKI, E.; WALDECK, D. H. Controlling Chemical Selectivity in Electrocatalysis with Chiral CuO-Coated Electrodes. *J. Phys. Chem. C.* 2019, vol. 123, no. 5, pp. 3024–3031. Available from: https://doi.org/10.1021/acs.jpcc.8b12027.

[15] NAAMAN, R.; PALTIEL, Y.; WALDECK, D. H. Chiral Induced Spin Selectivity Gives a New Twist on Spin-Control in Chemistry. *Acc. Chem. Res.* 2020, vol. 53, no. 11, pp. 2659–2667. Available from: https://doi.org/10.1021/acs.accounts.0c00485.

[16] LI, C.; JIANG, B.; CHEN, H.; IMURA, M.; SANG, L.; MALGRAS, V.; BANDO, Y.; AHAMAD, T.; ALSHEHRI, S. M.; TOMINAKA, S.; YAMAUCHI, Y. Superior Electrocatalytic Activity of Mesoporous Au Film Templated from Diblock Copolymer Micelles. *Nano Res.* 2016, vol. 9, no. 6, pp. 1752–1762. Available from: https://doi.org/10.1007/s12274-016-1068-z.

THE STABILITY OF SILVER NANOWIRE COATED POPLAR NONWOVENS UNDER WASHING CONDITIONS

[1]Aybeniz SEYHAN, [1]Canan USTA, [1]Alper GURARSLAN

[1]*Istanbul Technical University, Faculty of Textile Technologies and Design, Istanbul, Türkiye,* seyhanay@itu.edu.tr

https://doi.org/10.37904/nanocon.2024.4987

Abstract

Poplar fibers, known for their high cellulose content and hollow structure, are primarily discarded as waste in the wood industry despite the tree's abundance. Recent advancements have enabled the utilization of poplar fibers in the textile industry, specifically in the production of nonwoven fabrics. This study explores the potential of valorizing poplar fibers by functionalizing them with silver nanowires (AgNWs) for technical applications. The focus is on enhancing the durability of conductive poplar nonwovens against washing procedures. Initially, silver nanowires were synthesized via the NaCl-mediated polyol method and subsequently coated onto a blend of poplar and polyester (PET) nonwoven fabric. The conductive fabric samples underwent a series of washing cycles to assess their durability. Post-washing analyses included morphological characterization and conductivity measurements to evaluate the performance and resilience of the coatings. This study provides insights into the wash durability of conductive poplar nonwovens, contributing to their potential application in electronic wearables, where compliance and comfort are critical.

Keywords: Poplar fiber, sustainable textiles, silver nanowire, washing durability, wearable electronics

1. INTRODUCTION

The development of conductive textiles has gained considerable attention in recent years, particularly in response to the growing demand for wearable, flexible electronics and smart fabrics. These textiles offer promising applications in areas ranging from healthcare monitoring to interactive garments, owing to their ability to conduct electricity through incorporating conductive elements such as metal nanowires, intrinsically conductive polymers (ICPs), or carbon-related materials [1]. Among the materials explored for conductive textiles, silver nanowires (AgNWs) have emerged as a prominent candidate due to their exceptional electrical conductivity, flexibility, and ease of integration into fibrous structures [2, 3]. Silver as a component and silver nanoparticles have found use in textiles due to their antimicrobial activity in various applications, recently AgNW-coated fabrics have demonstrated notable performance in wearable applications as well [3, 4].

Beyond the choice of conductive materials and fabrication techniques, the selection of substrate plays a crucial role in the development of conductive surfaces, particularly when sustainable materials are considered for achieving both performance and environmental responsibility. Therefore, functionalization of natural fibers, particularly for technical and smart applications, represents a growing area of interest due to the increasing demand for sustainable and high-performance materials. The environmental impact of synthetic fibers, coupled with the vast quantities of natural fibers discarded as waste, has led to increased interest in valorizing renewable, biodegradable fibers. Poplar fibers belonging to the family of Populus genus, with their high cellulose content, large hollow lumen combined with a thin cell wall [5, 6], offer a promising alternative in this regard. Despite their abundance, poplar fibers are largely discarded as waste in the wood industry. Characterized by its large hollow lumen occupying about 80-90% of the cross-section, it offers a lightweight, fluffy texture with excellent thermal and sound insulation properties [7]. As a renewable, biodegradable resource, poplar fiber has gained increasing attention for various applications, including thermal [7] and acoustic insulation [8], oil sorbents for water cleanup [5, 9], composite reinforcement [10], aerogel formation [11] and as a natural substrate for wearable electronics [12, 13, 14]. Having the focus on the incorporation of poplar fibers for electronic wearables, this area of use requires materials that combine conductivity with softness, breathability, and durability, particularly in environments where washing and repeated handling are inevitable. However, the challenge of maintaining electrical conductivity after exposure to mechanical stresses, such as washing and repeated use, remains a significant hurdle in bringing these smart textiles to mass uses [15]. Washing can degrade the conductive coating by removing or altering the distribution of nanowires, which diminishes the fabric's electrical performance. The wash durability of such conductive fabrics is critical for their practical use in smart textiles and wearable devices. Research has thus focused on optimizing both the materials used and the processes involved in manufacturing conductive textiles to enhance their durability and washability.

The aim of this study is to explore the potential of poplar fibers as a sustainable substrate for conductive textiles. Using the NaCl-mediated polyol method, silver nanowires were synthesized and coated onto poplar-polyester (PET) nonwoven blends. The durability of these conductive nonwoven fabrics was evaluated by subjecting the samples to a series of washing cycles, followed by conductivity assessments to determine the performance of the AgNW coatings post-wash. This work aims to contribute to the growing body of research on sustainable, multifunctional textiles by providing insights into the wash durability of conductive poplar nonwovens, with particular relevance to electronic wearables where both comfort and compliance are essential.

2. EXPERIMENTAL STUDY

2.1. Materials

Poplar fibers belonging to nigra italica trees were collected from ground using a modified leaf-blower during the blooming season in Türkiye. Prior to the preparation of the nonwoven samples, seeds, dust, leaves, and other contaminants were meticulously removed manually to ensure clean poplar fibers. Hollow polyethylene terephthalate (PET) fibers (~64 mm in length) were generously provided by SASA Polyester Sanayi A.Ş., Türkiye. Silver nitrate ($AgNO_3$, ≥99.8%, Sigma-Aldrich), polyvinylpyrrolidone (PVP, average molecular weight of 40,000, Sigma-Aldrich), ethylene glycol (EG, Merck), ethanol (Merck), and acetone (99.5% purity, TEKKİM) were sourced from Labor-Teknik Laboratuvar Malzemeleri Sanayi Ve Ticaret A.Ş., Türkiye. All chemical reagents were used without further purification.

2.2. Nonwoven production

The cleaned poplar fibers were blended with PET fibers in three distinct weight ratios (0:100, 30:100, and 60:100) without any chemical pre-treatment. The resulting nonwoven fabrics were designated as sPET, popPET-30, and popPET-60, corresponding to blend ratios of 0/100, 30/100, and 60/100 (w/w) for poplar and PET fibers, respectively. A two-step process was employed to fabricate these blends. Initially, the fibers were processed through a carding machine (Mesdan 337A laboratory carding machine) to achieve uniform distribution. Subsequently, the carded webs were subjected to needle-punching (DILO needle-punching machine) to mechanically bond the fibers and produce stable nonwoven fabrics. The produced fabrics have similar basis weight of 243, 263 and 279 gsm along with 3 mm of thickness for sPET, popPET-30, and popPET-60, respectively as stated in the previous study [6].

2.3. Silver nanowire synthesis and coating

A NaCl-mediated polyol synthesis was carried out to produce silver nanowires. The capping agent, PVP, was dissolved in ethylene glycol in a three-necked glass flask. The flask was placed in an oil bath to provide homogenous heating over a hot plate at 180°C. The solution was stirred until homogenization for 45 minutes. Preparation of silver nitrate solution was carried out by dissolving silver nitrate in ethylene glycol at room temperature under continuous stirring in a aluminum foil covered beaker, simultaneously. After PVP solution is mixed, a 1M NaCl-EG solution was added and it was allowed to react for an additional 10 minutes. The silver nitrate solution was then fed into the system at a rate of 1 mL/min by a syringe pump. The resulting solution was stirred at 180°C for an additional 20 minutes and then allowed to cool to room temperature. After cooling, acetone was added to dilute the mixture, and silver nanowires (AgNWs) were precipitated through centrifugation.

The precipitated AgNWs were then re-dispersed in ethanol at a concentration of 50 mg/mL to prepare the coating solution. Nonwoven fabrics were cut into pieces having dimensions of 2.5 cm x 5 cm to prepare samples for coating procedure. They were then immersed in 15 mL of the AgNW dispersion to facilitate uniform coating, as schematized in **Figure 1**. Each cycle of dip-coating lasted for 10s. After each cycle an oven- drying step was performed at 60 ℃ for 30 minutes. Consecutive coatings were performed to decrease the electrical resistance of the samples. The number of dip-coating cycles was completed at 5 after a resistance value of below 20 Ohm was achieved for each fabric. AgNW functionalized poplar blend nonwovens were produced.

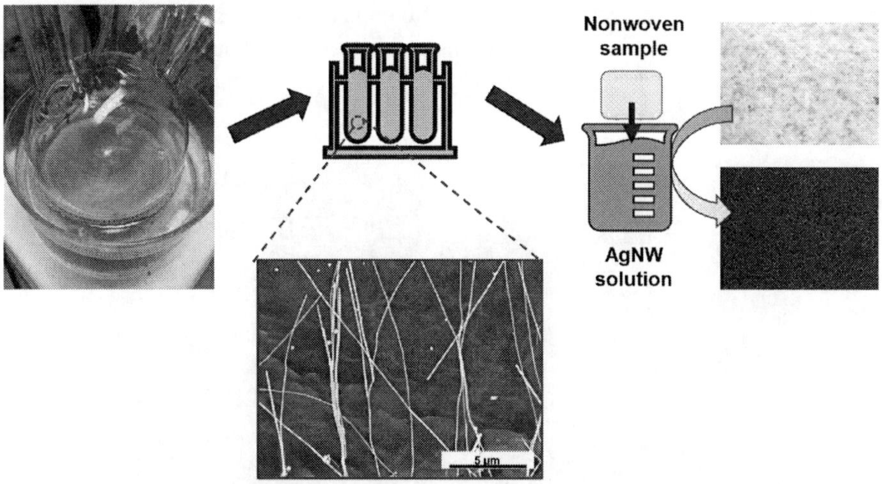

Figure 1 NaCl-mediated silver nanowire synthesis and dip-coating application on poplar blend nonwoven samples

2.4. Washing of AgNW coated poplar blend nonwovens

A gentle washing procedure was conducted as described [16]. Each fabric sample of 2.5 cm x 5 cm was placed in a beaker containing 50 mL of a tap water. The AgNW-coated sPET, popPET-30 and popPET-60 nonwoven fabrics underwent five washing cycles. Each cycle was performed at 40°C for 30 minutes. The washing was conducted under continuous stirring at 40 rpm with help of a magnetic stirrer, as demonstrated in **Figure 2**. After each washing cycle, the samples were rinsed using distilled water. Following each washing cycle, samples were oven-dried at 60 °C for 45 minutes and were subjected to resistance measurement in-between cycles.

2.5. Characterization

The surface electrical resistance values of the nonwoven samples were measured using a Digital Multimeter (FLUKE 17B+) having an accuracy of ±0.5%. The morphology of the silver was imaged using scanning electron microscopy (SEM, TESCAN VEGA3). The length of the nanowires was assessed with help of the ImageJ software using the SEM images as base for measurements. An average of 50 samples were taken into account. A Perkin Elmer Spectrum Two IR was used for the spectral analysis. Spectra were recorded with a resolution of 4 cm^{-1} across the range of 4000–50 cm^{-1}, with an average taken over 16 scans.

3. RESULTS AND DISCUSSION

The washing performance of conductive nonwovens was evaluated as focus of this study. Conductive nonwoven substrates were produced by a common dip-coating method. To impart conductivity, AgNWs were fabricated through polyol synthesis. After synthesis, a small fraction of nanowire dispersion was deposited on a smooth substrate and the excess ethanol was removed by oven-drying. This substrate was then subjected by scanning electron microscopy to analyze the produced silver nanowires. As represented in **Figure 1,** AgNWs having an average length of 10.9±5 μm has been synthesized and used for coating of nonwovens.

The dip-coating procedure lasted for 5 consecutive cycles. Increasing the dip-coating cycles to five significantly reduced electrical resistance, reaching below 20 Ω/cm, as each cycle imparted more silver nanowires onto the fabric surface. Although increased deposition has its advantages, the drawbacks may include increased cost, excessive coating without adhesion, which may be released during mechanical resistance or washing. Therefore, it is critical to achieve low resistivity at fewer deposition steps. The original fabrics have the colors white and a soft yellow for virgin PET (sPET) and poplar blends (popPET-30 and popPET-60), respectively. After full impregnation of the nonwoven fabrics with the AgNW ethanol suspension, the fabric colors shifted from their original appearance to a dark grey, signifying successful and complete AgNW deposition to create conductive network on the surface, as demonstrated in **Figure 1**.

Figure 2 Silver nanowire-coated nonwoven fabric samples before (a) and after (b) being subjected to washing procedure 5 times (c)

The initial electrical resistance values of the fabric samples were given in **Table 1.** The slight difference in resistivity of synthetic and blend samples may be attributed to physical and chemical characteristics of the poplar fiber. For which, the surface roughness and abundant hydroxyl groups deriving from cellulose content have significant role. After first resistance measurements, all samples have been subjected to washing in a laboratory environment, as demonstrated in **Figure 2.** A gentle washing procedure has been conducted to observe the washing stability of the samples, which have been neither pre-treated nor post-cured. As seen in the color difference between **Figure 2a** and **Figure 2b,** a total washing procedure having 5 cycles have significantly affected the conductive network created by AgNW coating. The disruption of this network became also apparent in terms of electrical resistance activity recorded by multimeter measurement in **Table 1.** It reveals a clear trend in the washing durability of silver nanowire-coated nonwoven fabrics composed of varying proportions of poplar and PET fibers. Initially (Cyc.0), all samples exhibit relatively low electrical resistance, with sPET having a resistance of 18 Ω/cm, popPET-30 at 7.5 Ω/cm, and popPET-60 at 10.4 Ω/cm. As the fabrics are subjected to washing cycles, the sPET sample shows a sharp and dramatic increase in resistance, rising to 80 Ω/cm after just one cycle and reaching a significant 1100 Ω/cm by the third cycle. In contrast, the poplar-blended fabrics (popPET-30 and popPET-60) show much slower increases in resistance, rising by 9.3% and 5.7%, respectively. This indicates that poplar fibers contribute to greater stability of the silver nanowire coating after initial washing, particularly when compared to the pure polyester fabric. After five washing cycles, the electrical resistance of popPET-30 reaches 88 Ω/cm, and popPET-60 stabilizes at 20 Ω/cm, while the sPET sample is no longer measurable after three cycles due to its extremely high resistance.

Table 1 Electrical resistance measurements of nonwoven fabric samples at initial stage and after each washing cycle

Washing Cycle	Electrical Resistance (Ω/cm)		
	sPET	popPET-30	popPET-60
Cyc.0	18	7.5	10.4
Cyc.1	80	8.2	11
Cyc.2	138	9	12.2
Cyc.3	1100	16	13
Cyc.4	-	33	15.8
Cyc.5	-	88	20

The trend continues over subsequent washing cycles. After two washing cycles, sPET's resistance further surges to 138 Ω/cm, reflecting a 666% total rise from the initial value. Meanwhile, popPET-30 increased slightly to 9 Ω/cm and popPET-60 to 12.2 Ω/cm. By the third cycle, sPET experienced a dramatic increase to 1100 Ω/cm, indicating a near-total loss of conductivity. Conversely, popPET-30 reached 16 Ω/cm and popPET-60 only 13 Ω/cm showing much better retention of conductive properties. After five washing cycles, popPET-30 and popPET-60 had final resistances of 88 Ω/cm and 20 Ω/cm, respectively. Overall, popPET-60 demonstrated the best wash resistance, with only a 92.3% increase over five cycles, while sPET became non-conductive after three cycles. These findings suggest that increasing the poplar fiber content significantly improves the fabric's durability and conductivity retention after repeated washing, offering substantial advantages over pure polyester fabrics.

4. CONCLUSION

Conductive nonwoven fabrics were produced from hollow PET and poplar fibers decorated with silver nanowires. PET and poplar fibers have been carded and needle-punched to achieve a mechanically stable nonwoven fabric. In order to impart conductivity to these fabrics, AgNWs were synthesized according to NaCl-mediated polyols synthesis. Nanowires were deposited on the fabric substrates by consecutive dip coating. Their ability to conduct electricity has been investigated in context of washing durability. The results suggest that adding poplar fibers to polyester nonwovens significantly improves the wash durability of silver nanowire coatings, with higher poplar content offering even better performance. This makes poplar-blended nonwovens a more promising material for applications requiring conductive textiles.

ACKNOWLEDGEMENTS

This work was supported by the Scientific and Technological Research Council of Türkiye (TÜBİTAK) under Grant No. 121M308 and the European Commission Horizon Marie Sklodowska-Curie Actions Program, SMARTWASTE project, under Grant Agreement ID: 101086258.

REFERENCES

[1] TSEGHAI, G. B., MALENGIER, B., FANTE, K. A., NIGUSSE, A. B., VAN LANGENHOVE, L. Integration of conductive materials with textile structures, an overview. *Sensors*. 2020, vol. 20, no. 23, 6910.

[2] CHENG, Y., WANG, R., SUN, J., GAO, L. Highly conductive and ultrastretchable electric circuits from covered yarns and silver nanowires. *ACS nano*. 2015, vol. 9, no. 4, pp. 3887-3895.

[3] YAO, S., YANG, J., POBLETE, F. R., HU, X., ZHU, Y. Multifunctional electronic textiles using silver nanowire composites. *ACS applied materials & interfaces*. 2019, vol. 11, no. 34, pp. 31028-31037.

[4] CUI, H. W., SUGANUMA, K., UCHIDA, H. Highly stretchable, electrically conductive textiles fabricated from silver nanowires and cupro fabrics using a simple dipping-drying method. *Nano Research*. 2015, vol. 8, pp. 1604-1614.

[5] LIKON, M., REMŠKAR, M., DUCMAN, V., ŠVEGL, F. Populus seed fibers as a natural source for production of oil super absorbents. *Journal of Environmental Management*. 2013, vol. 114, pp.158-167.

[6] USTA, C., SEYHAN, A., GURARSLAN, A. Utilization of poplar fibers in needle punched nonwovens. *Cellulose*. 2024, vol. 31, pp. 8351-8367.

[7] CHEN, H. L., CLUVER, B. Assessment of poplar seed hair fibers as a potential bulk textile thermal insulation material. *Clothing and Textiles Research Journal*. 2010, vol. 28, no. 4, pp. 255-262.

[8] LIU, Y., LYU, L., XIONG, X., QIAN, Y., LI, H., GUO, J. Structural characteristics and sound absorption properties of poplar seed fibers. *Textile Research Journal*. 2020, vol. 90, no. 21-22, pp. 2467-2477.

[9] XU, Y., SU, Q., SHEN, H., XU, G. Physicochemical and sorption characteristics of poplar seed fiber as a natural oil sorbent. *Textile Research Journal*. 2019, vol. 89, vol. 19-20, pp. 4186-4194.

[10] ZHAO, X., TEKINALP, H., MENG, X., KER, D., BENSON, B., PU, Y., RAGAUSKAS, A.J., WANG, Y., LI, K., WEBB, E., GARDNER, D.J., ANDERSON, J., OZCAN, S. Poplar as biofiber reinforcement in composites for large-scale 3D printing. *ACS Applied Bio Materials*. 2019, vol. 2, no. 10, pp. 4557-4570.

[11] LI, L., HU, T., SUN, H., ZHANG, J., WANG, A. Pressure-sensitive and conductive carbon aerogels from poplars catkins for selective oil absorption and oil/water separation. *ACS Applied Materials & Interfaces*. 2017, vol. 9, no. 21, pp. 18001-18007.

[12] GURARSLAN, A., NARINÇ, M.E. Investigating the rose oil and toluene absorption of populus fiber. *Textile Research Journal*. 2019, vol. 89, no. 10, pp. 1952–63.

[13] ZHANG, X., XU, G., WANG, H., CUI, H., ZHAN, X., WANG, X. Enhanced acetone sensing properties of hollow SnO2 fibers using poplar catkins as a bio-template. *Powder Technology*. 2019, vol. 344, pp. 183-189.

[14] SADI, M.S., NAHAR, N., KUMPIKAITĖ, E. Multifunctional Cu/poplar-PLA conductive nonwoven fabric for electromyography (EMG) sensing, pressure sensing, and thermal heating applications. *Journal of Material Science*. 2024, vol. 59, no. 16, pp. 6861–78.

[15] SADI, M.S., NAHAR, N., KUMPIKAITĖ, E. Advances in the robustness of wearable electronic textiles: Strategies, stability, washability and perspective. *Nanomaterials*. 2022, vol. 12, no 12, 2039.

[16] ZHOU, Y., LI, W., LI, L., SUN, Z., JIANG, L., MA, J., CHEN, S., NING, X., ZHOU, F. Lightweight and highly conductive silver nanoparticles functionalized meta-aramid nonwoven fabric for enhanced electromagnetic interference shielding. *Journal of Materials Science*. 2021, vol. 56, pp. 6499-6513.

HIGH PERFORMANCE FLEXIBLE CAPACITIVE PRESSURE SENSOR FOR WEARABLE TECHNOLOGIES

[1]Meltem TEKCIN, [2]Esma Nur NEBIPASAGIL, [3]Senem KURSUN

[1]*İstanbul Technical University, Textile Engineering Department, İstanbul, Türkiye, tekcin@itu.edu.tr*

[2]*İstanbul Technical University, Textile Engineering Department, İstanbul, Türkiye, gulluoglu@itu.edu.tr*

[3]*İstanbul Technical University, Mechanical Engineering Department, İstanbul, Türkiye, kursuns@itu.edu.tr*

https://doi.org/10.37904/nanocon.2024.4985

Abstract

Flexible capacitive pressure sensors have applications in a wide variety of fields, especially soft robotics, bioelectronics, wearable devices and healthcare applications. In this study, a capacitive pressure sensor has been designed and fabricated using silver, nickel, and carbon nanoparticle conductive inks. Pad printing method was used to print nanoparticle inks on a polyamide-based taffeta label fabric. Capacitive pressure sensor underwent testing through three methodologies; these are pressing different fingers on the sensor surface, dropping water on the sensor and applying bending-relaxation process to the pressure sensor to determine sensor performance measuring capacitance variations. The capacitance of the fabricated pressure sensor changed according to the type of finger pressed and the number of water droplets dropped. As the amount of water droplets and the force of the pressed finger increase, the capacitance changes in the printed sensor also increase. On the other hand, there was no significant change in the capacitance values of the sensor after the bending-relaxation application. This shows that the proposed flexible pressure sensor can be used in wearable electronics, human-robot interactions, pressure mapping or detection of finger touch.

Keywords: Pad printing, pressure sensor, conductive ink, flexible, capacitance measurement

1. INTRODUCTION

Flexible pressure sensors with various sensing mechanisms, especially piezoelectric [1], triboelectric [2], resistive [3] and capacitive [4], have become widely used in robotics, bioelectronics, human motion and wearable health monitoring applications. Compared to traditional and rigid pressure sensors, flexible sensors are distinguished by their ability to be bent and deformed, as well as their high sensitivity at low pressures and rapid response times [5], [6]. Furthermore, the stable and straightforward structure of capacitive pressure sensors, coupled with their low power consumption and cost-effective manufacturing, are reasons for their frequent preference [7], [8]. Flexible pressure sensors are essentially capacitors composed of two electrodes and a dielectric layer interposed between them. The application of pressure causes variations in the thickness of the dielectric layer and the distance between the electrodes, resulting in changes in the sensor's capacitance [5], [9].

In the production of capacitive pressure sensors, structures such as fibers, fluids, microstructured or porous elastomers, foam, composites are used as dielectric materials, and the sensitivity of these sensors is proportional to the compressibility of these dielectric materials [5], [7], [10]. A high sensitivity capacitive pressure sensor fabricated using double layer polydimethylsiloxane (PDMS) coated with silver nanowire electrodes was proposed in the study by Wang et al [11]. In another study, a capacitive pressure sensor obtained using hemispherical porous PDMS and carbon nanotube (CNT) used in monitoring pulse and joint movement was presented [12]. Qiu et al. produced a high-sensitivity, capacitive, and flexible tactile sensor from a 3D porous conductive composite using a polyurethane sponge substrate by dip-coating method [13]. The most critical parameters in sensor manufacturing processes are the ease of operation and low-cost. Additionally, possessing high sensitivity and being thin and miniature are also sought-after characteristics of sensors.

In this work, we report on the development of a high sensitivity printed flexible capacitive pressure sensor fabricated through pad printing method utilizing silver/carbon/nickel conductive ink. We evaluated the performance of the pressure sensors by measuring the capacitance variations of sensors produced without an elastic dielectric layer. Based on the obtained results, we believe that the proposed capacitive and flexible pressure sensor holds significant potential across various applications, ranging from wearable technologies to robotic implementations and human-robot interactions.

2. EXPERIMENTAL STUDY

2.1 Chemicals and Materials

In this study, silver, carbon, and nickel nanoparticle conductive inks were used for producing the electrodes of the pressure sensor. Silver, carbon, and nickel nanoparticle inks were mixed in a magnetic stirrer at a weight ratio of 1:1:1 until it becomes homogeneous. A polyamide-based taffeta label fabric was selected as a substrate. Waterproof polyurethane welding tape was used to cover the capacitive pressure sensor. The thickness of the welding tape is 44 μm.

2.2 Capacitive Pressure Sensor Fabrication

Interdigital electrodes were used for the design of the capacitive pressure sensor. The electrodes, which can be designed in many different shapes according to different gap patterns, have a square wave air gap [5]. The silver/carbon/nickel ink mixture prepared for the fabriation of pressure sensor was printed on polyamide taffeta label fabric using a pad printing machine, which has a fast and easy application. The image of the pressure sensor, which was printed 5 times to increase conductivity, is shown in **Figure 1**. Since the conductivity of nanoparticles increases with heat treatment, the printed sensor was subjected to a sintering process in an oven at 150°C for 30 minutes after the printing process. Except for the uncoated sensor, the sensor printed was covered with polyurethane welding tape using a hot press machine in order to find out whether the coating of the sensor surface has an effect on the capacitance performance of the sensor. The important point here is the softening temperature of the welding tape. Based on the datasheet of the welding tape, the hot press machine was operated at 90°C. Another important point is that while the sensor is being coated, the ends of the sensor are left open in order to take measurements with the multimeter.

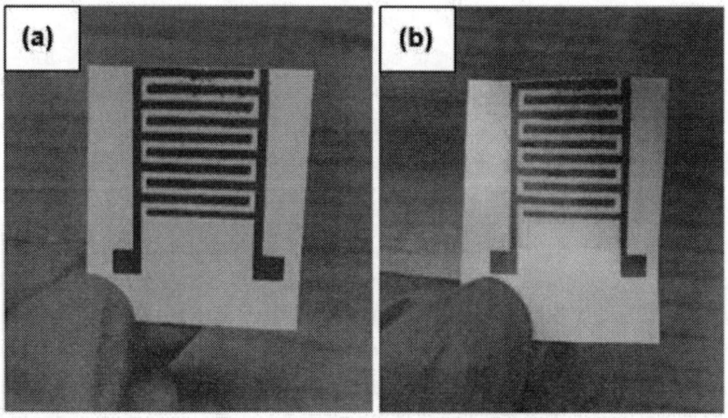

Figure 1 (a) Printed pressure sensor without coating, (b) Printed pressure sensor coated with welding tape

2.3 Capacitance Measurement

To evaluate the performance of coated and uncoated silver/carbon/nickel printed sensors, tests including finger pressing, water droplet application, and bending were conducted. Capacitance measurements were performed by direct contact of the probe tips of the Fluke 17B+ digital multimeter to the pressure sensor.

3. RESULTS AND DISCUSSION

Finger pressing was first applied to coated and uncoated silver/carbon/nickel printed sensors. Higher finger pressing force increases the relative capacitance. Increasing finger pressing force increases the contact angle between the finger and the sensor and creates a smaller parallel contact resistance in the circuit, thus producing a larger capacitance [5]. In this study, since fingers have different pressing forces, the middle finger, index finger and thumb were pressed separately on the pressure sensors and the capacitance changes were measured with a multimeter. The capacitance response graph of uncoated and coated sensors when pressed with three different fingers is given in **Figure 2**. Based on the capacitance changes shown in **Figure 2**, the highest capacitance change in both sensors according to finger pressing force and finger areas was obtained when the thumb was pressed, while the lowest capacitance change was obtained when the index finger was pressed. At the same time, the capacitance change in the uncoated pressure sensor is higher than the coated sensor. This shows that the uncoated sensor is more sensitive to the pressing force.

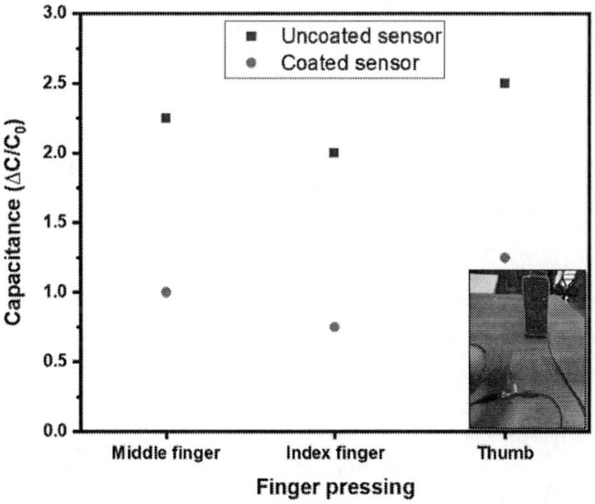

Figure 2 Capacitance response of the uncoated and coated sensors when pressing middle finger, index finger and thumb

Additionally, distilled water was dropped onto coated and uncoated sensors to measure the effect of the weight of water drops on capacitance changes in the sensors. Distilled water was dropped onto the sensors from 1 to 10 drops, each drop volume being 0.01 mL. Following the application of each water drop, the capacitance response of the sensors was measured with the help of a multimeter. Capacitance measurements were first started with the uncoated sensor. The capacitance change on the uncoated sensor could not be measured with the multimeter after the second drop of water. It is estimated that this situation is due to the absence of any dielectric layer between the electrodes.Therefore, in the continuation of the study, water droplets were dropped only onto pressure sensors coated with polyurethane welding tape. The capacitance change graph obtained when distilled water droplets were dropped onto the coated sensor is presented in **Figure 3**. The capacitance values read from the multimeter when each water droplet falls onto the sensor surface are also given in the figure. According to the capacitance responses in **Figure 3**, as the number of water droplets on the sensor surface increases, both the capacitance and the capacitance change in the sensor increases.

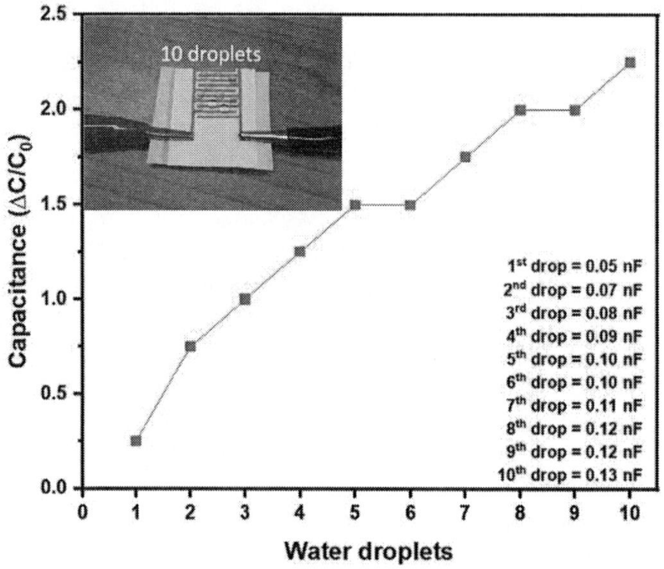

Figure 3 Capacitance response of the coated sensor when dropping distilled water

In addition to the water dropping and finger pressing tests, a bending test was applied to the printed sensor to test the suitability of the proposed pressure sensor especially for wearable technologies. To measure the stability and performance of the sensor against bending, the sensor was bent and released 10 times. The capacitance measurements were made by finger pressing the sensor before and after bending and compared these values in order to understand the performance of the sensor. **Figure 4** shows the capacitance change after applying force with the thumb to the coated pressure sensor before bending, while **Figure 5** indicates the capacitance results after the same application after bending process. When the graphs in **Figure 4** and **Figure 5** are examined, the values of the capacitance changes of the pressure sensor before bending and after the bending-relaxation application vary between 0.9 and 1.3. When the graphs are compared, the capacitance changes after the bending-relaxation process are quite similar to the capacitance changes of the sensor before bending. Therefore, the proposed flexible pressure sensor shows high resistance to bending.

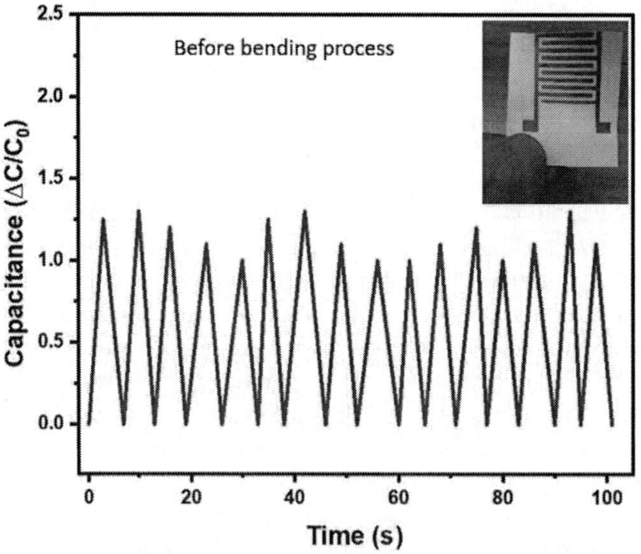

Figure 4 Capacitance change of the sensor before bending process

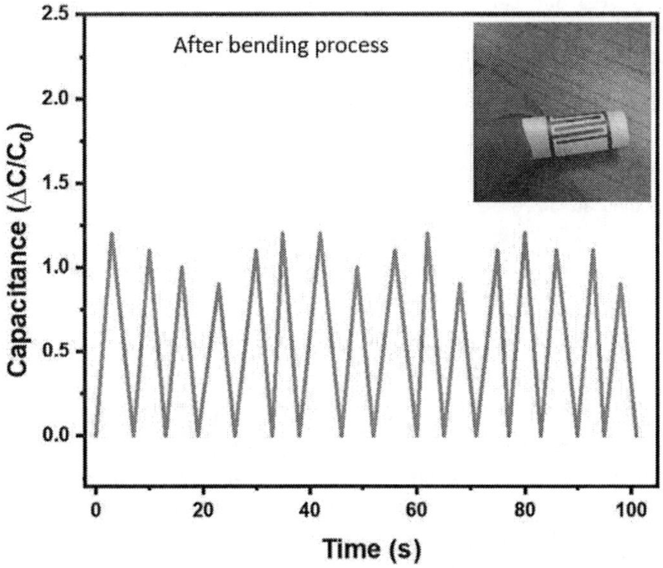

Figure 5 Capacitance change of the sensor after bending-relaxation process.

4. CONCLUSION

As a result, in this study, a flexible capacitive pressure sensor was fabricated by pad printing machine using silver/carbon/nickel mixture ink, especially targeting wearable technologies and robotic applications. To investigate the performance characteristics of the sensors, capacitance responses were examined using a multimeter by applying force with three different fingers on sensors coated with polyurethane welding tape and those without any coating. Since the capacitance change in the uncoated pressure sensor is greater as a result of finger pressing, the sensitivity of the uncoated sensor to pressure was found to be better. On the other hand, no meaningful results were obtained in the capacitance values of the uncoated sensor in the capacitance changes measured for the pressure formed by the weight of the water droplets. This situation highlighted the advantage of the coated sensor. Repeated bending-release process was applied to demonstrate the flexibility of the proposed sensor. The capacitance changes of the sensor after bending give very close results to the capacitance changes before bending. As a result, it is foreseen that the capacitive printed pressure sensor is flexible, bendable, highly sensitive and will be advantageous in areas such as human monitoring systems, robotics, wearable technologies, and bioelectronics.

REFERENCES

[1] A. AHMED *et al.* "Development of Highly Flexible Piezoelectric PVDF-TRFE/Reduced Graphene Oxide Doped Electrospun Nano-Fibers for Self-Powered Pressure Sensor." *Polymers (Basel).* 2024, vol. 16, no. 13.

[2] T. SUN *et al.* "Machine learning-coupled vertical graphene triboelectric pressure sensors array as artificial tactile receptor for finger action recognition." *Nano Energy*. 2024, vol. 123, no. December 2023, p. 109395.

[3] H. ZHU, H. LUO, M. CAI, and J. SONG. "A Multifunctional Flexible Tactile Sensor Based on Resistive Effect for Simultaneous Sensing of Pressure and Temperature." *Adv. Sci.* 2024, vol. 11, no. 6, pp. 1–10.

[4] X. GUO *et al.* "Bamboo-Inspired, Environmental Friendly PDMS/Plant Fiber Composites-Based Capacitive Flexible Pressure Sensors by Origami for Human-Machine Interaction." *ACS Sustain. Chem. Eng.* 2024, vol. 12, no. 12, pp. 4835–4845.

[5] S. CHEN, S. XIN, L. YANG, Y. GUO, W. ZHANG, and K. SUN. "Multi-sized planar capacitive pressure sensor with ultra-high sensitivity." *Nano Energy*. 2021, vol. 87, no. March, p. 106178.

[6] M. SU, P. LI, X. LIU, D. WEI, and J. YANG. "Textile-Based Flexible Capacitive Pressure Sensors: A Review," *Nanomaterials.* 2022, vol. 12, no. 9, pp. 1–24.

[7] R. B. MISHRA, N. EL-ATAB, A. M. HUSSAIN, and M. M. HUSSAIN. "Recent Progress on Flexible Capacitive Pressure Sensors: From Design and Materials to Applications." *Adv. Mater. Technol.* 2021, vol. 6, no. 4, pp. 1–34.

[8] S. R. A. RUTH, V. R. FEIG, H. TRAN, and Z. BAO. "Microengineering Pressure Sensor Active Layers for Improved Performance." *Adv. Funct. Mater.* 2020, vol. 30, no. 39, pp. 1–31.

[9] H. WANG *et al.* "Flexible capacitive pressure sensors for wearable electronics." *J. Mater. Chem. C.* 2022, vol. 10, no. 5, pp. 1594–1605.

[10] J. XIONG, J. CHEN, and P. S. LEE. "Functional Fibers and Fabrics for Soft Robotics, Wearables, and Human–Robot Interface." *Adv. Mater.* 2021, vol. 33, no. 19, pp. 1–43.

[11] S. WANG *et al.* "High sensitivity capacitive flexible pressure sensor based on PDMS double wrinkled microstructure." *J. Mater. Sci. Mater. Electron.* 2024, vol. 35, no. 1, pp. 1–10.

[12] J. YAO, C. QU, Z. CHEN, Y. ZHANG, and Y. XU. "High-Sensitive Wearable Capacitive Pressure Sensor with Hemispherical Porous Electrode." *ACS Appl. Electron. Mater.* 2024, vol. 6, no. 4, pp. 2649–2658.

[13] J. QIU *et al.* "Rapid-Response, Low Detection Limit, and High-Sensitivity Capacitive Flexible Tactile Sensor Based on Three-Dimensional Porous Dielectric Layer for Wearable Electronic Skin." *ACS Appl. Mater. Interfaces.* 2019, vol. 11, no. 43, pp. 40716–40725.

INNOVATIVE TEXTILE-BASED FORCE SENSOR

[1]Esma Nur NEBIPASAGIL, [2]Ahad KHABBAZ, [1]Meltem TEKCIN, [3]Senem KURŞUN

[1]*ITU – Textile Engineering Department - Istanbul Technical University, Beyoglu, Istanbul, Turkey, gulluoglu@itu.edu.tr*

[2]*ITU – Mechatronics Engineering Department - Istanbul Technical University, Beyoglu, Istanbul, Turkey*

[3]*ITU – Mechanical Engineering Department - Istanbul Technical University, Beyoglu, Istanbul, Turkey*

https://doi.org/10.37904/nanocon.2024.4986

Abstract

In this study, a flexible force sensor with silver and carbon mixed conductive nanoparticles was designed and tested for use in the robotics field. Conductive inks composed of Ag:C nanoparticles were printed on polyamide-based taffeta label fabric by pad printing method in order to fabricate a capacitive force sensor. The sensitivity of this capacitive sensor structure was tested and evaluated by loading different weights. The results showed that as the concentration of carbon nanoparticles in the conductive ink increases, a decrease in capacitive energy loading is observed. Hence, the production of a capacitive force sensor form was carried out with the appropriate number of printing layers and Ag:C concentration. The focus of research lies in the discovery and development of durable flexible force sensors for motion tracking in robotic applications.

Keywords: Force sensor, nano ink, pad printing, capacitive measurement, robotic applications

1. INTRODUCTION

The term robotics is succinctly defined as the operation, design, and utilization of robotic systems for production purposes. Generally, robots are engineered for applications that involve unsafe, hazardous, highly repetitive, or undesirable tasks [1]. Contemporary robotic systems also provide significant assistance in various aspects of daily life. Sensors capable of measuring forces, moments, or their combinations in multiple directions are widely utilized in robotics-related domains. Force sensors are increasingly finding applications in this field. Most flexible force sensors not only detect normal contact forces but also provide benefits in applications such as smart robotic grasping, e-skin, prosthetics, and wearable devices requiring comprehensive tactile force analysis. In addition to offering advantages such as high precision, flexibility, low hysteresis, and rapid response time, these sensors can be manufactured using simple and cost-effective methods [2] [3] [4]. The design of the sensor structure is critical in defining key characteristics, including stiffness, measurement range, isotropy, and any undesirable coupling between sensor outputs. To ensure the accuracy, reliability, and durability of a force sensor, meticulous attention must be given to both the sensitive elements and the elastic structure during the design process [5].

Pad printing offers several advantages over alternative printing techniques, including cost-effectiveness, ease of use, speed, compatibility with high-viscosity inks, versatility in printing on diverse surfaces, and the capability for layer-by-layer application [6]. The materials employed in the fabrication of flexible force sensors can primarily be classified into three categories: carbon-based materials, metallic materials, and polymeric materials [7]. Nanoparticle inks employed in wearable electronics can be fabricated using the pad printing method. The field of printed wearable electronics represents a rapidly expanding segment of the electronics industry, driven by advancements in conductive inks that incorporate conductive nanoparticles [8].

When examining the working principles of sensing units within sensor structures based on the type of output signal, they can be categorized as piezoresistive, capacitive, piezoelectric, triboelectric, electromagnetic, and optical. Among the diverse array of sensing mechanisms, the capacitive working principle emerges as particularly advantageous for flexible force sensors [9]. Capacitive sensing facilitates high-precision measurements by detecting even minute deviations. Additionally, it supports the development of flexible sensors that are highly effective in capturing and processing tactile information [10], [11], [5]. For these reasons, the focus of this research is the exploration and development of durable, flexible force sensors fabricated through the pad printing method utilizing conductive nanoparticles, specifically for motion tracking in robotic applications. The results of the study indicate that the proposed force sensor structure offers promising potential across a wide range of fields, particularly in robotic applications.

2. MATERIAL AND METHOD

2.1 Ink concentration

Conductive silver and carbon nano inks were utilized to fabricate the textile-based sensor intended for integration into robotic arms. Various concentrations of these inks were formulated, specifically 100% Ag, 75% Ag - 25% C and 50% Ag - 50% C.

2.2 Sample preparation

Samples were prepared using polyamide-based taffeta label fabric. The Inkprint EAZY90 pad printing machine was employed to apply the sensor production onto the selected fabrics. Various concentrations of carbon and silver nano inks were formulated. The desired density was achieved by mixing the conductive inks with a thinner at a weight ratio of 5:1, resulting in a uniform ink composition. The viscosity of the ink was subsequently adjusted to attain the optimal parameters for the pad printing machine. The pad printing process was repeated to enhance conductivity and achieve a smoother surface. Samples consisting of 5, 7, and 9 layers were fabricated. Following this, the samples underwent a sintering process, wherein they were cured in an oven at 150 °C for 30 minutes after printing. The pad printing machine is shown in **Figure 1A**, and the samples produced with this machine are shown in **Figure 1B**.

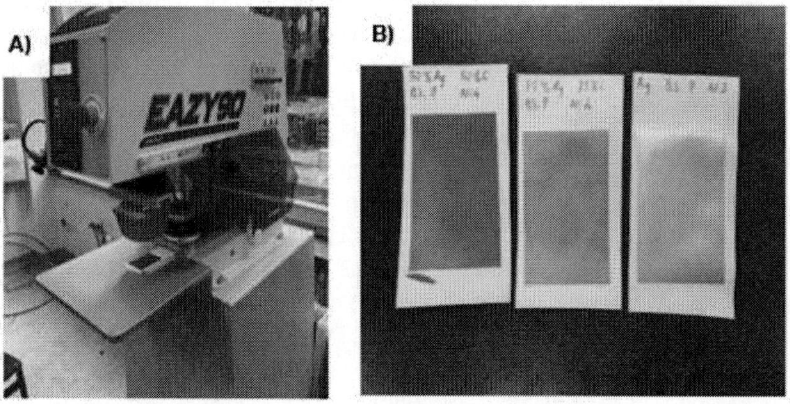

Figure 1 A) Pad printing machine, B) Prepared samples polyamide based taffeta label fabric

2.3 Measurements of force sensor

The Ossila four-point probe meter was utilized to investigate the conductivity of the samples. This device provides data on sheet resistance (Ω/square), conductivity (MS/m), and resistivity (nΩ·m). **Figure 2A** shows the Ossila four-point probe measuring system.

The force sensor developed consists of two layers. The produced samples were arranged in a face-to-face configuration. In the fabrication of capacitive sensors, welding tape was employed as an insulating layer positioned between two conductive surfaces. Consequently, the pressure exerted on the contact area can be inferred by measuring the difference in electrical resistance between the two contacting layers. As the applied weights were varied, resulting in changes in the pressure at the interface, the corresponding alterations in surface resistance due to the interaction between the two layers were recorded. Testing was conducted by applying loads ranging from 1 g to 10 kg. All measurements were obtained using a Fluke 17B+ digital multimeter. The pressure measurement system created in **Figure 2B** is located.

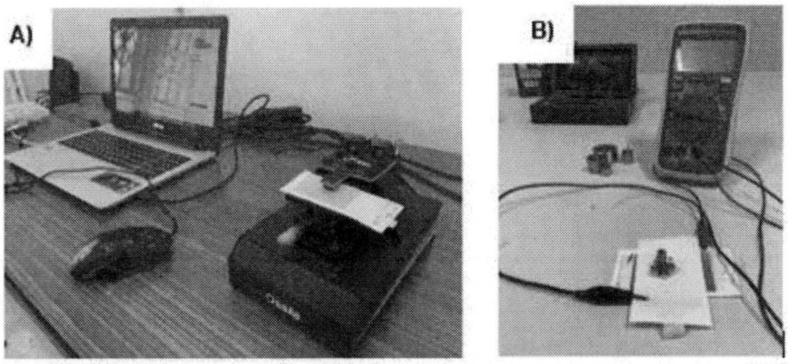

Figure 2 A) The Ossila four-point probe measuring device, B) Pressure measurement system

3. RESULTS AND DISCUSSION

3.1 Electrical properties

The results obtained from the samples measured using the Ossila device are presented in **Table 1**. Upon analysis, it is observed that an increase in the number of layers correlates with enhanced conductivity, while sheet resistance and resistivity demonstrate a corresponding decrease. Conversely, an increase in the concentration of carbon nano ink within the produced samples results in diminished conductivity, accompanied by increased sheet resistance and resistivity. Notably, the sample exhibiting the highest energy storage potential is identified as the one produced with 100% silver nano ink.

Table 2 Osilla test results

	Layer	Sheet Resistance (mΩ/square)	Resistivity (pΩ.m)	Conductivity (GS/m)
100% Ag	5	20,70	1034,85	0,97
	7	14,43	728,00	1,40
	9	11,89	594,58	1,70
75% Ag - 25% C	5	129,03	6452,00	0,16
	7	73,58	3679,00	0,27
	9	54,58	2716,50	0,37
50% Ag - 50% C	5	4301,00	215050,00	0,005
	7	1277,25	63862,50	0,02
	9	651,13	32555,00	0,03

3.2 Loading test

Figure 3, Figure 4 and **Figure 5** illustrates the charts of loading and unloading values for sensors with varying concentrations and layer configurations, up to a maximum of 10 kg. Additionally, values up to 200 grams are included in the table to investigate the limitations of our robotic arm in applying such pressure and to assess the behavior of the sensors under lower weight conditions.

When a load of 10 kg is applied, the 5-layer sensors composed of 50% Ag and 50% C (**Figure 5**) yield a capacitance value of 0.84 nF, while the 7-layer sensors also yield 0.84 nF, and the 9-layer sensors exhibit a capacitance of 0.85 nF. Although the linearity of the capacitance values is comparable across the different layer configurations, it is evident that increasing the number of layers does not significantly enhance the capacitance.

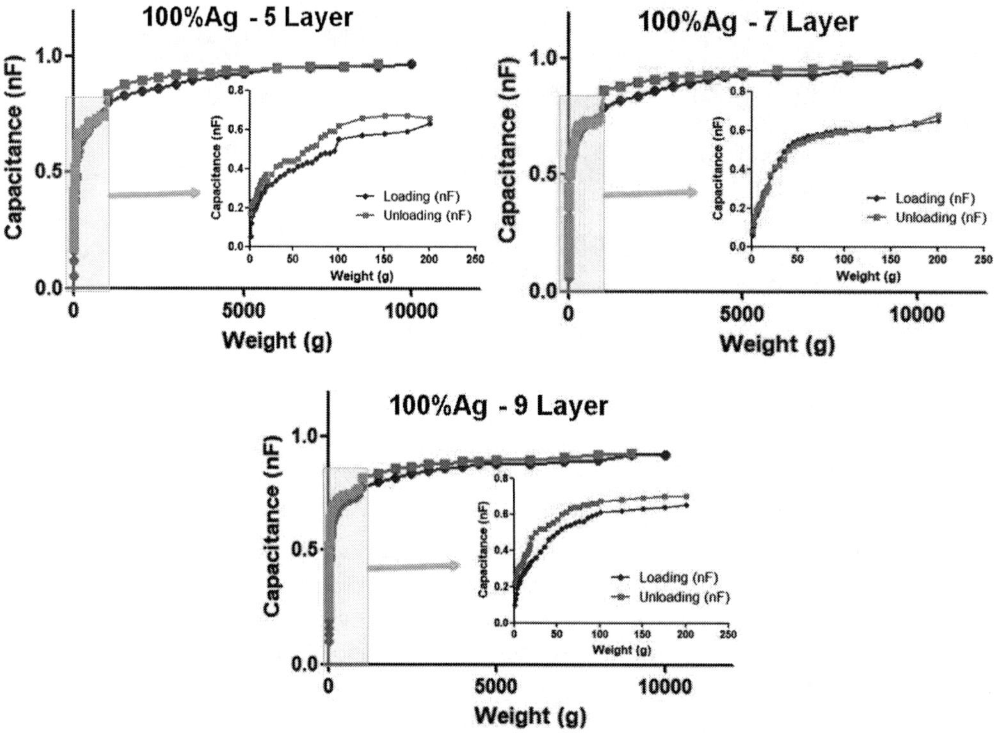

Figure 3 100%Ag concentration force sensor sample loading test results

Upon applying a load of 10 kg to sensors with a concentration of 75% Ag and 25% C (**Figure 4**), the 5-layer sensors exhibit a capacitance value of 0.89 nF, while the 7-layer sensors yield 0.87 nF, and the 9-layer sensors demonstrate a capacitance of 0.86 nF. Notably, the linearity of the capacitance values for the 7-layer and 9-layer configurations is similar; however, it is evident that increasing the number of layers does not positively influence the capacitance.

Finally, upon the application of a 10 kg load to sensors with a 100% Ag concentration (**Figure 3**), the 5-layer sensors yield a capacitance value of 0.97 nF, the 7-layer sensors yield 0.98 nF, and the 9-layer sensors demonstrate a capacitance of 0.92 nF. In comparison to the carbon-based samples, a significant increase in capacitance values is observed When examining linearity, both the 5-layer and 9-layer configurations exhibit favorable characteristics; however, the loading and unloading values of the 7-layer sensor structure align exceptionally well. Increasing the number of layers yields a marginal positive contribution to the capacitance value. However, for the 9-layer configuration, obtaining measurements at very high loads becomes increasingly challenging, leading to a compromise in the stability of the system. This instability is likely attributed to the deformation of the sensor structure under excessive load Considering all these results, it is evident that the 7-layer sensor structure with 100% Ag concentration is the most suitable choice for robotic applications.

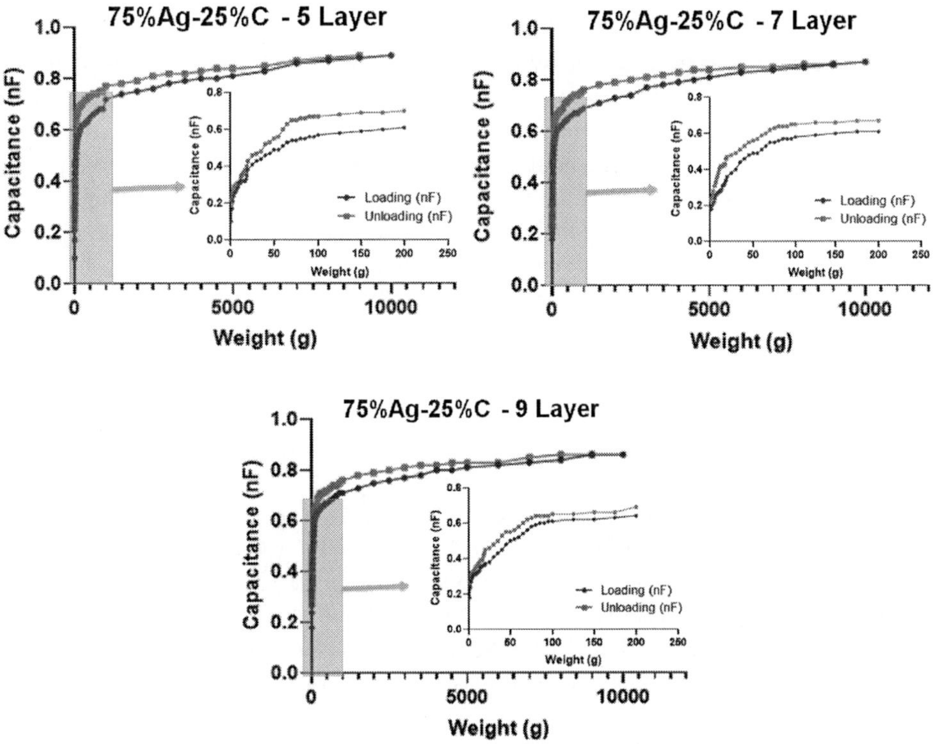

Figure 4 75%Ag-25% concentration force sensor sample loading test results

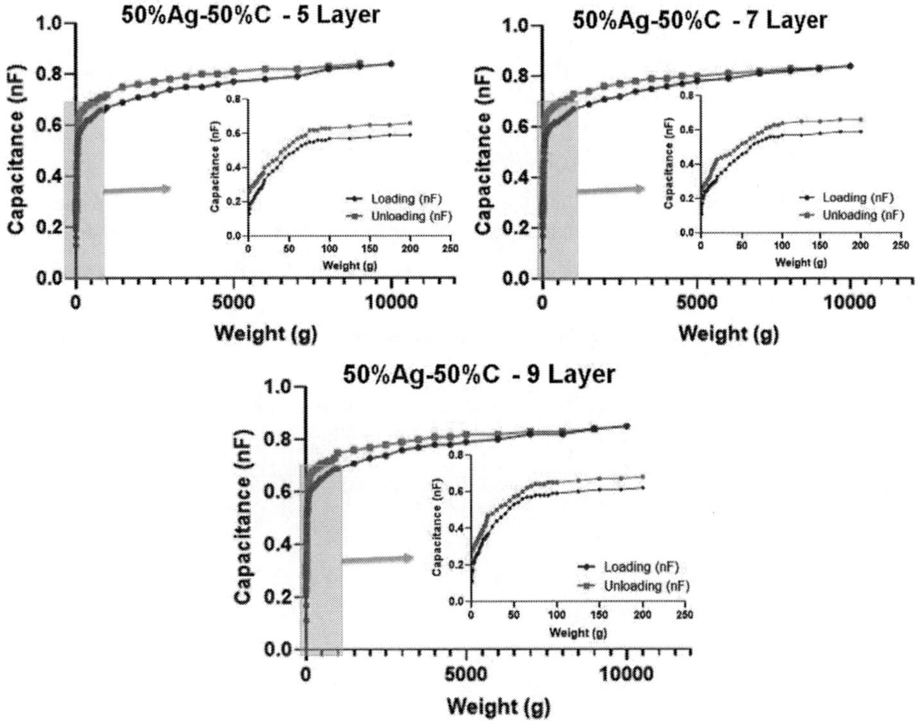

Figure 5 50%Ag-50% concentration force sensor sample loading test results

4. CONCLUSION

In this study, a flexible force sensor composed of conductive nanoparticles containing silver and carbon mixtures has been designed and tested for applications in the field of robotics. To produce a capacitive force sensor, conductive inks with nanoparticle concentrations of 100% Ag, 75% Ag - 25% C, and 50% Ag - 50% C were printed onto polyamide-based taffeta label fabric in 5, 7, and 9 layers using the pad printing method. Welding tape was incorporated between the printed samples to form the sensor structure. The sensitivity of this capacitive sensor configuration was tested and evaluated by applying various weights ranging from 1 g to 10 kg. The results indicate that an increase in the concentration of carbon nanoparticles within the conductive ink correlates with a decrease in capacitive energy storage. Furthermore, it was observed that a very high number of layers does not yield positive outcomes for the system. Consequently, it was determined that the capacitive force sensor configuration produced with 100% Ag concentration and 7 printed layers represents the most suitable structure.

REFERENCES

[1] ELFASAKHANY, A., YANEZ, E., BAYLON, K. & SALGADO, R. Design and Development of a Competitive Low-Cost Robot Arm with Four Degrees of Freedom. *Mod. Mech. Eng.* 2011, vol. 01, pp. 47–55.

[2] ZHANG T., LIU H., JIANG L., FAN S., YANG J. Development of a flexible 3-D tactile sensor system for anthropomorphic artificial hand. *IEEE Sensors J.* 2013, vol. 13, no. 2, pp. 510–518.

[3] CHEN X., SHAO J., TIAN H., LI X., TIAN Y., WANG C. Flexible three-axial tactile sensors with microstructure-enhanced piezoelectric effect and specially-arranged piezoelectric arrays. *Smart Mater. Struct.* 2018, vol. 27, no. 2.

[4] XU, D., HU, B., ZHENG, G., WANG, J., LI, C., ZHAO, Y., YAN, Z., JIAO, Z., WU, Y., WANG, M., LI, H., GUO, X., Sandwich-like flexible tactile sensor based on bioinspired honeycomb dielectric layer for three-axis force detection and robotic application. *J. Mater. Sci., Mater. Electron.* 2023, vol. 34, no. 11, p. 942.

[5] TEMPLEMANA J. O., SHEIL B. B., SUNB T. Multi-axis force sensors: A state-of-the-art review. *Sensors and Actuators A: Physical.* 2020, vol 304, 111772.

[6] JAAFAR A., SCHOINAS S., PASSERAUB P. Pad-printing as a fabrication process for flexible and compact multilayer circuits. *Sensors.* 2021, vol. 21, no. 20-

[7] CHENG M., ZHU G., ZHANG F., TANG W, JIANPING S., YANG J, ZHU L., A review of flexible force sensors for human health monitoring. *Journal of Advanced Research.* 2020, vol. 26, pp. 53–68

[8] HTWE Y. Z. N., MARIATTI M. Printed graphene and hybrid conductive inks for flexible, stretchable, and wearable electronics: Progress, opportunities, and challenges. *Elsevier B.V.* 2022, doi: 10.1016/j.jsamd.2022.100435.

[9] DONG T., WANG J., CHEN Y., LIU L., YOU H., LI T. Research Progress on Flexible 3-D Force Sensors: A Review *IEEE Sensors Journal.* 2024, vol. 24, no. 10.

[10] PUANGMALI P., ALTHOEFER K., SENEVIRATNE L.D., MURPHY D., DASGUPTA P. State-of-the-Art in force and tactile sensing for minimally invasive surgery. *IEEE Sens. J.* 2008, vol. 8, pp. 371–381.

[11] TREJOS A.L., PATEL R.V., NAISH M.D. Force sensing and its application in minimally invasive surgery and therapy: a survey. *Proc. Inst. Mech. Eng. C.* 2010, vol. 22, pp. 41435–1454.

PREPARATION OF $Ti_3C_2T_X$ NANOSHEETS FROM A LARGE-SIZE MAX PHASE PRECURSOR FOR THE ELECTROCONDUCTIVE FUNCTIONALISATION OF CELLULOSE FABRIC

[1]Alenka OJSTRŠEK, [2]Silvo HRIBERNIK, [1]Laura JUG

[1]University of Maribor, Faculty of Mechanical Engineering, Institute of Engineering Materials and Design, Maribor, Slovenia, EU, alenka.ojstrsek@um.si, laura.jug@um.si

[2]University of Maribor, Faculty of Electrical Engineering and Computer Science, Institute of Automation, Maribor, Slovenia, EU, silvo.hribernik@um.si

https://doi.org/10.37904/nanocon.2024.5005

Abstract

Textile material, which has always been perceived as a protective layer against weather conditions, nowadays plays a vital role in wearable electronics (smart textile-based devices/ electronic textiles), contributing to the areas of energy harvesting, energy storage, sensors, real-time monitoring of healthcare, personal thermal management, etc. Thus, the main aim of the presented research was to synthesise $Ti_3C_2T_x$ nanosheets with proper structure, flake size and high capacitance using a top-down synthesis approach utilising a large-size MAX (Ti_3AlC_2) phase precursor (100 μm) for further electroconductive functionalisation of cellulose fabric. Thus, the synthesis parameters were changed, i.e., time, temperature and concentrations of selected etchants, preparing eight stable aqueous dispersions of $Ti_3C_2T_x$ nanosheets plus reference $Ti_3C_2T_x$ (from small-size 40 μm MAX precursor). As-synthesised products were characterised using various analytical techniques such as Scanning Electron Microscopy (SEM), X-ray powder Diffraction (XRD) and Dynamic Light Scattering (DLS). In addition, prepared $Ti_3C_2T_x$ nanosheets were applied to cellulose fabric by an optimised dip-coating procedure, and fabrics' surface morphologies, as well as electrical resistances, were inspected. The obtained SEM and XRD results showed successful preparation of the $Ti_3C_2T_x$ nanosheets, wherein the synthesis parameters influenced their size and morphology. $Ti_3C_2T_x$ applied to cellulose fabric exhibited high electrical conductivity, revealing effective cellulose fabric' functionalisation and indicating the potential use of $Ti_3C_2T_x$ in wearable electronic applications.

Keywords: $Ti_3C_2T_x$ nanosheets, electroconductive functionalisation, cellulose fabric, characterisation

1. INTRODUCTION

A new family of two-dimensional (2D) materials, the transition metal carbides and nitrides, called MXenes (more than 30 types), have shown great promise as potential compounds for the electroconductive functionalisation of textiles; especially the most attractive $Ti_3C_2T_X$ due to its highest conductivity (up to 20,000 S/cm), tuneable surface chemistry, higher strength and stiffness compared to other solution-processed 2D materials, good hydrophilicity, high thermal conductivity and efficient photothermal conversion behaviour, excellent ion intercalation behaviour, scalable solution synthesis (kg batches), sufficient environmental stability, biocompatibility and aqueous solution processing without surfactants addition [1-3].

It is well-known that the electroconductivity of $Ti_3C_2T_X$-coated textiles is strongly influenced by the size of synthesised $Ti_3C_2T_X$ nanosheets [4,5]. Functionalisation with larger nanosheets gives higher electrical conductivity and mechanical strength than smaller ones. This can be explained by the fact that the smaller the $Ti_3C_2T_X$ flake size is, the greater the interfacial resistance between the flakes, reducing overall conductivity [6,7]. One of the possible approaches to increase the size of $Ti_3C_2T_X$ is to use a large-size MAX phase precursor (Ti_3AlC_2) and, further, to choose the adequate etching conditions during the synthesis of $Ti_3C_2T_x$ to maintain its large flake size [8,9].

In the presented study, the top-down synthesis of $Ti_3C_2T_x$ nanosheets (from a large-size MAX phase precursor) was carried out, changing the etching conditions, i.e., the selection of proper type and concentration of chemicals, as well as the process temperature, since all these factors influence the structure (defects) of $Ti_3C_2T_x$, and consecutively, its electrical conductivity. In addition, freshly synthesised $Ti_3C_2T_x$ nanosheets were applied on cellulose fabric by dip-coating, with the aim of its electroconductive functionalisation. Herein, the synthesised $Ti_3C_2T_x$ products were comprehensively analysed using SEM, XRD and DLS, as well as the $Ti_3C_2T_x$-coated cotton fabrics were characterised by means of SEM and measurement of electrical resistivity.

2. EXPERIMENTAL PART

2.1 Synthesis of $Ti_3C_2T_x$ nanosheets

Ten different $Ti_3C_2T_x$ nanosheets were synthesised from the 1 g of 40 μm (reference, R) and 1 g of 100 μm (S1-S9) MAX precursor powder (Sigma-Aldrich) by selective etching of A-element layers (top-down approach) utilising various process parameters (**Table 1**). Lithium fluoride (LiF; 99.99 wt%) and hydrochloric acid (HCl; 37 wt%) as etchants were purchased from Apollo Scientific Ltd. and Honeywell, respectively.

Table 1 Process parameters for the synthesis of $Ti_3C_2T_x$

Sample	Mass of LiF (g)	Volume of HCl (mL)	Concentration of HCl (M)	T (°C)
R	1.6	20	9	40
S1	1.6	20	9	20
S2	1.6	20	9	40
S3	2.2	30	9	20
S4	2.2	30	9	40
S5	2.2	20	12	20
S6	2	30	9	40
S7	2	40	9	40
S8	3	40	9	40
S9	4	50	9	40

2.2 Application of $Ti_3C_2T_x$ onto cellulose fabric

An industrially bleached plain-waived cotton fabric (with a mass of 92.7 ± 0.6 g/m^2, warp density of 51 threads/cm, weft density of 44 threads/cm, and thickness of 0.18 mm) was sectioned into samples of sizes 2×2 cm^2 and coated by differently synthesised $Ti_3C_2T_x$ employing a dip-padding procedure. The individual fabric sample was dipped three times into 10 mL $Ti_3C_2T_x$ aqueous dispersion (10 mg/mL) for 5 min, followed by an intermediate/final vacuum-assisted drying at 60 °C.

The mass load of $Ti_3C_2T_x$ on an individual sample was calculated (in mg/cm^2) as the difference between the mass of the $Ti_3C_2T_x$-functionalised cotton sample and the mass of the untreated sample divided by the area of the functionalised sample.

2.3 Analytical procedures

Scanning Electron Microscopy (SEM) was applied for the morphological characterisation of synthesised $Ti_3C_2T_x$ nanosheet and $Ti_3C_2T_x$-functionalised fabrics by means of a Zeiss Gemini Supra 35 VP Scanning

Electron Microscope (Carl Zeiss NTS GmbH, Germany), with a maximum scan resolution of up to 1.5 nm at 1 kV. To determine the hydrodynamic diameters of $Ti_3C_2T_x$ dispersions as a function of different synthesis parameters, Dynamic Light Scattering (DLS) measurement was accomplished using Litesizer 500 (Anton Paar, Austria). X-Ray Diffraction (XRD) spectrum of $Ti_3C_2T_x$ was obtained on a high-resolution diffractometer D2 phaser (Bruker Siemens) with CuKα1 radiation (λ1 = 1.54184 Å) in the range from 5 to 70° 2θ and a step size of 0.03° per 100 s using the LYNXEYE XE-T detector. The electrical resistivity of the $Ti_3C_2T_x$-functionalised fabrics was measured across the entire surface area between two consistently sealed connections using a two-point probe multimeter 34410A 6 1/2 Digit (Agilent technologies). A total of 10 measurements were conducted for an individual sample, and the average value of the specific electrical resistivity was calculated according to equation (1):

$$\rho = R \cdot \frac{A}{l} \tag{1}$$

where:

ρ - the specific electrical resistivity (Ω·cm)

R - the measured electrical resistivity (Ω)

A - the cross-sectional area of the fabric (cm^2)

l - the length between two consistently sealed probe connectors (cm)

3. RESULTS AND DISCUSSIONS

3.1 Analysis of synthesised $Ti_3C_2T_x$ nanosheets

The morphological structures of different $Ti_3C_2T_x$ nanosheets (parameters of synthesis in **Table 1**) are presented in **Figure 1**, the size distribution in **Figure 2** and detailed information about the crystallographic structure of $Ti_3C_2T_x$ in **Figure 3**.

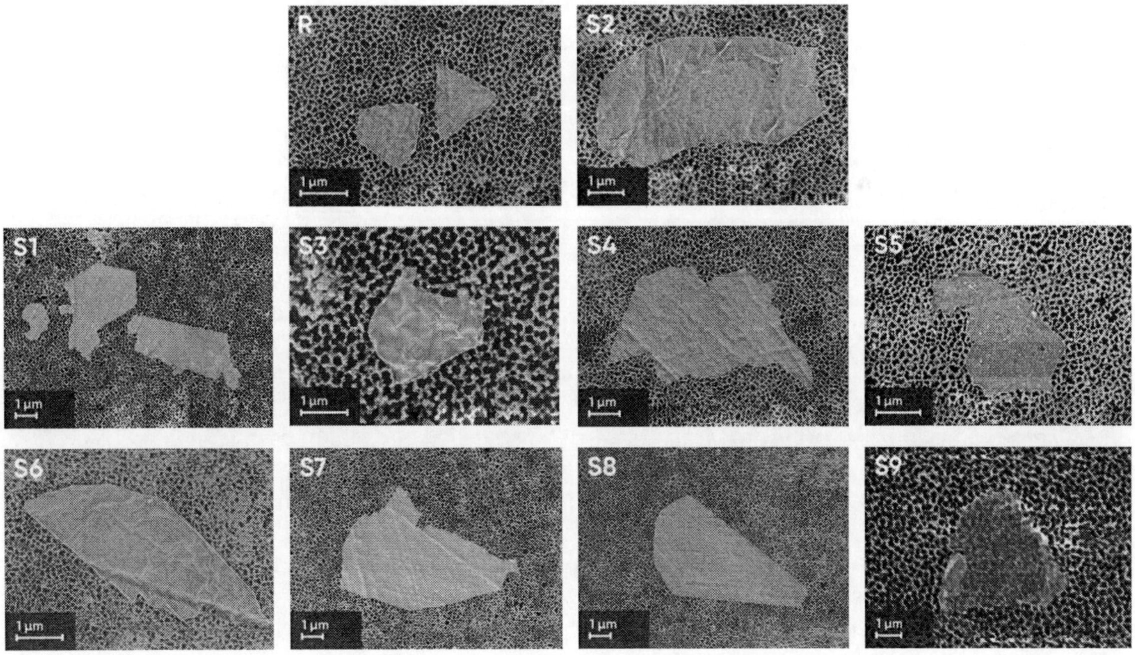

Figure 1 SEM micrographs of synthesised $Ti_3C_2T_x$ nanosheets

All SEM images in **Figure 1** reveal ultrathin flake-like $Ti_3C_2T_x$ nanostructures of different shapes and sizes with well-defined and sharp edges regarding the process parameters during synthesis. Harsh etching conditions (S5) resulted in the formation of structural defects. Inadequate etching, together with washing and delamination (S1 and S3), can lead to smaller flake sizes, which significantly change the surface properties of the particles.

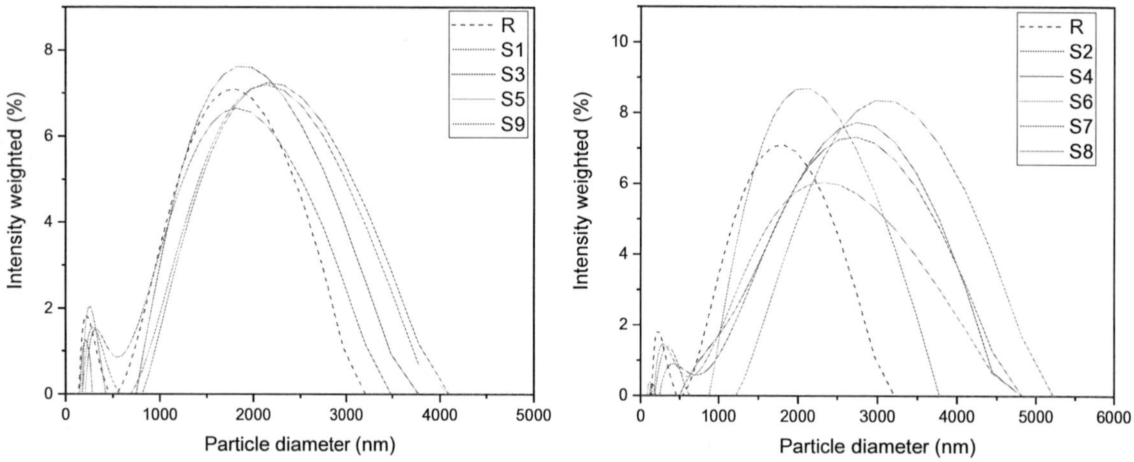

Figure 2 DLS analysis of synthesised $Ti_3C_2T_x$: R, S1, S3, S5 and S9 (left); R, S2, S4, S6, S7 and S8 (right)

DLS results in **Figure 2** confirmed that the process conditions influence the $Ti_3C_2T_x$ size. The left diagram depicts samples that produce particles of sizes like the reference sample (1,422 nm), either smaller or with a distribution that closely resembles it. The most favourable outcomes were observed for S6 with a hydrodynamic particle diameter of 2,723 nm. Unfavourable outcomes were found at S1 and S3, where the etching temperature was not raised, as well as at S5 and S9, where a higher concentration and volume of acid were employed (harsher etching conditions), influencing the particle breakup, as also proven by SEM.

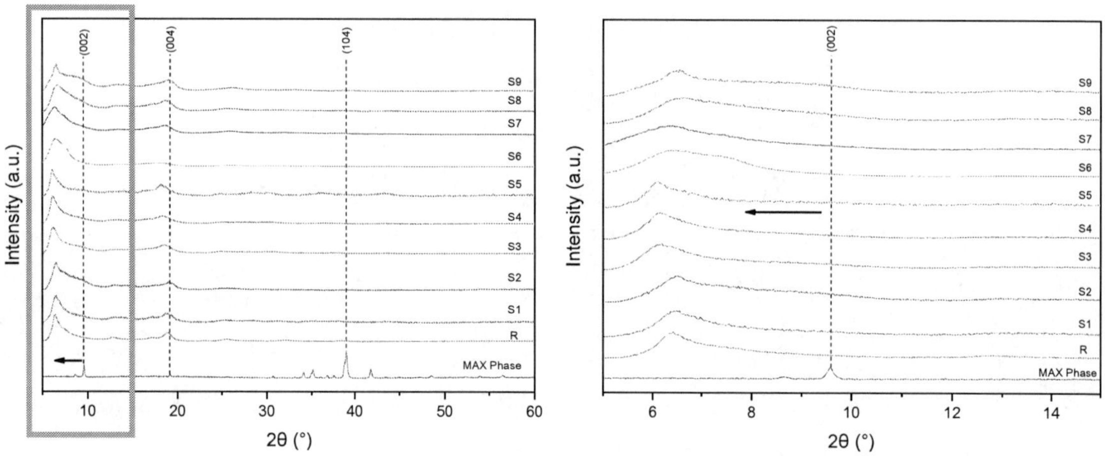

Figure 3 XRD of $Ti_3C_2T_x$; enlarged detail in the green square on the right

As can be seen from **Figure 3**, Ti_3AlC_2 MAX phase precursor exhibits two distinct diffraction peaks that serve as representative indicators of its crystal structure, i.e., (002) crystal plane at 9.5° 2θ and (104) crystal plane at 39.0° 2θ. After synthesis, the (104) peak diminished, and the (002) peak shifted to a lower angle (6.5-6.8°) and became broader compared to the MAX phase, regardless of the process conditions, indicating the effective removal of aluminium from the MAX phase and $Ti_3C_2T_x$ sheet delamination, respectively.

3.2 Characterisation of $Ti_3C_2T_x$-functionalised cellulose fabric

The surface morphology of the fabric before and after $Ti_3C_2T_x$ functionalisation was detailly inspected using the SEM technique, and the selected micrographs are shown in **Figure 4**. In addition, the surface resistivity of textile samples (TR and TS1-TS9 coated with the $Ti_3C_2T_x$ dispersions in **Table 1**) was calculated according to equation 1, and the obtained results were presented in **Figure 5**, together with the $Ti_3C_2T_x$ mass load.

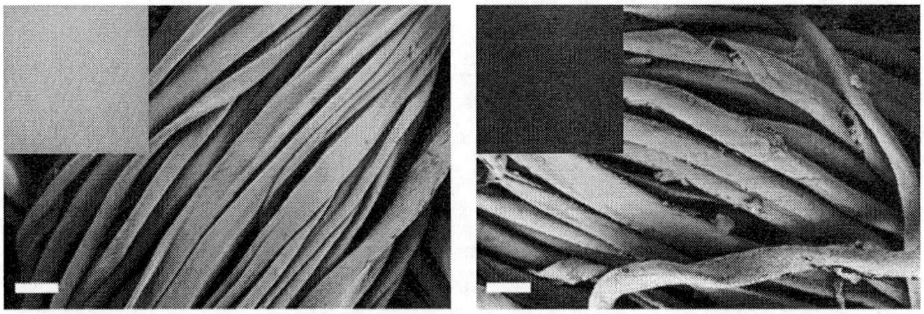

Figure 4 The selected SEM micrographs of pristine cotton fabric (left); and $Ti_3C_2T_x$-functionalised fabric, TS2 (right); the scale bar is 20 µm. In the left corners are photographs of untreated (white) and $Ti_3C_2T_x$-coated fabrics (black)

The SEM micrograph in **Figure 4a** showed the clear and clean cellulose fibres' surface of the untreated sample that was further successfully coated with $Ti_3C_2T_x$ by an optimised dip-padding procedure, changing the visual appearance (colour changed from white to black) and the surface morphology (**Figure 4b**) as well as the electroconductive properties. Overall, it appears that the employed $Ti_3C_2T_x$ functionalisation procedure had no deleterious effects on the morphology of the pristine cellulose fibres.

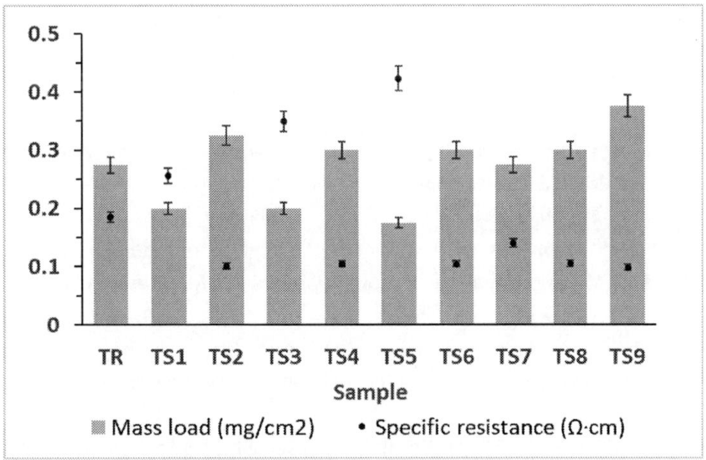

Figure 5 Specific electrical resistance of functionalised cotton fabric *versus* mass load

The specific electrical resistivities of the $Ti_3C_2T_x$-functionalised samples were in the range from 0.1 up to 0.4 Ω·cm, as can be observed in **Figure 5**, suggesting the excellent conductivity of samples, which is comparable with the electrical conductivity of metallised textiles [10]. The electrical resistivity varied between the samples, depending on the uniformity of the $Ti_3C_2T_x$ deposit (synthesis parameters). However, the relation between mass load and specific resistivity is not linear, since textiles are not homogeneous and anisotropic products like some other engineering materials. At a similar mass load of 0.275 mg/cm^2, sample TS7 functionalised with $Ti_3C_2T_x$ synthesised from a larger Ti_3AlC_2 MAX phase precursor has a lower specific resistivity (higher conductivity) in comparison with the reference sample (TR) functionalised with $Ti_3C_2T_x$ synthesised from a smaller MAX phase precursor.

CONCLUSION

In the presented study, the $Ti_3C_2T_x$ nanosheets were successfully synthesised using a top-down approach, as confirmed by SEM and XRD results. Different synthesis parameters influenced the sizes and morphologies of $Ti_3C_2T_x$ as well as the nanosheets' disintegration. $Ti_3C_2T_x$ applied to cellulose textiles employing dip-padding procedure exhibited meagre specific resistance (superior conductivity), depending on the synthesis parameters and the $Ti_3C_2T_x$ loading mass. This research reveals effective textiles' electroconductive functionalisation and, thus, indicates the potential use of $Ti_3C_2T_x$ in wearable electronic applications. Further work will focus on the optimisation of dip-padding parameters as well as the wash and wear durability of applied $Ti_3C_2T_x$.

ACKNOWLEDGEMENTS

The results leading to this work were funded by the Slovenian Research and Innovation Agency (ARIS) in the frame of research project no. J2-50087 and a research core program group for Textile Chemistry and Advanced Textile Materials P2-0118 within the Young Researchers Programme.

REFERENCES

[1] WANG, Qi-W., ZHANG, H.-B., LIU, J., ZHAO, S., XIE, X., LIU, L., YANG, R., KORATKAR, N., YU, Z.-Z. Multifunctional and Water-Resistant MXene-Decorated Polyester Textiles with Outstanding Electromagnetic Interference Shielding and Joule Heating Performances. *Advanced Functional Materials*. 2019, vol. 29, no. 7, pp. 1806819.

[2] ZHENG, X., NIE, W., HU, Q., WANG, X., WANG, Z., ZOU, L., HONG, X., YANG, H., SHEN, J., LI, C. Multifunctional RGO/$Ti_3C_2T_x$ MXene fabrics for electrochemical energy storage, electromagnetic interference shielding, electrothermal and human motion detection. *Materials and Design*. 2021, vol. 200, pp. 109442.

[3] ANASORI, B., GOGOTSI, Y. MXenes: trends, growth, and future directions. *Graphene and 2D Materials*. 2022, vol. 7, pp. 75-79.

[4] UZUN, S., SEYEDIN, S., STOLTZFUS, A. L., LEVITT, A. S., ALHABEB, M., ANAYEE, M., STROBEL, C. J., RAZAL, J. M., DION, G., GOGOTSI, Y. Knittable and Washable Multifunctional MXene-Coated Cellulose Yarns. *Advanced Functional Materials*. 2019, vol. 29, no. 45, pp. 1905015.

[5] AHMED, A., HOSSAIN, M. M., ADAK, B., MUKHOPADHYAY, S. Recent Advances in 2D MXene Integrated Smart-Textile Interfaces for Multifunctional Applications. *Chemistry of Materials*. 2020, vol. 32, no. 24, pp. 10296-10320.

[6] WANG, X., FAN, X., LI, M., ZHU, W., XUE, J., YE, F., CHENG, L. Structure and electromagnetic properties of $Ti_3C_2T_x$ MXene derived from Ti_3AlC_2 with different microstructures. *Ceramics International*. 2021, vol. 47, no. 10, pp. 13628-13634.

[7] MALESKI, K., REN, C. E., ZHAO, M.-Q., ANASORI, B., GOGOTSI, Y. Size-Dependent Physical and Electrochemical Properties of Two-Dimensional MXene Flakes. *ACS Applied Materials & Interfaces*. 2018, vol. 10, no. 29, pp. 24491-24498.

[8] ZHANG, J., KONG, N., UZUN, S., LEVITT, A., SEYEDIN, S., LYNCH, P. A., QIN, S., HAN, M., YANG, W., LIU, J., WANG, X., GOGOTSI, Y., RAZAL, J. M. Scalable Manufacturing of Free-Standing, Strong $Ti_3C_2T_x$ MXene Films with Outstanding Conductivity. *Advanced Materials*. 2020, vol. 32, no. 23, pp. 2001093.

[9] CHEN, N., YANG, W., ZHANG, C. Perspectives on preparation of two-dimensional MXenes. *Science and Technology of Advanced Materials*. 2021, vol. 22, no. 1, pp. 917-930.

[10] OJSTRŠEK, A., PLOHL, O., GORGIEVA, S., KUREČIČ, M., JANČIČ, U., HRIBERNIK, S., FAKIN, D. Metallisation of Textiles and Protection of Conductive Layers: An Overview of Application Techniques. *Sensors*. 2021, vol. 21, pp. 3508.

ENHANCED WASHING DURABILITY OF ELECTROCONDUCTIVE $Ti_3C_2T_x$ MXENES ON COTTON FABRIC USING VARIOUS PROTECTIVE COATINGS

Darinka FAKIN, Laura JUG, Alenka OJSTRŠEK

University of Maribor, Faculty of Mechanical Engineering, Institute of Engineering Materials and Design, Maribor, Slovenia, EU, darinka.fakin@um.si, laura.jug@um.si, alenka.ojstrsek@um.si

https://doi.org/10.37904/nanocon.2024.5006

Abstract

The incorporation of electroconductive compounds onto the textiles' surfaces leads to the fabrication of flexible electronic textiles (e-textiles) for numerous application fields, i.e., sport, medical, protective, fashion, military, etc. Their long-term functionality during daily usage depends on the sufficient durability of applied electroconductive compounds against washing, which can be significantly improved by the employment of protective layers. Thus, this study investigated the efficacy of different coatings for the protection of $Ti_3C_2T_x$ MXenes dip-coated onto cotton fabric against laundering without essentially influencing fabrics' electroconductive properties. Two Mxene-functionalised samples were protected with three different compounds, i.e., modified acrylate resin (MAR), waterborne polyurethane resin (WPR) and biantennary oligoglycine peptide (tectomer - 2 tailed) utilising spray-coating, dip-coating and impregnation procedures, respectively. MXene-functionalised/protected samples were washed up to 20 washing cycles and characterised by the determination of their electrical resistance, mass loss, changes in surface morphologies, water contact angle and optical properties. Based on these results, WPR provided the most suitable protection of MXenes on cotton fabric, followed by MAR, balancing high conductivity with long-term stability. Tectomer is not recommended for MXene protection due to its low durable performance against harsh washing conditions.

Keywords: $Ti_3C_2T_x$ MXenes, protective coatings, cellulose fabric, washing durability

1. INTRODUCTION

The application of MXenes often does not lead to permanent effects on the textile surface during daily use (limited wash durability) due to the inhomogeneity and roughness of textiles as well as the small sheet size and weak interlayer interaction between MXene sheets [1]. Consequently, textiles lose their electroconductive functionalities after laundering. Moreover, it is necessary for MXene to avoid its restack and oxidisation, which will hinder its relative capacitance and stability.

Several researchers enhanced the durability of MXenes on textiles and simultaneously added some advanced functionalities by employing different adhesives or polymers. Luo et al. [2] coated, firstly, an interface polydopamine layer on the elastic textile substrate to smooth the surface and enhance adhesion and, secondly, a protective polydimethylsiloxane (PDMS) layer over the Mxene network to obtain the superhydrophobic, breathable and elastic smart textile device. Ma et al. [3] also applied a thin layer of PDMS as an adhesive for MXene to prepare a hydrophobic and multifunctional textile composite. Wang et al. [4] prepared highly conductive and hydrophobic textiles with exceptional electromagnetic interference (EMI) shielding efficiency and excellent Joule heating performance by depositing in-situ polymerised polypyrrole (PPy) modified MXene sheets onto polyethylene terephthalate textiles followed by a silicone coating. Yan et al. [5] electrochemically deposited PPy on the surface of MXen-decorated plain cotton textiles to further improve the capacitive performance of the MXene-based electrode and avoid the oxygen oxidation of MXene.

In this research, $Ti_3C_2T_x$ MXene nanosheets applied on cotton fabric according to an optimised dip-coating procedure were further protected against peeling-off due to harsh washing conditions, i.e., water, temperature and physical-mechanical forces, by employing three different compounds, modified acrylate resin (MAR), waterborne polyurethane resin (WPR) and tectomer (TEC). MXene-functionalised/protected fabrics were characterised before and after washing (up to 20 washing cycles) by measuring the electrical resistivity and determining the mass loss and water contact angle. Moreover, the changes in optical properties and surface morphologies were inspected.

2. EXPERIMENTAL PART

2.1 Materials

Trials were performed using an industrially bleached plain-waived cotton fabric with a mass/unit area of 92.7 ± 0.6 g/m^2, warp density of 51 threads/cm, weft density of 44 threads/cm, and thickness of 0.18 mm. Two titanium aluminium carbide Ti_3AlC_2 MAX precursor powders, in sizes of 40 μm and 100 μm, for the synthesis of MXene nanosheets were supplied by Sigma-Aldrich. Lithium fluoride (LiF; 99.99 wt%) and hydrochloric acid (HCl; 37 wt%) as etchants were purchased from Apollo Scientific Ltd. and Honeywell, respectively.

Two types of $Ti_3C_2T_x$ Mxene flakes, small- (S) and large-sized (L), were synthesised from 1 g of MAX precursor powders (two sizes) by a top-down minimally intensive layer delamination (MILD) approach [6]. Briefly, LiF was initially dissolved in a 250 mL polytetrafluoroethylene beaker containing an HCl solution. The solution was continuously stirred at 35 °C for 30 minutes and 400 rpm, resulting in total dissolution of the LiF. Then, 1 g of Ti_3AlC_2 (MAX powder) was gradually added to the above mixture and left to etch the Al layer under continuous stirring. After completion of the reaction, the obtained mixture was washed with deionised (DI) water and centrifuged numerous times at 3,500 rpm for 5 minutes per cycle until a stable dark green supernatant solution with pH 6 and a swelling clay precipitate was produced.

2.2 Dip-coating of MXenes and application of protective coatings

Cotton fabric (Co) was sectioned into samples of sizes 2×2 cm^2 and modified by two synthesised $Ti_3C_2T_x$ MXene nanosheets (S and L) employing a dip-padding procedure. The individual fabric sample was dipped three times into 10 mL MXene aqueous dispersion (10 mg/mL) for 5 min, followed by an intermediate/final vacuum-assisted drying at 60 °C, preparing two samples (CoS and CoL).

Three different protective compounds were applied onto MXene-decorated samples, i.e., modified acrylate resin using a pre-optimised spray-coating (CoS-MAR and CoL-MAR), waterborne polyurethane resin by pad-drying (CoS-WPR and CoL-WPR) and tectomer by impregnation (CoS-TEC and CoL-TEC).

2.3 Washing

To evaluate the washing durability of the prepared MXene-modified and MXene-modified/protected cotton fabrics, samples were washed up to 20 times (washing cycles) at a temperature of 40 °C for 30 min. The washing solution consisted of a standard reference detergent without optical brighteners, with a concentration of 1 g/L and a liquor-to-fabric weight ratio of 50:1. After each washing cycle, the samples were rinsed in tap water for 1 min and then dried at 60 °C for 20 minutes. After each washing cycle, the mass loss of the samples was calculated.

2.4 Analytical procedures

Fourier Transform Infra-Red (FTIR) spectroscopic measurements of samples were performed using a Spectrum GX spectrophotometer (Perkin Elmer) with a Golden Gate ATR attachment and a diamond crystal. The absorbance spectra were obtained within the range of 4.000-400 cm^{-1}, with 32 scans and a resolution of 4 cm^{-1}. By using a two-point probe multimeter 34410A 6 1/2 Digit (Agilent technologies), the electrical resistance of selected samples was measured across the entire surface area between two consistently sealed connections before and after 5,10,15 and 20 washing cycles. In addition, the specific electrical resistivity (in $\Omega \cdot cm$) was calculated as a product of the measured electrical resistivity (in Ω) and the cross-sectional area of the fabric (in cm^2) divided by the length between two consistently sealed probe connectors (in cm). The Water Contact Angle (WCA) of MXene-coated/protected samples before and after washings was performed by means of the goniometer OCA 35 (DataPhysics Instruments) using the sessile drop technique. Optical properties were evaluated by measuring the reflectance of samples before and after each set of wash cycles (5, 10 and 20) using a two-ray spectrophotometer Spectraflash SF600 Plus (Datacolor, Trenton, NJ, USA), from which the CIE L*a*b* colour differences in all three directions of colour space were calculated.

3. RESULTS AND DISCUSSIONS

3.1 FTIR analysis of MXene-coated/protected fabrics

Figure 1 depicts FTIR results, confirming the successful application of protective compounds onto MXene-coated samples.

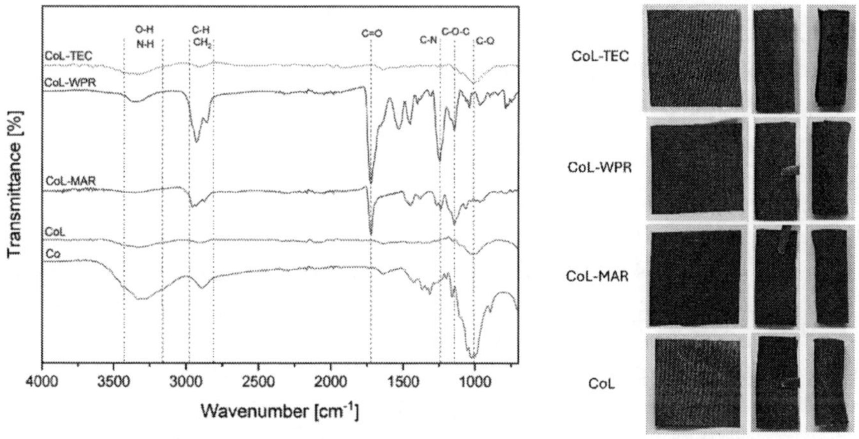

Figure 1 FTIR spectra of pristine (Co), MXene-coated (CoL) and MXene-coated/protected (CoL-TEC, CoL-WPR and CoL-MAR) cotton fabrics (left); and digital photographs of their foldability and rollability (right)

As noticed from **Figure 1**, the pristine cotton fabric shows a broad peak around 3200-3400 cm^{-1}, which was attributed to the stretching of the O–H groups; the peak near 2890 cm^{-1} was assigned to the band of C–H stretching vibrations, and the peak at 1210 cm^{-1} was the stretching of the C–O bond [7]. After the application of MXenes, the characteristic cotton peaks are diminished. A similar was noticed for the TEC-protected sample, implying that the MXene peak dominates over characteristic peptide peaks of TEC. Additionally, the MXene-modified/WPR-protected cotton spectrum has completely attenuated cotton peaks and shown absorption bands corresponding to N–H (3340 cm^{-1}), C–H (2920 cm^{-1}), C=O (1720 cm^{-1}), C–N (1240 cm^{-1}), and C–O–C (1100–1250 cm^{-1}) bonds, significant for WPR. For the sample protected with MAR, the most prominent peak appears at about 1720 cm^{-1} (C=O), at ~1140 cm^{-1} (C–O in the ester groups), and at 2950–2860 cm^{-1} (C–H) [8].

Moreover, the MXene-modified/protected fabrics maintained flexibility regardless of the protective coating used, demonstrating their potential in wearable electronic devices.

3.2 Washing durability of Mxene-functionalised/protected cotton fabric

With the aim to evaluate the efficiency of the selected compounds to protect MXenes onto cotton fabric against peeling off and oxidation, MXene-functionalised/protected samples were washed up to 20 cycles and comprehensively characterised in terms of measuring the electrical resistance (**Figure 2**), determination of mass loss after 20 washings, WCA analysis (**Figure 3**) and investigation of changes of hue in a CIE L*a*b* colour space after 5, 10, and 20 washing cycles (**Table 1** and **Figure 4**).

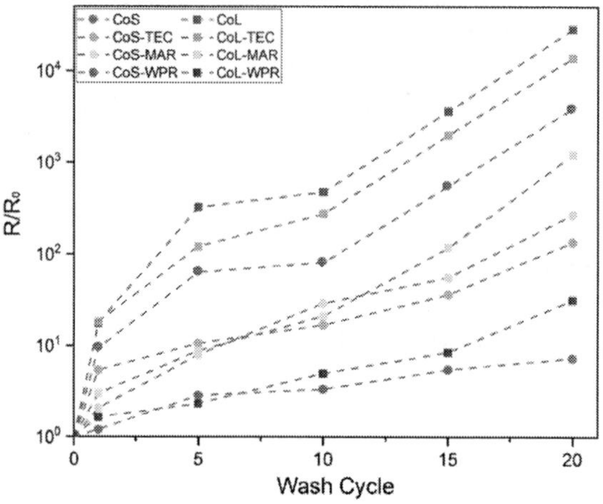

Figure 2 The electrical resistance relative to the initial value of the MXene-coated and MXene-coated/protected samples, up to 20 wash cycles

After five washes, both samples, CoS-PUR and CoL-PUR, exhibited a minimal raising of resistance, i.e., for 4.99% and 3.34 %, respectively, compared to their starting values, as seen in **Figure 2**. A similar trend was observed in MAR-treated samples with adequate protection during the early washing cycles, with a resistivity of 12.81 Ω·cm (CoS-MAR) and 6.34 Ω·cm (CoL-MAR) after five washes. Moreover, the CoS-MAR had a mass loss of only 0.92 % after the 20 washing cycles, while the CoL-MAR had a slightly higher mass loss of 1.30 %. Nevertheless, if the number of washing cycles exceeds 10, the resistivity rapidly escalates and can eventually reach the resistance level of untreated raw samples, losing electrical conductivity. The effectiveness of TEC protection varied significantly depending on the size of the applied MXenes.

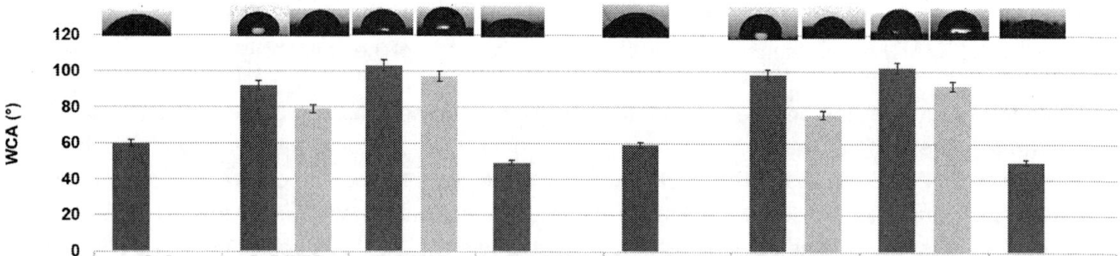

	CoS	CoS-WPR	CoS-MAR	CoS-TEC	CoL	CoL-WPR	CoL-MAR	CoL-TEC
before	60	92	103	49	59	98	102	50
20 wash	0	79	97	0	0	76	92	0

Figure 3 WCA of MXene-coated and MXene-coated/protected samples before and after 20x washings

The WCA analysis (**Figure 3**) revealed excellent hydrophobic properties of the MXene-functionalised/MAR- or WPR-protected samples with high contact angles of around 100 °, regardless of MXene nanosheets lateral size. Unprotected samples had contact angles of ~60 °, presumably due to the inherent hydrophilic nature of both MXenes and the cotton. The protected samples remained hydrophobic even after 20 washing cycles. For MAR-protected samples, the WCA decreased for 6 ° (CoS-MAR) and 10 ° (CoL-MAR), and for WPR-protected samples for 13 ° (CoS-WPR) and 22 ° (CoL-WPR), demonstrating the robustness and efficacy of the coating in preserving the surface's hydrophobic properties. Meanwhile, TEC was the least effective compound for the protection of MXenes against washings since the water drop spread immediately after surface contact after 20 washes. Consequently, water quickly penetrates through the pores during repeated washings, causing a loss of conductivity, as previously confirmed by an enlargement in electrical resistance.

Table 1 CIE colour differences of samples after 5, 10 and 20 washing cycles

Sample	CIE L*a*b* values			CIE colour differences											
				after five washing cycles				after ten washing cycles				after 20 washing cycles			
	L*	a*	b*	ΔL	Δa	Δb	ΔE	ΔL	Δa	Δb	ΔE	ΔL	Δa	Δb	ΔE
CoS	36.53	3.17	-4.05	-2.07	-1.98	1.29	3.14	-1.60	-2.80	2.00	3.79	0.33	-3.55	3.05	4.69
CoS-WPR	29.01	2.42	-1.54	-0.60	-0.46	-0.27	0.80	-1.53	-0.88	-0.41	1.81	-2.16	-1.58	-0.69	2.76
CoS-MAR	28.86	5.04	-9.47	0.55	-1.05	0.43	1.26	1.08	-1.09	-0.40	1.59	0.17	-1.91	0.84	2.09
CoS-TEC	33.17	2.44	-4.40	1.55	-1.28	1.49	2.50	1.06	-1.89	2.00	2.95	2.96	-2.70	3.14	5.09
CoL	38.77	3.06	-3.96	-1.63	-2.19	1.89	3.32	-1.13	-2.89	2.63	4.07	1.33	-3.49	3.57	5.17
CoL-TEC	36.24	2.22	-3.91	-0.81	-1.33	1.38	2.08	1.84	-1.81	2.01	3.27	6.60	-2.56	3.31	7.81
CoL-MAR	28.74	3.79	-6.62	0.38	-1.09	0.33	1.20	1.06	-0.74	-0.58	1.42	1.06	-1.52	0.13	1.86
CoL-WPR	28.95	2.11	-1.43	0.19	-0.16	-0.16	0.30	0.32	-0.36	-0.25	0.54	-0.14	-1.06	-0.36	1.13

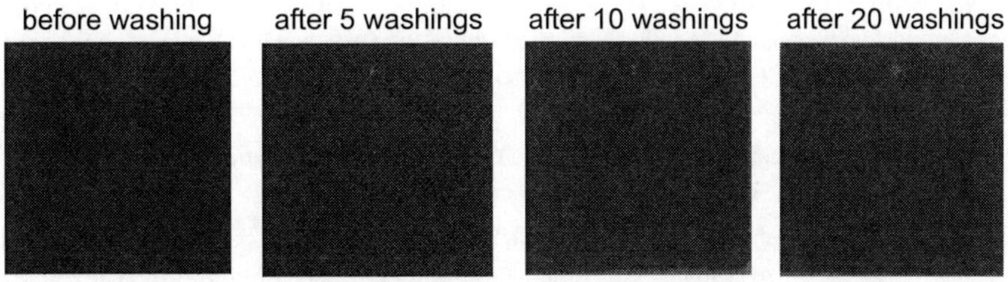

Figure 4 Photos of the CoL sample before and after washings

It is visually perceivable from **Figure 4** that with an increasing number of washing cycles, the sample CoL without protection becomes lighter and progressively exhibits a more greenish hue, i.e., the total colour change (ΔE) is 5.17. Values on the b-axis shift towards positive, and values on the a-axis towards negative (**Table 1**). Similar can be noted for all other samples. However, the changes are not linear and depend on the uniformity of the MXene deposition, the type of protective coatings, and the measurement location on each sample. The total colour change (ΔE) was significantly higher in samples without protective coatings. On the other hand, the ΔE values of WPR- and MAR-protected samples increased from 0.30 after 5 cycles up to 1.13 after 20 cycles (CoS-WPR) and from 1.20 after 5 cycles up to 1.86 after 20 cycles for the CoS-AR sample. The highest colour difference was observed in the CoS-TEC sample, with an ΔE of 7.81. The TEC was thus found unsuitable for the protection of MXenes, as also confirmed by low WCA and the high electrical resistance measured after washings.

CONCLUSION

In the presented study, the $Ti_3C_2T_x$ MXene nanosheets synthesised from small or large MAX precursors were successfully applied onto cotton fabric by the dip-padding procedure. The larger MXene flakes (in lateral size) imply higher electrical conductivity but lower adhesion and a greater tendency to oxidation. Therefore, three different protective compounds were applied to MXene-functionalised fabric, and two of them (WPR and MAR) significantly increased its hydrophobicity and, thus, protection against water accessibility. Moreover, the adhesion of MXenes was enlarged even after 20 washing cycles. This approach significantly expands the potential applications of MXene-decorated fabrics in smart textile applications by providing good durability of electroconductive functionalisation. TEC is not recommended for MXene protection due to its low adhesion and hydrophobic performance.

ACKNOWLEDGEMENTS

The results leading to this work were funded by the Slovenian Research and Innovation Agency (ARIS) in the frame of research project no. J2-50087 and a research core program group for Textile Chemistry and Advanced Textile Materials P2-0118 within the Young Researchers Programme.

REFERENCES

[1] ZHANG, H., JI, H., CHEN, J., WANG, N., XIAO, H. A multi-scale MXene coating method for preparing washable conductive cotton yarn and fabric. Industrial Crops & Products. 2022, vol. 188, pp. 115653.

[2] LUO, J., GAO, S., LUO, H., WANG, L., HUANG, X., GUO, Z., LAI, X., LIN, L., LI, R.K.Y., GAO, J. Superhydrophobic and breathable smart MXene-based textile for multifunctional wearable sensing electronics *Chemical Engineering Journal.* 2021, vol. 406, pp. 126898.

[3] MA, C., YUAN, Q., DU, H., MA, M.-G., SI, C., WAN, P. Multiresponsive MXene ($Ti_3C_2T_x$)-Decorated Textiles for Wearable Thermal Management and Human Motion Monitoring. *ACS Applied Materials & Interfaces.* 2020, vol. 12, pp. 34226-34234.

[4] WANG, Q.-W., ZHANG, H.-B., LIU, J., ZHAO, S., XIE, X., LIU, L., YANG, R., KORATKAR, N., YU, Z.-Z. Multifunctional and Water-Resistant MXene-Decorated Polyester Textiles with Outstanding Electromagnetic Interference Shielding and Joule Heating Performances. *Advanced Functional Materials.* 2019, vol. 29, pp. 1806819.

[5] YAN, J., MA, Y., ZHANG, C., LI, X., LIU, W., YAO, X., YAO, S., LUO. S. Polypyrrole–MXene coated textile-based flexible energy storage device. *RCS Advances.* 2018, vol. 9, pp. 39742.

[6] ALHEBEB, M., MALESKI, K., ANASORI, B., LELYUKH, P., CLARK, L., SIN, S., GOGOTSI, Y. Guidelines for Synthesis and Processing of Two-Dimensional Titanium Carbide ($Ti_3C_2T_x$ MXene). *Chemistry of Materials.* 2017, vol. 29, no. 18, pp. 7633-7644.

[7] LV, N., WANG, X., PENG, S., LUO, L., R. ZHOU. Superhydrophobic/superoleophilic cotton-oil absorbent: preparation and its application in oil/water separation. *RSC advances.* 2018 vol. 8, no. 53, pp. 30257-30264.

[8] SALIH, A.M., AHMAD, M.B., IBRAHIM, N.A., DAHLAN, K.Z.H.M., TAJAU, R., MAHMOOD, M.H., YUNUS, W.M.Z.W. Synthesis of Radiation Curable Palm Oil–Based Epoxy Acrylate: NMR and FTIR Spectroscopic Investigations. *Molecules.* 2015, vol. 20, no. 8, pp. 14191-14211.

THERMAL ANALYSIS OF PRISTINE AND CAFFEINE LOADED PA 4,6 NANOFIBROUS YARNS

[1,2]Juan Pablo PÉREZ AGUILERA, [1,2]Jaroslav MIKULE, [3]Luboš BĚHÁLEK, [2]Věra JENČOVÁ, [2]David LUKÁŠ

[1]*Technical University of Liberec, Faculty of Mechatronics, Informatics and Interdisciplinary studies, Liberec, Czech Republic, EU, juan.pablo.perez.aguilera@tul.cz*

[2]*Technical University of Liberec, Faculty of Science, Humanities and Education, Department of Chemistry, Liberec, Czech Republic, EU*

[3]*Technical University of Liberec, Faculty of Mechanical Engineering, Department of Engineering Technology, Liberec, Czech Republic, EU*

https://doi.org/10.37904/nanocon.2024.5013

Abstract

This work focuses on the research and development of composite multilayered AC electrospun materials for applications in technical, medical, and bioengineering fields. It investigates the effects of AC electrospinning on the thermal properties and morphology of Polyamide 4,6 (PA 4,6), both in its base pellet form and as nanofibrous materials with and without caffeine as a model drug. The high voltage applied during AC electrospinning can influence the crystallinity and thermal behavior of the material. Differential scanning calorimetry (DSC) and thermogravimetric analysis (TGA) were employed to analyze thermal transitions, crystallinity, stability, degradation profiles, and as a tool to properly asses the loading of the model drug. Scanning electron microscopy (SEM) was used to examine fiber morphology and material structure. DSC analysis revealed that, compared to PA 4,6 pellets, the pristine nanofibrous material exhibited a decrease in normalized melting enthalpy and a slight increase in melting temperature, while the caffeine-loaded nanofibrous material showed an increase in normalized melting enthalpy and a decrease in melting temperature. TGA confirmed the proper loading of the model drug with initial weight loss corresponding to the PA 4,6:caffeine ratio of 8:2 in solution. The inflection point decreased for both electrospun materials when compared to the pellets. SEM analysis showed smooth, randomly oriented nanofibers, with a reduction in fiber diameter for the caffeine-loaded sample. This study provides clear insights into the effects of AC electrospinning and the addition of a model drug on the thermal properties and fiber morphology of PA 4,6. This serves as a solid base for the next steps in analyzing the release profiles of biologically active substances from composite/layered AC electrospun nanofibrous materials.

Keywords: AC electrospinning, polyamide 4,6, nanofibrous materials, caffeine-loaded nanofibers, thermal properties, nanofibrous yarns, fiber morphology, drug loading

1. INTRODUCTION

The increasing demand for advanced materials with tailored properties has driven significant research into nanofibrous yarns, particularly those based on polyamide polymers. Among these, polyamide 4,6 (PA 4,6) has garnered attention due to its excellent mechanical strength, thermal stability, and chemical resistance. The unique morphology of nanofibers, characterized by a high surface area-to-volume ratio, enhances their potential applications in various fields, including textiles, filtration, and biomedical engineering [1].

Incorporating bioactive compounds into polymer matrices is a promising strategy to impart additional functionalities to nanofibrous structures [2]. Caffeine has been selected as a model drug for this study due to its well-documented pharmacological properties and its suitability for investigating drug release profiles [3]. Loading caffeine into PA 4,6 nanofibrous yarns could provide valuable insights into the mechanisms of drug release and the influence of fiber morphology on the release kinetics, which are critical for developing effective drug delivery systems.

Thermal analysis plays a crucial role in understanding the thermal behavior of polymeric materials and their composites. Techniques such as differential scanning calorimetry (DSC) and thermogravimetric analysis (TGA) allow for the evaluation of thermal transitions and stability, which are essential for assessing the performance of nanofibrous yarns under varying temperature conditions [4], [5]. This study aims to investigate the thermal properties of pristine and caffeine-loaded PA 4,6 nanofibrous yarns through comprehensive thermal analysis.

By elucidating the effects of caffeine loading on the thermal behavior of PA 4,6 nanofibers, this research seeks to contribute valuable insights into the design and optimization of multifunctional polymeric materials. Understanding these interactions will pave the way for developing innovative applications that leverage both the mechanical properties of PA 4,6 and the bioactive potential of caffeine as a model drug for future studies [6].

2. MATERIALS AND METHODS

2.1 Solution preparation

Polyamide 4,6 Stanyl® (PA 4,6) from Envalior was used as received for the unprocessed samples. For the non-functionalized solution, 14% wt PA 4,6 was dissolved in a 1:1 mixture of glacial acetic acid and formic acid (PENTA chemicals) and magnetically stirred for 24 hours. In the functionalized solution, 3.5% wt caffeine (C0750, Sigma-Aldrich) was added after dissolving the polymer, and the mixture was stirred under the same conditions.

2.2 AC Electrospinning

The solutions were electrospun to create 100% nanofibrous yarns from PA 4,6 (Pristine NY) and from PA 4,6 functionalized with caffeine (Caffeine NY). The AC electrospinning was carried out using the set-up described in patent number 2022-370 [7] as depicted in **Figure 1**. The parameters for the electrospinning process are described in **Table 1**.

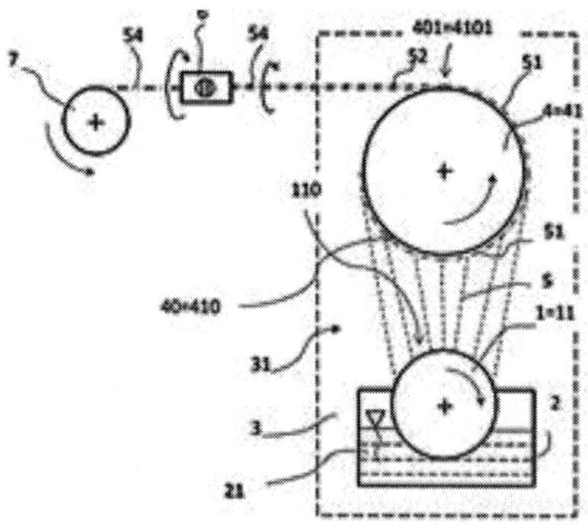

Figure 1 AC electrospinning device for the fabrication of 100% nanofibrous yarns utilizing a rotating disk electrode, rotating drum collector, a twirling device for yarn formation and a bobbin winder *[7]*

Table 1 AC electrospinning parameters for the production of Pristine NY and Caffeine NY.

Sample	Voltage (kV)	Frequency (Hz)	Winding speed (rpm)	Twirling speed (rpm)	Collector speed (rpm)	Ambient temperature (°C)	Relative humidity (%)
Pristine NY	40	100	0.85	10000	0.8	24	65
Caffeine NY	40	100	0.72	8000	0.7	25	55-60

2.3 Differential Scanning Calorimetry (DSC)

DSC is a thermal analysis technique that measures the heat flow into or out of a sample as a function of temperature or time. This technique is selected as it allows to analyze the thermal properties of the polymer materials being studied. More specifically their melting temperatures (Tm and Tc), specific heat capacity (Cp), purity, crystallinity, thermal conductivity, and thermal stability. DSC works by comparing the heat flow of a sample to that of a reference material, which is typically a thermally stable material. The difference in heat flow between the sample and the reference material indicates changes of state within the sample [8], [9], [10].

2.4 Thermogravimetric Analysis (TGA)

TGA is a thermal analysis technique that measures the change in mass of a sample as it is heated or cooled in a controlled environment. This technique provides invaluable insights into the thermal stability, composition, moisture content, and decomposition kinetics of materials [11], [12].

2.5 Scanning Electron Microscopy (SEM)

The SEM images were taken with a VEGA3 SBU – EasyProbe microscope, Tescan (Czech Republic) equipped with a Tungsten heated cathode as the electron gun and the secondary electron detector Everhart-Thornley type (YAG Crystal). The images were taken at High Vacuum Mode (SE) at a resolution of 8 nm at 15 kV.

3. RESULTS

The AC electrospinning set-up described in the patent successfully produced 100% nanofibrous yarns from PA 4,6 solutions, both with and without caffeine as an additive (at an 8:2 ratio in the dry material). The following section presents the results of the thermal and morphological analyses conducted on these samples.

3.1 Differential Scanning Calorimetry (DSC)

Table 2 DSC results for PA 4.6 in pellet form, pure caffeine, PA 4.6 pristine NY, and PA 4.6:Caffeine NY.

Sample	T_m N-Enthalpy 1(J/g)	T_m Peak 1(°C)	T_m N-Enthalpy 2(J/g)	T_m Peak 2(°C)
PA 4,6 Pellet	-	-	81.8 ± 2.5	288.3 ± 2.1
Caffeine	18.9	160.8	98.5	235.2
Pristine NY	-	-	65.8 ± 12.3	289.4 ± 0.9
Caffeine NY	26.4 ± 9.7	220.6 ± 11.7	100.2 ± 5.7	283.5 ± 1.7

Table 2 compares the thermal properties of the Pristine NY and Caffeine NY with the PA 4,6 pellets. The melting temperature of Pristine NY (289.4°C) is slightly higher than that of the pellet form by 1.1°C, suggesting that the electrospinning process may have enhanced molecular alignment in the nanofibers. However, the normalized enthalpy of Pristine NY decreases to 65.8 J/g, a 19.6% reduction compared to the pellet. This indicates a slight loss in crystallinity during electrospinning, as fewer polymer chains participate in the crystalline phase in the nanofibers.

For Caffeine NY, the melting temperature decreases to 283.5°C, 4.8°C lower than the pellet. This suggests that the addition of caffeine disrupts the molecular alignment of PA 4,6 in the nanofibers. Interestingly, the enthalpy of Caffeine NY's second thermal transition increases to 100.2 J/g, which is 22.5% higher than that of the pellet. This suggests that, despite the lower melting temperature, the overall crystallinity of the caffeine-loaded nanofibers may be higher than that of the pure polymer. This could be attributed to the interaction between caffeine and the polymer matrix, which may promote more orderly packing of polymer chains.

Moreover, Caffeine NY displays two distinct thermal peaks: the first at 220.6°C, corresponding to caffeine, and the second at 283.5°C, representing PA 4,6. The shift in caffeine's melting point from its pure form (160.8°C and 235.2°C) suggests a significant interaction between caffeine and the polymer matrix, resulting in an increase in caffeine's melting temperature. The normalized enthalpy of the caffeine peak in Caffeine NY (26.4 J/g) is higher than that of pure caffeine (18.9 J/g), indicating effective integration of caffeine into the nanofiber matrix. This earlier thermal transition of caffeine suggests that it could be released at lower temperatures, while PA 4,6 remains structurally intact until its higher melting point. This has important implications for drug release applications, as it demonstrates the potential for controlled release of caffeine while maintaining the stability of the polymer matrix.

3.2 Thermogravimetric Analysis (TGA)

Table 3 illustrates the thermal degradation behavior of PA 4,6 in its pellet form, with the degradation onset (TD,2%, or temperature at 2% weight loss) occurring at 380.4°C. The polymer exhibits high thermal stability, with midpoint and inflection point temperatures at 431°C and 436.9°C, respectively, indicating a gradual degradation at elevated temperatures. In contrast, caffeine degrades much earlier, with a TD,2% of 218.2°C, and significantly lower midpoint (293°C) and inflection point (311.2°C) temperatures. This highlights the thermal instability of pure caffeine compared to PA 4,6.

For the nanofibrous yarn without caffeine (Pristine NY), the thermal stability is slightly reduced, with a TD,2% of 365°C, compared to 380.4°C for the PA 4,6 pellets. This reduction can be attributed to the higher surface area of the nanofibers, which makes them more susceptible to thermal degradation. The midpoint and inflection point temperatures for the Pristine NY are also slightly lower (414°C and 419°C), further reflecting its slightly reduced stability compared to the pellet form.

Table 3 Pure PA 4.6 pellet, pure caffeine, PA 4.6 pristine NY, and PA 4.6:Caffeine NY

	Sample	$T_{D,2\%}$ (°C)	$T_{midpoint}$ (°C)	$T_{infl.point}$ (°C)	
	PA 4,6 Pellet	380.4 ± 0.9	431 ± 2.3	436.9 ± 3.4	
	Caffeine	218.2 ± 3.5	293 ± 6.6	311.2 ± 6.9	
	PA 4.6 Pristine NY	365 ± 10.1	414 ± 3.5	419 ± 2.6	
Sample	$T^*_{midpoint}$ (°C)	$T^*_{infl.point}$ (°C)	$T_{midpoint}$ (°C)	$T_{infl.point}$ (°C)	**First step wt loss [%]**
PA 4.6:Caffeine NY	253 ± 6.6	260.7 ± 6.8	410 ± 3.5	415.7 ± 1.5	23.2 ± 0.5

In the case of the caffeine-loaded nanoyarn (Caffeine NY), the first-step weight loss of 23.2% aligns well with the intended 20% caffeine loading, confirming successful caffeine incorporation. The first degradation step, with a Tmidpoint of 253°C and a Tinfl.point of 260.7°C, corresponds to the degradation of caffeine, consistent with its lower thermal stability. The second degradation step (Tmidpoint: 410°C, Tinfl.point: 415.7°C) is associated with the degradation of PA 4,6. These values are slightly lower than those of the pristine nanofibers, indicating that the addition of caffeine impacts the polymer's thermal stability, likely due to interactions between caffeine and the polymer matrix.

3.3 Scanning Electron Microscopy (SEM)

Figure 1 clearly shows that the addition of caffeine results in a significant reduction in the average fiber diameter, from 347 nm in the pristine nanoyarn to 251 nm in the caffeine-loaded nanoyarn, a decrease of approximately 28%. This indicates that caffeine influences the electrospinning process, leading to the production of thinner fibers. Although there is some variability in fiber diameter, the standard deviations suggest that the caffeine-loaded nanoyarn (±98 nm) has a more consistent fiber distribution compared to the pristine nanoyarn (±125 nm). This points to caffeine possibly promoting more uniform fiber formation during electrospinning, even as the average diameter decreases.

One explanation for the reduction in fiber diameter is that caffeine may alter the viscosity of the PA 4,6 solution. Typically, lower viscosity solutions tend to yield thinner fibers during electrospinning. If caffeine reduces the overall solution viscosity, this could account for the smaller fiber diameter observed. Additionally, caffeine may affect the electrical properties of the polymer solution, as changes in conductivity can influence the stretching and elongation of the electrospun jet, leading to finer fibers. The interaction between PA 4,6 and caffeine might also induce phase separation or alter polymer chain interactions, contributing to the observed differences in fiber formation and size.

The caffeine-loaded nanoyarn displayed thinner fibers, as seen in **Figure 1**, which could have several implications. Thinner fibers result in a higher surface area-to-volume ratio, potentially advantageous for applications like drug delivery, where increased surface area may enhance drug release or interaction with biological environments. A reduction in fiber diameter might also impact the mechanical strength and flexibility of the nanoyarn. Thinner fibers can sometimes reduce mechanical robustness, but this will depend on how caffeine affects the crystallinity and overall fiber structure.

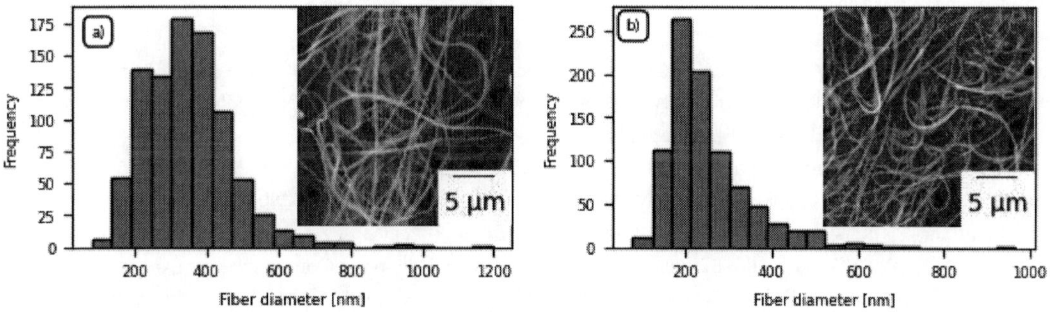

Sample	Average [nm]	St. deviation [nm]
Pristine NY	347	125
Caffeine NY	251	98

Figure 2 Fiber diameter histogram and SEM image for a) Pristine NY and b) Caffeine NY

4. CONCLUSION

The custom AC electrospinning setup successfully produced PA 4,6 nanofibrous yarns, both in pristine form and with caffeine as an additive. The DSC analysis showed that the electrospinning process slightly increased the melting temperature of the pristine nanoyarn, suggesting enhanced molecular alignment. However, the reduction in enthalpy values indicates a decrease in crystallinity, likely caused by the electrospinning process itself. For the caffeine-loaded nanoyarn, DSC revealed a lower melting temperature of PA 4,6, which points to some disruption in the polymer's crystalline structure. Despite this, the increased enthalpy suggests that caffeine may promote a secondary crystalline phase or improve the overall stability of the nanofibers. The two distinct thermal transitions observed for caffeine in the nanoyarn demonstrate its potential for controlled drug release, with caffeine melting at a lower temperature than the PA 4,6 matrix.

TGA confirmed the successful incorporation of caffeine into the nanofibers, with a degradation profile reflecting both caffeine and PA 4,6. The addition of caffeine slightly reduced the thermal stability of the nanoyarn compared to pristine PA 4,6. This decrease in stability is likely due to the interaction between caffeine and the polymer matrix. Finally, SEM revealed that caffeine loading significantly impacted the fiber morphology, reducing the average fiber diameter and leading to thinner, more uniformly distributed fibers. This reduction in fiber size could enhance surface-related properties, such as drug release, while potentially affecting the mechanical and thermal behavior of the nanoyarn. The observed changes suggest that caffeine influences the electrospinning process, likely through modifications to the viscosity, conductivity, or interaction with the polymer solution. These findings provide a comprehensive understanding of the thermal and morphological behavior of PA 4,6 nanofibers with caffeine. This foundation is valuable for future research on drug release mechanisms and the thermal performance of nanofibrous materials.

ACKNOWLEDGEMENTS

The study was supported by the Czech Health Research Council, project no. NW24-08-00133 "Nanofibrous sutures functionalized by bioactive compounds", provided by the Ministry of Health of the Czech Republic.

This study was supported by the National Competence Centres II program, project no. TN02000018/015 "Technology for the continuous production of multi-layer nanofibrous composite material using the AC electrospinning", provided by the Technology agency of the Czech Republic.

REFERENCES

[1] MALEKI, H., GHAREHAGHAJI, A.A., TOLIYAT, T., and DIJKSTRA, P.J. Drug release behavior of electrospun twisted yarns as implantable medical devices. *Biofabrication.* 2016, vol. 8, no. 3, p. 035019. Available from: https://doi.org/10.1088/1758-5090/8/3/035019.

[2] SAYIN, S. *et al.* Electrospun Nanofibers With pH-Responsive Coatings for Control of Release Kinetics. *Front Bioeng Biotechnol.* 2019, vol. 7, p. 309. Available from: https://doi.org/10.3389/fbioe.2019.00309.

[3] PINA-VIDAL, C., BERNED-SAMATÁN, V., PIERA, E., CABALLERO, M. Á., and TÉLLEZ, C., Mechanochemical Encapsulation of Caffeine in UiO-66 and UiO-66-NH2 to Obtain Polymeric Composites by Extrusion with Recycled Polyamide 6 or Polylactic Acid Biopolymer. *Polymers.* 2024, vol. 16, no. 5, Art. no. 5. Available from: https://doi.org/10.3390/polym16050637.

[4] DRAGAR, Č., ROŠKAR, R., AND KOCBEK, P. The Incorporated Drug Affects the Properties of Hydrophilic Nanofibers. *Nanomaterials.* 2024, vol. 14, no. 11, Art. no. 11. Available from: https://doi.org/10.3390/nano14110949.

[5] CHOU, S.-F., CARSON, D., AND WOODROW, K.A. Current strategies for sustaining drug release from electrospun nanofibers. *J Control Release.* 2015, vol. 220, no. Pt B, pp. 584–591. Available from: https://doi.org/10.1016/j.jconrel.2015.09.008.

[6] VALTERA, J. *et al.* Fabrication of dual-functional composite yarns with a nanofibrous envelope using high throughput AC needleless and collectorless electrospinning. *Scientific Reports.* 2019, vol. 9, p. art.#1801. Available from: https://doi.org/10.1038/s41598-019-38557-z.

[7] BERAN, J. *et al.* A method of the production of nanofiber yarn by alternating electrical spinning of the polymer solution or melt and an equipment to perform the method. Czech patent No. PV2022-370, 2022.

[8] GABBOTT, P., Fast Scanning DSC, in *Principles and Applications of Thermal Analysis.* John Wiley & Sons, Ltd, 2008, pp. 51–86. Available from: https://doi.org/10.1002/9780470697702.ch2.

[9] HAINES, P.J. *Thermal Methods of Analysis: Principles, Applications and Problems.* Berlin: Springer Science & Business Media, 2012.

[10] WENDLANDT, W.W. *Thermal analysis.* New Jersey: Wiley-Interscience, 1986.

[11] COATS, A.W., REDFERN, J.P., Thermogravimetric analysis. A review. *Analyst.* 1963, vol. 88, no. 1053, pp. 906–924. Available from: https://doi.org/10.1039/AN9638800906.

[12] LIU, X., YU, W. Evaluating the thermal stability of high performance fibers by TGA. *Journal of Applied Polymer Science*. 2006, vol. 99, no. 3, pp. 937–944. Available from: https://doi.org/10.1002/app.22305.

PREPARATION AND CHARACTERIZATION OF SPARTIUM JUNCEUM L FIBER FILLED POLYCARBONATE COMPOSITES APPLICABLE IN THE AUTOMOBILE INDUSTRY

Emi GOVORCIN BAJSIC[1], Mario MEHES[1], Ljerka KRATOFIL KREHULA[1], Ana PERSIC[1], Sanja PERINOVIC JOZIC[2]

[1]*Faculty of Chemical Engineering and Technology, University of Zagreb, Zagreb, Croatia, EU,* egovor@fkit.unizg.hr

[2]*Faculty of Chemistry and Technology, University of Slit, Split, Croatia, EU*

https://doi.org/10.37904/nanocon.2024.5017

Abstract

The aim of this study is the modification of the polycarbonate (PC) as synthetic polymer with the natural polymer - Spartium junceum L (SJL) fibers to obtain a new biodegradable polymer material. Spanish broom (Spartium junceum L) was chosen as reinforcement for PC composites because of her low cost, light weight, renewability, biodegradability and easily availability in nature. The effect of different content of SJL fiber on the properties of PC/SJL composites was investigated. The PC composites reinforced with randomly oriented, short Spartium junceum L. fibers were prepared in Brabender mixer and hydraulics press. Mechanical properties of prepared PC/SJL fiber composites were analysed using dynamic mechanical analysis (DMA) and tensile tests. The morphology of the composites was analysed using scanning electron microscopy (SEM). The chemical properties were characterized by Fourier transform infrared spectroscopy (FTIR). The obtained results show that the SJL fibers increased the tensile strength and loss modulus as well as decreased the elongation of composites. The SEM analysis revealed that good adhesion between SJL fiber as filler and PC matrix was observed. The resulting PC/SJL fiber composites have the potential to be used in the automotive industry.

Keywords: Polycarbonate, Spartium junceum L (SJL) fibers, DMA, mechanical properties, SEM

1. INTRODUCTION

Synthetic polymer materials are mostly used as disposable packaging which makes them major environmental pollutants. Because of the large consumption of polymer materials, the large amounts of waste are created, which cause the creation of large landfills of synthetic polymers which are not biodegradable. Therefore, interest in biodegradable materials stems from the growing concern of people about the potential harm and the problem of waste disposal arising from the use of materials based on non-renewable sources of energy. Due to all mentioned above, a good waste management system is important, which includes prevention of waste generation, minimization of the amount of waste, reuse whenever is that possible and different disposal procedures, of which various recycling procedures bring exceptional importance and environmental benefit. In the last few years, there has been a growing interest in the commercial use of composites based on natural fibers, i.e. biocomposites, especially in the automotive industry. A significant amount of research has been dedicated to the use of development of biocomposites due to recyclable, environmentally friendly, and compostable alternatives to petrochemical based conventional plastics such as polypropylene (PP) and polyethylene (PE). Natural fibers represent sustainable materials, easily available in nature, with several advantages such as low cost, light weight, renewability, biodegradability and high specific properties [1]. The introduction of reinforcements of natural fibers is a practice for improving the properties of polymer materials. Also, the use of natural fibers as reinforcements for polymers is increasingly being investigated due to the environmental acceptability of natural fibers compared to synthetic fibers. Some of the most used natural fibers are barely, oats, rye, sugar cane, bamboo, rice husks, hemp, etc. [2]. Polycarbonate (PC) is a thermoplastic polymer that is easily processed, moulded, and thermoformed. It is highly transparent to visible light, with better light transmission than many kinds of glass and because of this property PC finds many applications in optical devices. This polymer is a durable material, but although it has high impact resistance, its scratch resistance is poor and so a hard coating is generally applied to polycarbonate eyewear lenses and polycarbonate exterior automotive components. The biggest application of PC in automotive industry is for headlights, instrument panels and wheel cover as well as windows. In the Mediterranean region of Croatia Spanish broom are of great agricultural crop. It is composed of cellulose, hemicellulose and lignin. Lignocellulosic materials have a lot of positive features, such as low density, low requirements on processing equipment, no abrasion during processing, abundance and biodegradability.

2. EXPERIMENTAL

2.1 Materials

In this study Polycarbonate (PC) Makrolon 2805 was supplied in pellet form by Bayer, Germany. As indicated by the manufacturer, the material has a melt flow index of 10 g / 10 min at 300°C. The Spartium junceum L fibers were obtained from the Spartium junceum L plant which was harvested around town Šibenik, Croatia.

2.2 Preparation of composites

To determine the influence of the Spartium junceum L fibers on the properties of the PC matrix, a series of composites was manufactured with different content of randomly oriented, short Spartium junceum L fibers (0, 5, 10, 15, and 20 % by weight).The samples were prepared using Brabender mixer at a temperature of 200°C and a screw speed of 60 min^{-1} for 5 min, moulded in a hydraulic hot press Dake Model 44-226 (Holland) at 200 °C and a pressure of 25 kPa for 5 min and then cooled down between thick metal blocks at room temperature. **Figure 1** shows a photo of the hot pressed samples of investigated PC/SJL fiber composites used for further characterization.

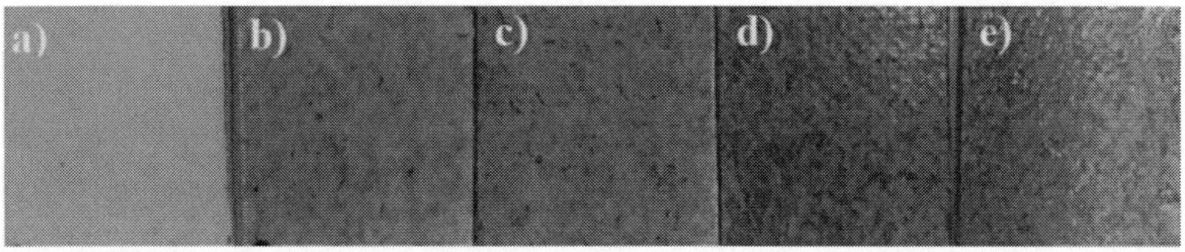

Figure 1 Photograph of PC/SJL fiber composites with a) 0; b) 5; c) 10; d) 15 and e) 20 wt.% of SJL fiber

2.3 Characterization

Dynamic mechanical measurements are performed with a TA Instruments model DMA 983 Analyzer. The measurements are conducted over temperature range of 25 °C to 200 °C at a constant frequency of 1Hz with amplitude of 0.3 mm. The heating rate was 3 °C min^{-1}. Tensile test was conducted on a universal testing machine (Zwick Roel Z020) with a testing speed 50 mm/min. At least five specimens were repeated. Infrared spectroscopy (FT-IR) spectra were obtained with a Perkin Elmer Spectrum 100 FTIR spectrometer (Waltham, MA, USA) using attenuated total-reflection (ATR) method. The analyses were carried out at room temperature and ambient humidity. All spectra were registered from 4000 cm^{-1} to 650 cm^{-1} with a resolution of 4 cm^{-1}. A scanning electron microscope (SEM) was used to evaluate the fracture surfaces, respectively, the interphase adhesion. Images of individual samples were taken on a JOEL JSM 7610F scanning electron microscope at different magnifications. The applied voltage was 1,00 kV.

3. RESULTS AND DISCUSSION

3.1 Viscoelastic properties

To investigate the viscoelastic properties (loss modulus, E'' and storage modulus, E') of the neat PC and PC/SJL fibers composites, we tested the dynamic mechanical properties using DMA. Generally, in biocomposites in which fibers are used as filler, the dynamic mechanical properties depend on the nature of the polymer and the interactions between the filler and the polymer matrix, as well as of the dispersion and orientation of the filler. **Figure 2a** shows the storage modulus (E') of the neat PC and PC/ SJL fiber composites with different SJL fiber content. From **Figure 2a**, the storage modulus below the T_g (glassy region) of the neat PC and PC/SJL fiber composites maintains plateau and shows no significant difference among the samples. Above the T_g (rubbery plateau region), the storage modulus of the neat PC and PC/SJL fiber composites decreases with increasing the temperature. For the rubbery plateau region, the storage modulus of the composites is significantly increased with increasing the SJL fibers content, except for the composite with a content of 20 wt. % of SJL fiber (**Table 1**). This increases in storage modulus probably due to the increase in the crystal phase fraction derived from the crystal structure of cellulose present in SJL fiber. **Figure 2b** shows the curves of the loss modulus (E'') vs temperature for the neat PC and PC/SJL fiber composites. According to **Figure 2b** loss modulus curve of the neat PC exhibit one broad relaxation maximum at 55 °C which is related to the β-relaxation in PC due to the mobility of shorter segments in the polymer chain [3]. At a higher temperature (153,3 °C), a relaxation maximum of high intensity corresponding to the T_g of the PC was obtained. The T_g 's obtained for PC and PC/SJL fiber composites are shown in **Table 1**. The T_g values of the PC/SJL fiber composites decrease with increasing the SJL fiber content compared with neat PC. Similar results have been obtained for nanocellulose/PC composites [4]. This decrease in the T_g may be due to increased chain mobility of the amorphous PC with higher amount of lignin which also has an amorphous structure.

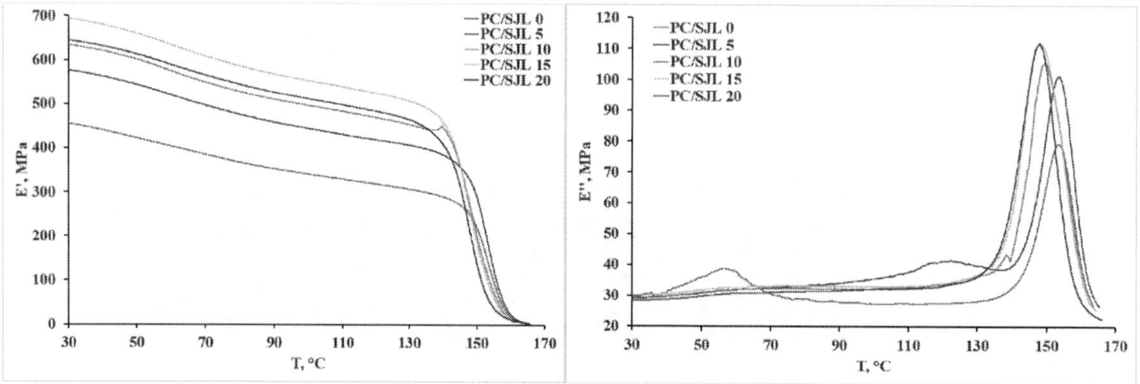

Figure 2 a) storage modulus (E') and b) loss modulus (E'') as a function of temperature for the neat PC, and PC/SJL fiber composites

3.2 Mechanical properties

The stress/strain curves of composites at different SJL fiber content were shown in **Figure 3. Table 1** present the results of tensile strength (σ) and elongation at break (ε) for the neat PC and PC/SJL fiber composites. It can be seen from **Table 1** that tensile strength of SJL fiber reinforced PC composites was higher when compared with the neat PC. Tensile strength of fiber reinforced PC composite increased with the increasing of SJL fiber content [5] and only slightly decreased when SJL fiber content reached 20 wt.%, as shown in **Table 1**. The reduction of tensile strength at higher SJL fiber content may possible due to the increasing of the probability of fiber agglomeration. On the other hand, the improvement in tensile strength was believed to be due to the better interfacial adhesion between SJL fibers and PC matrix. The interfacial bonding between fiber and matrix played an important role for the mechanical property of fiber reinforced composite. The increased of tensile strength agrees with the increase of storage modulus value observed from DMA measurements. However, the elongation at break of SJL fiber reinforced PC composites were lower than the neat PC and decreased with increasing of fiber content.

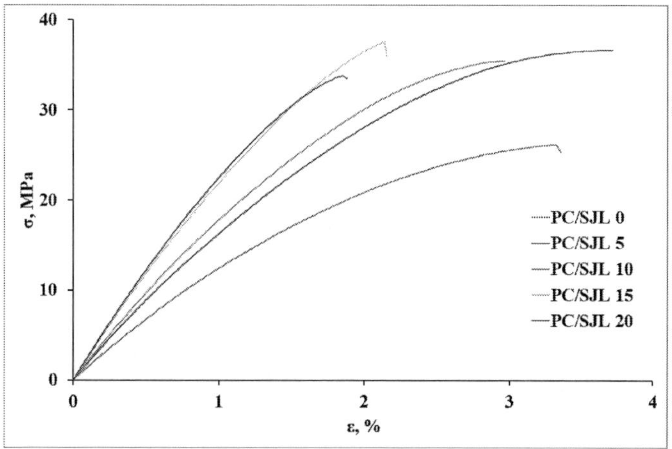

Figure 3 Stress/strain curves for the neat PC, and PC/SJL fiber composites

Table 1 DMA and mechanical properties of the neat PC and PC/SJL fiber composites

Sample	$T_{g\,E''}$ (°C)	**$E'_{25°C}$ (MPa)**	σ (MPa)	**ε (%)**
PC/SJL 0	153.3	461.6	25.2±0.4	3.7±1.3
PC/SJL 5	153.5	580.9	36.6±0.2	3.1±0.1
PC/SJL 10	149.5	634.7	36.4±3.3	2.9±0.2
PC/SJL 15	147.9	698.6	39.2±5.2	2.1±0.6
PC/SJL 20	147.0	646.0	33.1±8.1	1.8±0.6

3.3 FTIR results

The characteristic absorption bands for the neat PC are observed in **Figure 4** and summarized in **Table 2.** PC is characterized by absorption bands at 2968 cm^{-1}, 2920 cm^{-1}, 2873 cm^{-1} and 2841 cm^{-1} which can be attributed to the C-H stretching vibrations of the CH_3 groups while absorption bands at 1768 cm^{-1} are derived from molecular vibrations of C=O groups [6]. The absorption band at 1502 cm^{-1} is associated to stretching C=C vibrations in the aromatic ring, the absorption bands at 1409 cm^{-1} and 1363 cm^{-1} deformation OH, and 1218 cm^{-1}, 1185 cm^{-1}, 1157 cm^{-1} and 1109 cm^{-1}, are attached to the deformation of acetyl C-O group [7]. The absorption peak at 1012 cm^{-1} corresponded to the stretching vibrations of the C–O–C group. **Table 2** shows the IR spectra of SJL fibers which are usually characterized by several absorption bands. The broad absorption band at 3334 cm^{-1} represents vibration of free OH groups of the cellulose while bands at 2915 cm^{-1} and 2899 cm^{-1} are attributed to the C-H symmetrical stretching of cellulose, hemicellulose, pectin, fats and waxes. The absorption band at 1604 cm^{-1} corresponds to the C=C stretching from the amorphous lignin. The absorption band at 1025 cm^{-1} corresponds to the C-C, C-OH, C-H ring and side group vibrations from cellulose. The band at 1100 cm^{-1} is related to the C-O-C glycosidic ether from cellulose and band at 896 cm^{-1} of the COC, CCO and CCH deformation and stretching in the cellulose. The typical band of O-H at 3334 cm^{-1} was visible in the FTIR spectrum of the PC/SJL fiber composites, indicating the presence of SJL fibers. The results in **Figure 4** show that was not significantly different of shape and position in FTIR spectra with increasing of SJL fiber content. The overlap of some absorption bands of SJL and PC was observed.

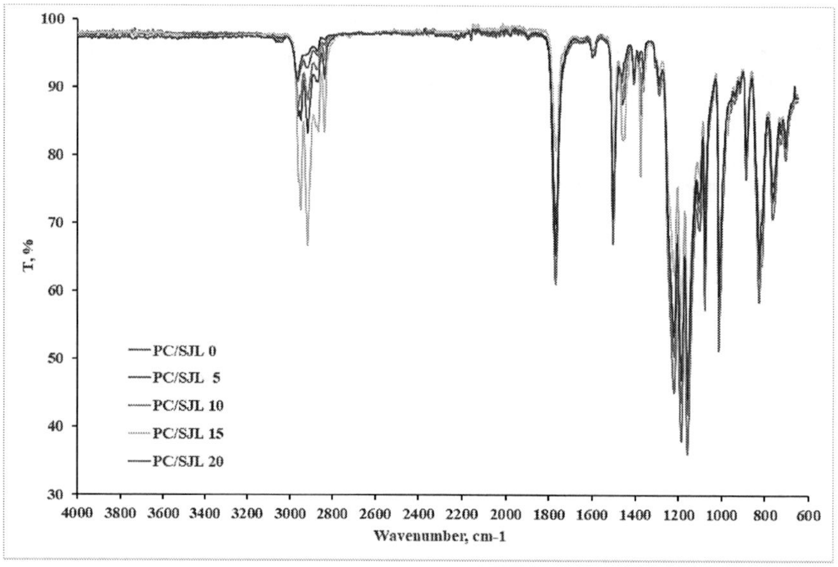

Figure 4 FTIR spectra for neat PC, and PC/SJL fiber composites

Table 2 FTIR data peak band assignments for PC and SJL fiber

PC	
Assignment	**Peak position (cm^{-1})**
CH stretching	2968 ,2920, 2873, 2841
C=O stretching	1768
C=C stretching	1502
OH deformation	1409, 1363
C-O deformation	1218, 1185,1157, 1109
C–O–C stretching	1012
SJL fiber	
Assignment	**Peak position (cm^{-1})**
OH stretching (Cellulose, Hemicellulose)	3334
C-H symmetrical stretching (Cellulose, Hemicellulose)	2915,2899
C=O stretching vibration (Pectin, Waxes)	1730
C=C stretching (lignin)	1604
C-O-C glycosidic ether (cellulose)	1100
C-C, C-OH, C-H ring and side group vibrations (cellulose)	1025
COC, CCO and CCH deformation and stretching (cellulose)	896

3.4 Morphology of the neat PC and PC/SJL fibers composites

The morphology of the neat PC and PC/SJL fiber composites was measured by scanning electron microscopy (SEM). The matrix-fillers interface plays critical role for ensuring that the properties of each component contribute optimally to the properties of the final product. The SEM micrograph of fracture surface after tensile test of the neat PC and PC/SJL fiber composites are shown in **Figure 5**. As observed from SEM images a good dispersion of fibers is observed for composites with content of fiber up to 15 wt.%. Also, it is possible to note that the better the interfacial adhesion between SJL fiber and PC matrix was observed in composites with less fiber content (**Figures 5b-5d**). We can conclude that a good adhesion between reinforcing fiber and PC matrix interphase resulted in tensile strength improvement up to 15 wt.% of SJL fiber content (**Table 1**).

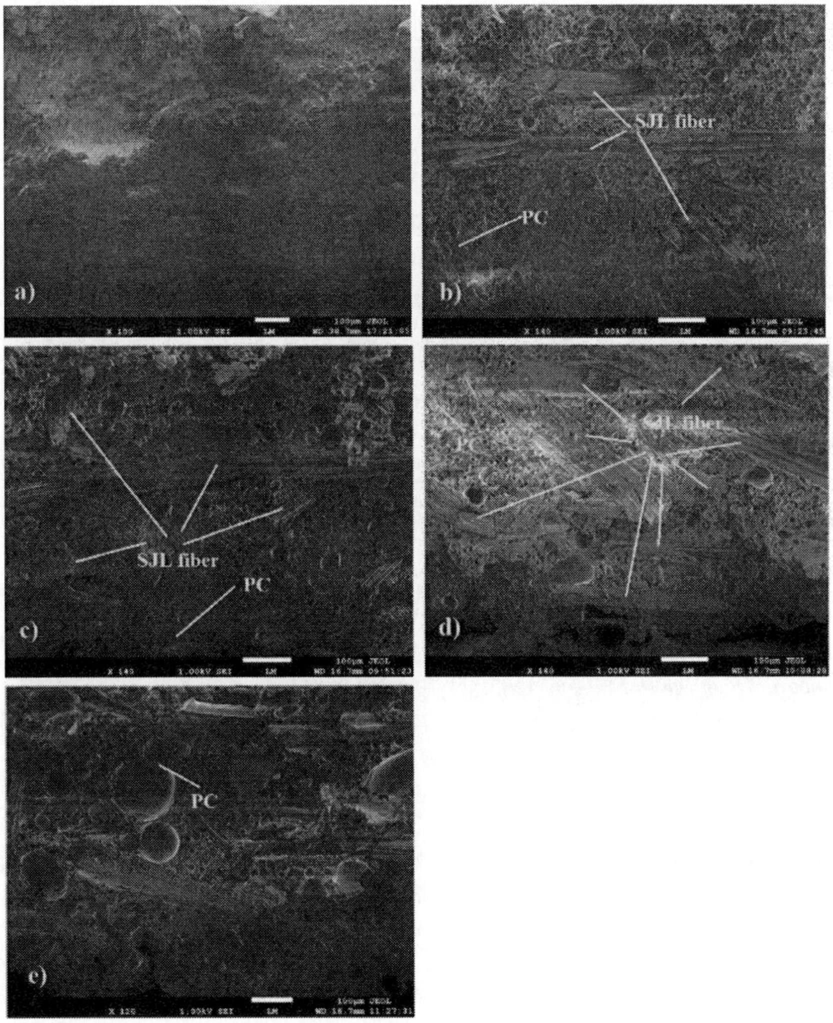

Figure 5 SEM images of the fracture surface of (a) neat PC; (b) PC/SJL 5; (c) PC/SJL 10; (d) PC/SJL 15 and (e) PC/SJL 20 composites

4. CONCLUSION

In this work, the polymer composite samples of polycarbonate (PC) matrix with Spartium junceum L (SJL) fibers were successfully prepared via hot melt in Brabender mixer. The results of dynamic mechanical analysis (DMA) according to storage modulus (E') and the tensile strength (σ) of PC/SJL fiber composites showed that E' and σ increased with increasing of fiber content. Good adhesion between SJL fiber and PC matrix resulted in tensile strength improvement reinforced PC composites. However, the elongation at break ($\boldsymbol{\varepsilon}$) and glass transition temperature (T_g) decreased with increasing fiber content compared with the neat PC due to the increased chain mobility of the amorphous PC. The FTIR analysis results showed the presence of SJL fibers in PC/SJL composites and the overlap of some absorption bands of SJL and PC. Upon SJL fiber addition, there are not major changes of shape and position of the characteristic absorption bands, and it can be concluded that FTIR spectra of the composites are practically identical to the spectra of neat PC, except of the absorption bends intensity. The results of morphological studies of PC/SJL fiber composites clearly showed that a good dispersion of fibers is observed for composites with content of fiber up to 15 wt.% and this resulted in tensile strength improvement. Considering the results of mechanical, chemical and morphological studies, it is evident that the SJL fiber as reinforcement might improve the investigated properties of PC. Based on these results we assumed that SJL fibers represent good candidate as potential application in the automobile industry.

ACKNOWLEDGEMENTS

The research in this work was supported by grant no.121174 University of Zagreb, Croatia.

We would like to acknowledge to the "Functional integration of the University of Split, PMF-ST, PF-ST and KTF-ST through the development of scientific research infrastructure in the building of three faculties" (KK.01.1.1.02.0018), a project co-financed by the Croatian Government and the European Union through the European Regional Development Fund—the Competitiveness and Cohesion Operational Programme.

REFERENCES

[1] VENKATESH, N., MOHAN, K., VIJAYANANDA, K. A Review on natural fiber composite materials in automotive applications. *Engineered Science.* 2022, vol. 18, pp. 1-10.

[2] TAJ, T., MUNAWAR, M.A., KHAN, S., Natural fiber – reinforced polymer composites. *Proceedings of the Pakistan Academy of Science.* 2007, vol. 44, pp. 129-144.

[3] VARDARAJAN, K., BOYER, R.F. Secondary relaxation processes in bisphenol-A polysulphone. *Journal of Polymer Science: Polymer Physics Edition.* 1982, vol. 20, pp. 141-154.

[4] MARIANO, M., EL KISSI, N., DUFRESNE, A. Melt processing of cellulose nanocrystal reinforced polycarbonate from a masterbatch process. *European Polymer Journal.* 2015, vol. 69, pp. 208–223.

[5] PANTHAPULAKKAL, S., SAIN, M. Preparation and characterization of cellulose canofibril films from wood fiber and their thermoplastic polycarbonate composites. *Int. Journal of Polymer Science.* 2012, vol. 2012, pp. 1–6.

[6] SACKEY, J., FELL, A., NGILIRABANGA, J., RAZANAMAHANDRY, L., NTWAMPE, S., NKOSI, M. Antibacterial effect of silver nanoparticles synthesised on a polycarbonate membrane. *Materials Today: Proceedings.* 2021, vol. 36, pp. 336–342.

[7] WANG, H., WANG, J., ZOU, Q., LIU, W., WANG, C., HUANG, W. Surface treatment using potassium ferrate for separation of polycarbonate and polystyrene waste plastics by froth flotation. *Applied Surface Science.* 2018, vol. 448, pp. 219–229.

WET CHEMISTRY TREATMENT AND SURFACE MODIFICATION OF RECYCLED GLASS PARTICLES FOR SUSTAINABLE CONSTRUCTION APPLICATIONS

[1]Oleg BABČENKO, [1]Klára BERANOVÁ, [1]Zdeněk REMEŠ, [1]Kateřina KOLÁŘOVÁ, [1,2]Jakub LÖRINC, [1,3]Jakub VOLF, [2]Zdeněk PROŠEK, [2]Pavel TESÁREK

[1]Institute of Physics, Czech Academy of Sciences, Prague, Czech Republic, EU, babcenko@fzu.cz

[2]Faculty of Civil Engineering, Czech Technical University in Prague, Prague, Czech Republic, EU

[3]Department of Inorganic Chemistry, University of Chemistry and Technology, Prague, Czech Republic, EU

https://doi.org/10.37904/nanocon.2024.5016

Abstract

The surface contamination or bulk impurities in the matrix of soda-lime glass can significantly alter glass properties and via leaching influenced the alternative glass waste utilization as a partial replacement for sand or binder in concrete-based materials for sustainable construction applications. Therefore, in this work, we perform wet chemistry (different pH) treatment of the recycled glass particles obtained from the commercial source. The resulting changes in glass particle properties were analyzed using microscopic and spectroscopic techniques. Changes in the concentrations of Mg, Al, Ca and Na near the particle surfaces were detected, indicating potential elements release. For instance, treatment in HCl resulted in a decrease of all observed non-silica ions in the glass subsurface region while NaOH treatment resulted in a slight increase of Ca amount in the subsurface region which potentially can positively affect the calcium-silicate-hydrate formation in concrete-based constructions. The observed results can be helpful in the explanation of processes at interfacial transition zones in waste glass containing concrete-based materials.

Keywords: Soda-lime glass, wet chemistry, surface analysis, metal ions diffusion, civil engineering

1. INTRODUCTION

The natural resources exhausting, and issues of environmental pollution leads to increased demands for sustainable management and recyclable materials use. For instance, glass waste recycling in the glass industry, especially container glass, significantly reduces the need for raw materials and lowers energy consumption [1–3]. Moreover, since silica-based glass can be recycled many times retaining its properties it contributes positively to its overall environmental impact when considered in lifecycle assessments. On the other hand, the strict industry requirements for recycled glass quality/separation have resulted in large amounts of glass waste being dumped in landfills, creating environmental problems. With the aim of reducing glass waste landfilling, a variety of alternative recycling strategies was considered. It was proposed to use the waste glass for foam glass or glass wool fabrication, manufacturing of sandblasting media or filtration media for water treatment, road base material, partial replacement for sand or binder in concrete-based materials for sustainable construction applications [1]. Nevertheless, in many cases the contaminations or additives in waste glass that prohibit its recycling in glass industry may also limit the alternative recycling strategies.

Indeed, in the estimation of environmental pollution, glass is generally considered as non-biodegradable material with the landfilling as main environmental impact. However, the reaction of glass surface with water, especially if cracks in the glass are present, is known [4–6]. Therefore, diffusion of metal ions from glass waste, e.g. through reaction of glass in the form of shards or small fragments with rainwater, with leachate into the environment can be an important pollution issue as well [6–8]. Thus, the risk of pollutants release from the waste glass via leaching must be taken into account as from the point of environmental pollution as from the point of alternative recycling strategies [8–11]. Moreover, glass treatment in alkaline or acidic liquids or melts can affect the composition of glass subsurface and thus its surface properties. In particular, the wet chemistry treatment of the waste glass and surface properties (chemistry or morphology) modification can be interesting for applications in the concrete-based materials and calcium-silicate-hydrate formation.

In the current study we perform the wet chemistry treatment of the recycled waste glass (WG) in the form of glass powder from the commercial source. We investigate the effect of WG interaction with mild acidic and/or alkaline solutions and study WG powder properties before and after wet chemistry treatment by microscopy and spectroscopy techniques. The observed WG surface and subsurface composition changes can be helpful in the explanation of processes in high alkalinity mixtures for sustainable concrete-based constructions.

2. EXPERIMENT

Ground waste glass powder from ordinary, recycled municipal glass waste (Recifa a.s.) was used in the experiments [12]. The deionized (DI) water and three different solutions with 0.25 M concentration of NaOH, KOH, and HCl were used for wet chemistry treatment. The WG powder was placed in the amount of 10 g to 100 ml of each treatment medium/solution and mixed by magnetic steer for 1 h. It follows by the 24 h sedimentation of WG powder. After sedimentation treatment solutions were removed from vessels and each glass powder was washed by DI water (short mixing by glass stick) with following 24 h sedimentation. The washing cycle was repeated 5 times after which treated WG powders were dried by natural water evaporation.

The result of WG powder interaction with used solutions was studied using Zetasizer Nano ZS – ZEN3600 equipped with multi-purpose titrator MPT-2 (Malvern Panalytical). The pH measurements of the treatment solutions and washing media (all after sedimentation) were performed. The measurements were supported by the conductometry data obtained using PWT conductivity tester (Hanna instruments).

Dry powder microstructure analysis was realized by scanning electron microscopy (SEM) using TESCAN MAIA3 microscope (TESCAN GROUP, a.s.). Analyzed powders were spread in a dense layer on sticky copper tape on solid support. The size, shape, and surface morphology of WG particles were analyzed at acceleration voltage 10 kV with secondary electrons detector. Additionally, mounted to SEM setup the energy dispersive X-ray (EDX) detector with IDFix analytical software was used to perform the basic elemental analysis of WG.

Fourier Transform Infrared (FTIR) spectroscopy was realized using the dry-air ventilated commercial Nicolet IS50 FTIR Spectrometer (Nicolet CZ, Ltd.). The FTIR absorption spectra (64 scan/spectrum with 4 cm^{-1} spectral resolution) were collected using a built-in attenuated total reflectance (ATR) accessory with a diamond ATR prism, DTGS detector and solid substrate beamsplitter in 200-1800 cm^{-1} spectral range (SiO_2 fingerprint region). The spectra in 400-4000 cm^{-1} spectral range were collected additionally using a KBr beamsplitter. The absorbance spectra were spectrally corrected using built-in software (OMNIC Spectra Software).

The X-ray photoelectron spectroscopy (XPS) characterization of dry WG powders was performed using the AxisSUPRA photoelectron spectrometer (Kratos Analytical). The monochromatized Al Kα (1486.6 eV) with emission of 10 mA, activated charge-compensation and resolution 20 was used for all measurements. Regions of the survey spectra containing main peaks of all elements were measured with passing energy of 80 eV for purposes of quantitative analysis. The atomic concentrations were calculated using KolXPD software. Correction was performed according to Smith et al. [13], while 3.5 nm was used for the electron attenuation length of C 1s photoelectrons.

The excitation-emission photoluminescence mapping of dried WG powder was performed using Agilent Cary Eclipse Fluorescence Spectrophotometer. Maps were measured in 300–800 nm excitation range with 10 nm step and 320–1100 nm emission range with 1 nm step. Measured spectra were processed using OriginPro 2016 software (OriginLab Corporation) under the academic license.

3. RESULTS AND DISCUSSION

The measured pH values (within ±0.5) and conductivity (2% accuracy) for WG powders (several minutes sedimentation after mixing) are summarized in **Table 1**. As it can be seen, adding WG powder to deionized water (H_2O) resulted in an increase in pH and conductivity which indicates on induced exchange of sodium (mainly) and hydrogen ions in water [6, 9, 14]. A slight decrease of these values was observed after 24 h sedimentation, which continues with further washing. Nevertheless, the values remain high even after 5th water washing that indicates continued sodium and hydrogen ions exchange. Next, both alkaline solutions have comparable high pH values that are not changes after WG addition and even increase after 24 h sedimentation. With water washing pH and conductivity of solutions decreases till reach comparable values with water-treated WG. The 0.25 M HCl solution demonstrates low pH typical for acidic medium that again not changed after adding the WG powder or 24 h sedimentation. The following water washing purified the powder from acid rests, whilst the "final" pH value became as high as 8.4. The conductivity measurements of HCl treated WG after purification was also lower than for other solutions that probably can be related to decreased reactivity of HCl treated WG comparing to other (lower alkali-hydrogen ions exchange).

Table 1 Measured pH and conductivity of treated WG powders

Liquid	H_2O		NaOH		KOH		HCl	
Quantity	**pH**	**σ (μS/cm)**	**pH**	**σ (μS/cm)**	**pH**	**σ (μS/cm)**	**pH**	**σ (μS/cm)**
Initial	5.3	4	13.3	165	13.4	160	1.6	155
WG add	10.5	155	13.3	165	13.4	160	1.6	155
1d stay	10.1	145	14.4	>200	14.4	>200	1.6	>200
1st wash	9.8	115	13.2	145	13.0	145	1.8	130
5th wash	9.2	80	9.8	85	9.6	80	8.4	70

In our earlier studies [12], we demonstrate that used untreated WG powder consist of particles with large size distribution (mean values 36.0 ±0.2 μm) with its large (up to 50 μm), irregular pieces with sharp edges decorated by the small fragments (tens of nanometers). Therefore, in this study the shown SEM images (**Figure 1**) were taken with high magnification and are focused on the analysis of WG surface morphology changes. The detailed SEM of untreated WG powder (**Figure 1a**) agrees with previous observation [12] and demonstrates flat surfaces of large particles with "scratches" or cracks related to grinding process as well as smaller WG fragments on surfaces.

The performed EDX analysis of powders composition reveals expected high amount of Si and O elements (two of the strongest signal). Also, strong signal from Na and presence of Ca was observed, but also presence of Mg, Al, C and K were detected. However, since EDX depth resolution is in order of a few μm and the intensity of response signal from elements is energy sensitive and requires precise quantitative correction for estimation of samples composition, the observed data were taken as orientation for more surface precise analysis and here shown for the untreated sample only (**Figure 1b**). It must be added that presence of carbon has been directly or indirectly observed earlier [12, 15] and was related to WG surface contamination by organic residues which is not distinguishable by SEM.

Similar particles morphology (with "scratches" and cracks on large particles) as for untreated WG powder were found on water treated powder (**Figure 1c**). In the case of NaOH and KOH treated WG particles the high magnification SEM images (**Figures 1d, 1e**) demonstrate that the surface of particles is not smooth but slightly eroded (small cavities formed). These observations are in agreement with other works where alkaline solutions of high molarity decolorize or etch glass surfaces [14, 16]. This indicates the dissolving of WG matrix even in mild alkaline medium and its potential participation in calcium-silicate-hydrate formation in the concrete-based materials. The last used solution was 0.25 M HCl with no evident changes of WG particles morphology comparing to untreated samples (**Figure 1f**).

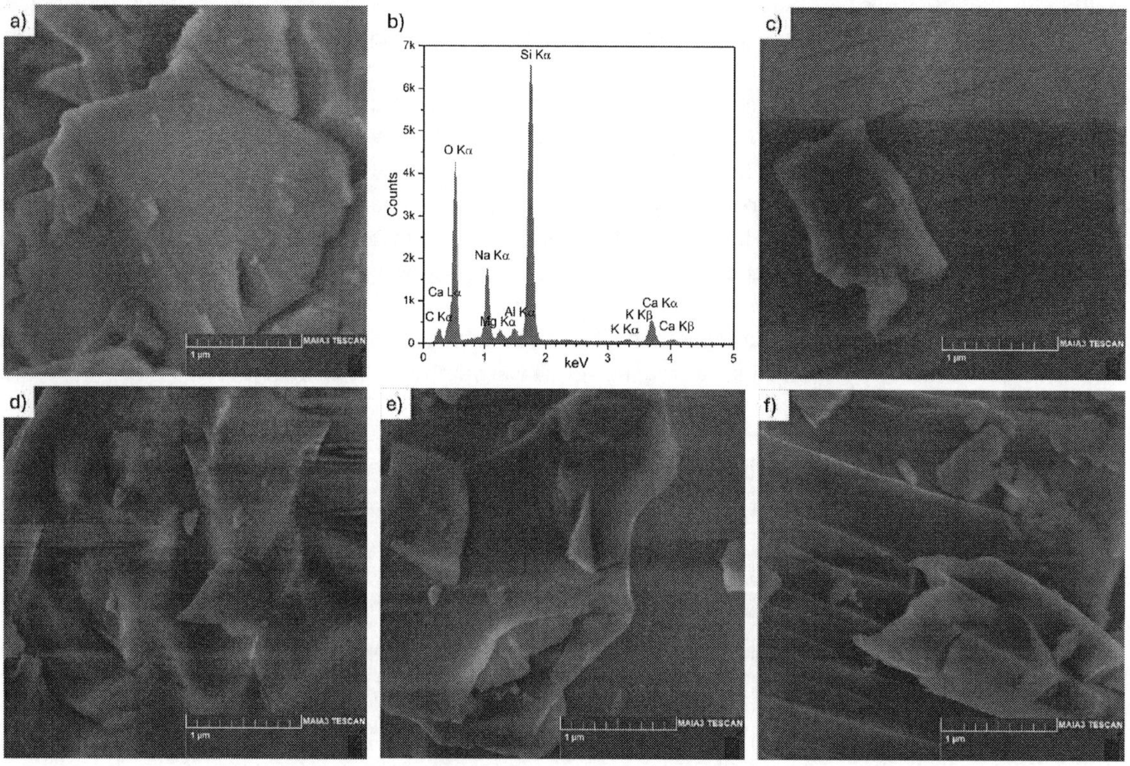

Figure 1 SEM image (a) and EDX spectrum (b) of untreated WG powder, and SEM images of WG powder treated in DI water (c), NaOH solution (d), KOH solution (e) and HCl solution (f).

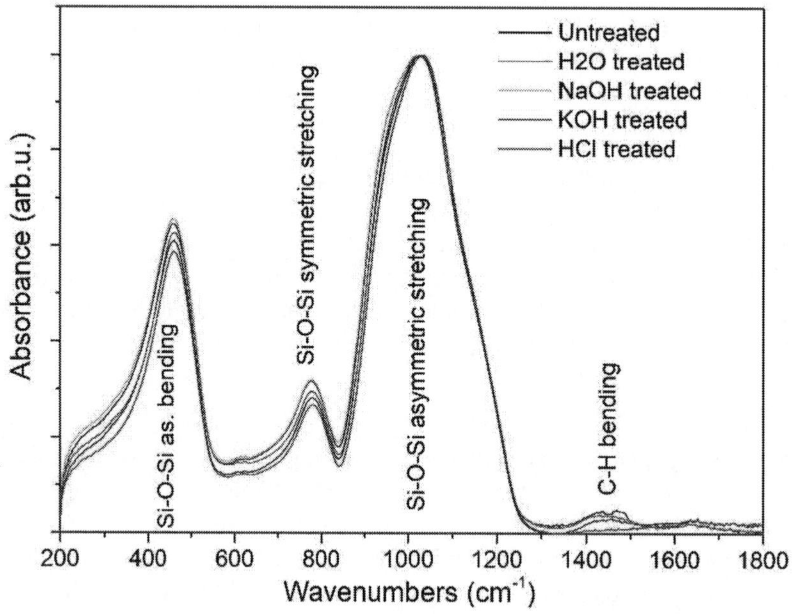

Figure 2 Infrared absorbance spectra of investigated dry WG powders

The normalized ATR FTIR spectra of analyzed powders are shown in **Figure 2**. As it can be seen, the measured IR absorbance spectra for untreated and wet chemistry treated WG powders are similar. The broad bands found in the spectra were attributed to Si-O-Si (in some cases noted just as Si-O) vibrations: asymmetrical bending at 463 cm^{-1}, symmetrical stretching at 788 cm^{-1}, and asymmetrical stretching at 1082 cm^{-1} characteristic for amorphous (non-crystalline) SiO_2 [5, 6]. The weak band at 1450 cm^{-1} was attributed to C-H bending vibration that for HCl treated WG significantly decreases. This indicates possible removing of organic contaminates [15] from the WG particles surface.

The XPS analysis of WG powders was performed to investigate changes in WG subsurface composition and clarify the observations from other analytic methods. The calculated atomic concentration of chemical elements for studies samples is summarized in **Table 2**. Due to surface sensitivity of XPS method the highest elements concentration was observed for oxygen and carbon while all other elements (including Si) have lower signal. The treatment in DI water (H_2O) or HCl resulted in a decrease of the carbon on WG surface (which corresponds to FTIR spectra shown in **Figure 2**) and increase of signal from oxygen. Contrariwise, treatment in KOH resulted in a higher percentage of carbon (probably adsorbed CO_2 from air) and lower of oxygen in the total composition. The estimation of carbon layer thickness (d carbon in **Table 2**) as well as measured results correction can be performed according [13]. The calculations show that the estimated carbon layer thickness varies from 2.4 nm for KOH treated WG to 1.55 nm for HCl treated WG. As it was shown earlier, such layer can be efficiently detected by photoluminescence [15].

The significant change of chemical element amount after treatment was found for sodium that seems to be intensively washed out from the WG matrix (especially for HCl treated powder). These data correspond to pH and conductivity measurements that indicate a decrease of chemical reactivity of WG with treatment (especially in HCl case) and water washing (**Table 1**). The amount of Mg or Al in subsurface region slightly increases after DI water or alkaline solution treatment. Interestingly, that after NaOH treatment the amount of Ca on surface also slightly increases comparing to other treatment media, which can be beneficial for application in sustainable concrete-based constructions. Finally, it is important to note that potassium in the WG subsurface composition was below the detection limit of the spectrometer even after KOH treatment.

Table 2 Atomic concentration of chemical elements derived from XPS and estimated carbon layer thickness.

WG treatment	Mg (at%)	Na (at%)	O (at%)	Ca (at%)	C (at%)	Si (at%)	Al (at%)	d carbon (nm)
Untreated	0.2	5.8	36.8	1.4	43.5	12	0.3	2.00
H_2O	1	3.7	40.3	1.6	37.9	14.7	0.8	1.67
NaOH	0.7	3.6	36.4	2.4	44.2	12.2	0.4	2.04
KOH	0.6	3	32.8	1.6	49.7	11.6	0.6	2.41
HCl	0.1	2.5	42.9	0.8	35.8	17.6	0.2	1.55

Concerning the photoluminescence mapping, the lines related to first, second and third-order scattering maxima were clearly observed in the measured maps (**Figure 3**). The broadening of first order scattering line was observed in the region from 320 to 560nm excitation for wet chemistry treated samples. This broadening was stronger for NaOH and HCl pre-treated powder and probably should be related to eroded WG surface in the NaOH case (**Figure 1d**) and surface cleaned from carbon in the HCl case. The emission peaks around 414 nm at λ_{ex} =470 nm and 426 nm at λ_{ex} = 370 nm were attributed to carbon layer of organic contaminants on the WG particles surface calculated from XPS measurement (**Table 2**) which is in agreement with [15].

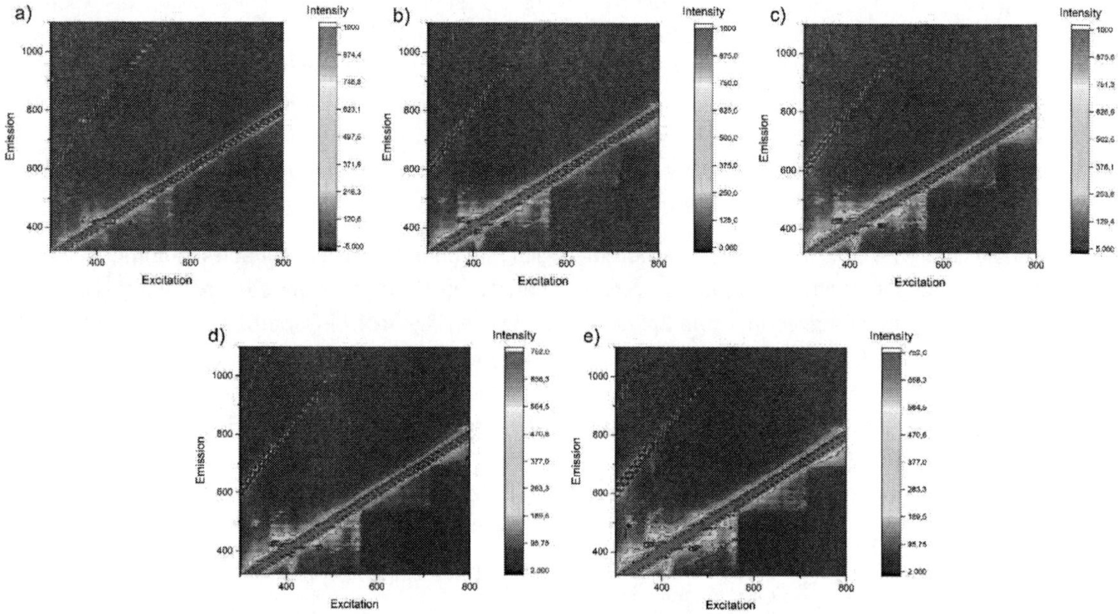

Figure 3 The excitation-emission photoluminescence maps of investigated WG powders: untreated (a), treated in DI water (b), NaOH solution (c), KOH solution (d) and HCl solution (e)

4. CONCLUSION

Commercially available ground waste soda-lime glass powder was treated in different solutions. We observe that treatment in NaOH and KOH resulted in slight erosion of used non-crystalline WG particles due to surface etching that also affects the light scattering. Next, all treatment solutions resulted in a decrease of sodium amount in the WG subsurface region and thus decrease of sodium and hydrogen ions exchange in solutions. Treatment in DI water or 0.25 M HCl resulted in a decrease of organic residues (carbon) contamination of the WG surface. Moreover, treatment in HCl resulted in a decrease of all observed non-silica ions in the WG subsurface region reducing its reactivity while morphology of the WG particles is not changed. Treatment in KOH resulted in a higher amount of carbon on WG particles while no increase in potassium amount on the surface was observed. Treatment in NaOH, besides WG surface etching, resulted in a slight increase in amount of Ca in the subsurface region which potentially can positively affect the calcium-silicate-hydrate formation in concrete-based constructions. These findings create a good basis for investigation of treated WG particles interaction processes in cement-based mixtures used in civil engineering.

ACKNOWLEDGEMENTS

The work was supported by the Czech Science Foundation project no 23-05500S. This work has used large research infrastructure CzechNanolab supported by the Ministry of Education, Youth and Sports of the Czech Republic through the project no. LM2023051.

The assistance of J. Libertínová with the SEM imaging is appreciated.

REFERENCES

[1] ADEKOMAYA, Oludaisi a Thokozani MAJOZI. Mitigating environmental impact of waste glass materials: review of the existing reclamation options and future outlook. *Environmental Science and Pollution Research* [online]. 2021, vol. 28, no. 9, pp. 10488–10502. ISSN 0944-1344, 1614-7499. Available from: https://doi.org/ 10.1007/s11356-020-12263-0

[2] BERGER, Gregor, Zlatko RAONIC, Daniel FORTHUBER, Harald RAUPENSTRAUCH a Robert HERMANN. Energy and Material Flow Evaluation with CO2 Emissions in the Glass Production Process. *Advances in Materials Physics and Chemistry* [online]. 2022, vol. 12, no. 05, pp. 82–105. ISSN 2162-531X, 2162-5328. Available from: https://doi.org/10.4236/ampc.2022.125007

[3] BUTLER, John H. a Paul D. HOOPER. Glass Waste. In: *Waste* [online]. B.m.: Elsevier, 2019 [vid. 2023-08-18], pp. 307–322. ISBN 978-0-12-815060-3. Available from: https://doi.org/10.1016/B978-0-12-815060-3.00015-3

[4] ZANINI, Roberta, Giulia FRANCESCHIN, Elti CATTARUZZA a Arianna TRAVIGLIA. A review of glass corrosion: the unique contribution of studying ancient glass to validate glass alteration models. *npj Materials Degradation* [online]. 2023, vol. 7, no. 1, 38. ISSN 2397-2106. Available from: https://doi.org/10.1038/s41529-023-00355-4

[5] KIM, Jong Min, Sang Mok CHANG, Sung Min KONG, Kyo-Seon KIM, Jinsoo KIM a Woo-Sik KIM. Control of hydroxyl group content in silica particle synthesized by the sol-precipitation process. *Ceramics International* [online]. 2009, vol. 35, no. 3, pp. 1015–1019. ISSN 02728842. Available from: https://doi.org/10.1016/j.ceramint.2008.04.011

[6] TORRES-CARRASCO, M., J. G. PALOMO a F. PUERTAS. Sodium silicate solutions from dissolution of glasswastes. Statistical analysis. *Materiales de Construcción* [online]. 2014, vol. 64, no. 314, e014. ISSN 1988-3226, 0465-2746. Available from: https://doi.org/10.3989/mc.2014.05213

[7] TSAI, C.L., U. KROGMANN a P.F. STROM. Handling leachate from glass cullet stockpiles. *Waste Management* [online]. 2009, vol. 29, no. 4, pp. 1296–1305. ISSN 0956053X. Available from: https://doi.org/10.1016/j.wasman.2008.10.013

[8] IMTEAZ, Monzur Alam, Mm Younus ALI a Arul ARULRAJAH. Possible environmental impacts of recycled glass used as a pavement base material. *Waste Management & Research: The Journal for a Sustainable Circular Economy* [online]. 2012, vol. 30, no. 9, pp. 917–921. ISSN 0734-242X, 1096-3669. Available from: https://doi.org/10.1177/0734242X12448512

[9] BUNKER, B.C., G.W. ARNOLD, D.E. DAY a P.J. BRAY. The effect of molecular structure on borosilicate glass leaching. *Journal of Non-Crystalline Solids* [online]. 1986, vol. 87, no. 1–2, pp. 226–253. ISSN 00223093. Available from: https://doi.org/10.1016/S0022-3093(86)80080-1

[10] TOGNONVI, Monique Tohoue, Arezki TAGNIT-HAMOU, Léon Koffi KONAN, Ablam ZIDOL a Wilfried Cyrille N'CHO. Reactivity of Recycled Glass Powder in a Cementitious Medium. *New Journal of Glass and Ceramics* [online]. 2020, vol. 10, no. 03, pp. 29–44. ISSN 2161-7554, 2161-7562. Available from: https://doi.org/10.4236/njgc.2020.103003

[11] MARIAKOVÁ, Diana, Klára Anna MOCOVÁ, Kristina FOŘTOVÁ, Tereza PAVLŮ a Petr HÁJEK. Waste Glass Powder Reusability in High-Performance Concrete: Leaching Behavior and Ecotoxicity. *Materials* [online]. 2021, vol. 14, no. 16, 4476. ISSN 1996-1944. Available from: https://doi.org/10.3390/ma14164476

[12] BABČENKO, O, Z REMEŠ, K BERANOVÁ, K KOLÁŘOVÁ, J ČERMÁK, A KROMKA, Z PROŠEK a P TESÁREK. Characterization of different types of silica-based materials. *Journal of Physics: Conference Series* [online]. 2024, vol. 2712, no. 1, 012010. ISSN 1742-6588, 1742-6596. Available from: https://doi.org/10.1088/1742-6596/2712/1/012010

[13] SMITH, Graham C. Evaluation of a simple correction for the hydrocarbon contamination layer in quantitative surface analysis by XPS. *Journal of Electron Spectroscopy and Related Phenomena* [online]. 2005, vol. 148, no. 1, pp. 21–28. ISSN 03682048. Available from: https://doi.org/10.1016/j.elspec.2005.02.004

[14] MARAGHECHI, Hamed, Farshad RAJABIPOUR, Carlo G. PANTANO a William D. BURGOS. Effect of calcium on dissolution and precipitation reactions of amorphous silica at high alkalinity. *Cement and Concrete Research* [online]. 2016, vol. 87, 1–13. ISSN 00088846. Available from: https://doi.org/10.1016/j.cemconres.2016.05.004

[15] REMEŠ, Zdeněk, Oleg BABČENKO, Vítězslav JARÝ a Klára BERANOVÁ. Enhanced Photoluminescence of Plasma-Treated Recycled Glass Particles. *Nanomaterials* [online]. 2024, vol. 14, no. 13, 1091. ISSN 2079-4991. Available from: https://doi.org/10.3390/nano14131091

[16] SUN, Lianfang, Xingji ZHU, Minjae KIM a Goangseup ZI. Alkali-silica reaction and strength of concrete with pretreated glass particles as fine aggregates. *Construction and Building Materials* [online]. 2021, vol. 271, 121809. ISSN 09500618. Available from: https://doi.org/10.1016/j.conbuildmat.2020.121809

ISSUES IN MACHINING EPOXY RESIN - NANO REINFORCEMENT COMPOSITE SYSTEMS WITH RESPECT TO SURFACE INTEGRITY AND TOOL WEAR

Artur KNAP

TUL – Technical University of Liberec, Liberec, Czech Republic, EU, artur.knap@tul.cz

https://doi.org/10.37904/nanocon.2024.5008

Abstract

Polymer nanocomposites, reinforced with nanoparticles such as SiO_2, are gaining attention due to their improved mechanical properties. However, the impact of SiO_2 nanoparticle content on machining processes, particularly tool wear and surface quality, remains insufficiently explored. This study investigates the effect of different SiO_2 concentrations (0–8 wt.%) on tool wear (V_B), surface roughness (R_a), and chip formation during the micromachining of epoxy-based composites. Samples were fabricated with varying SiO_2 content and machined using a micro-milling process. The tool wear and surface roughness were evaluated at regular intervals, with the wear predominantly occurring within the first few minutes of cutting. Results show that as the SiO_2 content increases, tool wear accelerates, leading to higher surface roughness. In particular, the addition of SiO_2 contributed to the formation of non-dispersed clusters, which exacerbated tool wear. The chip morphology shifted from worm-like to crumbly as tool wear increased, reflecting the progressive degradation of the machining process. These findings suggest that while higher nanofiller content may improve material strength, it significantly impacts the machining efficiency and surface quality, particularly when filler dispersion is poor. The study highlights the importance of optimizing the dispersion of nanofillers to maintain machining stability and surface integrity. This research provides insights into the micromachining behavior of SiO_2-reinforced nanocomposites and serves as a foundation for further investigations into the machining of advanced composite materials.

Keywords: Nanocomposites, nanoreinforcement, micromilling, tool wear, surface roughness

1. INTRODUCTION

Polymer systems are widely used due to their unique properties, such as ease of manufacture, light weight, high structural strength and more. They are used in a wide range of applications – from the aerospace and automotive industries to objects of everyday use. However, polymers have a lower modulus of elasticity and strength compared to metal and ceramic materials. One way to improve their mechanical properties is to reinforce polymers with inclusions (fibers, platelets or particles). Currently, developments are underway in the use of nano-reinforcements and the production of nanocomposites. Experimental investigations point to the occurrence of new phenomena capable of noticeably influencing the properties of the resulting composite - an increase in strength when using relatively small amounts of reinforcement [1, 2]. Also, reducing the size of these components to the nanometer level causes an improvement in the mechanical and physical properties of the composite due to the presence of a large number of nanoparticles (large specific surface area of the reinforcement) [3]. One of the most desirable nanoparticles is SiO_2, which is inexpensive, non-toxic, biocompatible, highly thermally and mechanically stable, and particularly effective as a mechanical reinforcement. All these aspects make it an ideal choice for filler in composites. The size of spherical SiO_2 nanoparticles alone is around 20 nm. The disadvantage of this filler is the high hydrophilicity of the SiO_2 surface, which causes the aggregation of nanoparticles that are then difficult to disperse in the polymer matrix [4, 5, 6, 7].

With the development of polymer nanocomposites and their applications in various industries, it was necessary to be able to tailor the nanocomposites themselves to the design specifications. Regardless of where these products are ultimately used, some of the finishing machining operations associated with the growing production of composite parts cannot be dispensed with. However, unlike metal parts, composite systems behave differently in machining due to their material structure [8, 9]. These properties mainly affect tool wear, which has an impact on the quality of the machining process and the resulting quality (roughness) of the machined surface [10]. The machining of composite systems is now well described. Nevertheless, new studies are still appearing that address the issues of tool wear, optimization of machining conditions, delamination and the character of the workpiece surface. A noticeable drawback of the application of nano-strengthening is the still insufficient understanding of the synergy of the individual phases [11].

2. MATERIALS AND METHODS

As a matrix to produce composite samples, the two-component low molecular weight epoxy resin ChS-EPOXY 520 was chosen. To cure the resin, P11 hardener was used, added in a weight ratio of 100:11. To produce the samples, a discontinuous process of casting the material mixtures into a pre-prepared mould was chosen, which was made of two-component silicone rubber Silastic T1. Silicone rubber was chosen for the mould production due to its ease of production and easy handling. The mould did not need to be separated, the samples could be easily removed, and the mould did not require complicated cleaning. The dimensions of the samples were chosen according to the requirements for the subsequent micromachining operations. The filler used was the pyrogenic amorphous silica LM- 150 Cab-O-Sil. The material is amorphous, highly pure, with a large specific surface area (150 m^2/g). It was added to the epoxy resin in proportions of 0, 2, 4, 6, 8 wt.%. A hardener is then added. The mixture thus prepared is mixed well and poured into the mould. The mould is placed in a pressure vessel and the mixture is cured for 24 hours at a laboratory temperature of 22 °C and a pressure of 0,7 MPa. To achieve the final values of the mechanical parameters it is desirable to cure at 50 °C for 10 hours. The basic sample dimensions were 30x10x10 mm.

The DMG MORI CMX 600 V machine was used to machine the samples. Before starting the machining process, the samples were clamped longitudinally to the direction of the tool path in a clamp and angled. The cornering was followed by end-plane micro-milling, which was carried out according to the defined cutting conditions, see below. The control criterion for evaluating the amount of wear incurred was a tool time interval of 5 minutes in the cut corresponding to a machined area of 300 mm^2.

The tool used for the experiment was 2ALE-005-005-S04, a double-edged cylindrical sharp cutter with a diameter of 0.5 mm and a length of 40 mm suitable for machining soft plastics and composite materials (**Figure 1**). The machining experiments were carried out using cutting conditions as recommended by the cutting tool manufacturer. The main parameters were: spindle speed n=12 000 [$rev.min^{-1}$], feed per tooth f_z= 0.01 [mm], depth of cut a_p=0.033 [mm], width of cut a_e=0.20 [mm].

a) b)
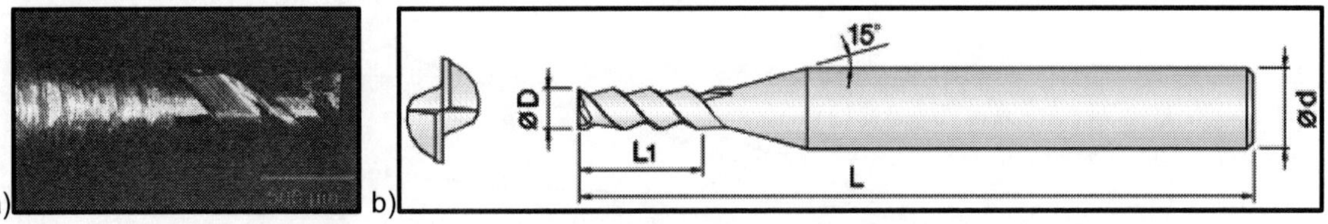

Figure 1 Milling cutter 2ALE-005-005-S04. (**a**) Detail of cutting geometry. (**b**) Scheme of basic parameters of the cutter

For the evaluation of the monitored parameters - tool wear (V_B), surface roughness (R_a), surface integrity (burrs, cavities, etc.) and the nature of the resulting chips, a confocal microscope Keyence VK-X1100 was used using a set of objectives with a zoom range of 5x to 100x. The controlled parameters were monitored after 5 minutes of the instrument in the cut. A total of 6 comprehensive measurements (30 minutes of instrument in section) were performed. In the course of evaluation of the achieved quality of the machined surface, the occurrence of cavities, unmachined areas, filler clusters and other phenomena associated with machining of nanocomposite materials based on particle reinforcement was observed. A summary of the occurrence of cavities, clumps and others is shown within the sample subsections along with the final surface at the end of the experiment and the chips taken.

3. RESULTS

The sample without the addition of reinforcing nanoparticles, Sample 1, acts as a basic reference for the observation of the observed parameters. It is therefore a pure epoxy resin made according to the methodology presented in Chapter 2. **Figure 2** shows the processed values of the observed tool wear parameter V_B for machining samples with 0, 2, 4, 6 and 8 wt.% SiO_2. Similar is the case for **Figure 3** for the parameter R_a. The individual plots will be described within the sub-sections corresponding to the sample under examination.

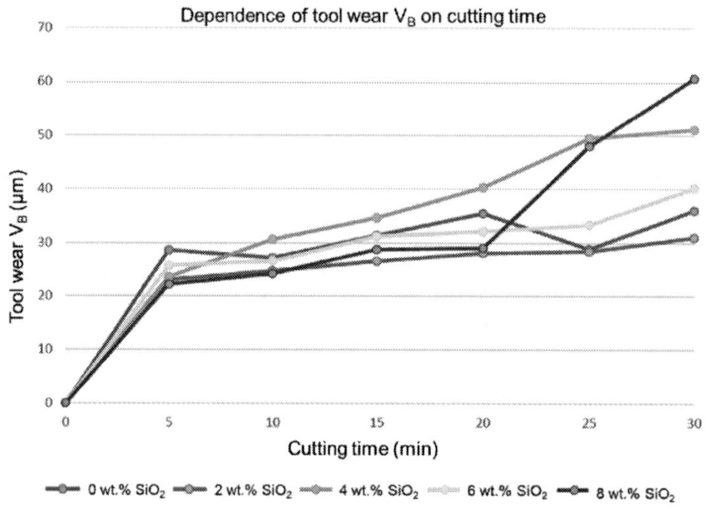

Figure 2 Summary graph of achieved V_B parameter values for individual samples

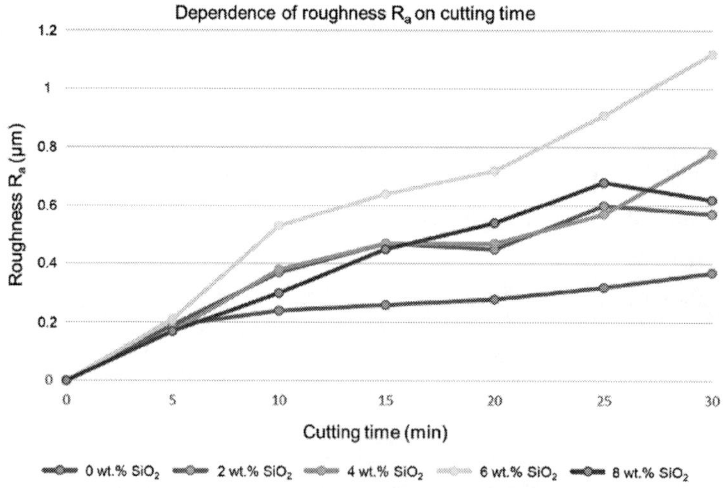

Figure 3 Summary graph of achieved R_a parameter values by each sample

3.1 Sample 1 – 0 wt.%

In **Figure 2** the wear development of the tool can be observed. When the sample was machined without the addition of nanofiller, almost 80% of the total wear occurred within the first 5 minutes of milling. The wear trend was of increasing character, except at the points corresponding to 10 and 25 minutes in the cut. These decreases were due to the difficulty in positioning the microtool to the initial measuring positions and the presence of post-machining debris (chip clusters) that can visually act as cutting tool material. The dependence of roughness on tool time in cut for Sample 1 (**Figure 2**) shows a slight increasing trend along with increasing magnitude of V_B wear. This can be explained by the area of the sample in which the material may have been ploughed resulting in an increase in the surface roughness, by the deposition of a larger amount of very fine particles of chips or by the greater presence of cavities in the sample. The machining cycles revealed the presence of cavities in the sample (**Figure 4a**), that were created during the production of the blanks for the experimental part, e.g. by insufficient venting during casting. These cavities may increase tool wear due to the interruption of smooth tool engagement. Furthermore, a material surface can be observed where, due to tool wear, ploughing of the material has started to occur and the chip thus separated has then attached to the machined surface (**Figure 4b**). The machining of the specimen, which contains only epoxy resin, produced worm-like chips (**Figure 4c**). The same chip morphology was present during the last machining cycle.

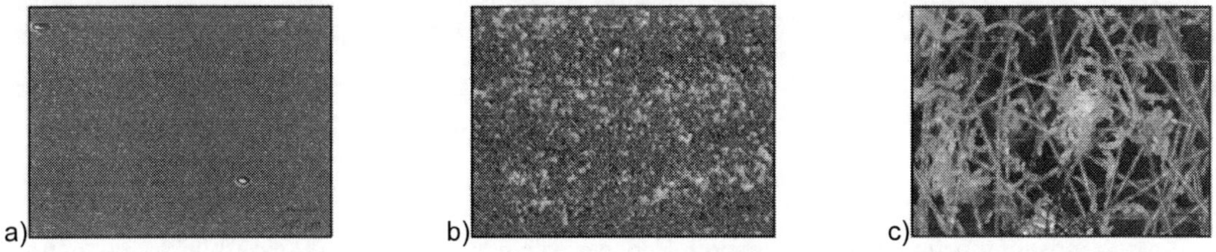

Figure 4 Overall overview of Sample 1. (**a**) Occurrence of cavities on the surface.; (**b**) Surface after 30 minutes of the tool in the cut.; (**c**) Nature of the chips formed – white areas.

3.2 Sample 2 – 2 wt.%

As with Sample 1, the greatest wear occurred during the first machining cycle, with approximately 74 % of the total wear (**Figure 2**). The trend of the curve is increasing and the differences between the measured sections are almost constant. Contrary to expectations, the total wear is approximately 5 μm less than the tool that milled the sample without nano-reinforcement. A regular increase in R_a values can be observed in the roughness plot (**Figure 3**). However, it is clear from the roughness measurements that the resulting surface of Sample 2 has lower R_a values than Sample 1. When machining the samples with SiO_2 addition, it was possible to observe clusters of undispersed nanofiller (white areas) in addition to cavities (black areas) (**Figure 5a**). The quality of the machined surface and the recognizability of the toolpaths also rapidly decrease due to the increasing magnitude of tool wear (**Figure 5b**). After adding 2 wt.% SiO_2, the character of the resulting chips did not change much. Their shape is still worm-like. After the last machining cycle, i.e. 30 min, there was a relatively large change in the nature of the chips. Due to tool wear, the shape of the chips changed from worm-like to a crumble shape, which forms accumulations of chips (**Figure 5c**).

Figure 5 Overall overview of Sample 2. (**a**) Occurrence of cavities and clusters of undispersed nanofiller on the surface.; (**b**) Surface in the end of machining.; (**c**) Changed nature of the chip's formation.

3.3 Sample 3 – 4 wt.%

During the machining of Sample 3, the tool also experienced the steepest increase in wear magnitude during the first milling cycle. Further, the rate of wear increase slowed down over the machining cycles and a steady increase in tool wear continued. A very slight increase in R_a parameters was observed in the machining of this sample. Specimen 3 was made without observable cavities, nanofiller clusters, or other undesirable structures that would contribute excessively to tool wear. In **Figures 6a** and **6b**, the difference between the visibility of the toolpaths after the first and last machining cycle can be observed. Due to the addition of 4 wt.% SiO_2, the resulting chips are regular flake shaped. These chips are close to 200 µm in size. As with the previous sample, the chips were strongly influenced by the gradual increase in tool wear and their character gradually changed from flake-like (**Figure 6c**) to crumbly again.

a) b) c)

Figure 6 Overall overview of Sample 3. (**a**) Surface of Sample 3 after 5 minutes of milling.; (**b**) Surface of Sample 3 after 30 minutes of milling.; (**c**) Flake character of chips of Sample 3.

3.4 Sample 4 – 6 wt.%

For Sample 4, most of the wear can be observed to occur within the first 5 minutes of machining. In the case of this sample with 6 wt.% SiO_2, this represents the formation of approximately 64 % of the final measured V_B size. The rest of the graph shows a slightly increasing trend (**Figure 2**). For this sample, it is possible to observe regularly increasing values of roughness R_a (**Figure 3**). In **Figure 7a,** many white regions representing non-dispersed clusters of nanofiller revealed by the 4th machining cycle can be observed. The size of said clumps relative to the toolpaths can be observed in **Figure 7b**. These randomly distributed clusters contribute significantly to the increase in the magnitude of the V_B wear and the magnitude of the roughness parameters. As the amount of nanofiller increases in the sample structure, it is evident that the filler contributes significantly to the friability of the chip. In **Figure 7c**, it can be observed that even though the chips are flake shaped, there are areas around and between the chips that are crumbly in nature. As the tool wears, the chips change from a flake shape to an almost completely friable shape, forming very small particles.

a) b) c)

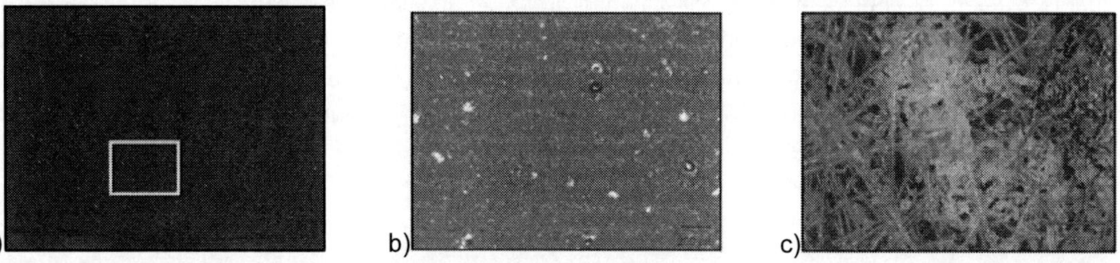

Figure 7 Overall overview of Sample 4. (**a**) Macro view at machined surface-yellow rectangle represents area of interest.; (**b**) Detail view at the surface structure with clusters.; (**c**) Flake shaped with higher number of areas with crumbly shaped nanocomposite chips.

3.5 Sample 5 – 8 wt.%

The tool wear during machining of sample 5 shows different characteristics. During the first cycle there is more wear, as with the other samples, then this trend stabilizes after the next cycle. This is followed by a very gradual increase in the magnitude of V_B until the 4th cycle. During the last two cycles, however, the wear increases rapidly to almost three times the measured value after the first five minutes of machining. This sample shows an atypical trend in the dependence of the roughness parameters. The parameter R_a does not show the expected gradual increase. The increase of the observed roughness parameter shows a linear character until the sudden decrease caused by possible surface homogenisation after the removal of surface defects (**Figure 3**). Cavities of considerable size appear on the surface (**Figure 8a)**, often extending across the entire width of the tool. These cavities interrupt the smoothness of the tool stroke and cause unwanted additional stress on the tool. Furthermore, it is possible to observe the difference in surface quality achieved by the experiment (**Figure 8a** and **8b**). Due to the greater amount of impurity in the sample, there was more tool wear, which is evident in the comparison between the 1st and 6th milling cycle (**Figure 2**). At the end of machining, toolpaths are not discernible on the sample surface (**Figure 8b**). Here, already after the first milling cycle, a relatively large representation of crumbly chips can be observed, which in the previous samples was only visible at the end of the experiment. The chips after the last machining cycle show primarily a crumbly chip character with very fine particles (**Figure 8c**).

Figure 8 Overall overview of Sample 5. (**a**) Surface structure with cavities across toolpaths.; (**b**) Quality of finished surface after 30 minutes of milling.; (**c**) Crumbly chips on the fibers of the cloth used for the chips collection.

4 CONCLUSION

The trend shows that a higher nanofiller content in the sample structure is associated with a deterioration in surface roughness and associated surface quality. The abrasive nature of the nanofiller increases tool wear and this has an immediate effect on the chip removal mechanism, which in turn affects the quality of the machined surface. This finding must be supplemented by another input factor - the degree of mixing between the nano-reinforcement and the support element, which distorts the assumption of higher wt.% reinforcement = higher tool wear rate. The occurring reinforcement clusters contribute significantly to the influence of tool cut stability and the mechanisms associated with chip formation and resulting surface roughness. The progressive tool wear was already visible after the second machining cycle, where the traces of the chip separation mechanism (visible toolpaths) disappeared. Due to further tool wear, by the end of the experiment, there was almost no chip shearing anymore, but rather ploughing, resulting in a further deterioration of the surface quality. In all the samples examined, the chips were worm-like or flake-shaped at first, but when the chips formed during the last machining cycle were examined, their character always tended towards a crumbly structure with particles smaller than those at the beginning of the experiment.

This study advances knowledge in the field of micromachining of nanocomposite materials. The reason for this study was to assess the effect of wt.% SiO_2 on the magnitude of V_B tool wear, R_a roughness and the resulting surface quality. The degree and quality of dispersion of the nanofiller in the matrix appears to be a critical factor for process stability. This is an initial study that will be a starting data set for further research and to expand the knowledge in this area of research.

ACKNOWLEDGEMENTS

This research study was funded by the Student Grant Competition of the Technical University of Liberec under project number SGS-2023-5387. "Research in the field of micromachining - micromilling".

REFERENCES

[1] JORDAN, Jeffrey, Karl I JACOB, Rina TANNENBAUM, Mohammed A SHARAF a Iwona JASIUK. Experimental trends in polymer nanocomposites—a review. Materials science and engineering: A. 2005, vol. 393, no. 1–2, pp. 1–11.

[2] MALICK, P.K. *Fiber-Reinforced Composites: Materials, Manufacturing and Design*, 3rd ed.; CRC Press: Boca Raton, Florida, USA, 2007.

[3] OMANOVIĆ-MIKLIČANIN, Enisa, Almir BADNJEVIĆ, Anera KAZLAGIĆ a Muhamed HAJLOVAC. Nanocomposites: A brief review. *Health and Technology*. 2020, vol. 10, no. 1, pp. 51–59.

[4] LANDOWSKI, Michał, Gabriel STRUGAŁA, Michał BUDZIK a Krystyna IMIELIŃSKA. Impact damage in SiO2 nanoparticle enhanced epoxy–Carbon fibre composites. *Composites Part B: Engineering*. 2017, vol. 113, pp. 91–99.

[5] MALLAKPOUR, Shadpour a Mina NAGHDI. Polymer/SiO2 nanocomposites: Production and applications. *Progress in Materials Science*. 2018, vol. 97, pp. 409–447

[6] ZENERINO, Arnaud, Sonia AMIGONI, Elisabeth Taffin DE GIVENCHY, Denis JOSSE a Frédéric GUITTARD. Homogeneous dispersion of SiO2 nanoparticles in an hydrosoluble polymeric network. *Reactive and Functional Polymers*. 2013, vol. 73, no. 8.

[7] LIU, Chun-Hua a Cai-Yuan PAN. Grafting polystyrene onto silica nanoparticles via RAFT polymerization. *Polymer*. 2007, vol. 48, no. 13, pp. 3679–368

[8] UHLMANN, E.; SAMMLER, F.; RICHARZ, S.; HEITMÜLLER, F.; BILZ, M. Machining of carbon fibre reinforced plastics. *Procedia CIRP*. 2014, vol. 24, pp. 19-24.

[9] KARATAS, M.; GÖKKAYA, H. A review on machinability of carbon fiber reinforced polymer (CFRP) and glass fiber reinforced polymer (GFRP) composite materials. *Defence Technology*. 2018, vol. 14, pp. 318-326.

[10] THAKUR, RK a KK SINGH. Influence of fillers on polymeric composite during conventional machining processes: a review. *Journal of the Brazilian Society of Mechanical Sciences and Engineering*. 2021, vol. 43, no. 2, pp. 1–20.

[11] HANEMANN, Thomas a Dorothée Vinga SZABÓ. Polymer-nanoparticle composites: from synthesis to modern applications. *Materials*. 2010, vol. 3, no. 6, pp. 3468– 3517.

HUMIDITY AND TEMPERATURE INFLUENCE ON THE SELF-CLEANING PERFORMANCE OF BUILDING MATERIALS CONTAINING THE PHOTOCATALYSTS

Bohdan POLISHCHUK, Kryštof FONIOK, Vlastimil MATĚJKA

[1] *VSB - Technical University of Ostrava, Faculty of Materials Science and Technology, Ostrava, Czech Republic, EU, bohdan.polishchuk.st@vsb.cz, krystof.foniok@vsb.cz, vlastimil.matejka@vsb.cz*

https://doi.org/10.37904/nanocon.2024.5018

Abstract

Titanium dioxide is a well-known substance for its photocatalytic properties. The aim of this article is to describe the influence of humidity and temperature on the self-cleaning properties of construction surfaces covered with a layer of titanium dioxide (further labeled as TiO_2). Its bandgap allows for the use of its photocatalytic properties strictly when exposed to the UV irradiation. The physico-chemical characteristics of the used TiO_2 were investigated and described using analytical methods: x-ray diffraction spectroscopy (XRD), diffuse reflectance spectroscopy (DRS) and scanning electron microscopy (SEM). The experiments carried out aimed to identify the effect of humidity and temperature on the self-cleaning performance of TiO_2 coated surfaces. The effect of humidity was tested at 20, 40, 60 and 80 % of relative humidity in testing chambre, the temperature was kept constant at 20 °C. The effect of temperature was tested for the 10, 20 and 30 °C, and the relative humidity was kept constant at 60 %. It was observed that increasing of relative humidity always led to higher self-cleaning performance. With respect to the temperature effect, a more complicated relationship was observed between the temperature and self-cleaning performance. The lowest self-cleaning ability was observed for the samples tested at 20 °C. The highest self-cleaning effect was observed for samples tested at 10 °C; this effect was slightly lower for samples tested at 30 °C. The best self-cleaning performance was observed for the experimental conditions of 80% RH and 10°C. Relative humidity was identified as a key factor influencing the self-cleaning ability of TiO_2 coated cement pastes.

Keywords: Photocatalysis, titanium dioxide, cement pastes, self-cleaning properties

1. INTRODUCTION

Nowadays, TiO_2 is the most studied substance among all photocatalysts [1]. Its accessibility and chemical stability provide humanity with a wide range of application of TiO_2 in nanotechnology [2], medicine [3], wastewater treatment [4]. and manufacturing industries [5]. Over the last 30 years, the construction industry has also begun to show its interest in photocatalysts application in building materials. ISO UNI 22197:2007 and UNI 11247:2007 standards [6,7] were developed specially for the testing of the photodegradation properties of TiO_2 if used in building materials. Since a suitable form of TiO_2 is one of the most accessible photocatalysts that also show a self-cleaning effect, TiO_2 has quickly started to be used as a cement additive or building surface coating, which also reduces construction maintenance costs [8].

The most stable TiO_2 phase modifications are anatase and rutile with band gaps of 3.2 eV and 3.02 eV [1,5]. These energies correspond with wavelengths at 385 and 410 nm of light radiation, which means that pure photocatalytically active TiO_2 shows its self-cleaning capabilities exclusively when exposed to the UV irradiation without being doped with other additives [9]. Nevertheless, if solar light may provide enough UV radiation, photocatalysts still require water as a necessary resource for the photodegradation process. In that case, active centers of TiO_2 use humidity contained in the surrounding atmosphere to produce hydroxyl $\cdot OH$ and superoxide $\cdot O^2$ radicals, which are well-known for their high oxidizing properties [10]. It was figured out, that during different periods of a year, photocatalytic coatings show different photodegradation activities as well [11]. In our work, we focused on the study of the effect of relative humidity and temperature on the self-cleaning ability of surfaces coated with photocatalytically active TiO_2.

2. MATERIALS AND METHODS

2.1 Sample preparation

TiO_2 photocatalyst used for the experimental part is an AEROXIDE® P25 (further labeled as P25), containing its two crystal phases - anatase and rutile. In the first step, the suspension of P25 with exact concentration of 4 $g \cdot dm^{-3}$ was prepared. In order to eliminate inaccuracies during application process, the suspension was properly homogenized with ultrasonicator OmniSonic Ruptor 400.

In the second step, samples of cement pastes were prepared. In sum, 3 samples of cement paste (one reference sample, and two samples for coating with TiO_2) were prepared for testing at selected temperature and humidity. The cement paste was made with 150 g of cement (CEM I 42,5R) and 55 ml of water, the resulting mixture was then poured into three equal steel forms (100 x 50 x 10 mm) and put into the humidity box. After the 24 hours of hardening, fresh hardened samples were unmold and stored in a moist environment for one week. After that period, the samples were let to dry 24 hours in laboratory conditions with aim to eliminate the surface moisture as it can undesirably affect the application process of P25 suspension on the surface of pastes.

In the last step, the P25 suspension was evenly applied on the surface of samples with a lab sprayer. The surface of each sample was coated with three approximately equal layers, applied one by one after the previous one completely dries. Such prepared samples were stored in a dark at laboratory conditions for 24 hours before the initiation of self-cleaning test. The rest of the suspension volume prepared was not used.

2.2 Characterization methods

X-ray diffractometer Miniflex600 (Rigaku, Japan) was used for registration of diffraction patterns of P25. An X-ray lamp, which has a cobalt anode, was operated at 40 kV and 15 mA. Primary optics included a 2.5° divergence Soller aperture and a 1.25° divergence aperture. Secondary optics included a 2.5° Soller aperture and a K_β filter made of iron foil. D/tex Ultra 250 operated in 1D mode was used as a detector. The measurement was performed in the range of 10-100 °2 theta, with a step of 0.2 ° and at a speed of 3 °/min.

Scanning electron microscope Aspex Explorer (Thermofisher Scientific, USA) was used for SEM analysis of the surface of P25 particles. Cerium hexaboride used as a thermoemission source of electrons in this microscope operated at an accelerating voltage of 15 keV. Secondary and backscattered electrons were detected using SE and BSD detectors, the chemical composition was studied using an energy dispersive detector (EDS).

UV-VIS Diffuse reflectance spectroscopy was performed on spectrometer UV-2600 (Shimadzu, Japan). The instrument in the DRS configuration is equipped with an IRS-2600Plus integrating sphere. Registration of DRS spectra was performed in the range of 220 - 800 nm. The indirect bandgap energy of P25 sample was evaluated with Tauc's plot.

The self-cleaning test was conducted by the discoloration of rhodamine B applied on the surface of TiO_2 coated cement pastes. The modified procedure, based on standard UNI 11259, was used for this purpose. Our procedure included application of rhodamine B solution with concentration of 50 $g \cdot dm^{-3}$ on the sample surfaces using lab sprayer. The amount of RhB on each sample was calculated gravimetrically giving approximately 80 mg of P25 deposited on the surface of each sample. Such prepared samples were left 24 hours in Memmert ICH110 L climate chamber at conditions defined for given test (temperature and relative humidity). After this period, the own self-cleaning test was initiated by switching-on the UV light and lasted for 26 hours. To monitor the discoloration process (RhB degradation), L*a*b* coordinates were measured with MiniScan EZ spectrometer (HunterLab, USA) at the 0 h (before the irradiation was initiated), 4 h, and 26 h of 26 h of irradiation. Self-cleaning performance of TiO_2 coating at given combination of temperature and relative humidity was determined based on the change of the a* coordinate in a given time period (4, and 26 h) relative to its original value obtained before the irradiation was started (0 h).

3. RESULTS AND DISCUSSION

3.1 Characterization of P25

X-ray diffraction pattern of P25 photocatalyst is shown in **Figure 1** and confirms the presence of two TiO_2 modifications, anatase and rutile, respectively.

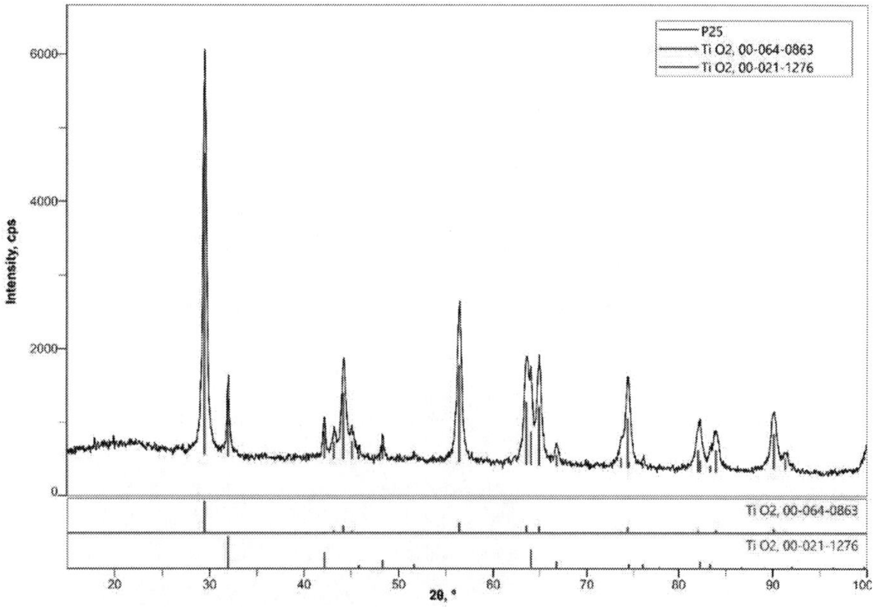

Figure 1 X-ray diffraction analysis of P25

Figure 2 describes the DRS spectrum of the P25 sample. **Figure 3** describes the Tauc's plot [12]. It was used to evaluate the optical bandgap of this sample, and the value of 3.3 eV was achieved. This value is closely similar to the value 3.2 eV typically reported for anatase and reflects the dominant contribution of anatase phase to Eg values. It should be noted that a slightly higher value of Eg than 3.2 eV was observed also by other researchers, for example, by Feng et al. [13].

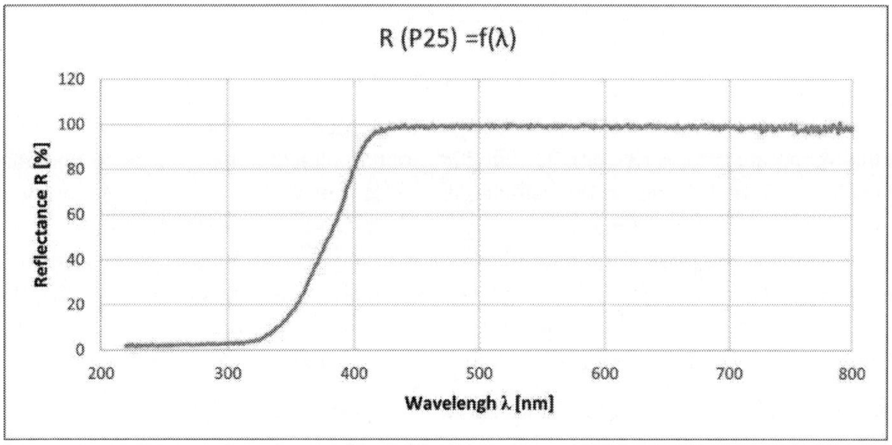

Figure 2 DRS spectra of P25 sample

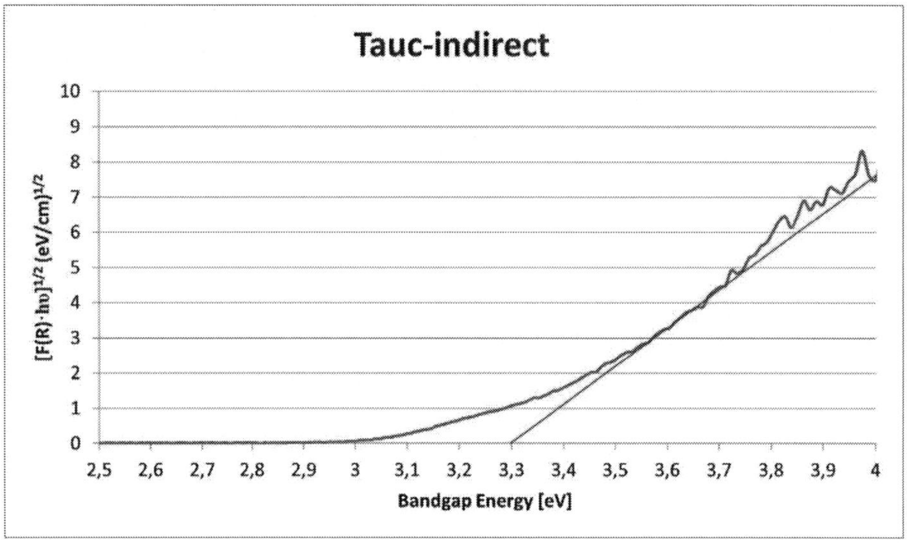

Figure 3 Tauc's indirect plot

Figure 4 shows the morphology of P25 observed with SEM technique and it is evident, that the individual TiO_2 nanoparticles form larger aggregates.

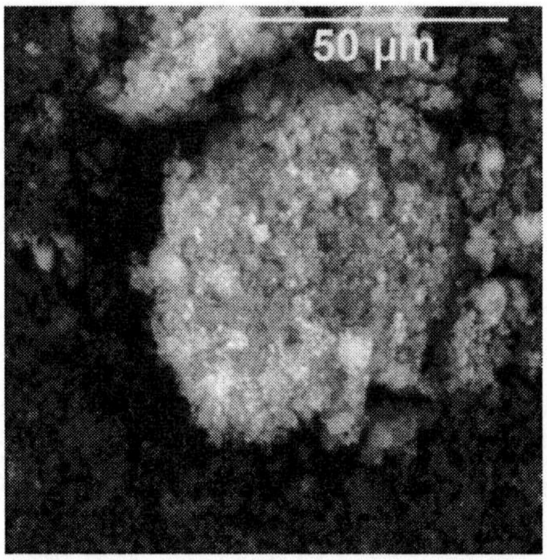

Figure 4 SEM image of P25 particle

3.2 Testing of the effect of humidity and temperature on self-cleaning performance

Figure 5 demonstrates the influence of humidity on the self-cleaning performance of TiO_2 coated cement pastes, while the temperature remained constant at 20 °C during all experiments. As can be seen, increasing the relative humidity led to an increase in the self-cleaning performance. Higher relative humidity is associated with a higher number of water molecules in the testing chambre. The self-cleaning performance of the surfaces of the samples is based on the photodegradation activity of TiO_2, which is related to the formation of hydroxyl and superoxide radicals from H_2O molecules. In general, the higher the humidity is, the higher the number of water molecules are available for the formation of hydroxyl and superoxide radicals. **Figure 5** shows that there is a linear relationship between photodegradation activity and humidity.

Figure 6 shows the impact of temperature on self-cleaning performance at 60% of relative humidity. The temperature relationship of self-cleaning performance is more complicated than was observed in the case of the relative humidity. The highest self-cleaning performance was observed at a temperature of 10 °C, while the minimum performance was observed for a temperature of 20 °C. This relation is attributed to the fact that with the growing temperature the number of water molecules absorbed on the surface is decreased because of their larger evaporation. On the other hand, a higher self-cleaning performance registered at 30 °C can be connected with the effect of temperature on the RhB degradation kinetics.

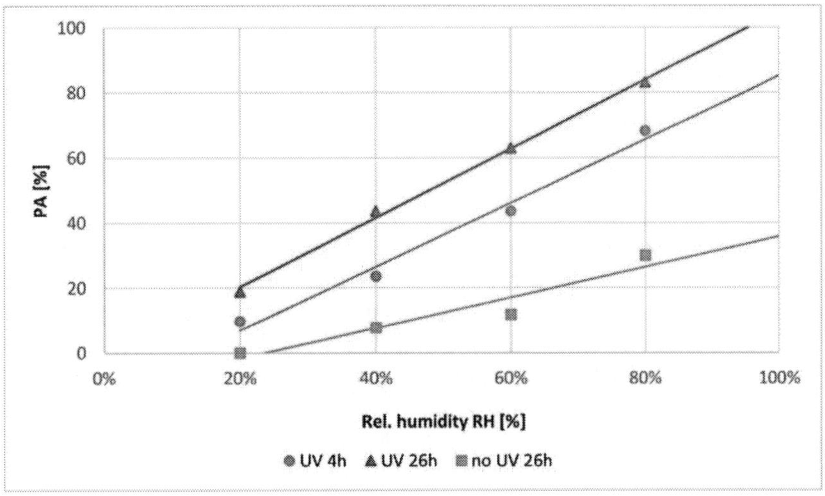

Figure 5 Relative humidity impact on photodegradation activity at 20 °C

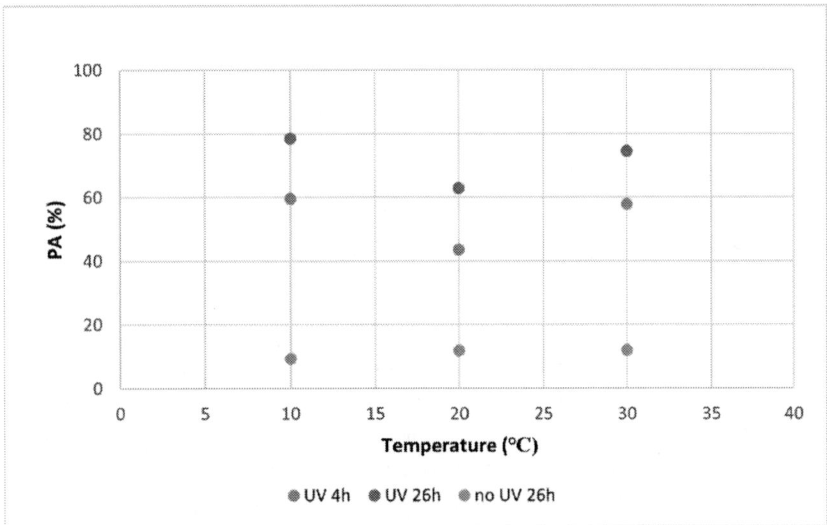

Figure 6 Temperature impact on self-cleaning performance

4. CONCLUSION

This work was devoted to studying the influence of relative humidity and temperature on self-cleaning capabilities of building materials coated with TiO_2 as photocatalyst. In the case of humidity, the positive effect of its increasing value was observed and explained by higher amount/number of water molecules present on the tested surface. It was observed that the dependence of the self-cleaning performance on humidity has linear character.

The second important finding is the effect of temperature on the self-cleaning ability of the samples. The test was carried out at temperatures of 10, 20 and 30 °C and a constant relative humidity 60 %. The decolorization of the tested surfaces caused by the photodegradation of RhB dye was the least effective at a temperature of 20 °C, while the most effective was at 10 °C. This effect could be attributed to the fact that the lower temperature leads to the condensation of a higher number of water molecules (larger volume of water vapor) and higher retention of condensed molecules on the surface. The water molecules retained on the surface of the TiO_2 coating can produce hydroxyl and superoxide radicals after irradiation with light of a suitable wavelength. A higher number of originated radicals caused a higher extent of RhB degradation and thus a higher level of surface discoloration. Increasing the temperature to 30 °C resulted in an increase in the self-cleaning ability of the surfaces of cement pastes compared to the experiment at 20°C, which can generally be explained by the positive effect of the higher temperature on the rate of chemical reactions.

ACKNOWLEDGEMENTS

This research was supported by the Czech Science Foundation (project 24-10949S), while the infrastructure has been utilized in the frame of project No. CZ.02.01.01/00/22_008/0004631 Materials and technologies for sustainable development within the Jan Amos Komensky Operational Program financed by the European Union and from the state budget of the Czech Republic. Authors also thank to project SP2024/025 (VSB – Technical University of Ostrava).

DATA AVAILABILITY

10.5281/zenodo.13918261

REFERENCES

[1] FUJISJIMA, A., WATANABE, T., HASHIMOTO. TiO2 photocatalysis, fundamentals and application. Tokyo: BKC Inc, 2001.

[2] FATTAKHOVA-ROHLFING, D., ZALESKA, A., BEIN, T. Three-Dimensional Titanium Dioxide Nanomaterials. *Chemical Reviews.* 2014, vol. 114, no. 19, pp. 9487-9558.

[3] LU, Y., GUAN, S., HAO, L., YOSHIDA, H., NAKADA, S., TAKISAWA, T., ITOI, T. Inactivation of SARS-CoV-2 and photocatalytic degradation by TiO2 photocatalyst coatings. *Scientific Reports.* 2022, vol. 12, no. 1, pp. 16038.

[4] LEE, S.-Y., PARK, S.-J. TiO2 photocatalyst for water treatment applications. *Journal of Industrial and Engineering Chemistry.* 2013, vol. 19, no. 6, pp. 1761-1769.

[5] NAKATA, K., FUJISHIMA, A. TiO2 photocatalysis: Design and applications. *Journal of Photochemistry and Photobiology C: Photochemistry Reviews.* 2012, vol. 13, no. 3, pp. 169-189.

[6] ISO 22197-1:2016. *Fine ceramics (advanced ceramics, advanced technical ceramics) — Test method for air-purification performance of semiconducting photocatalytic materials. Part 1: Removal of nitric oxide.* Geneva, Switzerland: International Organization for Standardization, 2016.

[7] UNI 11247:2007, *Determination of the catalytic degradation of nitrogen oxides in air by photocatalytic inorganic materials.* Roma, Italy: Ente Nazionale Italiano di Unificazione (UNI), 2007.

[8] HASHIMOTO, K., IRIE, H., FUJISHIMA, A. TiO2 Photocatalysis: A Historical Overview and Future Prospects. *Japanese Journal of Applied Physics.* 2005, vol. 44, no. 12, pp. 8269–8285

[9] NAH, Y., PARAMASIVAM, I., SCHMUKI, P. *Doped TiO 2 and TiO 2 Nanotubes: Synthesis and Applications.* ChemPhysChem. 2010, 2010-09-10, vol. 11, no. 13, pp. 2698-2713.

[10] DAOUD, W. A. *Self-cleaning materials and surfaces: a nanotechnology approach*, 2013.

[11] JANUS, M., ZAJAC, K. Concretes with Photocatalytic Activity. *In: High Performance Concrete Technology and Applications.* InTech, 2016, pp. 141-162.

[12] TAUC, J. Optical properties and electronic structure of amorphous Ge and Si,Materials. *Research Bulletin.* 1968, vol. 3, no. 1, pp. 37-46.

[13] FENG, S., LI, F. Photocatalytic dyes degradation on suspended and cement paste immobilized TiO2/g-C3N4 under simulated solar light. *Journal of Environmental Chemical Engineering.* 2021, vol. 9, no. 4, pp. 105488.

NANOCOMPOSITES OF ZNO-DECORATED HPHT NANODIAMOND AS A PERSPECTIVE GREEN PHOTOCATALYSTS

[1]Julia MICOVA, [1]Natalia KOSUTOVA, [2,3]Monika JERIGOVA, [4]Zdenek REMES

[1]Slovak Academy of Sciences, Institute of Chemistry, Bratislava, Slovakia, EU, chemjumi@savba.sk

[2]International Laser Centre, Bratislava, Slovakia, EU

[3]Department of Physical and Theoretical Chemistry, Faculty of Natural Sciences, Comenius University, Mlynska dolina, Bratislava, Slovakia, EU

[4]Czech Academy of Sciences, Institute of Physics, Prague, Czech Republic, EU

https://doi.org/10.37904/nanocon.2024.5014

Abstract

Electrical and optical properties of zinc oxide, in particular the relatively wide direct band gap of ~3.3 eV predetermines ZnO as a photocatalytic material at room temperature. ZnO itself is non-toxic and the hydrothermal growth method we use to prepare its nanostructures is environmentally friendly. These facts contribute to the attractiveness of this material. Recently, we have focused on studying the influence of doping or of plasma surface treatment on the photocatalytic efficiency of ZnO. However, the main drawbacks are the high recombination rate of charge carriers and the poor in capturing visible light in a narrow spectral range. The solution is the use of ZnO as part of nanocomposites. In the past, ZnO-detonation nanodiamons (DND) composites were studied. The disadvantage of DND is the formation of clusters of different sizes, therefore we focused our attention on the use of HPHT nanodiamonds (NDs). We modified the surface of commercial HPHT NDs by oxidation in order to increase the concentration of hydroxyl-, carbonyl- and carboxyl- functional groups. The resulting ZnO-HPHT NDs composites were synthesized using a dehydration-condensation process between oxygen-containing functional groups on the NDs surface and hydroxyl groups on the ZnO surface. The prepared nanocomposites were characterized by SEM, EDX spectroscopy and secondary ion mass spectrometry (SIMS) and tested by photocatalytic dye degradation.

Keywords: HPHT NDs, ZnO, nanocomposites, photocatalyst, photocatalytic oxidation

1. INTRODUCTION

Zinc oxide (ZnO) represent semiconducting nanomaterial, which has been studied for its wide range of application beyond conventional optoelectronics. Its unique properties can be used for cultivation of mammalian cells, wound healing, inhibition of bacterial growth or in cosmetics for absorbing ultra-violet (UV) light. In addition, ZnO is an effective photocatalyst for the degradation of inorganic and organic pollutants [1]. The nanoparticles ZnO-NPs are the particles with the largest variety of synthesized structure. including one-dimensional (1D), two-dimensional (2D), and three-dimensional (3D) structures. The largest faction of ZnO represent one-dimensional structures such as needles, helixes, nanorods, ribbons, belts, wires and combs. 2D-ZnO can make up nanopellets or nanosheet. The last group of ZnO are 3D structures, whore include snowflakes, dandelion, flower [2]. This nanomaterial can be relative easily synthetized using different methods including sol-gel method, microwave heating synthesis, RF plasma synthesis or by hydrothermal growth method [3]. The latter is a simple and environmentally friendly method for the preparation of these nanoparticles since this method does not include any organic solvents or additional processing of the product [2,3]. This approach for the synthesis of ZnO brings many advantages such as the possibility of carrying out the synthesis at low temperatures or the high purity and the degree of crystallinity of the prepared material [4,5]. However, disadvantage of this method can requirement for expensive autoclaves and the difficulty of observing the crystal as it grows [6]. In our recent study, we investigated the photocatalytic properties of doped and plasma treated ZnO [1]. The main disadvantages of this materials are the high recombination rate of charge carriers and the poor use of visible light. The possible solution of this problem could be the use of ZnO as part of nanocomposites.

Hybrid zinc oxide (ZnO)/detonation nanodiamond (DND) composites have been the interest of study for many researchers over the years. DNDs form particles with the size around 5 ± 2 nm, while the diamond grains are unstable and spontaneously aggregate to clusters of different sizes [7]. Therefore, in this work, ZnO- HPHT NDs composites were prepared and studied. The use of HPHT nanodiamonds brings possible solution of the problem with DND. Detonation and High-pressure high-temperature (HPHT) methods are the two main methods applied for the synthesis of the commercially available nanodiamonds [8]. However, the final NDs products significantly differ in their structure (concentration of defects and embedded nitrogen) and also in their surface quality and reactivity [7]. In general, HPHT NDs create bigger and stable structures. Their other advantage over DNDs is certainly its monocrystalline character with low concentration of lattice defects. This is very essencial especially for their spintronic and sensing applications [9].

The studied ZnO-HPHT NDs composites were synthesized using a dehydration-condensation process between oxygen-containing functional groups on the NDs surface and hydroxyl groups on the ZnO surface. The prepared nanocomposites were then characterized by SEM, EDX, SIMS and tested by photocatalytic dye degradation (methylene blue MB).

2. EXPERIMENTAL PART

2.1 Photocatalyst preparation

2.1.1 Treatment of HPHT nanodiamonds

The commercial type Ib HPHT (high-pressure, high-temperature) diamond powder of size about 30 nm was purchased from Microdiamant AG., Switzerland. The powder was oxidized by air at 510 °C for 6 h followed by the wet chemical treatment in a mixture of concentrated H_2SO_4 and HNO_3 at 75 °C for 3 days to increase the carbonyl surface coverage. After dilution of the reaction mixture was removed by centrifugation. Subsequently the nanodiamonds were washed with 0.1 M NaOH and 0.1 M HCl. Finally, the particles were washed five times with MilliQ water. The solution was lyophilized.

2.1.2 Preparation ZnO nanostructures

The process of synthesis zinc oxide by hydrothermal growth method was previously reported [3]. Briefly, equimolar aqueous solutions of 25 mM $Zn(NO_3)_2 \cdot 6H_2O$ and HMTA at 90 °C for 3 h. The precursor salt residue was removed from the sample by washing 3 -times with deionized water followed by centrifugation at 18 000 rpm (RCF: 23542g) for 20 min. Finally, the powder was dried by lyophilization.

2.1.3 Preparation of ZnO-NDs composites

The ND-ZnO nanocomposite was synthesized via a simple dehydration condensation process between the oxygen-containing functional groups on NDs and the hydroxyl groups on the surface of ZnO nanostructures 10 mg ND was first added to 20 ml water and ultrasonicated for 30 min. Then 100 mg ZnO powder was added into the suspension of NDs and stirred for 1 h and allowed to react overnight at r.t. The obtained light gray powder was dried at 130 °C for 6 h.

2.2 Structural analysis of the photocatalyst ZnO-ND

2.2.1 SEM analysis and EDX spectroscopy

Structural observations on ZnO-ND nanocoposite were performed with field emission scanning electron microscopy (SEM – JEOL 7600 FEG), and chemical compositions were analysed using energy dispersive (EDS) X-ray spectroscopy (Oxford Instruments INCA microanalysis system with X-Max 50 detector) at 10 kV. Original JOEL PC SEM software Ver. 2.1.0.9 (Tokyo, Japan) was used to collect SEM images and process the results, and INCA software was used for EDS-chemical composition analysis.

2.2.2 SIMS spectrometry

Mass spectrometry measurements were performed using a ToF-SIMS IV (ION-TOF, Germany) reflectron type time-of-flight mass spectrometer equipped with a bismuth ion source. Pulsed 25 keV Bi^+ were used as primary ions with a current of 1 pA with total primary ion dose density below the static limit of 10^{13} ions cm^{-2}. Secondary ion mass spectra were measured by scanning over a selected 100x100 μm^2 analysis area. Imaging was performed by selecting masses of interest and recording their intensities with respect to the position of the primary ion beam in the fields of view 250 x 250 μm^2. The lateral resolution of measured 2D images was about 1 µm. Measurements were performed in positive and negative polarities. The samples in a powder form were deposited on silicon substrate with double-sided graphite tape without any other procedures.

2.2.3 Characterisation of photocatalytic properties

Testing the efficiency of the ZnO-ND photocatalyst was realized through the photodegradation of MB. 20 mg of photocatalyst was added to 15 ml of MB aqueous solution (MB concentration was 10 ppm). The suspension was allowed to stir in a completely dark environment for half an hour before photocatalytic degradation to reach adsorption-desorption equilibrium. Subsequently, the suspension was exposed to UV radiation THORLAS LED 365 nm, 880 mW (min), 1000 mA. At regular 10-minute intervals, 2 mL samples were taken and filtered through a PVDF filter (diameter 13 mm, with a pore size of 0.22 µm. MB concentration was monitored by measuring the absorbance by a Beckman Coulter DU 730 UV/Vis spectrophotometer at a wavelength 662 nm. (For comparison, the photocatalytic efficiency of the prepared ZnO nanostructures was tested in an identical way).

3. RESULTS AND DISCUSSION

The SEM image as synthesized ZnO-ND nanocoposite is presented in **Figure 1**. The SEM image at 5 000 times magnification, it shows the homogeneous distribution of ZnO on the NDs surface. The elemental compositions of ZnO-ND nanocomposite were determined by EDX spectroscopy. The results are presented in **Table 1**.

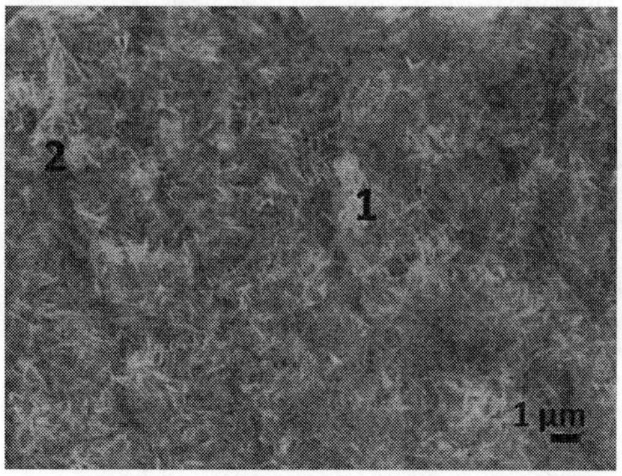

Figure 1 The SEM image of the ZnO-ND nanocomposite with marking of the positions for measuring the EDX spectrum

Table 1 The table of the elemental composition of the ZnO-ND nanocomposite in the marked SEM positions (**Figure 1**) in mass percentage

Position in the SEM image	**wt%**		
	C	**Zn**	**O**
1	19.10	57.79	23.11
2	15.84	60.20	23.97

Secondary ion mass spectrometry of the ZnO-ND nanocomposite sample identified the following ions and fragments with the corresponding isotopic distribution in positive polarity: C+, Zn+, ZnOH+. The following ions and fragments originating from the nanodiamond structure and ZnO coating were identified in negative polarity a: C, O-, OH-, CO_3-, ZnO-, ZnOH-, ZnO_2-. Due to the high sensitivity of SIMS at the ppm level, impurities originating probably from the synthesis process, such as Na+, F-, Cl, were identified.

The images were measured over the area of 250 x 250 µm2. The brighter is the yellow colour, the higher is the emission of selected secondary ion from given area. The darker or black colour means the low or no emission of selected secondary ions. The numbers below each image denote the maximum number of detected ions as counts in the single shot measurement (MC) and the total number of detected ions as counts (TC). The total ion image represents the distribution of all secondary ions detected during SIMS analysis in selected polarity. The **Figure 2** represents the 2D distribution of the selected species in the positive polarity. The distributions of Zn+ and ZnOH+ represent the homogeneous coverage of nanodiamond with ZnO.

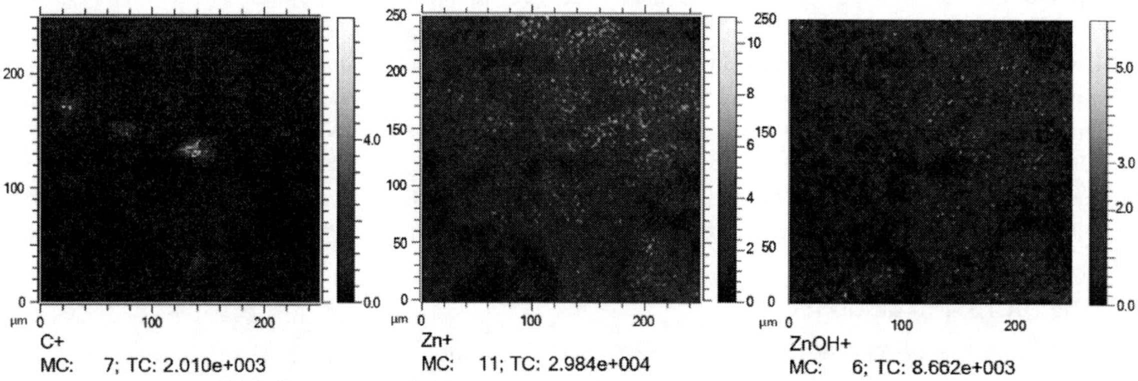

Figure 2 SIMS 2D images obtained in positive polarity for the homogeneous coverage of nanodiamond ZnO

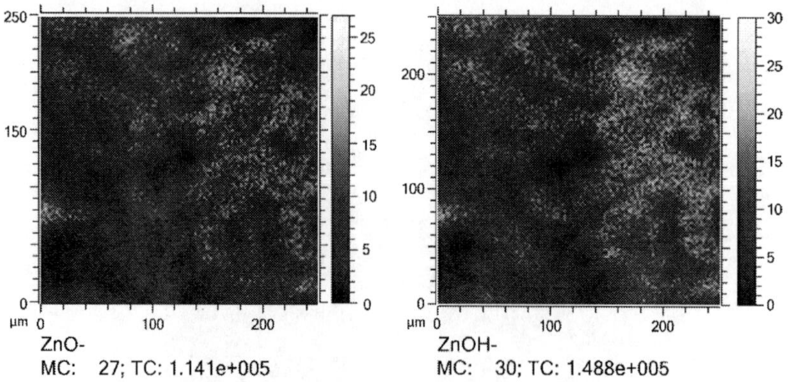

Figure 3 SIMS 2D images obtained in negative polarity for the homogeneous coverage of nanodiamond ZnO

The **Figure 3** represents the 2D distribution of the selected species in the negative polarity. ZnO^- and $ZnOH^-$ distributions representing the ZnO coverage of the nanodiamond sample are homogeneous. The intensity distributions of ZnO species in both polarities correspond to the total ion distribution and the intensities variations within one image are given by topographical effect of the sample.

Results of degradation processes were further fitted to first-order rate kinetic based on the Langmuir-Hinshelwood model. Langmuir–Hinshelwood (LH) kinetics is the most commonly used kinetic expression to explain the kinetics of the heterogeneous catalytic processes. The kinetics of the photodegradation MB follows an apparent first order, like the Langmuir-Hinshelwood mechanism.

$$ln(c_0 / c) = k_s \cdot t \quad (1)$$

where:

k_S- the rate constant (min^{-1})

c_0- the initial concentration of MB (mg/l)

c-the concentration of MB at reaction time of t (mg/l)

The values of the degradation rate constants indicate the rate of MB degradation (detected by MB decolorization). The larger the rate constant, the faster the reaction takes place, i.e. degradation of MB.

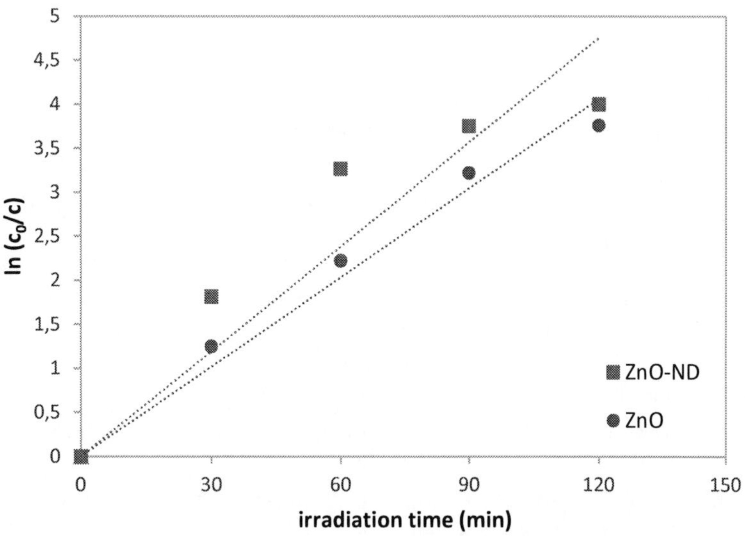

Figure 4 Graph of function of time and photocatalytic degradation MB kinetics, the straight lines represent the linear fit from the Langmuir-Hinshelwood model

All the plots of $ln(c_0/c)$ vs t follow a linear relationship with a correlation coefficient (R^2) and the slope (**Figure 4**) deduced from each fit provides an estimate of k_S, as presented in **Table 2**.

Table 2 Comparison of rate constants k_S and correlation coefficients R^2 for photocatalysts ZnO and ZnO-ND

Photocatalyst	R^2	k_s
ZnO	0.9931	0.0338
ZnO-ND	0.9595	0.0396

4. CONCLUSION

Zinc oxide has unique optoelectronic properties. One of them is the band gap region (3.2 ev), which predetermines its use as a photocatalyst. Its indispensable advantage is cheap and environmentally friendly preparation. However, it exhibits a high recombination rate of charge carriers. One of the solutions is the use of ZnO as a component of composites. In our study, attention was first focused on the ZnO-NDs nanocomposite, while HPHT nanodiamonds were used as NDs. ZnO was prepared by the method of hydrothermal growth. By oxidizing commercial NDs, we increased their surface concentration of hydroxyl, carbonyl and carboxyl functional groups. This allowed us to prepare a ZnO-ND nanocomposite by dehydration condensation process. We examined the morphology of the composite using SEM microscopy. We confirmed the elemental composition using EDX and SIMS spectrometry was used to study the structure. We tested the photocatalytic efficiency of the composite through the photodegradation of methylene blue. By comparing the degradation of MB under identical conditions with the use of ZnO as a photocatalyst, we found that the rate of degradation when using ZnO-ND was 1.17 times faster.

ACKNOWLEDGEMENTS

This work was supported by the Slovak Research and Development Agency under the Contract no. DS-FR-22-0035 and APVV -23-0083 and Ministry of Education, Youth and Sports of Czech Republic (MEYS) MSTC Danube project 8X23025, and the Mobility project CAS-SAS-2022-08, SAV-23-13 and by the Slovak Scientific Grant Agency (VEGA: 2/0132/23).

REFERENCES

[1] ONG, Ch. B., NG L. Y., MOHAMMAD, A. W. A review of ZnO nanoparticles as solar photocatalysts: Synthesis, mechanisms and applications. *Renewable and Sustainable Energy Reviews.* 2018, vol. 81, pp. 536-551.

[2] KOŁODZIEJCZAK-RADZIMSKA, A., JESIONOWSKI, T. Zinc Oxide—From Synthesis to Application: A Review. *Materials.* 2014, vol. 7, no. 4, pp. 2833-2881.

[3] MIČOVÁ, J., BURYI, M., ŠIMEK, D., DRAHOKOUPIL, J., NEYKOVA, N., CHANG, Y., REMEŠ, Z., POP-GEORGIEVSKI, O., SVOBODA, J., IM, C. Synthesis of zinc oxide nanostructures and comparison of their crystal quality. *Applied Surface Science.* 2018, vol. 461, pp. 190-195.

[4] DJURIŠCIĆ, A.B., CHEN, X.Y., LUNG, Y.H. Recent progress in hydrothermal synthesis of zinc oxide nanomaterials. *Recent Pat. Nanotechnol.* 2012, vol. 6, no. 2, pp. 124–134.

[5] TSUZUKI, T., DAWKINS, H., DUNLOP, J., TROTTER, G., NEARN, M., MCCORMICK, P.G. Nanotechnology and cosmetic chemist. *Cosmet. Aeorosol Toilet. Aust.* 2002, vol. 15, no. 5, pp. 10–24.

[6] SONAWANE, G.H., PATIL, S.P., SONAWANE, S.H. *Chapter 1 - Nanocomposites and Its Applications* in *Applications of Nanomaterials.* Sawston: Woodhead Publishing, 2018, p. 1-12.

[7] KOLAROVA, K., BYDZOVSKA, I., ROMANYUK, O., SHAGIEVA, E., UKRAINTSEV, E., KROMKA, A., REZEK, B., STEHLIK, S. Hydrogenation of HPHT nanodiamonds and their nanoscale interaction with chitosan. *Diamond & Related Materials.* 2023, vol. 134, art. no. 109754.

[8] NUNN, N., TORELLI, G., MCGUIRE, G., SHENDEROVA, O. Nanodiamond: a high impact nanomaterial. *Curr. Opinion Solid State Mater. Sci.* 2017, vol. 21, no. 1, pp. 1-9.

[9] STEHLIK, S., VARGA, M., LEDINSKY, M., JIRASEK, V., ARTEMENKO, A., KOZAK, H., ONDIC, L., SKAKALOVA, V., ARGENTERO, G., PENNYCOOK, T., MEYER, J.C., FEJFAR, A., KROMKA, A., REZEK, B. Size and purity control of HPHT nanodiamonds down to 1 nm. *J. Phys. Chem. C.* 2015, vol.119, no.49, pp. 27708–27720.

PULSED LASER DEPOSITION OF NOVEL SPINEL MATERIALS FOR PHOTOVOLTAIC CONVERSION

[1,2]Jakub HOLOVSKÝ, [1,2]Lucie LANDOVÁ, [3]Lukáš HORÁK, [3]Milan DOPITA, [1]Naini JAIN, [1]Eva HORYNOVÁ, [1,2]Neda NEYKOVA, [2]Jitka LIBERTÍNOVÁ, [1]Rupendra Kumar SHARMA

[1]Centre for Advanced Photovoltaics, Faculty of Electrical Engineering, Czech Technical University in Prague, Prague, Czech Republic, EU, jakub.holovsky@fel.cvut.cz

[2]Institute of Physics, Czech Academy of Sciences, Prague, Czech Republic, EU

[3]Faculty of Mathematics and Physics, Charles University, Prague, Czech Republic, EU

https://doi.org/10.37904/nanocon.2024.5007

Abstract

While all traditional photovoltaic materials are covalent compounds with coordination number 4, currently the most progressive materials - hybrid halide perovskites - have coordination number 6 and are ionic. This can be the key to success, especially regarding surface electronic properties, which are essential for polycrystalline semiconductors. This motivates us to explore completely new materials - cubic lithium thiospinels - and compare them with the former. Li thiospinels are prepared from stoichiometrically mixed precursors as a pellet in a nitrogen-filled glovebox. This tablet serves as a target in Pulsed Laser Deposition (PLD) and a special chamber is constructed that avoids air exposure of the precursor tablet during the loading. Thin films of lithium thiospinels are deposited on glass substrates and they are characterized by X-ray diffraction and optically by Photothermal Deflection Spectroscopy for accurate determination of absorption edge. Results are promising and demonstrate the validity of this approach.

Keywords: Thiospinels, methodology, hybrid halide perovskites, novel material, pulsed laser deposition

1. INTRODUCTION

To meet the Paris agreement, the implementation of low carbon emission energy sources must be quick and widespread. Photovoltaics, being almost a synonym for renewable energy has not yet completely convinced the society. To do so, efficiency has to be further increased and reliability has to be maintained. For the former, the tandem solar cell is the only option [1,2]. For the latter, the technology of crystalline silicon is an obvious choice for the bottom cell, but the question about the top cell remains open as the currently mostly explored material hybrid perovskite is not stable [3] and is potentially toxic due to lead content and water solubility [4]. Among many new potential materials there are also lithium thiospinels with formula ($Li_2MSn_3S_8$) with M = Mn, Fe, Mn, Ni, Co or with formula $Li_{1.66}CuSn_{3.33}S_8$ [5]. These Li thiospinels have similarly as halide perovskites also tuneable optical band gaps between 0.91 and 2.19 eV, they are ionic crystals and have slightly indirect gap [4]. They are however less toxic and more stable in water. On the other hand, there are several challenges related to the preparation and processing of this material from powder form to thin film. These materials have so far been prepared by high temperature synthesis [5] and mechano-chemical synthesis [6]. In this work, we demonstrate the first successful preparation of Li thiospinel $Li_{1.66}CuSn_{3.33}S_8$ thin film by pulsed laser deposition (PLD). In order to avoid air exposure of the material precursors and to limit the contamination of PLD chamber, we build a small portable chamber, that can enter nitrogen-filled glove box and also fits inside PLD setup. In this way the material precursors are kept in inert atmosphere until the PLD deposition is finished.

2. EXPERIMENTAL SECTION

The material precursors were mixed into stoichiometric mixture corresponding to $Li_{1.66}CuSn_{3.33}S_8$ and milled in agate mortar and pestle. The powder was then pressed in hydraulic press Specac into small pellets of 5 mm diameter (**Figure 1a**) in a nitrogen-filled glovebox. This tablet then serves as a target in Pulsed Laser Deposition (PLD), and a special chamber (**Figure 1b**) is constructed to avoid the precursor tablet's air exposure during the loading into the PLD setup and unloading. The chamber has quartz dome cover that is transparent for PLD laser (KrF 248 nm) and is equipped with vacuum valve and vacuum flange so that it can be connected to a turbomolecular vacuum pump. The pellet together with glass substrate (2 × 2 cm) was placed inside the chamber in nitrogen-filled glovebox.

Figure 1 (a) Pellet of Li thiospinels **(b)** Special portable chamber designed **(c)** Special chamber placed inside of the original PLD chamber to avoid contamination and connected to a turbomolecular pump.

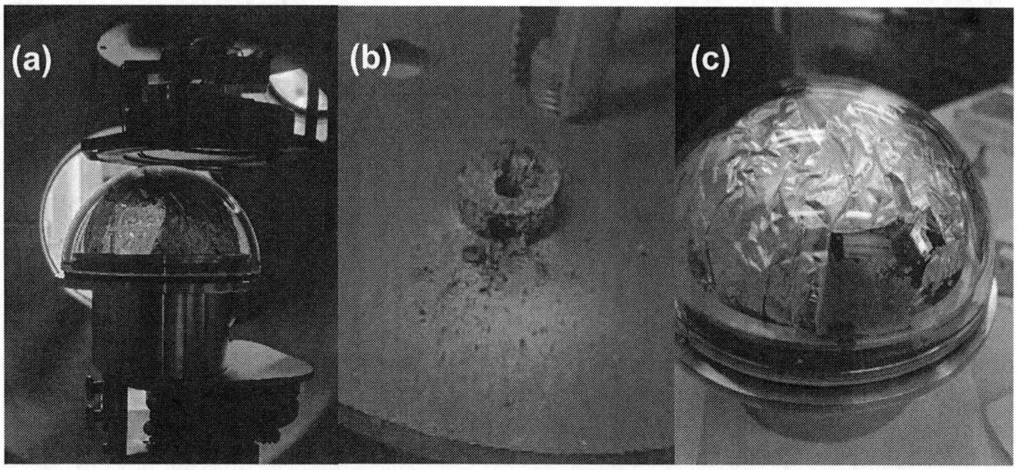

Figure 2 The violet colour of plasma plume visible at the uncovered part of the quartz dome **(a)**. Pellet of Li thiospinels after first successful deposition of thin film by PLD **(b)**. Contamination of quartz dome during deposition **(c)** of the very first thin film of Li thiospinel successfully deposited by PLD.

Before the deposition, the chamber was connected to an external vacuum pump and pumped to vacuum lower than 10^{-4} mbar (**Figure 1c**). Chamber was placed inside PLD setup made by TSST B.V. equipped with KrF (λ = 248 nm) excimer laser COMPex 50 with 150 mJ pulses. The laser was focused 8 times to a spot of theoretical area of ca. 2 mm^2. The alignment of the laser beam is very important because of the small size (5 mm) of the pellet. For the deposition from the Li thiospinels target, the pulse energy (measured inside the chamber) was kept at 24 mJ giving the fluence of approx. 1.2 J/cm^2. In **Figure 2a** the violet colour of plasma inside the chamber during ablation is visible. The first sample was deposited at a repetition rate of 30 Hz and the number of pulses was 5000. We observed that because of a higher repetition rate of 30 Hz and a large number of pulses, the target was fully ablated in the region of the beam hitting the target (**Figure 2b**) as no movement of the laser or the pellet was implemented and the laser was hitting always one spot. The window glass of the special chamber was also contaminated during deposition (**Figure 2c**). This contamination can be removed by ethanol. For the next experiments we reduced the number of pulses as well as the frequency in the subsequent depositions. The pellet was flipped for the next deposition. All the depositions were made at room temperature, as no substrate heating is implemented in the portable chamber. **Table 1** shows the laser frequency and number of pulses used.

Table 1 List of samples and PLD parameters

Sample no.	Frequency (Hz)	No. of pulses	Thickness (nm)
1	30	5000	51
2	20	1000	30
3	20	2000	150

3. RESULTS AND DISCUSSIONS

The deposited thin films were characterized by X-ray diffraction (**Figure 3a**) and by scanning electron microscope Tescan MAIA (**Figure 3b**). We can observe that the thin layers remain amorphous, however higher number of pulses leads to thicker layers, that eventually feature crystalline peaks. Thicker layers are unstable, likely due to poor adhesion, and tend to recrystallize. Morphology of such thicker layers is not uniform (**Figure 3b**).

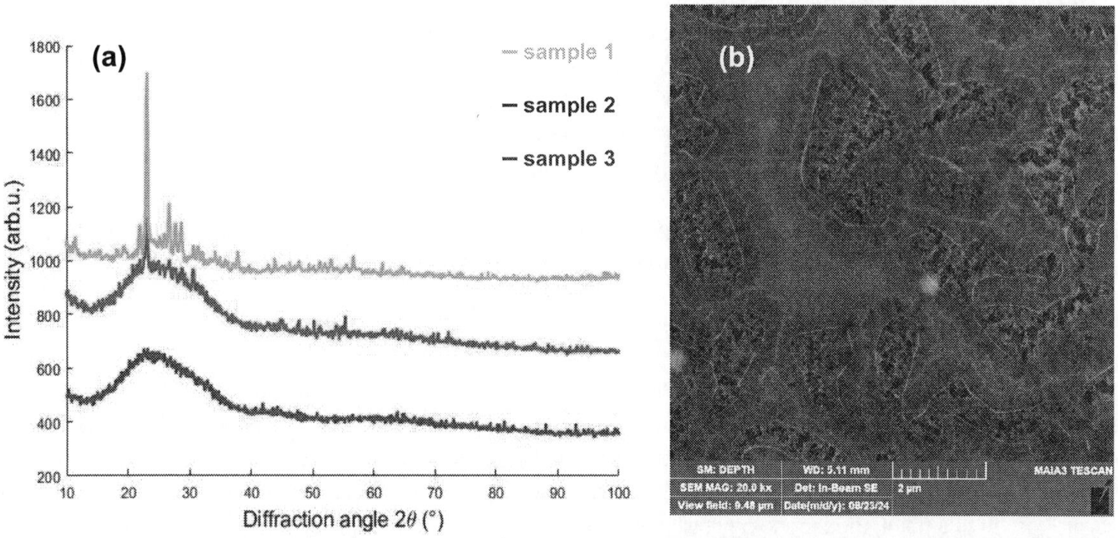

Figure 3 XRD diffraction patterns collected approx. 1 month after the deposition **(a)**. SEM image of sample no 1 approximately one week after the deposition showing probably the recrystallized regions **(b)**.

Table 2 Atomic composition (without lithium).

Atom	Ideal stoichiometry (%)	Experimental - EDX (%)
S	65	73
Cu	8	1.5
Sn	27	25.5

We evaluated also the atomic composition by Energy Dispersive X-ray spectroscopy (EDX) as presented in **Table 2**. As the lithium is giving X-ray line outside of the detection limits of the EDX, we can only assess the ratio of atomic composition of sulphur, copper and tin. We see that the stoichiometry is not perfectly maintained and has to be further tuned, e.g. by laser fluence or repetition rate, or adding low amount of argon into the chamber to slow down the ions.

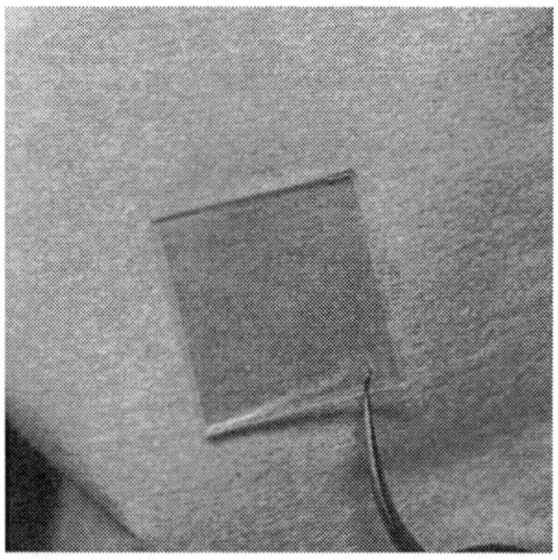

(a)

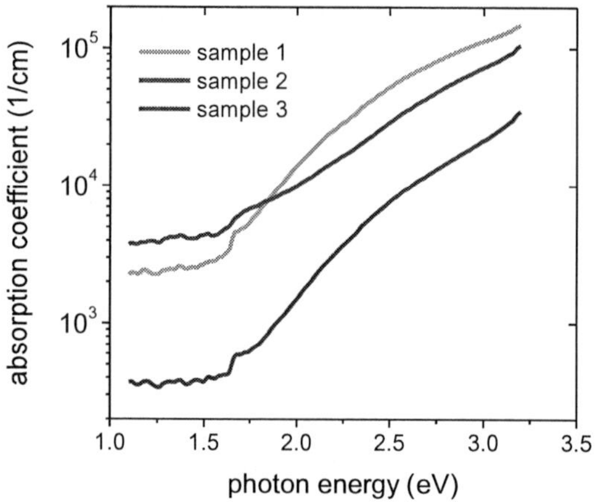

(b)

Figure 4 Visual appearance of thin film (sample no 2) of Li thiospinels successfully deposited by PLD **(a)**. Spectra of absorption coefficient evaluated from PDS **(b)**.

Optically, it can be seen that the layers possess absorption edge in near infrared-to-visible spectrum, yielding typical brownish appearance (**Figure 4a**). For more accurate evaluation of absorptance Photothermal Deflection Spectroscopy (PDS) in Fluorinert FC-72 liquid (refractive index 1.25) was performed. Our PDS setup simultaneously acquires also transmittance and reflectance and allows for an absolute and more accurate evaluation of the absorption coefficient and refractive index in the range from 1.1 eV to 3.2 eV. Refractive index varied in the range from 2.1 to 2.3. Thickness variation measured optically by PDS is shown also in **Table 1**. Absorption coefficient is shown in **Figure 4b**. We can observe clear indication of bandgap, however not sufficiently sharp to yield high open circuit of solar cell [7,8]. Interestingly the absorption coefficient is varying not so much in shape, but especially in absolute value. We attribute this, as well as the variation of optical thickness, to the variable density of the material.

4. CONCLUSION

For the first time, we prepared thin films of Li thiospinels using PLD. These films were characterized using numerous methods including X-ray diffraction, Energy Dispersive X-ray spectroscopy and Photothermal deflection spectroscopy. The first attempts led to layers featuring absorption edge and medium sub-bandgap absorptance. At this point, the technology still needs optimization of deposition parameters, but the first results are very promising and served as excellent proof of the concept of the idea of preparing photosensitive thin films of novel materials by PLD. Also, we demonstrated the validity of our approach of using small portable chamber to avoid air exposure of the precursor materials including sulphides.

ACKNOWLEDGEMENTS

This work was supported by the Czech Science Foundation grant no. 23-06285S and by Czech Ministry of Education, Youth and Sports grant no. CZ.02.01.01/00/22_008/0004617 – "Energy conversion and storage" and Czech Technical University in Prague student grant no SGS24/135/OHK3/3T/13.

REFERENCES

[1] E. AYDIN, T.G. ALLEN, M. DE BASTIANI, A. RAZZAQ, L. XU, E. UGUR, J. LIU, S. De WOLF. Pathways toward commercial perovskite/silicon tandem photovoltaics. *Science*. 2024, vol. 383, eadh3849. Available from: https://doi.org/10.1126/science.adh3849.

[2] LONGi sets a new world record of 33.9% for the efficiency of crystalline silicon-perovskite tandem solar cells, (2023). Available from: https://www.longi.com/en/news/new-world-record-for-the-efficiency-of-crystalline-silicon-perovskite-tandem-solar-cells/ (accessed April 1, 2024).

[3] R. WANG, M. MUJAHID, Y. DUAN, Z. WANG, J. XUE, Y. YANG, A Review of Perovskites Solar Cell Stability. *Adv Funct Materials*. 2019, vol. 29, 1808843. Available from: https://doi.org/10.1002/adfm.201808843.

[4] C.E. TORRENCE, C.S. LIBBY, W. NIE, J.S. STEIN. Environmental and health risks of perovskite solar modules: Case for better test standards and risk mitigation solutions. *iScience*. 2023, vol. 26, 105807. Available from: https://doi.org/10.1016/j.isci.2022.105807.

[5] M.A. QUINTERO, S. HAO, S.V. PATEL, J.-K. BAO, X. ZHOU, Y.-Y. HU, C. WOLVERTON, M.G. KANATZIDIS, Lithium Thiostannate Spinels: Air-Stable Cubic Semiconductors. *Chem. Mater.* 2021, vol. 33, pp. 2080–2089. Available from: https://doi.org/10.1021/acs.chemmater.0c04651.

[6] E.M. HEPPKE, S. MAHADEVAN, M. LERCH, New compounds of the $Li_2MSn_3S_8$ type. *Zeitschrift Für Naturforschung B 75.* 2020, pp. 625–631. Available from: https://doi.org/10.1515/znb-2020-0050.

[7] U. RAU. Reciprocity relation between photovoltaic quantum efficiency and electroluminescent emission of solar cells, *Physical Review B.* 2007, vol. 76, 085303. Available from: https://doi.org/10.1103/PhysRevB.76.085303.

[8] J. HOLOVSKÝ, K. RIDZOŇOVÁ, A. PETER AMALATHAS, B. CONRAD, R.K. SHARMA, X.Y. CHIN, E. BASTOLA, K. BHANDARI, R.J. ELLINGSON, S. DE WOLF. Below the Urbach Edge: Solar Cell Loss Analysis Based on Full External Quantum Efficiency Spectra. *ACS Energy Lett.* 2023, vol. 8, pp. 3221–3227. Available from: https://doi.org/10.1021/acsenergylett.3c00951.

FABRICATION AND ANTIBACTERIAL ACTIVITY OF ALLYL ISOTHIOCYANATE-LOADED CHITOSAN NANOPARTICLES FOR ENHANCED FOOD PRESERVATION

[1]Hoang-Anh CAO, [1]Jana PEKÁRKOVÁ, [2]Eliška BIRGUSOVÁ, [2]Lukáš RICHTERA

[1]*Brno University of Technology, Central European Institute of Technology, Brno, Czech Republic, EU,* Hoang-Anh.Cao@ceitec.vutbr.cz, pekarkova@vut.cz

[2]*Mendel University in Brno, Department of Chemistry and Biochemistry, Brno, Czech Republic, EU,* xbirguso@mendelu.cz, richtera@mendelu.cz

https://doi.org/10.37904/nanocon.2024.5009

Abstract

Allyl isothiocyanate (AITC) has been recognized for its potent antibacterial and antioxidant properties, positioning it as a promising agent for food preservation. However, its practical application in the food industry is significantly hindered by its high volatility and strong, unpleasant odor. To address these limitations, we utilized the multifunctional attributes of chitosan, a polysaccharide derived from chitin, to stabilize and enhance the efficacy of AITC. The abundance of amino and hydroxyl groups in chitosan chain facilitates strong interactions with AITC, thus achieving high encapsulation efficiency. Additionally, the inherent antibacterial property of chitosan synergizes with that of AITC, potentially reducing the required dose for effective microbial inhibition. In this study, AITC-loaded chitosan nanoparticles (AITC-CSNPs) were synthesized using a two-step emulsion-ionic gelation method. The characterization of the nanoparticles was performed using Dynamic Light Scattering (DLS) to determine their hydrodynamic size, zeta potential, and polydispersity index, while Scanning Electron Microscopy (SEM) provided detailed insights into their morphology and distribution. Afterwards, the antibacterial properties of AITC-CSNPs were evaluated on prevalent food-borne pathogens. The findings indicated that the as-synthesized AITC-CSNPs possessed a spherical shape with nanosize and demonstrated satisfactory antibacterial effects, thereby presenting a viable method for extending the shelf life of perishable food products.

Keywords: Allyl isothiocynate, chitosan nanoparticles, antibacterial properties, food preservation

1. INTRODUCTION

One of the primary challenges in food manufacturing is the control of pathogenic microorganisms during production. These pathogens contribute significantly to food spoilage, resulting in health problems, food waste, and economic losses [1]. In this context, allyl isothiocyanate (AITC), derived from plants of the Brassicaceae family, has garnered significant attention as a natural agent for food preservation [2-4]. AITC, a volatile sulfur-containing compound, contributes mainly to the strong flavors of mustard, wasabi, and horseradish [5]. Since the early 20th century, AITC has been noted for its potent antimicrobial [6] and antifungal [7] properties even at low concentrations. Since 2005, the U.S. The Food and Drug Administration (USFDA) designated AITC as Generally Recognized as Safe (GRAS) [8], while the European Food Safety Authority (EFSA) regulated its use as a food additive in the EU in 2010 [9]. Despite these advantages, the practical use of AITC in active food packaging is hindered by challenges related to its volatility, solubility, and controlled release over time [10]. In addition, its strong odor can negatively affect consumer acceptance [3]. To address these issues, encapsulation techniques have been developed to stabilize AITC for use in food processing. Recent studies have explored nanotechnology as a means of enhancing AITC's stability and efficacy of AITC. For example, Aytac et al. [11] demonstrated the incorporation of AITC-loaded β-cyclodextrin into electrospun nanofibers, whereas Encinas-Basurto et al. [12] enhanced the functionality of AITC using poly(lactic-co-glycolic acid) nanocarriers.

In this study, chitosan (CS), a chitin-derived polysaccharide, was employed as a multifunctional biopolymer to entrap AITC. Firstly, CS exhibits remarkable characteristics, including non-toxicity, biocompatibility, and biodegradability, which are preferred for various applications including food preservation [13]. Secondly, CS possesses excellent antibacterial and antioxidant properties against a wide range of bacteria and fungi [14]. Furthermore, the abundance of amino and hydroxyl groups on the chain facilitates CS to interact with a plenty of active substances, thereby being promising to encapsulate productively AITC [15]. Indeed, Patil & Patel [16] described the successful fabrication of CS nanoparticles (CSNPs) loaded AITC with encapsulation efficiency of 88 % for breast cancer treatment. To the best of our knowledge, no study has demonstrated the synergistic antibacterial effect of CS and AITC in nanoformulations for food preservation.

Herein, this study aims to fabricate AITC-loaded chitosan nanoparticles (AITC-CSNPs) and explore their antibacterial properties. To achieve the research goal, the main objectives were specified as follows: (1) Synthesis of AITC-CSNPs using a two-step emulsion-ionic gelation method; (2) Characterization of AITC-CSNPs by Dynamic Light Scattering (DLS) and Scanning Electron Microscopy (SEM), and (3) Antibacterial assessment of AITC-CSNPs against common bacteria.

2. MATERIALS AND METHODS

2.1 Materials

Low molecular weight chitosan (CS) (50–190 kDa, 75–85 % degree of deacetylation), Sodium tripolyphosphate (TPP), Allyl isothiocyanate (AITC), Acetic acid, Tween 20 and all solvents were purchased from Sigma-Aldrich (Darmstadt, Germany) with quality of analytical grade. The following bacteria were provided by the Czech Collection of Microorganisms (Brno, Czech Republic): *S. aureus* (CCM 4223), Methicillin-resistant *S. aureus* (CCM 7110), and *E. coli* (CCM K12). Ultrapure water (18.2 MΩ·cm) was utilized in all steps of this study.

2.2 Preparation of unloaded and loaded-AITC Chitosan nanoparticles (CSNPs and AITC-CSNPs)

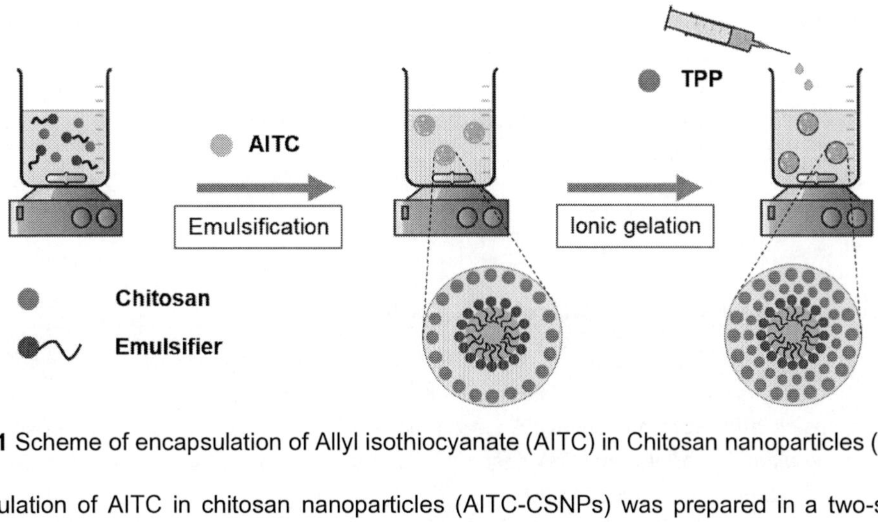

Figure 1 Scheme of encapsulation of Allyl isothiocyanate (AITC) in Chitosan nanoparticles (CSNPs)

The encapsulation of AITC in chitosan nanoparticles (AITC-CSNPs) was prepared in a two-step process (**Figure 1**) as reported by Patil and Patel [16]. In brief, 0.1 % CS solution (w/v) was stirred at 1000 rpm with 0.15 % Tween 20 and preheated to 50 °C. Then, AITC was added to achieve a ratio of CS:AITC by 1:0.5 w/w and emulsified under continuous stirring at 14000 rpm for 15 min at room temperature. Finally, a TPP aqueous solution of 0.1 % (w/v) was added using a microsyringe at 1 mL/min under continuous stirring to establish the

AITC-CSNPs. The mixture was kept for 1 hour under gentle stirring. The nanoparticles were collected by centrifugation and washed several times with ultrapure water. Afterwards, the AITC-CSNPs were stored at 4 °C for further analysis. The unloaded CSNPs were synthesized similarly as described above without presence of AITC. **Table 1** presents the detailed composition of CS, TPP, and AITC in each synthesis batch.

Table 1 Formulation of CSNPs and AITC-CSNPs

Code	AITC (mg/mL)	Ratio of CS:TPP (w/w)
F0 (CS:TPP 1:1)	0	1:1
F1 (CS:TPP 1:1)	0.5	1:1
F2 (CS:TPP 2:1)	0.5	2:1
F3 (CS:TPP 3:1)	0.5	3:1

2.3 Characterization of AITC-CSNPs

Dynamic light scattering (DLS)

The hydrodynamic size, polydispersity index (PDI), and zeta potential of CSNPs and AITC-CSNPs were provided by Dynamic light scattering (DLS) method using Zetasizer Nano ZS (Malvern Panalytical, UK) with the specifications as following: chitosan refractive index (RI) of 1.700, absorption index of 0.001 and water solvent RI of 1.33, with a viscosity of 0.8872 cP. All measurements were performed in triplicates.

Scanning Electron Microscopy (SEM)

SEM micrographs were obtained on Verios 460L (Thermo Fisher, USA) to explore morphology, shape and size of CSNPs and AITC-CSNPs. An accelerating voltage of 20 kV was applied at the working distance of 5 mm. Before imaging, all samples were coated with a 15-nm layer of gold using a coating device (EM ACE600, Leica, Germany) to achieve sufficient conductivity. The average particle size was estimated from at least 100 measurements using ImageJ from three SEM micrographs.

Determination of encapsulation efficiency (EE) of AITC-CSNPs

To evaluate the amount of encapsulated AITC, the supernatant was carefully separated from the AITC-CSNPs by centrifugation. A UV-Vis spectrophotometer was utilized to determine the absorbance of the supernatant at 248 nm. The quantity of AITC in the supernatant was calculated based on the calibration curve, which was constructed from the concentration of standard AITC solutions and their respective absorbances. As the initial amount of AITC added to the synthesis was defined, the encapsulation efficiency (EE) of AITC was calculated according to equation (1) as follows:

$$\text{EE (\%)} = \frac{\text{Initial AITC (mg)} - \text{AITC in supernatant (mg)}}{\text{Initial AITC (mg)}} \times 100\ \% \qquad (1)$$

2.4 Assessment of antimicrobial activity of AITC-CSNPs

The antimicrobial properties of CSNPs and AITC-CSNPs were assessed against common Gram-positive and Gram-negative bacteria, namely *S. aureus*, Methicillin-resistant *S. aureus*, and *E. coli.* Two commonly used methods, dics diffusion and broth microdilution, were employed in this study. For dics diffusion tests, filter paper discs were soaked in nanoformulation suspensions (10 mg/mL) and placed on bacterial inoculated Mueller-Hinton agar plates. The inhibition zones around the discs were measured after a 24-hour incubation period at 37 °C. In case of broth microdilution method, microbial suspensions (1.5×10^8 CFU/mL) were incubated with AITC, CSNPs, and AITC-CSNPs in concentration of 10 mg/mL to 0.156 mg/mL at 37 °C. The growth of bacteria were checked after 24h of the treatment.

3. RESULTS AND DISCUSSION

3.1 Hydrodynamic diameter and surface charge of AITC-CSNPs

Table 2 presents the hydrodynamic diameter and surface charge of the AITC-loaded and unloaded chitosan nanoparticle formulations. The size of the AITC-CSNP formulations ranged from 100.4 to 129.0 nm. The observed size reduction from formulations F1 to F3 correlated with the decreasing TPP amount, calculated as the mass ratio relative to that of chitosan. This trend aligns with the ionic crosslinking mechanism underlying the formation of the chitosan nanoparticles. The extent of this crosslinking network is primarily governed by the balance of negative charges from the amino groups on the chitosan chains and positive charges from the phosphate groups of TPP. Higher TPP concentrations promote a more extended crosslinking network, resulting in larger particle sizes. Additionally, a comparison between the unloaded (F0) and AITC-loaded (F1) formulations at the same CS/TPP ratio revealed that incorporating AITC increased particle size. This size increase was consistent with previous studies on the encapsulation of essential oils in chitosan nanoparticles [15]. The homogeneity of the nanoformulations was demonstrated by the polydispersity index (PDI). The nanoformulations in this study, with PDI values from 0.377 ± 0.010 to 0.571 ± 0.007 (**Table 2**), exhibited mid-range polydispersity. Zeta potential results of nanoformulations were in the range of +28.6 ± 0.9 mV to +36.4 ± 3.0 mV, which indicated a sufficient stability of them in aqueous solutions. Zeta potential values above +30 mV or below -30 mV are ideal, as they ensure adequate electrostatic repulsion, leading to a more stable suspension.

Table 2 Average hydrodynamic diameter, Polydispersity index (PDI), Zeta potential , Encapsulation efficiency (EE) of AITC loaded and unloaded CSNPs

Code	Average size (nm)	PDI	Zeta potential (mV)	EE (%)
F0	117.6 ± 1.2	0.377 ± 0.010	+28.9 ± 0.7	-
F1	129.0 ± 0.9	0.380 ± 0.008	+28.6 ± 0.9	7.08 ± 3.26 %
F2	124.4 ± 1.0	0.571 ± 0.007	+35.5 ± 0.7	4.32 ± 1.19 %
F3	100.4 ± 0.6	0.447 ± 0.012	+36.4 ± 3.0	1.87± 4.23 %

3.2 Encapsulation of AITC into Chitosan nanoparticles

In this study, UV spectrophotometry was utilized to evaluate the encapsulation efficiency of AITC in each formulation. The findings revealed promising results, with encapsulation values ranging from 1.87 % to 7.08 %. A clear trend emerged, showing that the encapsulation efficiency decreased as the TPP content in the nanoformulations was reduced. While the AITC-loaded chitosan nanoparticles (AITC-CSNPs) were small in size and exhibited good dispersion when prepared with lower TPP ratio, this advantage came at the cost of reduced encapsulation efficiency. This trade-off suggests that optimizing TPP levels is critical for balancing the particle size and encapsulation efficiency in the formulation process.

3.3 Morphology of AITC-CSNPs

SEM analysis was carried out to investigate the morphology and surface characteristics of the nanoformulations. The micrographs in **Figure 2** revealed nanoparticles with a spherical shape, which was typical for chitosan nanoparticles. The particle sizes for formulations F0 to F3, as determined by SEM, were 53.9 ± 7.5 nm, 54.1 ± 8.2 nm, 44.9 ± 6.8 nm, and 25.1 ± 2.1 nm, respectively. Although these values were smaller than those measured by DLS, the same trend in size was observed across the formulations. Specifically, the incorporation of AITC led to an increase in particle size (as seen in F0 and F1), whereas decreasing the proportion of TPP resulted in progressively smaller nanoparticles (as demonstrated in F1, F2, and F3). The difference in particle size obtained by DLS and SEM can be explained based on the particle aggregation phenomenon commonly encountered in polymeric nanoparticles. Unlike microscopy, the DLS technique cannot determine the size of individual particles in aggregates, resulting in hydrodynamic diameters that can be up to 10 times larger than the dimensions of the particles determined by microscopy, such as TEM or SEM [17].

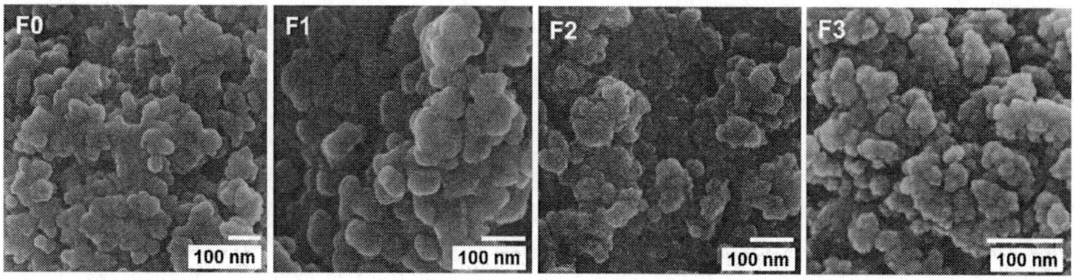

Figure 2 Representative SEM micrographs of CSNPs and AITC-CSNPs

3.4 Antimicrobial activity of AITC-CSNPs

Table 3 presents the minimum inhibitory concentrations (MIC) of AITC and nanoformulations against Gram-positive and Gram-negative bacteria. All formulations of AITC-CSNPs exhibited higher antimicrobial activity against *S. aureus* and Methicillin-resistant *S. aureus* than the free AITC and blank CSNPs. For *E. coli*, the lowest MIC was observed with nanoformulation F1, followed by higher values for free AITC and nanoformulations F2 and F3, respectively. Blank chitosan nanoparticles (CSNPs) showed no inhibitory effect on *E. coli* growth within the tested concentration range.

Table 3 Minimum inhibitory concentration (μg/mL) of AITC, CSNPs, and AITC-CSNPs against Gram-positive and Gram-negative bacteria

Microbial strains	AITC	F0	F1	F2	F3
S. aureus	625	1250	156.2	312.5	312.5
Methicillin-resistant *S. aureus*	1250	2500	312.5	625	625
E. coli	2500	>10,000	1250	5000	5000

The disc diffusion results showed limited inhibition in the growth of tested microorganisms. This outcome may be attributed to the limited diffusion of AITC and the nanoformulations into the agar medium.

CONCLUSION

In this study, AITC was successfully encapsulated in chitosan nanoparticles using a two-step ionic gelation technique, achieving a highest efficiency of 7.08 %. All nanoformulations exhibited characteristic spherical shapes, nanoscale sizes, and stability in aqueous media, as confirmed by the DLS and SEM results. Microbial assays demonstrated the enhanced antibacterial activity of AITC-loaded chitosan nanoparticles (AITC-CSNPs) against *S. aureus,* Methicillin-resistant *S. aureus,* and *E. coli.* The AITC-CSNPs showed higher efficacy compared with AITC or CSNPs alone at the same dosage, underscoring the synergistic effect of AITC integration within the CSNP matrix. However, further studies should be conducted to investigate the controlled release of AITC from these nanoformulations and to evaluate how the antibacterial efficacy varies with concentration and time.

ACKNOWLEDGEMENTS

This study was supported by the project CEITEC VUT-J-24-8615, Brno University of Technology and CzechNanoLab Research Infrastructure funded by MEYS CR (LM2023051)

REFERENCES

[1] AMIT, S.K., UDDIN, M.M., RAHMAN, R., ISLAM, S.M.R., KHAN, M.S. A review on mechanisms and commercial aspects of food preservation and processing. Agriculture & Food Security. 2017, vol. 6, no. 1, pp. 51.

[2] SHIN, J., HARTE, B., RYSER, E., SELKE, S. Active Packaging of Fresh Chicken Breast, with Allyl Isothiocyanate (AITC) in Combination with Modified Atmosphere Packaging (MAP) to Control the Growth of Pathogens. *Journal of Food Science*. 2010, vol. 75, no. 2, pp. M65-M71.

[3] KO, J.A., KIM, W.Y., PARK, H.J. Effects of microencapsulated Allyl isothiocyanate (AITC) on the extension of the shelf-life of Kimchi. *International Journal of Food Microbiology*. 2012, vol. 153, no. 1, pp. 92-98.

[4] KRAMER, B., WUNDERLICH, J., MURANYI, P. Impact of volatile allyl isothiocyanate on fresh produce. *Food Packaging and Shelf Life*. 2018, vol. 16, pp. 220-224.

[5] CORRALES, M., FERNÁNDEZ, A., HAN, J.H. *Innovations in Food Packaging (Second Edition)*. San Diego: Academic Press, 2014.

[6] LIN, C.-M., KIM, J., DU, W.-X., WEI, C.-I. Bactericidal Activity of Isothiocyanate against Pathogens on Fresh Produce†. *Journal of Food Protection*. 2000, vol. 63, no. 1, pp. 25-30.

[7] WINTHER, M., NIELSEN, P.V. Active Packaging of Cheese with Allyl Isothiocyanate, an Alternative to Modified Atmosphere Packaging. *Journal of Food Protection*. 2006, vol. 69, no. 10, pp. 2430-2435.

[8] USFDA. GRAS Claim notification for general Recognition of Safety of Allyl isothiocyanate in a food shelf-life extension and antio-sopilage system. GRN0180. 2005.

[9] EFSA. Scientific Opinion on the safety of allyl isothiocyanate for the proposed uses as a food additive. *EFSA Journal*. 2010, vol. 8, no. 12, pp. 1943.

[10] MOLLER, A., LEONE, C., KATARIA, J., SIDHU, G., RAMA, E.N., KROFT, B., THIPPAREDDI, H., SINGH, M. Effect of a carrageenan/chitosan coating with allyl isothiocyanate on microbial spoilage and quality of chicken breast. *Poultry Science*. 2023, vol. 102, no. 3, pp. 102442.

[11] AYTAC, Z., DOGAN, S.Y., TEKINAY, T., UYAR, T. Release and antibacterial activity of allyl isothiocyanate/β-cyclodextrin complex encapsulated in electrospun nanofibers. *Colloids and Surfaces B: Biointerfaces*. 2014, vol. 120, pp. 125-131.

[12] ENCINAS-BASURTO, D., IBARRA, J., JUAREZ, J., BURBOA, M.G., BARBOSA, S., TABOADA, P., TRONCOSO-ROJAS, R., VALDEZ, M.A. Poly(lactic-co-glycolic acid) nanoparticles for sustained release of allyl isothiocyanate: characterization, in vitro release and biological activity. *Journal of Microencapsulation*. 2017, vol. 34, no. 3, pp. 231-242.

[13] PRIYADARSHI, R., RHIM, J.-W. Chitosan-based biodegradable functional films for food packaging applications. *Innovative Food Science & Emerging Technologies*. 2020, vol. 62, pp. 102346.

[14] WANG, X., CHENG, F., WANG, X., FENG, T., XIA, S., ZHANG, X. Chitosan decoration improves the rapid and long-term antibacterial activities of cinnamaldehyde-loaded liposomes. *International Journal of Biological Macromolecules*. 2021, vol. 168, pp. 59-66.

[15] MONDÉJAR-LÓPEZ, M., CASTILLO, R., JIMÉNEZ, A.J.L., GÓMEZ-GÓMEZ, L., AHRAZEM, O., NIZA, E. Polysaccharide film containing cinnamaldehyde-chitosan nanoparticles, a new eco-packaging material effective in meat preservation. *Food Chemistry*. 2024, vol. 437, pp. 137710.

[16] PATIL, P.B., PATEL, J.K. Preparation, characterization, and in vitro cytotoxicity activity of allyl-isothiocyanate-embedded polymeric nanoparticles for potential breast cancer targeting. *Breast Cancer*. 2023, vol. 30, no. 6, pp. 1065-1078.

[17] FILIPPOV, S.K., KHUSNUTDINOV, R., MURMILIUK, A., INAM, W., ZAKHAROVA, L.Y., ZHANG, H., KHUTORYANSKIY, V.V. Dynamic light scattering and transmission electron microscopy in drug delivery: a roadmap for correct characterization of nanoparticles and interpretation of results. *Materials Horizons*. 2023, vol. 10, no. 12, pp. 5354-5370.

MAGNETICALLY MODIFIED BENTONITE AND LADLE FURNACE SLAG FOR STRONTIUM SORPTION

[1]Michaela TOKARČÍKOVÁ, [1]Julie SMIJOVÁ, [1]Roman GABOR, [1,2]Jana SEIDLEROVÁ

[1]VSB - Technical University of Ostrava, Centre for Energy and Environmental Technologies, Nanotechnology Centre, Ostrava, Czech Republic, EU, michaela.tokarcikova@vsb.cz

[2]VSB - Technical University of Ostrava, Faculty of Materials Science and Technology, Ostrava, Czech Republic, EU, jana.seidlerova@vsb.cz

https://doi.org/10.37904/nanocon.2024.5010

Abstract

The removal of pollutants, especially those with long half-lives, from water remains a current challenge. Commercial bentonite and waste materials - ladle furnace slags - were selected as suitable sorbents for Sr^{2+}. Bentonite and ladle furnace slags were modified with iron oxide (Fe_xO_y) nanoparticles (NPs) with magnetic properties using a microwave-assisted process, which allows easy removal of sorbents, including adsorbed pollutants from aqueous environments with a magnet. The sorption efficiencies of bentonite, ladle furnace slags and their magnetic modifications were compared. The determined sorption capacity of furnace slags and their magnetic modifications was higher than the sorption capacity of bentonite and magnetically modified bentonite. The release of Fe_xO_y NPs from the magnetically modified sorbents during the sorption experiments was monitored as well. The results of the experiments demonstrate the potential of using magnetically modified ladle furnace slags for the removal of Sr^{2+} from aqueous environments.

Keywords: Bentonite, ladle furnace slag, magnetic properties, iron oxide nanoparticles, sorption

1. INTRODUCTION

The presence of different types of metal ions in the environment is related to increasing anthropogenic activities such as the discharge of wastewater, or accidental leaks from storage tanks. Several methods including chemical precipitation [1], extraction [2], membrane separation [3], biological treatment processes [4] or sorption [5] have been investigated for the treatment of liquid solutions containing strontium. Adsorption has several advantages over the above processes, such as ease of operation, high selectivity, rapid removal of contaminants and effectiveness when a suitable adsorbent is selected. The solution of adsorbent separation from the aqueous environment is the modification of the adsorbent surface with iron oxide particles. Iron oxides with paramagnetic properties allow the fast separation of the adsorbent with a magnet.

Composite sorbents containing iron oxides in different polymer matrices [6], cellulose and alginate granules [7], chitosan [8], aluminosilicates [9] have been developed and tested. Especially, inorganic materials of aluminosilicate composition with the content of ion exchange cations Na^+, K^+, Ca^{2+} and Mg^{2+} are characterised by excellent absorption capabilities [10].

Bentonite ($Al_2O_3.4SiO_2.H_2O$) is a clay mineral mainly composed of montmorillonite (85–90%, $(Na, Ca)_{0.33}(Al, Mg)_2Si_4O_{10}(OH)_2{\cdot}nH_2O$), which exhibit a complex porous structure, high ion exchange capacity, large specific surface area, low permeability and no toxicity [10]. A similar chemical composition, particularly in terms of SiO_2 and Al_2O_3 content, is found in ladle furnace slag. Although the ladle furnace slag is a non-metallic waste output of steel production, it has found applications in building materials, soil rejuvenation or CO_2 capture [11]. Metals present in the FS are tightly bound to the matrix due to the high temperature production process, they are not easily leached and, therefore, the FS is considered environmentally safe [12]. The availability of FS, along with its polymineral composition and specific properties that determine its high sorption activity, makes FS an attractive adsorbent [12][15].

The strontium occurs natural in environment and by the anthropogenic activities belongs milling processes, coal burning, fertilizers or nuclear industry [16]. The radioactive isotope ^{90}Sr is one of the main fission products in wastes from nuclear power plant. Due to its long lifetime (half-life of 28 years), it is considered the most dangerous to human health and causes long-term pollution of ecosystems due to its high transferability, high solubility and easy assimilation in living organisms [8]. Sr^{2+} sorption has been tested in silica, iron oxide, aluminium oxides [16] or clays [9]. Therefore, two ladle furnace slags and their magnetically modified form were investigated as sorbents for elimination of Sr^{2+} from aqueous solution. The adsorption data were compared with the adsorption capacity of commercial bentonite and bentonite after magnetic modification. The stability of the prepared magnetically modified composites was also investigated. The release of Fe_xO_y NPs from the composites was monitored as well.

2. MATERIALS AND METHODS

2.1 Materials characterisation

Bentonite (denoted as B) (commercial material, < 40 μm) and two ladle furnace slags (denoted as FSI and FSII, < 100 μm) (they come from a metallurgical complex in the Moravia region of the Czech Republic) were used as substrates for Fe_xO_y NPs with magnetic properties. The chemical composition expressed in weight % was determined using X-ray fluorescence spectroscopy (SPECTRO Analytical Instruments Inc., Germany). The chemical composition of FS depends on the input materials, and is therefore more or less variable (see **Table 1**). CaO, Al_2O_3 and SiO_2 are the main phases of FSI and FSII. Lower contents of Zn, Ti, V, Cr, Sr, Zr oxides are also present and P, Ba, Co, Ni and Rb oxides are less than 0.001 wt.%. FSI and FSII also contain mixtures of iron oxides with magnetic properties: FeO (wüstite), Fe_2O_3 (hematite) and Fe_3O_4 (magnetite).

Table 1 The chemical content of bentonite (B), the main and minority chemical content of ladle furnace slags (FSI and FSII) expressed in weight % (wt.%).

The main oxides content (wt.%)				The content of the minority oxides (wt.%)		
	B	**FSI**	**FSII**		**FSI**	**FSII**
SiO_2	71.1	9.26	5.09	TiO_2	0.251	0.419
Al_2O_3	11.9	19.1	7.94	V_2O_5	0.057	<0.01
Fe_2O_3	1.89	3.72	5.11	Cr_2O_3	0.087	0.125
CaO	0.81	50.98	73.9	MnO	2.68	3.99
MgO	1.89	3.52	2.21	ZnO	<0.001	0.022
Na_2O	1.31	0.08	<0.001	SrO	0.046	<0.001
K_2O	0.56	0.02	0.02	ZrO_2	0.016	<0.001
SO_3	0.19	0.89	0.92	Nb_2O_5	0.069	<0.001

2.2 Chemicals

Fe_xO_y NPs were prepared from the precursor $FeSO_4.7H_2O$ (per analyses, MACH CHEMIKÁLIE s.r.o., Ostrava, Czech Republic). NaOH (per analyses, MACH CHEMIKÁLIE s.r.o., Ostrava, Czech Republic) was used to prepare of 1 mol/L solution. $Sr(NO_3)_2$ (per analyses, MACH CHEMIKÁLIE s.r.o., Ostrava, Czech Republic) was used to prepare solutions with various concentration of Sr^{2+} ions.

2.3 Preparation and characterisation of magnetically modified materials

Magnetically modified materials were prepared in two steps based on a verified Fe_xO_y NPs preparation method [16]. In the first step, Fe_xO_y NPs in suspension were precipitated from the iron hydroxide solution and synthesised in a microwave oven. SO_4^{2-} were washed from the suspension by decantation. In a second step, the suspension was mixed with deionised water and dried B (< 40 μm), FSI and FSII (< 100 μm) materials were separately added to the suspensions. The mixtures were filtered the following day. Magnetically modified B, FSI and FSII were denoted as MB, MFSI and MFSII, respectively. The total Fe content in MB, MFSI and MFSII was determined using atomic absorption spectrometer (AAS, UNICAM 969) after their decomposition in acid mixtures. The increase in total Fe content after magnetic modification procedure was as follows: 15.0 wt.% in MB (1.89 wt. % in B), 10.6 wt.% in MFSI (2.60 wt.% in FSI) and 11.0 wt.% in MFSII (3.57 wt.% in FSII).

2.4 Sorption experiment

0.1 g ± 0.01 g of material was mixed with 50 ml of a solution containing Sr^{2+} ions and was shaken in discontinuous rotation container (4.5 rpm) for 1 hour under laboratory conditions. The composite was then removed with a magnet and the concentration of Sr^{2+} in filtered (PRAGOPOR, 0.45 μm) solutions was determined using AAS. The adsorbed amount was calculated according to Equation 1:

$$q_e = \frac{c_0 - c_e}{m} \cdot V \qquad (1)$$

where:

- q_e – the equilibrium adsorption amount of Sr^{2+} (mg/g)
- c_0 - the initial concentration of Sr^{2+} (mg/l)
- c_e - the concentration of Sr^{2+} at the equilibrium after sorption experiment (mg/l)
- m - the weight of the sample (g)
- V - the volume of the sorption solution (l)

The experimental adsorption data were fitted using Freundlich and Langmuir isotherm models [18], [19].

The favourability of the adsorption process provides a separation parameter expressed by Equation 2 [20]:

$$R_L = \frac{1}{1 + b c_0} \qquad (2)$$

where:

- R_L - the separation factor (-)
- b - the Langmuir constant (l/mg)
- c_0 - the initial concentration of Sr^{2+} (mg/l)

The concentration of Fe released from composites during the sorption experiments, was also determined in the final extracts using AAS.

3. RESULTS AND DISCUSSION

A comparison of the Sr^{2+} adsorption capacities of B, MB, FS and MFS is shown in **Figures 1 (a-c)**.

The maximum adsorption capacity of Sr^{2+} determined by the Langmuir isotherm model (**Table 2**) was 23.75 mg/g for B and 22.32 mg/g for MB, which is higher compared to the study [5] (19.62 mg/g of Sr^{2+}). Although the Fe_xO_y NPs are participate in the sorption by forming the surface complexes between Sr^{2+} and Fe_3O_4 [21] [22], by ion exchange or complexation which allows Fe-R-OH and Fe-R-COOH groups on the MB surface [23], the Fe_xO_y NPs probably occupied some active sites on the MB surface and, therefore, the adsorption capacity of MB was slightly lower than the sorption capacity of B.

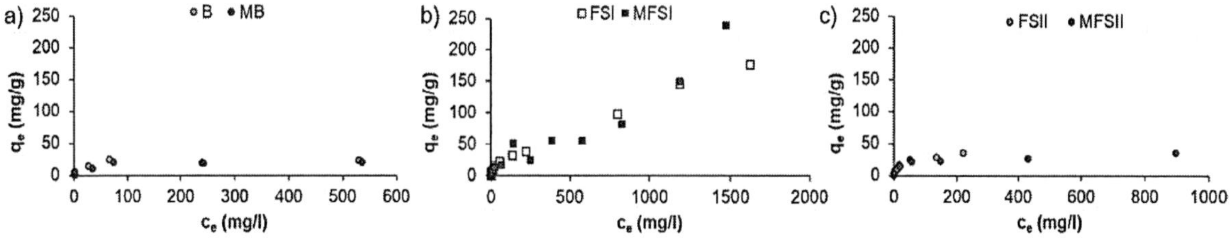

Figure 1 The amount of Sr^{2+} adsorbed on a) B, MB, b) FSI, MFSI, c) FSII, MFSII.

Maximum adsorption capacity was not reached for FSI, FSII, MFSI and MFSII, the amount of Sr^{2+}adsorbed on this sorbent increased with increasing initial Sr^{2+} concentration. The maximum adsorption capacities determined by Langmuir adsorption model were higher than the adsorption capacities on B and MB (see **Table 2**). **Figure 1 a)** shows, that the magnetic modification of FSI did not affect the adsorption capacity. Similar adsorption amount of Sr^{2+} 175.4 mg/g and 178.6 mg/g were adsorbed on the FSI and MFSI, respectively. The high uptake capacity of FSI and MFSI is a result of the alkalising ability of the FS, the Sr^{2+} probably forms hydrooxocomplexes and is related to the formation of poorly soluble compounds (silicates) on the sorbent surface. The presence of hydroxides allows the pH of the aqueous solution to be increased and catalyses the precipitation of dissolved heavy metal cations [15][24].

The amount of Sr^{2+} adsorbed on the FSII (37.45 mg/g) and MFSII (35.21 mg/g) was lower than the amount of Sr^{2+} adsorbed on the FSI and MFSI. The cause may be either a lower content of SiO_2 and Al_2O_3, a higher CaO content of the FSII and, therefore, the establishment of the chemical equilibria in the solutions during the sorption processes. Therefore, the kinetics of the adsorption process will be studied in detail in the future.

The adsorption capacity for FSI and MFSI, is better than the adsorption properties of FSII, MFSII, B and MB. The Freundlich isotherm constant 1/n was less than 1 and the R_L values were found to be $0 < R_L < 1$, which predicted the favourable adsorption of Sr^{2+} on all the tested sorbents.

Table 2 The Freundlich exponent (*n*), Freundlich isotherm constant (*k*), Langmuir constant (*b*), the maximum adsorption capacity of Sr^{2+} (q_{max}) of B, MB, FSI, MFSI, FSII, MFSII. The equilibrium adsorption amount of Sr^{2+} (q_e), the separation factor (R_L) and correlation coefficient (R^2) are also provided.

		B	MB	FSI	MFSI	FSII	MFSII
Freundlich	***n*** **(-)**	1.79	1.88	1.71	1.59	1.65	1.93
	k **(mg/g)·(l/mg)$^{1/n}$**	1.24	1.20	2.27	1.54	2.03	1.81
	R^2 (-)	0.884	0.934	0.945	0.950	0.866	0.838
Langmuir	***b*** **(l/mg)**	0.066	0.273	0.004	0.003	0.044	0.026
	q_{max} (mg/g)	23.75	22.32	175.4	178.6	37.45	35.21
	R^2 (-)	0.987	0.993	0.821	0.569	0.975	0.979
	$q_{e.}$ (mg/g)	23.9	20.6	176	240	35.8	26.0
	R_L (-)	0.028	0.027	0.009	0.010	0.065	0.017

The released amount of Fe determined in extracts after sorption experiment were lower than 0.01 mg/l, therefore, the prepared magnetically modified composites are stable. The Fe_xO_y NPs are firmly attached to the MB surface which is in agreement with our previous study [25].

4. CONCLUSION

The present study tested the elimination of strontium ions from aqueous solution using magnetically modified bentonite and two types of ladle furnace slag. Magnetic modification of the adsorbents did not significantly affect the adsorption capacity. The Sr^{2+} adsorption capacity of bentonite and magnetically modified bentonite determined by the Langmuir isotherm model was 23.75 mg/g and 22.32 mg/g, respectively. Ladle furnace slags and their magnetic modification were more efficient in elimination of strontium ions from solution than bentonite and magnetically modified bentonite. The amount of Sr^{2+} adsorbed, determined by the Langmuir isotherm model, reached 175.4 mg/g for FSI and 178.6 mg/g for MFSI. FSII and MFSII adsorbed 37.45 mg/g and 35.21 mg/g of Sr^{2+}. The adsorption amount of Sr^{2+} on the ladle furnace slags increased, therefore, the kinetics of the adsorption process will be a studied in detail the future. Iron oxide nanoparticles were not released during the sorption experiments, so the prepared composites are stable for use in an aqueous environment. In addition, ladle furnace slags and their modification are shown to be inexpensive, effective and viable sorbents for strontium removal.

ACKNOWLEDGEMENTS

The work was supported by ERDF/ESF New Composite Materials for Environmental Applications (No. CZ.02.1.01/0.0/0.0/17_048/0007399).

Research on the management of waste, materials and other products from metallurgical and related sectors (No. CZ.02.1.01/0.0/0.0/17_049/0008426).

REFERENCES

[1] LUO, X., ZHANG, G., WANG, X., GU, P. Research on a pellet co-precipitation micro-filtration process for the treatment of liquid waste containing strontium. Journal of Radioanalytical and Nuclear Chemistry. 2013, vol. 298, iss. 2, pp. 931–939.

[2] MAKRLÍK, E., VAŇURA, P., SELUCKÝ, P., HÁLOVÁ, J. Solvent extraction of microamounts of strontium and barium from water into nitrobenzene using hydrogen dicarbollylcobaltate in the presence of benzo-18-crown-6. *Journal of Radioanalytical and Nuclear Chemistry*. 2007, vol. 274, iss. 3, pp. 625–629.

[3] LI, C., LI, Z., WANG, Z., GUAN, K.C., CHIAO, Y.H., ZHANG, P.F., XU, P., GONZALES, R.R., HU, M.Y., MAI, Z.H., YOSHIOKA, T., MATSUYAMA, H. Fabrication of polydopamine/rGO Membranes for Effective Radionuclide Removal. ACS Omega. 2024, vol. 9, iss. 12, pp. 14187-14197.

[4] SINGH, S., EAPEN, S., THORAT, V., KAUSHIK, C.P., Raj, K., D'SOUZA, S.F. Phytoremediation of ^{137}cesium and ^{90}strontium from solutions and low-level nuclear waste by *Vetiveria zizanoides*. *Ecotoxicology and Environmental Safety*. 2008, vol. 69, iss. 2, pp. 306-311.

[5] LI, H., HAN, K., SHANG J., CAI, W., PAN, M., XU, D., DU, C., ZUO, R. Comparison of Adsorption Capacity and Removal Efficiency of Strontium by Six Typical Adsorption Materials. *Sustainability*. 2022, vol. 14, iss.13, no. 7723.

[6] EGORIN, A., TOKAR, E., MATSKEVICH, A., IVANOV, N., TKACHENKO, I., SOKOLNITSKAYA, T., ZEMSKOVA, L. Composite Magnetic Sorbents Based on Iron Oxides in Different Polymer Matrices: Comparison and Application for Removal of Strontium. *Biomimetics*. 2020, vol. 5, iss. 2, no. 22.

[7] TOLMACHEVA, V.V., APYARI, V.V., KOCHUK, E.V., DIMITRIENKO, S.G. Magnetic adsorbents based on iron oxide nanoparticles for the extraction and preconcentration of organic compounds. *Journal of Analytical Chemistry*. 2016, vol. 71, iss. 4, pp. 321–338.

[8] WANG, J., CHEN, C. Chitosan-based biosorbents: Modification and application for biosorption of heavy metals and radionuclides. *Bioresource Technology*. 2014, vol. 160, pp. 129-141.

[9] DRAN'KOV, A., SHICHALIN, A., PAPYNOV, E., NOMEROVSKII, A., MAYOROV, V., PECHNIKOV, V., IVANETS, A., BURAVLEV, I., YARUSOVA, S., ZAVJALOV, A., OGNEV, A., BALYBINA, V., LEMBIKOV, A., TANANAEV, I., SHAPKIN, N. Hydrothermal synthesis, structure and sorption performance to cesium and strontium ions of nanostructured magnetic zeolite composites. *Nuclear Engineering and Technology*. 2022, vol. 54, iss. 6, pp. 1991-2003.

[10] ZHANG, B., ZHU, W., HOU, R., YUE, Y., FENG, J., ISHAG, A., WANG, X., QIN, Y., SUN, Y. Recent advances of application of bentonite-based composites in the environmental remediation. *Journal of Environmental Management*. 2024, vol. 362, no. 121341.

[11] JACOB, L. Ladle Furnace Slag: Synthesis, Properties, and Applications. *ChemBioEng Reviews*. 2024, vol. 11, iss. 1, pp. 60-78.

[12] KANEL, S.R., CHOI, H., KIM, J.Y., VIGNESWARAN, S., SHIM, W.G. Removal of Arsenic(III) from Groundwater using Low-Cost Industrial By-products-Blast Furnace Slag. *Water Quality Research Journal*. 2006, vol. 41, iss. 2, pp. 130–139.

[13] ABDELBASIR, S.M., KHALEK, M.A.A. From waste to waste: iron blast furnace slag for heavy metal ions removal from aqueous system. *Environmental Science and Pollution Research*. 2022, vol. 29, pp. 57964–57979.

[14] TADIĆ, M., NIKOLIĆ, I., CUPARA, N., DJUROVIĆ, D., MILAŠEVIĆ, I. Strontium removal from aquatic solution using electric arc furnace slag: kinetic, equilibrium and thermodynamic approach. *Desalination and Water Treatment*. 2023, vol. 294, pp. 139-148.

[15] DIMITROVA, S.V. Metal sorption on blast-furnace slag. *Water Research*. 1996, vol. 30, iss. 1, pp. 228-232.

[16] MAYORDOMO, N., ALONSO, U., MISSANA, T. Effects of γ-alumina nanoparticles on strontium sorption in smectite: Additive model approach. *Applied Geochemistry*. 2019, Vol. 100, pp. 121-130.

[17] TOKARČÍKOVÁ, M., SEIDLEROVÁ, J., MOTYKA, O., ŽIVOTSKÝ, O., DROBÍKOVÁ, K., GABOR, R. Experimental verification of regenerable magnetically modified montmorillonite and its application for heavy metals removal from metallurgical waste leachates. *Journal of Water Process Engineering*. 2021, vol. 39, no. 101691.

[18] FREUNDLICH, H.M.F. Über die Adsorption in Lösungen. *Zeitschrift fur Physikalische Chemie*. 1906, vol. 57, pp. 385-470.

[19] LANGMUIR, J. The adsorption of gases on plane surfaces of glass, Mica and platinum. *Journal of American Chemical Society*. 1918, vol. 40, iss. 9, pp. 1361-1403.

[20] HALL, K.R., EAGLETON, K.R., ACRIVERS, L.C., VERMENLEM, T. Pore and solid diffusion kinetics in fixed adsorption constant pattern conditions. *Industrial & Engineering Chemistry Fundamentals*. 1966, vol. 5, iss. 2, pp. 212-223.

[21] EGORIN, A., TOKAR, E., KALASHNIKOVA, A., SOKOLNITSKAYA, T., TKACHENKO, I., MATSKEVICH, A., FILATOV, E., ZEMSKOVA, L. Synthesis and Sorption Properties towards Sr-90 of Composite Sorbents Based on Magnetite and Hematite. *Materials (Basel)*. 2020, vol. 13, iss. 5, no.1189.

[22] BOPARAI, H.K., JOSEPH, M., O'CARROLL, D.M. Cadmium (Cd^{2+}) removal by nano zerovalent iron: surface analysis, effects of solution chemistry and surface complexation modeling. *Environmental Science and Pollution Research*. 2013, vol. 20, pp. 6210–6221.

[23] YUAN, S., ZHANG, J., TAN, Z. Adsorption effect and the removal mechanism of silicate composite biochar particles on cadmium in soil. *Chemosphere*. 2022, vol. 303, no. 134970.

[24] MERCADO-BORRAYO, B.M., CONTRERAS, R., SÁNCHEZ, A., FONT, X., SCHOUWENAARS, R., RAMÍREZ-ZAMORA, R.M. Optimisation of the removal conditions for heavy metals from water: A comparison between steel furnace slag and CeO_2 nanoparticles. *Arabian Journal of Chemistry*. 2020, vol. 13, iss. 1, pp. 1712-1719.

[25] DROBÍKOVÁ, K., ŠTRBOVÁ, K., TOKARČÍKOVÁ, M., MOTYKA, O., SEIDLEROVÁ, J. Magnetically modified bentonite: Characterization and stability. *Materials Today: Proceedings*. 2021, vol. 37, pp. 53-57.

INKJET-PRINTED CHEMIRESISTIVE SENSOR ARRAY WITH INTEGRATED HEATER

[1]Vojtěch POVOLNÝ, [1]Alexandr LAPOSA, [1]Jiří KROUTIL, [1]Pavel HAZDRA

[1]*Faculty of Electrical Engineering, Czech Technical University in Prague, Prague, Czech Republic, EU,* *povolvoj@fel.cvut.cz*

https://doi.org/10.37904/nanocon.2024.5012

Abstract

An inkjet-printing method was used to fabricate a flexible gas sensor. This fabrication method is characterized by low material consumption, a fast design-to-fabrication process, and low fabrication cost. Due to these properties, this technology is increasingly popular in the fabrication of electronic circuits and devices. Gas sensors are devices generally used to detect the presence or, in more complex solutions, to measure the concentration of various gases. The fabrication technology and materials used determine sensor properties such as sensitivity, selectivity, repeatability, and stability for applications in general or industrial environments. In this research, we present the design preparation, fabrication, and characterization of the inkjet-printed gas sensor. The structures are based on silver nanoparticle ink and dielectric ink. The structures are printed on a flexible substrate and sintered using intense pulsed light and UV light, respectively. The printed sensors consist of an interdigital electrode (IDE) structure, a dielectric separation layer, and a heater pattern. The IDE structure is printed on the dielectric layer, which separates the heating and sensing structures. As a sensing layer, polyaniline (PANI) is drop-cast onto the IDE sensor structures. The inkjet-printed sensor was characterized based on its response to humidity and a test gas.

Keywords: Inkjet, gas sensor array, PET, IPL, microheater

1. INTRODUCTION

Monitoring of gaseous substances is crucial in many fields, such as industry, aviation, agriculture, healthcare, defense, and environmental monitoring [1], [2]. In response to meet the demand for sensors, a wide range of sensors, including metal oxide and conductive organic polymer-based devices, has been under intense research focus [3]. Flexibility, cost-effectiveness, light weight, and the ability to operate at room temperature are the key features that make chemical gas sensors based on polypyrrole (PPy), polyaniline (PANI), and poly(3,4-ethylene-dioxythiophene) (PEDOT) highly attractive [4]. Conducting polymers, as a class of organic materials, can be synthesized by chemical and electrochemical methods using a range of monomers in both aqueous and non-aqueous media [1]. Moreover, they provide a broad range of applications, such as energy storage [5]–[7], optoelectronic devices [8], and sensing of different gases and vapours like methanol, ethanol, acetone, benzene, NO, NO2, SO2, H2S, and some other toxic gases and vapours [9], [10].

Inkjet printing is a popular manufacturing technology today. It has gained popularity due to its low manufacturing costs, compatibility with a wide range of flexible and rigid substrates, low-temperature production process, and ease of use [11]. A wide range of electronic structures and devices can be produced with this technology, such as polymer light-emitting diodes (PLEDs) [12], nanoparticle MEMS (Micro Electro Mechanical Systems) [19], or chemiresistive gas sensors [14].

In this work, we report on the preparation of an inkjet-printed gas sensor matrix on a flexible substrate. We discuss the fabrication process, sensor's response to the detected gas, and the influence of temperature on the sensor's response.

2 MATERIALS AND METHODS

2.1 Substrate and ink

A combination of two inks was used to fabricate the presented gas sensor. For the preparation of the conductive part of the sensor, silver nanoparticle ink (ANP Silverjet DGP-40LT-15C, particle size <50 nm, solid content 35%, and resistivity 11 μΩ·cm) was used. A dielectric ink (Dycotec DM-INI-7003, Dycotec Materials Ltd, UK; volume resistivity $1.4 \cdot 10^{14}$ Ω·cm, UV curing wavelength 380 nm, UV curing energy 500 mJ·cm^{-2} – 1000 mJ·cm^{-2}) served as the isolation layer. Polyethylene terephthalate (PET) was selected as a suitable substrate for the fabrication of the sensor due to its advantageous properties for printed electronics, such as flexibility, dimensional stability, chemical resistance, and cost-effectiveness. As a result, PET is popular for low-cost, low-temperature applications in inkjet printing fabrication [15].

The printing conditions were set as follows: the metal structures were printed with the cartridge (DMC 11610), which was heated to 35 °C, at a resolution of 1016 dpi and a jetting frequency of 2 kHz. The substrate temperature was maintained at 40 °C. High-intensity pulsed light (Xenon X-1100, Xenon Corporation, USA) was used to sinter the printed structures, with a radiant energy of up to 9 J·cm^{-2} per broad-spectrum light pulse. Two flashes with durations of 300 μs and 200 J, with a 60-second interval, were applied to each sintered structure. The dielectric material was also printed with the DMC 11610 cartridge, heated to 32 °C. The printing resolution was set to 1016 dpi at 2 kHz, and the substrate was heated to 50 °C.

2.2 Synthesis of PANI ink

Polyaniline in the form of protonated emeraldine salt was prepared by the oxidation of 0.2 M aniline hydrochloride with 0.25 M ammonium persulfate at room temperature (**Figure 1**), as described in [10]. The temperature of the reaction mixture was monitored because the reaction is exothermic. The polymerization process lasted 15 minutes at a temperature of 37 °C. The resulting precipitate was filtered and washed with acetone and 0.2 M hydrochloric acid. Subsequently, it was dried on a hotplate at 60 °C for 2 hours, and then over silica gel in a desiccator for 24 hours. The PANI ink was prepared by mixing 2 mg of PANI in 1 mL xylene. A 0.5 μl aliquot of the prepared ink was deposited onto the interdigital electrode structures using a micropipette.

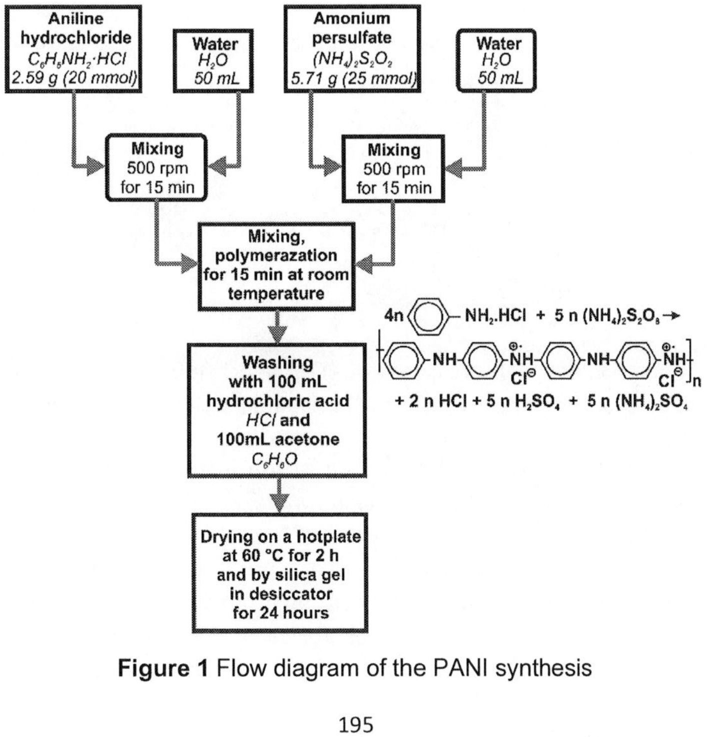

Figure 1 Flow diagram of the PANI synthesis

2.3 Sensor array fabrication

The prepared sensor array was initially divided into several individual parts, which were printed and characterized separately. Firstly, interdigital structures were printed and characterized, focusing on the appropriate resolution and the dimensions of the interdigital electrode structure. The results of the printed interdigital electrodes and dielectric mask are presented in [15]. Next, the heating elements were printed and characterized in terms of heat distribution and power consumption of the fabricated structures. The results of the prepared heaters are presented in [16]. Finally, heaters and IDE structures separated by a dielectric layer were printed and characterized. **Figure 2** shows the complete layout of the prepared sensor and each individual structure that was deposited.

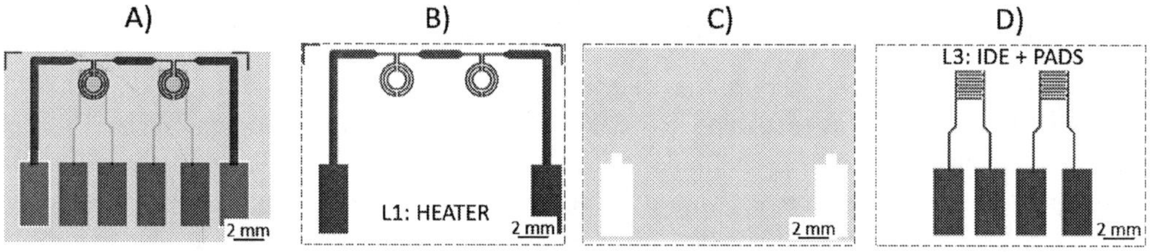

Figure 2 Layout of the prepared sensor multilayer structure: A) All layers combined together, B) First layer – heating structure printed with silver ink, C) Second layer – isolation printed with dielectric ink, D) Third layer – IDE structure and pads printed with silver ink

The sensor fabrication process began with the heater layer, which was printed in two layers. This resulted in a thickness of 1 µm, according to previous results in [15].]. The heating structure and connecting paths were printed with a resolution of 1016 dpi using 2 nozzles to achieve high uniformity in each printed layer. The connecting pads were printed with the same resolution but with four nozzles. After drying for 20 hours, the structures were sintered using an IPL system. Subsequently, the isolation layers were printed with 15 nozzles. Four layers were deposited to ensure a flat and smooth surface. Each layer was cured with a UV LED immediately after printing. After completing the printing of the dielectric structure, it was further sintered in a laboratory oven for 30 minutes at 120 °C to complete the curing process of the polymer material.

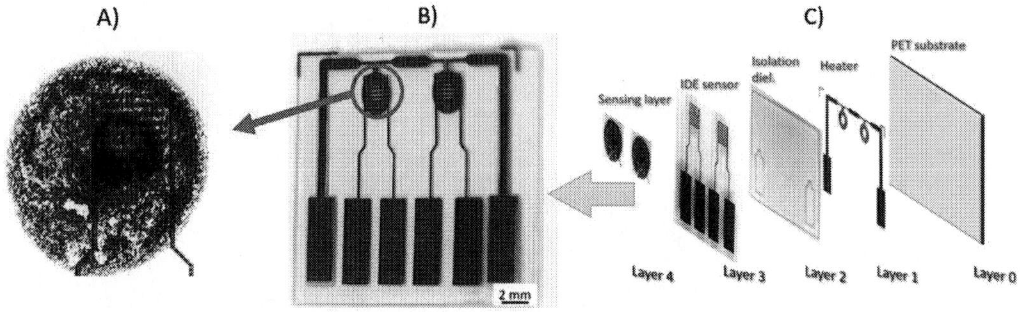

Figure 3 Platform of printed interdigital sensor array: A) PANI active layer drop cast on IDE structure, B) Complete sensor array with all layers, C) Exploded schema of the sensor array

Before the deposition of the IDE structure, the surface of the insulation layer was prepared by plasma treatment. The structures were exposed to plasma etching with SF_6 gas for 30 seconds, with the parameters set as follows: ICP power 100 W and RF generator power 30 W. After the treatment, the IDE structure was immediately printed. Two layers of silver ink were deposited. However, due to the slow drying of the printed layer, the structures were transferred to a laboratory oven and dried at 60 °C for 30 minutes. After the first layer was completely dried, the second layer was printed. Following 20 hours of drying under room conditions, the structures were sintered using the IPL system under the same conditions as before. Finally, the sensing material, PANI, was deposited using the drop-casting method. **Figure 3** shows the fully printed sensor with the deposited PANI layer and a schematic of the separately printed layers. The printed heater, IDE structure, and connecting pad with dimensions can be seen in **Figure 4**.

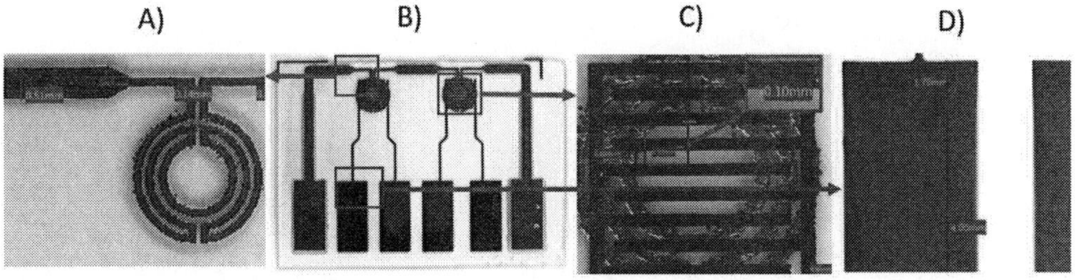

Figure 4 Details of printed sensor structures: A) Heating structure with dimensions, B) Completely printed sensor array, IDE with dimensions overprinted on dielectric and heating layers, C) Connecting pad with dimensions

3 RESULTS AND DISCUSSIONS

3.1 Temperature analysis and current-voltage characteristics of PANI

The temperature dependence of the prepared PANI layer is shown in **Figure 5a**. The temperature coefficient of resistance (TCR) of PANI was calculated in the temperature range from 23 °C to 80 °C using the formula TCR = $(\Delta R/R_0)/\Delta T$, where ΔR is the change in electrical resistance over a given temperature range, R_0 is the initial resistance, and ΔT is the temperature range. The TCR of PANI was found to be $-0.0023\ K^{-1}$. The temperature dependence of the polyaniline layers exhibits a negative temperature coefficient.

Figure 5b shows the current-voltage characteristics at various temperatures. These characteristics exhibit linear dependencies. The electrical resistivity of the PANI at 25 °C and 50 % relative humidity is 1934 Ω.

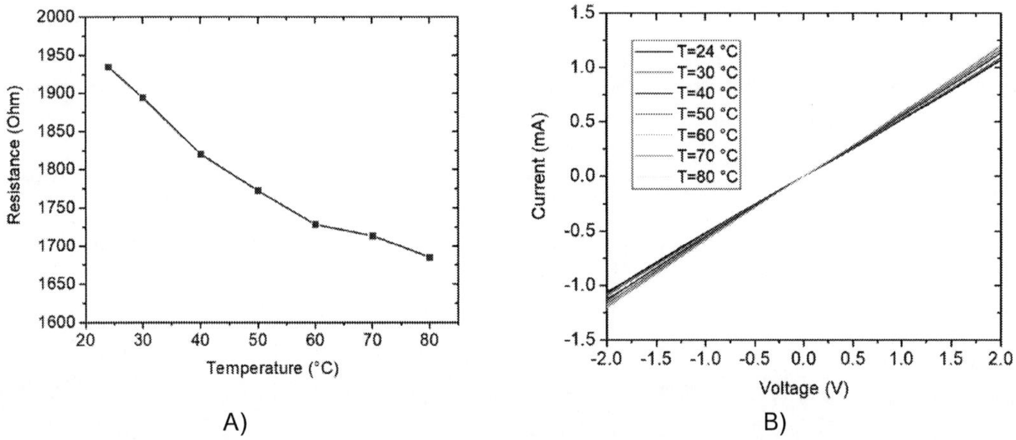

Figure 5 Temperature dependence of the PANI A) and current-voltage characteristics of PANI at various temperatures B)

3.2 Gas sensing analysis

The prepared sensor was tested in a custom-built apparatus for its response to ammonia (NH_3) and humidity (**Figure 6**). The exact concentration of ammonia was obtained by mixing it with the carrier gas. The desired humidity concentration was achieved by bubbling the carrier gas through water in the washer and then mixing it with the carrier gas. The flow rate was maintained at 200 ml·min⁻¹ using mass flow controllers. Synthetic air (SA - 21% O_2 and the remainder N_2) was used as the carrier gas and also served as the purge gas.

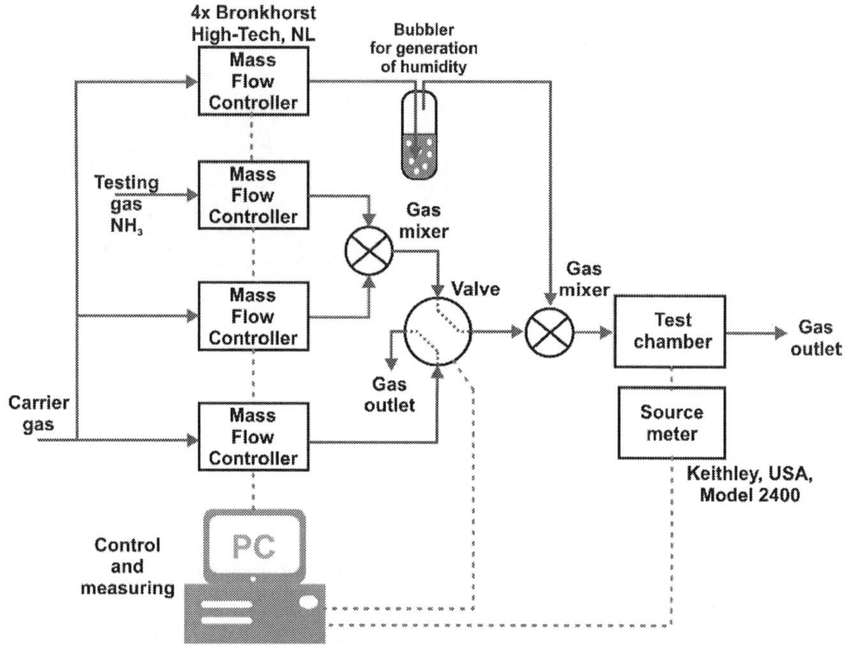

Figure 6 Schematic diagram of the customized sensing setup

The response of the PANI sensing layer was calculated using the formula for the relative resistance change, $\Delta R/R_0 = (R_s - R_0)/R_0$, where R_0 is the resistance at room temperature in synthetic air, and R_s is the resistance of the sensor in the presence of ammonia or humidity.

Figure 7 shows the response of the PANI layer to ammonia (12.5 ppm NH_3). The response of the sensor at room temperature and 50% relative humidity is shown in **Figure 7A**. Three cycles with the specified concentrations (10 minutes of exposure to the test gas, 10 minutes of synthetic air purging, flow rate 200 ml·min⁻¹) were performed during the testing with a constant voltage of 1 V.

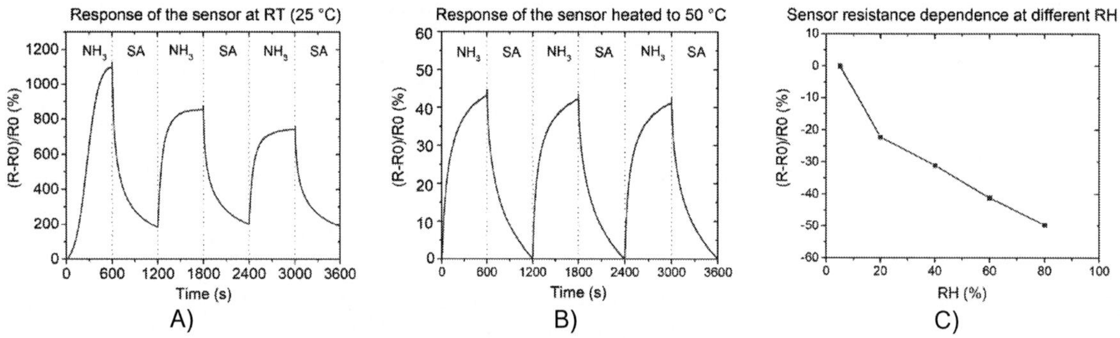

Figure 7 Gas characterization of the PANI layer: A) response to ammonia at room temperature, B) response to ammonia at 50 °C, C) dependence of PANI layer resistance on different levels of relative humidity

When polyaniline is exposed to ammonia at room temperature, it shows a relatively high response but limited recovery and stability. When compared to a conventionally manufactured sensor, we found that our sensor achieved a response that was three times higher [10]. Increasing the operating temperature to 50 °C

(**Figure 7B**) results in a decrease in response but improved stability and better recovery of the sensor resistance to the initial value. The sensing mechanism was explained in [10].

The PANI layer exhibits significant sensitivity to humidity (**Figure 7C**). Electrical resistance decreases with increasing relative humidity.

4 CONCLUSION

In this work, a fully inkjet-printed gas sensor was fabricated. Silver ink was used to print the heating (bottom) and sensing (top) layers, and dielectric ink was used to print the isolation layer. The investigation demonstrated that the interdigital structure can be printed on the dielectric layer with expected dimensions. Moreover, the active layer of polyaniline was drop-cast, and its response to the test gas was examined. The results showed that the sensor exhibits a similar response to that of conventionally produced sensors. When the sensor was exposed to the test gas, it was confirmed that heating the sensing structure is desirable to achieve a stable response and effective recovery to the baseline. Additionally, the sensor exhibits significant sensitivity to varying levels of humidity. When the PANI layer was exposed to increasing temperatures, it showed a decrease in electrical resistivity. Both the thermal and humidity responses are comparable to those of conventionally produced sensors.

ACKNOWLEDGEMENTS

This work was supported by the Czech Science Foundation grant No. 22-04533S and by the Czech Technical University Student Grant No. SGS23/182/OHK3/3T/13.

REFERENCES

[1] S. H. HOSSEINI, R. A. KHALKHALI, and P. NOOR. "Study of polyaniline conducting/electroactive polymer as sensor for some agricultural phosphorus pesticides." *Monatshefte für Chemie - Chem. Mon.* 2010, vol. 141, no. 10, pp. 1049–1053. Available from: https://doi.org/10.1007/s00706-010-0374-5.

[2] S. AKBAR, P. DUTTA, and C. LEE. "High-Temperature Ceramic Gas Sensors: A Review." *Int. J. Appl. Ceram. Technol.* 2006, vol. 3, no. 4, pp. 302–311. Available from: https://doi.org/10.1111/j.1744-7402.2006.02084.x.

[3] E. BESCHER and J. D. MACKENZIE. "Hybrid organic-inorganic sensors." *Mater. Sci. Eng. C.* Nov. 1998, vol. 6, no. 2–3, pp. 145–154. Available from: https://doi.org/10.1016/S0928-4931(98)00039-3.

[4] S. C. NAGARAJU, A. S. ROY, J. B. P. KUMAR, K. R. ANILKUMAR, and G. RAMAGOPAL. "Humidity Sensing Properties of Surface Modified Polyaniline Metal Oxide Composites." *J. Eng.* 2014, pp. 1–8. Available from: https://doi.org/10.1155/2014/925020.

[5] B. J. FELDMAN, P. BURGMAYER, and R. W. MURRAY. "The potential dependence of electrical conductivity and chemical charge storage of poly(pyrrole) films on electrodes." *J. Am. Chem. Soc.* 1985, vol. 107, no. 4, pp. 872–878. Available from: https://doi.org/10.1021/ja00290a024.

[6] H. KUZMANY, M. MEHRING, and S. ROTH. *Electronic Properties of Conjugated Polymers*, vol. 76. Berlin, Heidelberg: Springer Berlin Heidelberg, 1987.

[7] N. Li, J. Y. Lee, and L. H. Ong, "A polyaniline and Nafion composite film as a rechargeable battery," *J. Appl. Electrochem.* 1992, vol. 22, no. 6, pp. 512–516. Available from: https://doi.org/10.1007/BF01024090.

[8] A. BOYLE, E. M. GENIÈS, and M. LAPKOWSKI. "Application of the electronic conducting polymers as sensors: Polyaniline in the solid state for detection of solvent vapours and polypyrrole for detection of biological ions in solutions." *Synth. Met.* Jan. 1989, vol. 28, no. 1–2, pp. 769–774. Available from: https://doi.org/10.1016/0379-6779(89)90602-4.

[9] N. E. AGBOR, M. C. PETTY, and A. P. MONKMAN, "Polyaniline thin films for gas sensing," *Sensors Actuators B Chem.* Oct. 1995, vol. 28, no. 3, pp. 173–179. Available from: https://doi.org/10.1016/0925-4005(95)01725-9.

[10] J. KROUTIL *et al.* "Performance Evaluation of Low-Cost Flexible Gas Sensor Array With Nanocomposite Polyaniline Films." *IEEE Sens. J.* May 2018, vol. 18, no. 9, pp. 3759–3766. Available from: https://doi.org/10.1109/JSEN.2018.2811461.

[11] B. J. KANG, C. K. LEE, and J. H. OH. "All-inkjet-printed electrical components and circuit fabrication on a plastic substrate." *Microelectron. Eng.* 2012. Available from: https://doi.org/10.1016/j.mee.2012.03.032.

[12] T. R. HEBNER, C. C. WU, D. MARCY, M. H. LU, and J. C. STURM. "Ink-jet printing of doped polymers for organic light emitting devices." *Appl. Phys. Lett.* 1998. Available from: https://doi.org/10.1063/1.120807.

[13] S. B. FULLER, E. J. WILHELM, and J. M. JACOBSON. "Ink-jet printed nanoparticle microelectromechanical systems." *J. Microelectromechanical Syst.* 2002. Available from: https://doi.org/10.1109/84.982863.

[14] L. GE *et al.* "A fully inkjet-printed disposable gas sensor matrix with molecularly imprinted gas-selective materials." *npj Flex. Electron.* Jun. 2022, vol. 6, no. 1, p. 40. Available from: https://doi.org/10.1038/s41528-022-00168-6.

[15] V. POVOLNY, A. LAPOSA, J. KROUTIL, J. VOVES, M. DAVYDOVA, and P. HAZDRA. "Evaluation of printed interdigital electrodes array for chemo-resistive gas sensor application." In *Proceedings 14th International Conference on Nanomaterials - Research & Application*, 2022, pp. 45–51. Available from: https://doi.org/10.37904/nanocon.2022.4588.

[16] V. POVOLNY, A. LAPOSA, J. KROUTIL, J. VOVES, P. ASHCHEULOV, and P. HAZDRA. "Evaluation of inkjet printed heaters array for chemo-resistive gas sensor." 2023. Available from: https://doi.org/10.37904/nanocon.2023.4770.

SELF-CLEANING COATINGS FOR SAND STONES

[1]Kryštof FONIOK, [2]Blanka KOLINKEOVÁ, [2]Jiří ŠČUČKA, [1]Vlastimil MATĚJKA

[1]*VSB - Technical University of Ostrava, Faculty of Materials Science and Technology, Ostrava, Czech Republic, EU, krystof.foniok@vsb.cz, vlastimil.matejka@vsb.cz*

[2]*Institute of Geonics, Czech Academy of Sciences, Ostrava, Czech Republic, EU, blanka.kolinkeova@ugn.cas.cz, jiri.scucka@ugn.cas.cz*

https://doi.org/10.37904/nanocon.2024.5015

Abstract

In this research, we tested the performance of the self-cleaning coatings based on exfoliated g-C_3N_4 applied on the surface of sandstone. Research activities include: i) preparation of exfoliated g-C_3N_4, ii) preparation of exfoliated g-C_3N_4 water suspensions, iii) application of the suspensions on the surface of the sandstone samples, and iv) testing of the self-cleaning performance. Exfoliated g-C_3N_4 was prepared by thermal exfoliation of bulk g-C_3N_4 prepared by thermal polycondensation of melamine. Samples were characterized using elemental analysis, infrared spectroscopy, scanning electron microscopy, the band gap values were determined with help of reflectance spectra measured using UV-VIS DRS spectrometry. Rhodamine B dye was used a model substance to evaluate the photodegradation activity of prepared samples, and the results showed significantly better photodegradation performance of exfoliated sample, which was further used for application on the surface of sand stone. Water suspensions of g-C_3N_4 (2 and 4 $g \cdot dm^{-3}$) were prepared and applied on the surface of the sandstone samples using the spraying technique. Such prepared samples were subjected to the self-cleaning performance test based on the discoloration of rhodamine B under VIS irradiation and constant humidity and temperature. Changes in the a* coordinate in CIE L*a*b* color space were used to quantitatively monitor the discoloration of the tested samples.

Keywords: Graphitic carbon nitride, sand stones, photocatalysis, self-cleaning surfaces

1. INTRODUCTION

Photodegradation of environmental pollutants over photocatalytically active materials, for example, TiO_2, ZnO, and recently g-C_3N_4, has been widely studied topic for many years, and many scientific articles have been published on the use of photodegradation processes in the fields of water purification. Comprehensive reviews of these applications are summarized, for example, in the following papers [1, 2]. However, interest in research on the application of heterogeneous photocatalysis in the building industry has been noted relatively recently. The creation of photocatalytically active surfaces on the walls of buildings represents a promising way of cleaning the surrounding atmosphere, eliminating the growing of algae on the surfaces, and assuring the self-cleaning properties of the surfaces with deposited photocatalysts [3, 4]. As examples of the application of photocatalytically active TiO_2 in the building industry is the jubilee church Dives in Misericordia in Rome opened in 2003, the Hospital Manuel Gea Gonzalez in Mexico City opened in 2013 [5].

So far, the most commonly used photocatalytically active material used in building industry is titanium dioxide (TiO_2). In general, TiO_2 is considered as one of the most effective photocatalysts due to its strong oxidative ability and long-term photostability. The common crystallographic phases of TiO_2 are anatase, rutile and brookite, whereas the anatase is the TiO_2 phase showing higher photodegradation activity. Anatase is the semiconductor with large bandgap at around 3.2 eV [6]. This energy value determines that the wavelength 388 nm and lower is necessary to evoke the photocatalytic activity of anatase, and the radiation of this wavelength belongs to UV region. The day light contains around 5% of the light available for anatase activation, and thus alternatives to TiO_2 with lower band gap values are intensively searched.

Graphitic carbon nitride (g-C_3N_4) has attracted a great interest after Wang et al. [7] discovered its ability of photocatalytic water splitting. The band gap of g-C_3N_4 reaches the value 2.7 eV, what predetermines the possibility of its activation with the light of wavelength 460 nm and lower and therefore its ability to use a significantly higher proportion of daylight for its activation. G-C_3N_4 is often synthesized via thermal polymerization of melamine at the temperatures in the range of 550 – 600 °C, and the so-called bulk-g-C_3N_4 is obtained. One of the parameters limiting the bulk-g-C_3N_4 photodegradation activity is its low specific surface area, which is often reported around 10 $m^2 \cdot g^{-1}$. G-C_3N_4 has layered structure similar with graphite, and using the exfoliation, its surface area can be significantly increased, and the values in the range of 100 – 200 $m^2 \cdot g^{-1}$ are reported. Increasing of specific surface area leads to significant increase in photodegradation activity of g-C_3N_4. Despite the excellent photocatalytic properties of g-C_3N_4, only a few reports on the possibilities of using g-C_3N_4 in the field of inorganic building materials have been published in recent years, for example, [8, 9].

In our research we tested the application of both bulk-C_3N_4 and exfoliated g-C_3N_4 on the surface of sand stones with the aim to reveal its ability to ensure the self-cleaning properties of such treated surface. The self-cleaning performance of the surfaces was tested using discoloration of rhodamine B applied on the sand-stone samples.

2 MATERIALS AND METHODS

2.1 Sample preparation

Bulk g-C_3N_4 was prepared by thermal polycondensation of melamine precursor. In a typical procedure, 10 grams of melamine was put to ceramic crucible, which was covered with a ceramic lid and calcined for 4h at 550 °C in a muffle furnace. After the calcination step, the sample was let to cool down slowly in a furnace. The resulting bulk g-C_3N_4 was labelled as Bulk-gCN.

Sandstone from the locality of Mšené-lázně (Czech Republic) was chosen for the experiment because of its light and relatively homogeneous colour.

Exfoliation of g-CN was performed thermally at 600 °C. In this process, approximately 0.5 g of bulk gCN was placed on the bottom of ceramic dish to form thin layer. Such prepared sample was heated at 600 °C for 45 min and the sample was labelled TEX45.

From TEX45 sample, the suspensions were prepared in demineralized water, with concentrations of $2 g \cdot dm^{-3}$ and $4 g \cdot dm^{-3}$. The suspensions were mixed in the Omni Sonic Ruptor 400 Ultrasonic Homogenizer, for 15 minutes. The prepared suspensions were applied to the surface of test specimens of sand stones by spraying, and samples TEX45/2 and TEX45/4 were obtained.

2.2 Sample characterization

Carbon, hydrogen and nitrogen content of prepared samples was determined on instrument LECO 828 (LECO, USA). Typically, around 100 mg of sample was loaded in tin foil, in a next step the tin foil with sample was gently squeezed and such prepared sample was loaded to an autosampler.

X-ray diffraction (XRD) analysis was performed using MiniFlex600 theta/2theta diffractometer (Rigaku, Japan) equipped with Co tube (λ = 1,78897 Å) and a 1D silicon strip D/teX Ultra250 detector. The powder sample was pressed in a rotational holder, the patterns were recorded in the range 10 – 80 °2Theta with step 0.02 °, and scan speed 3 $° \cdot min^{-1}$. The obtained data were visualised using SW Origin 2020b.

Scanning Electron Microscope (SEM) eXPLORER (ASPEX, USA) was used to observe the morphology of bulk-gCN and TEX120 samples. Before the measurement the samples were coated with an Au/Pd film, and the SEM images were obtained using a backscattered electron (BSE) detector.

UV-VIS DRS spectra were registered using Shimadzu UV-2600 UV-VIS spectrometer equipped with integrating sphere 2600Plus (Shimadzu Ltd). Powder samples were placed inside a dedicated holder with quartz window and the spectra were recorded at room temperature in the range 200-800nm. Barium sulphate powder was used for base line registration. Tauc's plots [12] were used to evaluate indirect band gap energies.

Specific surface area of the samples was obtained with the method based on physisorption of nitrogen using an instrument Sorptomatic 1990 (Thermo Scientific, USA). The data of adsorbed volume of nitrogen was registered in the range of 0.05 – 0.25 of relative pressure. The obtained data were evaluated using Brunauer–Emmett–Teller (BET) method.

The photodegradation activity of the prepared samples was evaluated by photodegradation Rhodamine B (RhB). In a typical experiment 0.05 g of sample was weighted into 200 ml glass baker and 150 ml of RhB water solution (10 mg·dm^{-3}) was added. Such prepared suspension was subjected to irradiation with the light of wavelength 420 nm (FWHM 10 nm) under continuous stirring. In selected time intervals (0 and 120 min) 2ml of suspension was sampled using syringe and further filtered using syringe filter (CHROMAFIL GF/RC-20/25 filters, Macherey-Nagel, Germany) directly to 1 cm microcuvettes made of quartz glass. The absorbance of such collected filtrate was measured at 580 nm using UV-2600 spectrometer (Shimadzu, Japan). The obtained absorbance values at given irradiation time were recalculated to RhB concentration using the method of calibration curve.

The self-cleaning test was conducted using method derived from UNI 11259:2008 standard. This method is based on the discoloration of rhodamine B applied on the surface of tested samples. Our procedure included application of rhodamine B solution with concentration of 50 g·dm^{-3} on the sample surfaces using lab sprayer. After the RhB application, the samples were conditioned in a dark in climate chambre ICH110L (Memmert, Germany) for 24 h at a temperature of 20 °C and a relative humidity of 60 % The climate chamber is equipped with an illumination source simulating a day light and after 24h dark period, the relative humidity inside the chambre was raised to 70%, and the illumination was started. The colorimetric coordinates L*a*b* of the samples were measured after 4 and 26h of irradiation. The relative changes in coordinate a* at given time of irradiation in comparison with this value measured before the irradiation were calculated and expressed as R_4 and R_{26}.

Colorimetric measurements on the test specimens, with the aim to obtain the values of L*a*b* coordinates, were performed using MiniScan EZ 4500S spectrophotometer (HunterLab, VA, USA). The system uses xenon lamps, and intensity of the light reflection lies in the range of visible light spectrum (λ = 400–700 nm).

3 RESULTS AND DISCUSSION

Carbon, nitrogen and hydrogen content was determined using elemental analysis, and the results are compared for both samples in **Table 1**.

Table 1 Specific surface area and band gap energy of the prepared samples

Samples	C (wt.%)	H (wt.%)	N (wt.%)	C/N	SSA (m^2·g^{-1})	Eg (eV)
Bulk-gCN	33.6	1.81	59.6	0.56	10	2.67
TEX45	31.7	2.25	61.4	0.51	78	2.72

As evident in **Table 1**, the sum of these elemental contents does not reach 100 wt.% and for both samples, around 5 wt.% is missing and this content corresponds to oxygen, which was not determined. Carbon content decreased in the case of sample TEX45, while hydrogen and nitrogen contents increased in this sample. The C/N ratio calculated for Bulk-gCN reached the value 0.56 and indicates a non-fully polymerized gCN structure, which has this ratio 0.64. The decrease in C/N ratio observed for sample TEX45 (see **Table 1**) is in good agreement with the results observed by the other authors [10], indicating that the carbon of the tri-s-triazine building blocks was attacked and burn out during the thermal exfoliation process.

The morphology of the bulk-gCN and TEX45 particles was observed with SEM and the micrographs obtained are compared in **Figure 1**. Bulk-gCN particles (**Figure 1a**) have a compact flaky character, while TEX45 particles (**Figure 1b**) exhibit a spongy character, with distinctive pores of large size. The formation of pores in TEX45 documented with SEM supports the results of the chemical composition (**Table 1**), when the decrease in the carbon content was observed for TEX45 and was explained by burning of the carbon of the tri-s-triazine building blocks.

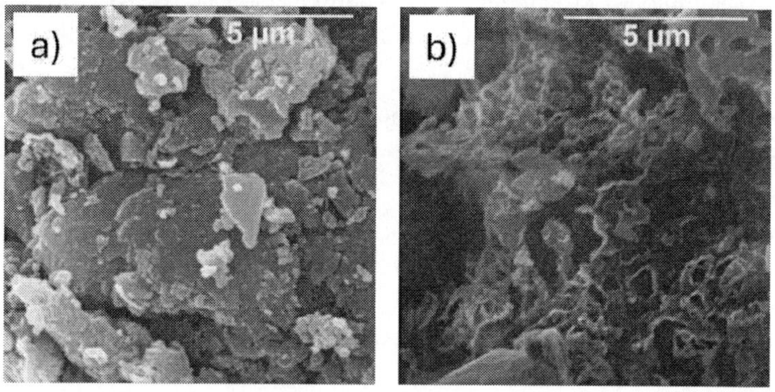

Figure 1 SEM micrographs of samples a) bulk-gCN, and b) TEX45

The structure of both samples was studied using the XRD method and the registered diffraction patterns are shown in **Figure 2a)**. The strong diffraction peak cantered at approximately 32°2θ (CoKα) corresponds to the (002) plane and is characteristic for the interlayer stacking of aromatic systems, while the low intensity diffraction peak centred at around 15°2θ (CoKα) belonging to the (100) planes corresponds to the interplanar separation [11]. The character of both diffraction patterns is closely similar and indicates that the character of $g\text{-}C_3N_4$ remained unchanged after exfoliation. A slight shift of the maxima of the TEX45 (002) diffraction peak towards higher angles is related to a denser packing of $g\text{-}C_3N_4$ layers.

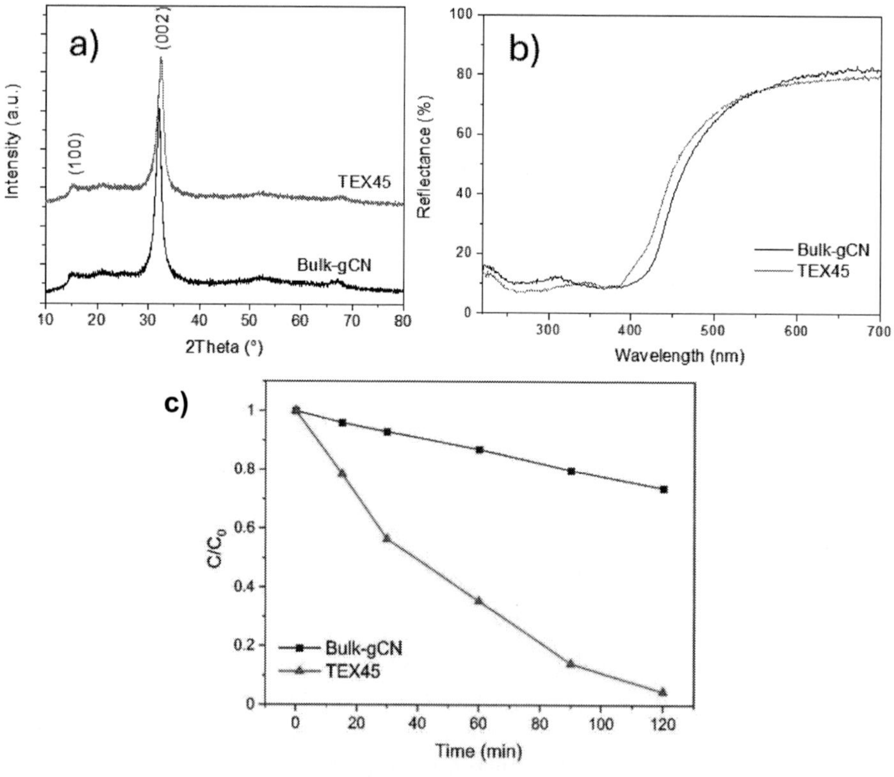

Figure 2 Comparison of photodegradation activity of tested samples bulk-gCN, and b) TEX45

Based on the photodegradation results, the TEX45 sample was used to coat of the surface of the sand stone samples and the results are summarized in **Table 2**. The pictures of the sand stone samples coloured with RhB show the progressive change in their colour appearance with irradiation time.

Table 2 Results of the self-cleaning test

Sample	R_x	Before test	After 4 h	After 26 h	Original colour
TEX45/2	R_4 = 10.5 R_{26} = 59.6		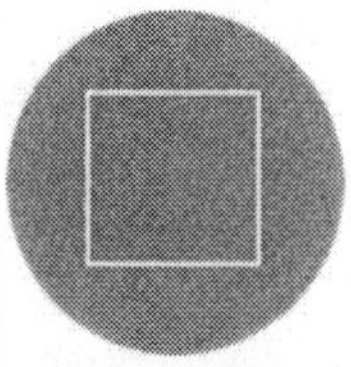		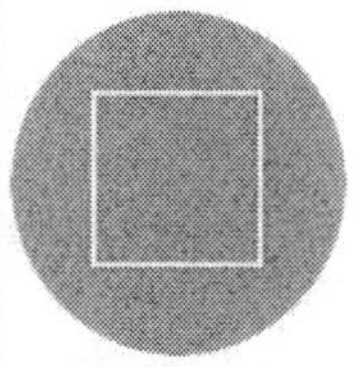
TEX45/4	R_4 = 31.5 R_{26} = 72		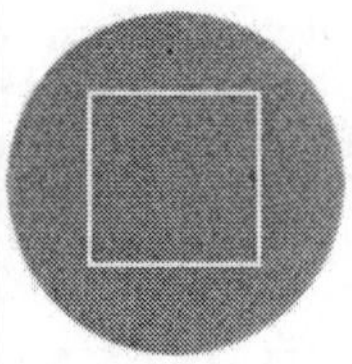	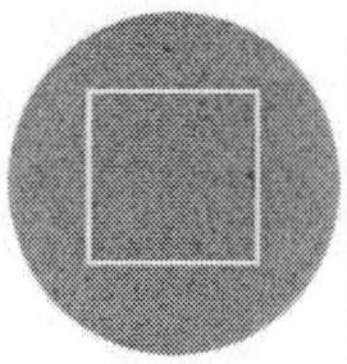	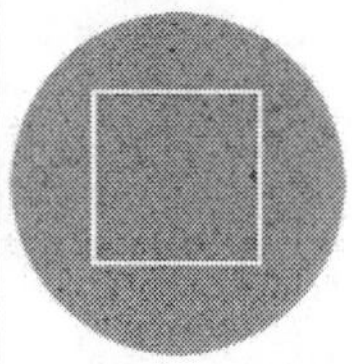

The calculated values of R4 and R26 clearly indicate that application of the TEX45 suspension of concentration of 4g·dm^{-3} (sample TEX45/4), showed significantly higher value R_{26}, and thus indicates higher self-cleaning performance of this sample.

4 CONCLUSION

The research proved that the application of exfoliated graphitic carbon nitride on the porous surface of sand stones imparted self-cleaning abilities to this important building material. The results proved a higher self-cleaning ability for the sand stone sample, whose surface was covered with a higher amount of TEX45. Sand stones are highly porous material and the solution of the RhB dye easily penetrates inside the pores. Next studies will focus on the search for the hydrophobic agent, which could easily be added to exfoliated gCN suspensions and which would block the penetration of the RhB solution to the pores of the sand stones.

ACKNOWLEDGEMENTS

This research was supported by the Czech Science Foundation (project 24-10949S), while the infrastructure has been utilized in the frame of project No. CZ.02.01.01/00/22_008/0004631 Materials and technologies for sustainable development within the Jan Amos Komensky Operational Program financed by the European Union and from the state budget of the Czech Republic. Authors also thank to project SP2024/025 (VSB – Technical University of Ostrava).

DATA AVAILABILITY

10.5281/zenodo.13927505

REFERENCES

[1] ZHAO, L., DENG, J., SUN, P., LIU, J,. JI, Y., et al. Nanomaterials for treating emerging contaminants in water by adsorption and photocatalysis: Systematic review and bibliometric analysis. Science of The Total Environment. 2018, vol. 627, pp. 1253-1263.

[2] BANG TRUONG, H., CUONG NGUYEN, X. and HUR, J. Recent advances in g–C3N4–based photocatalysis for water treatment: Magnetic and floating photocatalysts, and applications of machine-learning techniques. *Journal of Environmental Management*. 2023, vol. 345.

[3] CHEN, X., QIAO, L., ZHAO, R., WU, J., GAO, J., et al. Recent advances in photocatalysis on cement-based materials. Online. *Journal of Environmental Chemical Engineering*. 2023, vol. 11, pp. 109416

[4] KLIENCHEN DE MARIA, V.P., et al., Advances in ZnO nanoparticles in building material: Antimicrobial and photocatalytic applications – Systematic literature review. *Construction and Building Materials*. 2024, vol. 417. pp. 135337.

[5] PAOLINI, R., et al., Self-cleaning building materials: The multifaceted effects of titanium dioxide. *Construction and Building Materials*. 2018, vol. 182, pp. 126-133.

[6] ŽERJAV, G., et al., Brookite vs. rutile vs. anatase: What`s behind their various photocatalytic activities? *Journal of Environmental Chemical Engineering*. 2022, vol. 10, pp. 107722

[7] WANG, X., et al., A metal-free polymeric photocatalyst for hydrogen production from water under visible light. *Nature Materials*. 2009, vol. 8, pp. 76-80.

[8] PENG, F., et al., New g-C3N4 based photocatalytic cement with enhanced visible-light photocatalytic activity by constructing muscovite sheet/SnO2 structures. *Construction and Building Materials*. 2018, vol. 179, pp. 315-325.

[9] YANG, Y., et al., Efficiency and durability of g-C3N4-based coatings applied on mortar under peeling and washing trials. *Construction and Building Materials*. 2020, vol. 234, pp. 117438.

[10] Li, Y., et al., Tuning and thermal exfoliation graphene-like carbon nitride nanosheets for superior photocatalytic aktivity. *Ceramics International*. 2016, vol. 42, pp. 18521-18528.

[11] CAO, S., et al., Polymeric Photocatalysts Based on Graphitic Carbon Nitride. *Advanced Materials*. 2015, vol. 27, pp. 2150-2176.

THE DIFFERENCES IN THE EFFECTS OF PVP STABILIZED SILVER NANOPARTICLES ON THE MICROALGA *PORPHYRIDIUM CRUENTUM* AND THE CYANOBACTERIUM *ARTHROSPIRA PLATENSIS*

[1]Liliana CEPOI, [1]Ludmila RUDI, [1]Tatiana CHIRIAC, [2]Ion ROTARI, [1]Svetlana DJUR, [1]Anastasia CEPOI, [1]Liliana ZOSIM, [1]Iulia IATCO

[1]*Technical University of Moldova, Institute of Microbiology and Biotechnology, Chisinau, Republic of Moldova, liliana.cepoi@imb.utm.md*

[2]*Moldova State University, Doctoral School of Natural Sciences, Chisinau, Republic of Moldova, ioninhopld@gmail.com*

https://doi.org/10.37904/nanocon.2024.5011

Abstract

Nanoparticles are increasingly becoming part of microalgal biotechnologies. Certain types of nanoparticles are suggested as stimulators for lipid synthesis in microalgal biofuel production. The variety of responses from microalgal cultures depends on many factors, such as the type of nanoparticles, their size, concentration, origin, the duration of contact with microorganisms, the age of the culture, and the cultivation conditions. Another determining factor in the relationship between microorganisms and the presence of nanoparticles in their growth environment is the type of microorganisms. Eukaryotic organisms, such as microalgae, respond differently to nanoparticles than prokaryotic organisms, such as cyanobacteria. A study was conducted to determine the differences in the effects of PVP-stabilized silver nanoparticles on the microalga *Porphyridium cruentum* and the cyanobacterium *Arthrospira platensis*. Silver nanoparticles at concentrations ranging from 0.05 mg/L to 5 mg/L were added to the mineral medium from the first day of the cultivation cycle. The microorganism cultures were cultivated under autotrophic conditions. Biochemical tests were performed on the collected biomass. The cyanobacterium *Arthrospira platensis* reacted to AgNP-PVP by reducing biomass content and showing an insignificant change in structural compounds. Reduced antioxidant levels evidenced stress in cyanobacterial culture. For *Porphyridium cruentum*, biomass content increased, while lipid and carbohydrate content decreased. Higher malondialdehyde levels indicated stress in microalgal culture. The variety of responses from microalgae and cyanobacteria cultures to silver nanoparticles supports the idea that diverse mechanisms are involved in maintaining viability, which creates the possibility of remodeling biosynthetic activity depending on the intended purpose

Keywords: *Porphyridium cruentum*, *Arthrospira platensis*, PVP, silver, nanoparticles

1. INTRODUCTION

Silver nanoparticles (AgNPs) are recognized for their antimicrobial properties, which make them widely used in various products such as medical devices, textiles, cosmetics, and food packaging [1,2]. Recently, AgNPs have been exploited in biotechnologies for biofuel production, using microalgae and cyanobacteria as producer substrates [3]. Due to their stimulating effects, which involve reducing toxicity, AgNPs can be employed in phycological technologies to produce biologically active substances [4].

One key aspect of this effect is reducing AgNP aggregation and silver ion (Ag^+) release. Organic coatings and stabilizing substances address this issue [5]. To enhance colloidal stability and control the release of Ag+, nanoparticles are often coated with stabilizers such as polyvinylpyrrolidone (PVP). PVP is a polymer frequently used in nanoparticle synthesis because it prevents aggregation through steric repulsion. It forms a physical barrier on the surface of the nanoparticles, limiting their direct interactions and thus reducing the risk of forming large aggregates, otherwise decreasing nanoparticle bioavailability [2,5].

One of the most significant effects of PVP is reducing the release of Ag^+ ions, considered the primary cause of acute toxicity to aquatic organisms. Silver nanoparticles coated with PVP have shown a slower release of Ag+ ions, reducing toxicity in microalgae species tests [1]. AgNP-PVP particles have been detected in microalgae's periplasmic space and cytoplasm without significantly affecting their growth or photosynthesis [2]. The reduced aggregation of AgNP-PVP allowed for the maintenance of microalgal viability, lowering the generation of free radicals [6]. These findings suggest that PVP stabilizes AgNPs and moderates their toxicity by reducing oxidative stress.

The stimulating effect of AgNPs depends not only on the type of nanoparticles, their size, and concentration but also on the specific microalgae species with which they interact [7].

This research aimed to evaluate and compare the effects of AgNP-PVP on the cyanobacterium *Arthrospira platensis* (a prokaryotic organism) and the microalga *Porphyridium cruentum* (a eukaryotic organism), focusing on how they induce oxidative stress and influence biomass accumulation and biosynthetic activity.

2. EXPERIMENTAL DESIGN

The cyanobacterium *Arthrospira platensis* CNMN-CB-02 and the microalga *Porphyridium cruentum* CNMN-AR-01, stored in the National Collection of Non-Pathogenic Microorganisms at the Institute of Microbiology and Biotechnology of the Technical University of Moldova, were cultivated in mineral media under photoautotrophic conditions, with continuous illumination and constant temperature. The cultivation cycle lasted 6 days for *Arthrospira platensis* and 12 days for *Porphyridium cruentum*. Silver nanoparticles stabilized with PVP (AgNPs-PVP), with a size of 20 nm ± 2 nm, were added to the cultivation media at various concentrations.

Biomass content was determined spectrophotometrically. In the collected biomass, lipid content was measured using the phospho-vanillin reagent, and carbohydrate content was assessed using the anthrone reagent. The malondialdehyde (MDA) level was quantified using thiobarbituric acid reactive substances (TBARS). Antioxidant activity was evaluated using the ABTS (2,2'-azino-bis (3-ethylbenzothiazoline-6-sulfonic acid) assay on 55% ethanol extracts from the collected biomass. All experiments were performed in triplicate, and statistical data analysis was conducted using Microsoft Excel software (Microsoft 365 Excel version 2108).

3. RESULTS AND DISCUSSION

This study investigated the effects of PVP-stabilized AgNPs on the growth and biosynthetic activity of two representative species, the cyanobacterium *Arthrospira platensis* and the microalga *Porphyridium cruentum*. Changes in biomass content primarily indicated the presence or absence of the toxic effect of AgNPs. **Figure 1** shows the effects of AgNPs-PVP on growth and biomass accumulation in *Arthrospira platensis* and *Porphyridium cruentum* cultures.

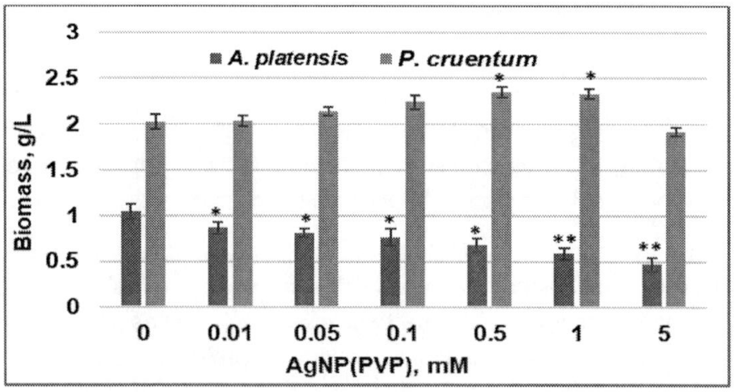

Figure 1 Changes in biomass content (g/L) of *Arthrospira platensis* and *Porphyridium cruentum* cultures under the action of AgNPs-PVP; *$p < 0.05$; $p < 0.01$

Applying AgNPs-PVP at the studied concentrations led to a significant reduction in biomass in the *Arthropsira platensis* culture. At a concentration of 0.01 mM AgNPs, biomass decreased by 17.3% ($p < 0.05$), while at 5 mM, the reduction was 55.4% ($p < 0.01$). A strong negative correlation was established between the concentration of nanoparticles in the cultivation medium and the final biomass values of the cyanobacterium, with an $r^2 = -0.8557$. In contrast, applying AgNPs-PVP to the *Porphyridium cruentum* culture resulted in a different response about biomass accumulation. Concentrations ranging from 0.01 mM to 0.1 mM did not affect biomass production or cause a slight, non-significant increase. At concentrations of 0.5 mM and 1.0 mM, the microalgal biomass content increased by 16.4% ($p < 0.05$) and 15.3% ($p < 0.05$), respectively. The correlation between AgNP concentrations and biomass values indicates a strong negative relationship, with an $r^2 = -0.6057$, suggesting a trend of decreasing *Porphyridium cruentum* biomass as AgNP concentration increases.

Figure 2 graphically presents the results regarding the accumulation of lipids, carbohydrates, and oxidative stress severity (MDA) in the biomass of the investigated cultures.

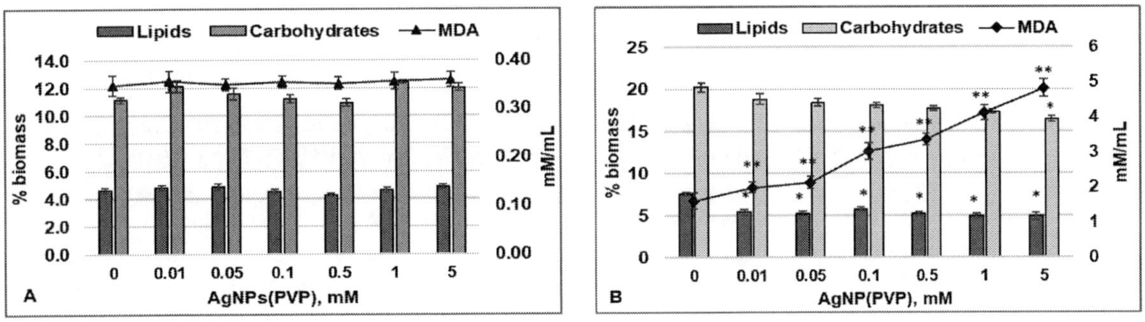

Figure 2 Changes in carbohydrate content (% biomass), lipid content (% biomass), and MDA (mM/mL) in the biomass of (**A**) *Arthrospira platensis* and (**B**) *Porphyridium cruentum* under the action of AgNPs-PVP; *$p < 0.05$; **$p < 0.01$

The carbohydrate and lipid content indices provide valuable information about maintaining cellular membrane integrity, reflecting the biosynthesis of structural and storage components. Malondialdehyde (MDA) is a marker of the severity of oxidative stress in cells. In the case of the cyanobacterium *Arthrospira platensis*, where AgNPs-PVP inhibited biomass production, the biosynthetic activity of the culture was not affected (**Figure 2A**). The lipid and carbohydrate content remained at the same level as the control. Only at the 5 mM concentration of AgNPs-PVP was a significant reduction of 18.8% ($p < 0.05$) observed in carbohydrate content. MDA values did not change, suggesting that, against a backdrop of reduced biosynthetic activity, excessive free radical production did not occur. A completely different results was established in the microalgae *Porphyridium cruentum* culture exposed to the action of AgNPs-PVP (**Figure 2B**). In all experimental variants, the carbohydrate content in the microalgal biomass did not change, remaining at the same level as in the control. In contrast, the lipid content significantly decreased due to AgNPs-PVP exposure, with this effect being observed at all applied concentrations. The reduction in lipid content ranged from 24.3% ($p < 0.05$) to 35.6% ($p < 0.05$) below control values. A strong negative correlation, $r^2 = -0.6204$, was determined between the nanoparticle concentrations and lipid content values. In the final biomass of *Porphyridium cruentum*, MDA test values were significantly higher than the control. AgNPs-PVP concentrations of 0.01 mM and 0.05 mM led to an increase in malondialdehyde content by 23.6% ($p < 0.01$) and 32.3% ($p < 0.01$), respectively. At a concentration of 1.0 mM AgNPs, the MDA value increased by 88.2% ($p < 0.01$). Additionally, concentrations of 0.5 mM and 1.0 mM caused a doubling of malondialdehyde content, which was significant. At 5 mM AgNPs, MDA values tripled compared to the control. Lipid biosynthesis supported microalgal biomass production, but the lipid content decreased due to oxidative degradation of cellular membranes. However, the maintenance of biomass productivity in the presence of AgNPs-PVP suggests that biosynthetic activity was not significantly affected. The reduction in lipid content appears to be a necessary mechanism to mitigate the oxidative stress induced by the nanoparticles. Notably, the decrease in lipid values is not critical for the *Porphyridium cruentum* culture, indicating some metabolic flexibility of this species in the face of oxidative stress induced by this type of nanoparticles.

Antioxidant tests can verify cell stress, which can either increase or decrease. **Figure 3** analyzes the ABTS antioxidant test values for the hydro-ethanolic extracts from the biomass of the studied cultures.

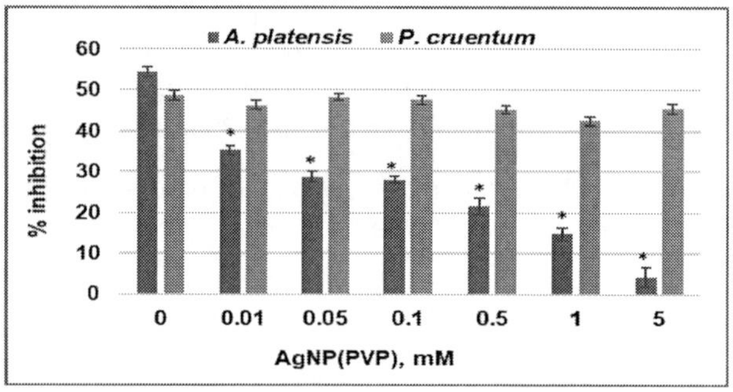

Figure 3 Changes in antioxidant activity (% ABTS inhibition) of *Arthrospira platensis* and *Porphyridium cruentum* biomass obtained from cultivation in the presence of AgNPs-PVP; *$p < 0.01$

In the samples obtained from *Arthrospira platensis* biomass, antioxidant activity was significantly reduced under the influence of AgNPs-PVP. Thus, the ABTS test value for the sample treated with 0.01 mM AgNPs decreased by 35% ($p < 0.01$). As the nanoparticle concentration increased, antioxidant activity continued to decline, reaching a 92% reduction ($p < 0.01$) at a concentration of 5 mM AgNPs. The relationship between AgNP concentrations and antioxidant test values indicates a strong negative correlation, $r^2 = -0.8829$. For the microalga *Porphyridium cruentum*, the stress experienced by the cells was not reflected in changes in the antioxidant test values, which remained at the control level.

The study highlighted clear differences in the effects of AgNPs-PVP on the cyanobacterium *Arthrospira platensis* and the microalga *Porphyridium cruentum*. In the case of *Arthrospira platensis*, biomass production was reduced, accompanied by stress indicated by a decrease in antioxidant activity without significant changes in structural compounds. On the other hand, *Porphyridium cruentum* showed increased biomass but with a reduction in lipid and carbohydrate content and an increase in oxidative stress, indicated by elevated malondialdehyde levels. These differentiated responses suggest that eukaryotic microalgae and prokaryotic cyanobacteria adopt distinct adaptive strategies in the presence of silver nanoparticles. These results open the possibility of optimizing the use of nanoparticles in microalgal biotechnologies, depending on the type of organism and the specific objectives pursued. The toxicity of AgNPs on aquatic microorganisms has been extensively studied and described in several works. It was reported that *Chlorella vulgaris* significantly loses viable cells at high concentrations of 200 mg/L AgNPs [8]. Another study showed that the freshwater species *Scenedesmus* sp. was more affected by AgNPs compared to the marine diatom *Thalassiosira* sp., suggesting that AgNPs exhibit different toxicity mechanisms for microalgae and are more toxic in freshwater environments than in marine ones [9]. The differences in toxicity between marine and freshwater environments of AgNP-PVP, where marine species seem to be less sensitive to the toxicity of silver nanoparticles compared to freshwater species, are attributed to the greater complexity of the marine environment, which reduces the availability of silver ions released from the nanoparticles. It has been observed that the toxicity of silver nanoparticles is lower than that of directly released silver ions. However, toxicity varies significantly depending on the initial concentration of AgNPs and the exposed microalgae species. It has been established that the toxicity of PVP-coated silver nanoparticles varies considerably depending on testing methodologies, environmental parameters, and microalgae species. Thus, for *Phaeodactylum tricornutum*, the EC50 was 0.06 mg/L, for *Isochrysis galbana* 0.039 mg/L, and for *Tetraselmis suecica* only 0.0052 mg/L [1]. PVP-stabilized silver nanoparticles are considered less toxic because PVP ensures their stability in the environment, preventing aggregate formation and reducing the release of large amounts of Ag+ ions. However, a positive correlation has been established between AgNP-PVP concentration and growth inhibition of the microalga *Raphidocelis subcapitata* [10], with a response similar to that obtained in this study for *Arthrospira platensis* and AgNPs-PVP. Another study demonstrates the toxicity of Polyvinylpyrrolidone coating substances in combination with Polyethyleneimine, which, when applied to 5 nm silver nanoparticles, significantly inhibited the growth of marine microalgae *Isochrysis galbana*, *Phaeodactylum tricornutum*, and *Tetraselmis suecica*. The coating products are said to promote the release of silver ions [11]. As can be seen, the results are very contradictory. The reactions of aquatic cultures to the action of AgNPs have been investigated depending on their size. Thus, the freshwater microalga *Chlamydomonas reinhardtii* and the marine microalga *Phaeodactylum tricornutum* were sensitive to small-sized AgNPs, resulting in growth inhibition at low concentrations. In contrast, for larger-sized nanoparticles, the cells exhibited a hormetic response [12]. Such an effect can be assumed for the microalga *Porphyridium cruentum*, which tended to reduce biomass production at the high concentrations of AgNPs-PVP applied in this study. Differences in the response of the cyanobacterium *Arthrospira platensis* and the microalga *Porphyridium cruentum* were also observed in the case of the action of citrate-stabilized AgNPs [7]. The lipid peroxidation effect has been described for the microalgae *Chlorella vulgaris* and *Dunaliella tertiolecta*, where exposure to AgNPs led to the formation of cell aggregates due to the amplification of the lipid peroxidation process. At a concentration of 1 mg/L, lipid peroxidation increased by 45% for *Chlorella vulgaris* and by 48% for *Dunaliella tertiolecta*. At 10 mg/L, lipid peroxidation increased fourfold for *Chlorella vulgaris* and fifteenfold for *Dunaliella tertiolecta* [6]. In contrast, AgNPs-PVP did not induce lipid peroxidation in *Chlorella vulgaris* [2]. In the present study, this effect was established for *Arthrospira platensis*, for which MDA values did not change.

4. CONCLUSIONS

The cyanobacterium *Arthrospira platensis* and the microalga *Porphyridium cruentum* exhibit different responses to AgNPs-PVP in their cultivation media. *Arthrospira platensis* experienced a decrease in biomass without significant cellular content changes. At the same time, *Porphyridium cruentum* maintained biomass productivity but showed signs of oxidative stress and a decrease in lipid content. Oxidative stress was confirmed based on different indicators. For *Arthrospira platensis*, oxidative stress manifested through a significant reduction in the antioxidant activity of the biomass, measured by the ABTS test. For *Porphyridium cruentum*, oxidative stress was confirmed by increased malondialdehyde (MDA) levels, a marker of lipid peroxidation. These differences suggest that microalgae and cyanobacteria employ different adaptation mechanisms to nanoparticle-induced stress, offering opportunities for optimizing nanoparticle use depending on the organism studied and the biotechnological goal pursued.

ACKNOWLEDGEMENTS

This research was funded by Government of Republic of Moldova, Ministry of Education and Research, Research Subprogram 020101, Institutional financing contract no. 4/FI of 22 February 2024.

REFERENCES

[1] MORENO-GARRIDO, I., PÉREZ, S., BLASCO, J. Toxicity of silver and gold nanoparticles on marine microalgae. *Marine Environmental Research*. 2015, vol. 111, pp. 60-73. Available from: https://doi.org/10.1016/j.marenvres.2015.05.008

[2] BIBA, R., KOŠPIĆ, K., KOMAZEC, B., MARKULIN, D., CVJETKO, P., PAVOKOVIĆ, D., PEHAREC ŠTEFANIĆ, P., TKALEC, M., BALEN, B. Surface coating-modulated phytotoxic responses of silver nanoparticles in plants and freshwater green algae. *Nanomaterials.* 2022, vol. 12, no. 24, pp.1-29. Available from: https://doi.org/10.3390/nano12010024.

[3] VARGAS-ESTRADA, L., TORRES-ARELLANO, S., LONGORIA, A., ARIAS, D.M., OKOYE, P.U., SEBASTIAN, P.J. Role of nanoparticles on microalgal cultivation: A review. *Fuel.* 2020, vol. 280, no. 118598. Available from: https://doi.org/10.1016/j.fuel.2020.118598.

[4] YUAN, X., GAO, X., LIU, C., LIANG, W., XUE, H., LI., Z., JIN, H. Application of Nanomaterials in the Production of Biomolecules in Microalgae: A Review. *Marine Drugs.* 2023, vol. 21, no. 11, 594. Available from: https://doi.org/10.3390/md21110594.

[5] RÓNAVÁRI, A., BÉLTEKY, P., BOKA, E., ZAKUPSZKY, D., IGAZ, N., SZERENCSÉS, B., PFEIFFER, I., KÓNYA, Z., KIRICSI, M. Polyvinyl-Pyrrolidone-Coated Silver Nanoparticles - The Colloidal, Chemical, and Biological Consequences of Steric Stabilization under Biorelevant Conditions. *International Journal of Molecular Sciences.* 2021, vol. 122, no. 16, 8673. Available from: https://doi.org/10.3390/ijms22168673.

[6] OUKARROUM, A., BARHOUMI, L., PIRASTRU, L., DEWEZ, D. Silver nanoparticle toxicity effect on growth and cellular viability of the green alga *Chlorella vulgaris. Environmental Toxicology and Chemistry.* 2012, vol. 31, no. 7, pp.1633-1642. Available from: https://doi.org/10.1002/etc.1887.

[7] CEPOI, L., RUDI, L., CHIRIAC, T., VALUTA, A., DJUR, S., RUDIC, V. Effect of gold and silver nanoparticles on cyanobacterium *Arthrospira platensis* and microalga *Porphyridium cruentum.* In: 15th International Conference on Nanomaterials - Research & Application (NANOCON 2023). Brno: Tanger Ltd, 2023, pp. 340-345. Available from: https://doi.org/10.37904/nanocon.2023.4759.

[8] HAZEEM, L.J., KUKU, G., DEWAILLY, E., SLOMIANNY, C., BARRAS, A., HAMDI, A., BOUKHERROUB, R., CULHA, M., BOUOUDINA, M. Toxicity Effect of Silver Nanoparticles on Photosynthetic Pigment Content, Growth, ROS Production and Ultrastructural Changes of Microalgae *Chlorella vulgaris. Nanomaterials.* 2019, vol. 9, no. 914. Available from: https://doi.org/10.3390/nano9070914.

[9] PHAM, T.L. Effect of Silver Nanoparticles on Tropical Freshwater and Marine Microalgae. *Journal of Chemistry.* 2019, 7 pages. Available from: https://doi.org/10.1155/2019/9658386.

[10] LEKAMGE, S., MIRANDA, A.F., ABRAHAM, A., BALL, A.S., SHUKLA, R., NUGEGODA, D. The toxicity of coated silver nanoparticles to the alga *Raphidocelis subcapitata. SN Applied Sciences.* 2020, vol. 2, no. 596. Available from: https://doi.org/10.1007/s42452-020-2430-z.

[11] SCHIAVO, S., DUROUDIER, N., BILBAO, E., MIKOLACZYK, M., SCHÄFER, J., CAJARAVILLE, M.P., MANZO, S. Effects of PVP/PEI coated and uncoated silver nanoparticles and PVP/PEI coating agent on three species of marine microalgae. *Science of the Total Environment.* 2016, vol. 565, pp. 102-112. Available from: https://doi.org/10.1016/j.scitotenv.2016.10.051.

[12] SENDRA, M, YESTE, M.P, GATICA, J.M., MORENO-GARRIDO, I, BLASCO, J. Direct and indirect effects of silver nanoparticles on freshwater and marine microalgae (*Chlamydomonas reinhardtii* and *Phaeodactylum tricornutum*). *Chemosphere.* 2017, vol. 179, pp. 279–289. Available from: https://doi.org/10.1016/J.CHEMOSPHERE.2017.03.123.

SESSION C
BIONANOTECHNOLOGY, NANOMATERIALS IN MEDICINE

Chairmen

Prof. Dr. Anton FOJTIK	Czech Technical University in Prague, Czech Republic, EU
Prof. Dr. Jiri HOMOLA	Institute of Photonics and Electronics of the CAS, Prague, Czech Republic, EU
Prof. Dr. Bohuslav REZEK	Czech Technical University in Prague, Czech Republic, EU

ENANTIOSELECTIVE SERS DETECTION USING CHIRAL AU HELICOIDS

[1]Andrea TOSOVSKA, [1]Anastasia SKVORTSOVA, [1]Vaclav SVORCIK, [1]Oleksiy LYUTAKOV

[1]Department of Solid State Engineering, University of Chemistry and Technology, Prague, Czech Republic, EU, tosovska@vscht.cz

https://doi.org/10.37904/nanocon.2024.5020

Abstract

The interaction between light and matter can be significantly enhanced through the excitation of surface plasmons and related nanospace-restricted focusing of light energy. The local enhancement of light energy is especially pronounced in the case of so-called plasmonic hotspots, which have found wide application in surface-enhanced Raman spectroscopy (SERS). This work focuses on the preparation and utilization of ordered array of chiral gold nanoparticles. The nanoparticles used have unique helical shapes with chirality encoded in each individual nanoparticle. An ordered layer of Au helices with a consistent gap between nanoparticles on a plasmon-active gold grating was fabricated. The created structure was able to support the excitation and propagation of surface plasmon-polaritons and local plasmons. Chiral molecules of the analyte(s) were placed between Au helicoids (i.e., in the place of the plasmonic hot spots). The influence of the chiral dielectric „spacer" (i.e., added chiral molecules) on the intensity of the local electric field and the SERS response of the enantiomers was investigated. A significant difference in the enhancement of SERS signal from enantiomers was observed, depending on the interplay of helicoids chirality and analytes chirality.

Keywords: Nanoparticles, gold nanoparticles, SERS, plasmon

1. INTRODUCTION

The chiral nature of molecules represents a fascinating aspect of chemistry and biochemistry, ranged from deeper understanding of molecular interactions to the potential for use in various practical applications. Considering the different impact and properties of opposite enantiomers the ability to distinguish between enantiomers is the key concept in analytical chemistry. Enantiomers can be distinguished using 3D nanostructures that interact differently with polarized light based on their shape.[1]

In this work, we used plasmon-active nanoparticles with specific shape, where chirality is encoded in each individual particle. The main advantage of such structures is their ability to support chiral plasmonic distribution in the near field, thanks to the unique shape of the particle, offering numerous benefits.[2]

Chiral plasmonic nanostructures are gradually being integrated into the field of Surface-enhanced Raman Spectroscopy (SERS).[3] In the pharmaceutical industry, SERS provide sensitive means for distinguishing between drugs and bio-active molecules, which in crucial for ensuring its efficacy and safety. In addition, enantioselective SERS discrimination offers an elegant method for rapidly obtaining structural and optical information related to molecular steric conformation, even in cases of low analytic concentrations or small sample volumes.[2]

The effort to precisely distinguish enantiomers using Raman spectroscopy has a relatively long history, beginning with the SEROA (Surface Enhanced Raman Optical Activity) technique. Recently, various strategies have been developed for the enantioselective capture of target analyte molecules on plasmon-active surface for subsequent SERS measurement and enantioselective discrimination.[3] Although these approaches are well-studied and sometimes yield satisfactory results, they have limitations mainly related to the lack of universality. In such instances, the use of chiral nanostructures capable of supporting localized plasmon should be considered.[4] It is well known that the intensity and strength of the local plasmonic wave depend on the coupling of plasmon-active nanostructures, which is influenced by the gaps between the nanostructures and the surrounding dielectric environment. Since plasmonic nanostructures create chiral plasmonic fields, it can be supposed that local values of plasmonic energy will be directed by the chirality of the surrounding dielectric environment. In this context, improved coupling and higher values of the local electric field are expected when the plasmonic chirality aligns with the chirality of the environment. However, this phenomenon may also have another effect – an optically active dielectric environment with appropriate chirality will exhibit a higher local electric field value, thereby enhancing its SERS response.[5]

2. EXPERIMENTAL

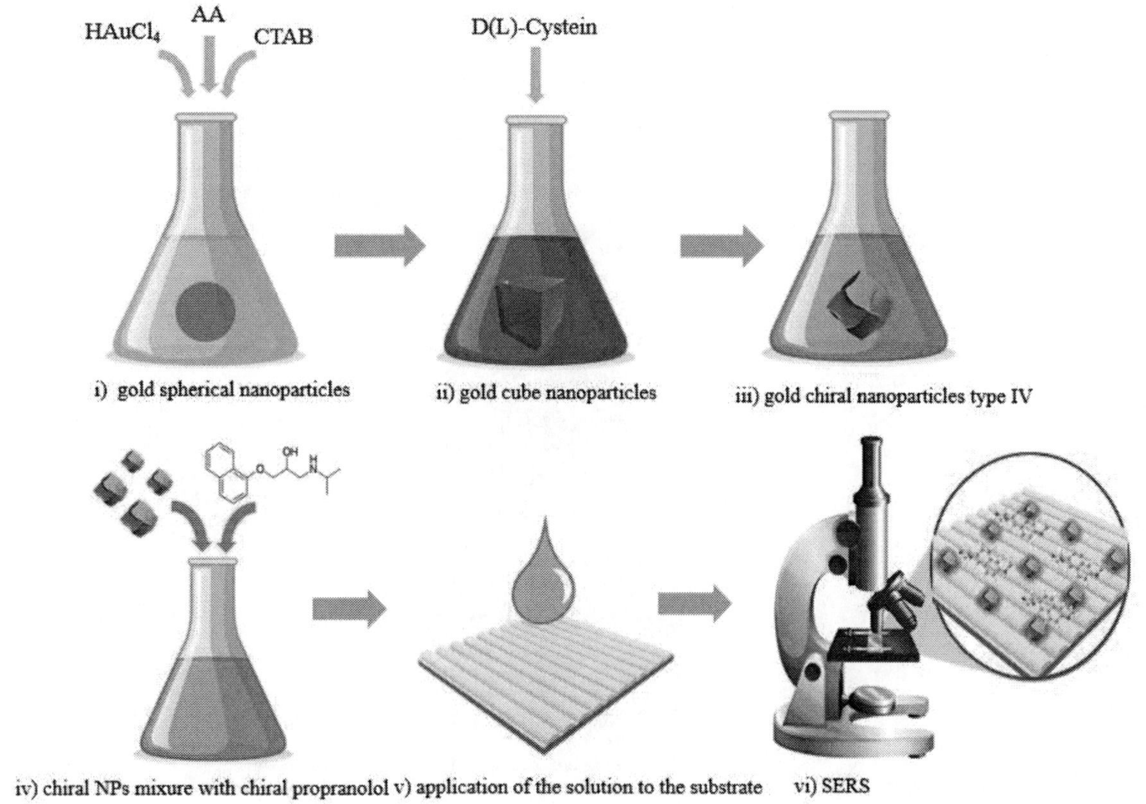

Figure 1 Schematic representation of the gold NPs synthesis and sample preparation for SERS measurement

In the initial phase of the synthesis (**Figure 1**, steps i–iii), chiral gold nanoparticles (type IV) were prepared. Step i) involved the synthesis of cubic gold nanoparticles by adding a growth solution, containing HAuCl4, ascorbic acid, and CTAB, to a solution of spherical gold nanoparticles. This step resulted in a colour change of the solution from brown to pink. Step iii) is the critical step in synthesizing chiral gold nanoparticles and forming type IV chiral structures. In this step, the growth solution along with one of the enantiomers of cysteine was added to the cubic nanoparticles. This reaction was accompanied by a colour change of the solution from pink to blue. Ascorbic acid acted as a reducing agent, converting Au3+ to Au+, while CTAB served as a stabilizer. By blocking the surface of the cubic nanoparticles, chiral cysteine facilitated the formation of chiral gold nanoparticles of type IV.

Step iv) details the preparation of the sample. The chiral analyte, specifically propranolol, was mixed with a solution containing chiral gold nanoparticles at a predetermined concentration.

Then, the solution prepared in step iv) was drop deposited on the substrate made from DVD disc (step v). Before this, the textured side of the disc was cleaned with ethanol, then coated with gold using vacuum sputtering (40 mA for 350 seconds). Once the drop had dried, the sample was ready for surface-enhanced Raman spectroscopy (SERS) measurements (step vi).

3. RESULTS AND DISCUSSION

3.1 Characterization of nanoparticles

3.1.1 Cubic nanoparticles

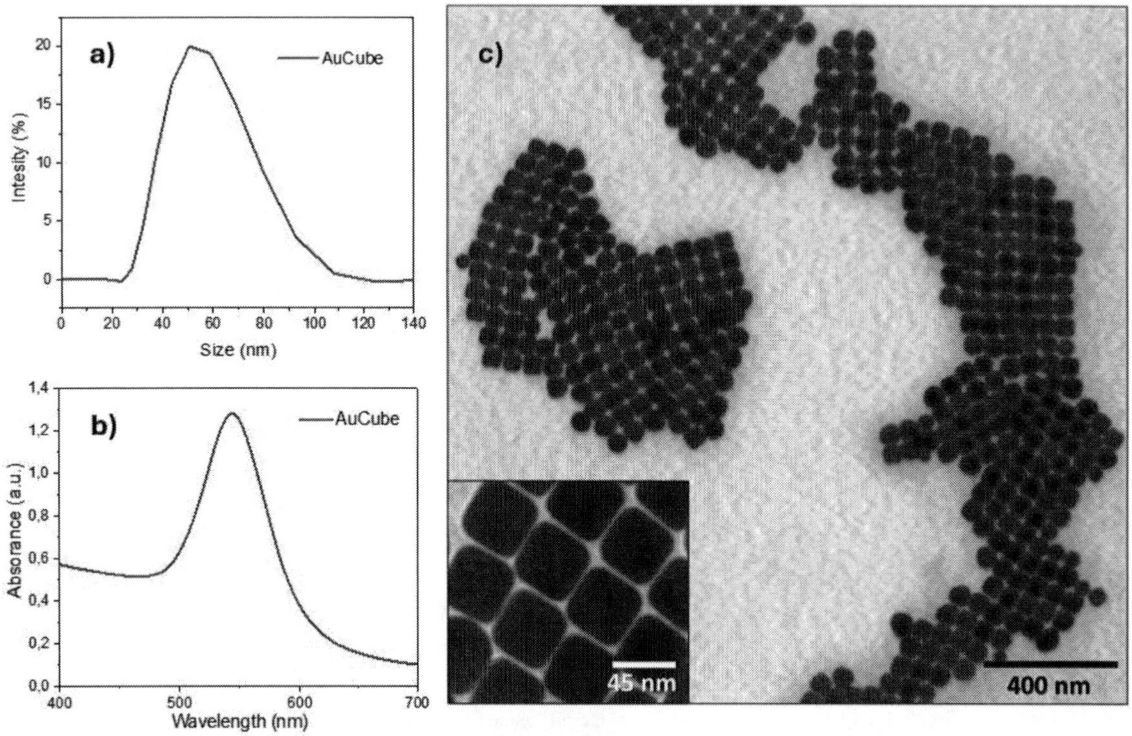

Figure 2 Characterization of cubic nanoparticles: a) DLS spectrum, b) UV-Vis spectrum, c) TEM image

Figure 2 illustrates the characterization of cubic gold nanoparticles (see **Figure 1**, step ii). The size of the cubic nanoparticles was analysed using dynamic light scattering (DLS). In **Figure 2a**, the DLS graph indicates that the predominant particle size in the solution ranges from 45 to 60 nm. **Figure 2b** presents the UV-Vis spectrum of the cubic nanoparticles, displaying an absorption maximum at 548 nm. Similar absorption maxima, such as 550 nm and 554 nm, are reported in other publications. [6]

The morphology of cubic gold nanoparticles was examined using transmission electron microscopy (TEM). **Figure 2c** displays the TEM image of the nanoparticles, revealing an approximate size of 55 nm. When comparing this to **Figure 2a**, it is evident that the sizes measured by DLS are very similar to those obtained from TEM.

3.1.2 Chiral nanoparticles type IV

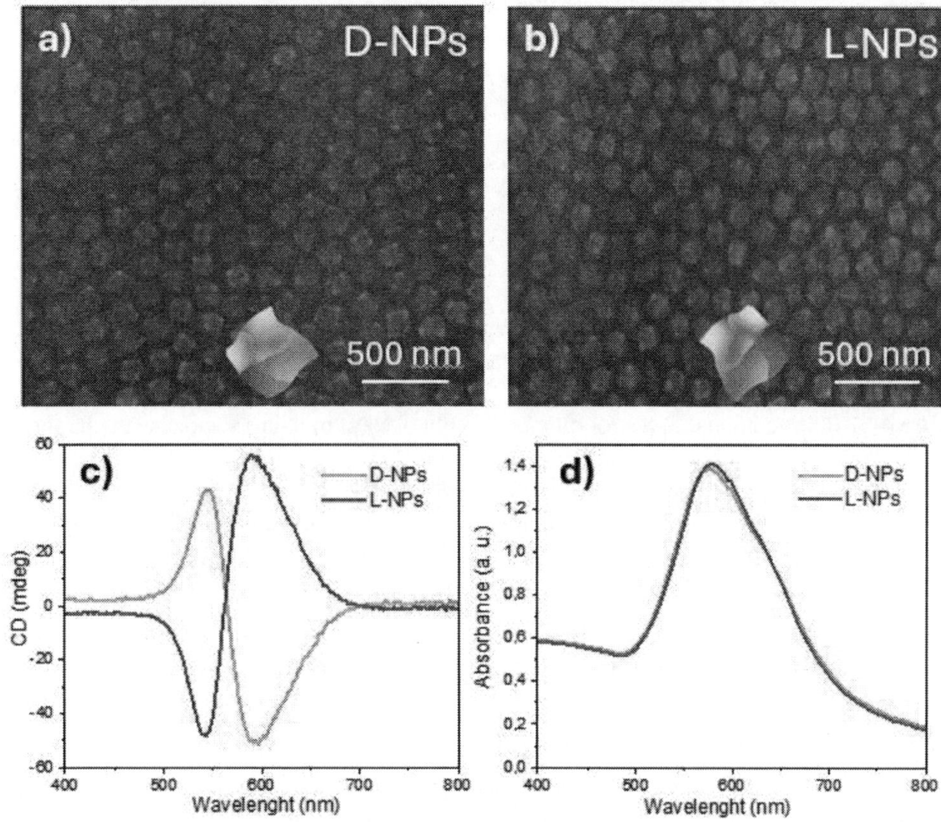

Figure 3 Characterization of chiral nanoparticles type IV: a) SEM image D-NPs, b) SEM image L-NPs, c) CD spectrum chiral NPs, d) UV-Vis spectrum chiral NPs

The characterization of chiral nanoparticles (prepared during step iii, depicted in **Figure 1**) is presented in **Figure 3**. The morphology of these nanoparticles was investigated using scanning electron microscopy (SEM). Notably, chirality is encoded in each particle due to their distinctive shapes. **Figure 3a** shows D-chiral nanoparticles, which exhibit a left-to-right twisting morphology. Conversely, **Figure 3b** illustrates L-chiral nanoparticles, characterized by a right-to-left twisting structure.

Chirality was confirmed using circular dichroism, as shown in **Figure 3c**. The signal strength for type IV chiral nanoparticles is in the range of 45-50 mdegs for both variants (D and L), indicating a high response. Furthermore, it can be observed that the graphs exhibit axial symmetry, suggesting that chiral nanoparticles efficiently absorb left or right handled light, depending on the nanoparticles shape.

Figure 3d presents the UV-Vis spectrum, showing an absorption maximum at 600 nm. In comparison to **Figure 2b**, this maximum has shifted from 548 nm to 600 nm. Similar absorption maxima can be found in other publications.[6]

3.2 SERS measure

The premise for SERS measurements was that chiral gold particles would react differently with various enantiomers. One enantiomer was expected to demonstrate a stronger response with D-NPs than with L-NPs, and vice versa.

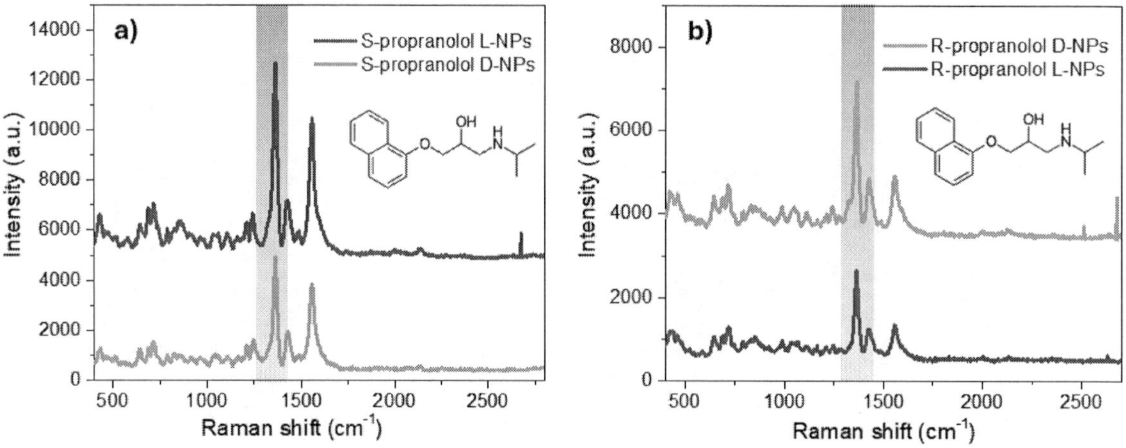

Figure 4 SERS spectrums: a) S-propranolol with chiral NPs, b) R-propranolol with chiral NPs

Figure 4a illustrates the spectra of the enantiomer (S)-(-)-propranolol combined with D and L nanoparticles. A significant signal enhancement is observed at the peak of 1350 cm^{-1}, with greater enhancement noted for the combination of L nanoparticles with (S)-(-)-propranolol. In contrast, the combination of D nanoparticles with (S)-(-)-propranolol results in a considerably lower SERS response.

Figure 4b presents the spectra of the enantiomer (R)-(+)-propranolol on substrates with L and D nanoparticles. Similarly, an increase in intensity at the same peak of 1350 cm^{-1} is observed, with more pronounced signal enhancement occurring in the combination of D nanoparticles with (R)-(+)-propranolol compared to the combination of L nanoparticles with (R)-(+)-propranolol.

Comparing the results, it is evident that the combination of (S)-(-)-propranolol with L nanoparticles and (R)-(+)-propranolol with D nanoparticles leads to higher peak intensity enhancement than the reverse combinations, allowing in this way enantioselective SERS discrimination.

4 CONCLUSION

This study investigated the use of chiral gold nanoparticles for enantioselective drug detection via surface-enhanced Raman spectroscopy (SERS). Enantioselective SERS detection was achieved by arranging chiral Au helicoids in grooves of a plasmon-active grating, ensuring homogeneous distances between the nanoparticles. The use of the plasmonic grating as a template for the self-assembly of nanoparticles also provided additional advantages from the excitation of surface plasmon polaritons (SPP) at the surface of the grating. Consequently, a significant difference in SERS signal intensities was observed for the enantiomers (S)-(-)-propranolol and (R)-(+)-propranolol using the arranged L or D gold helicoids. These differences were attributed to the varying strength of plasmonic coupling (under conditions of equal distances between the organized nanoparticles), which depends on the chirality of the dielectric gap in the plasmonic hot spot between the nanoparticles. Overall, it can be concluded that the combination of chiral nanoparticles and surface-enhanced Raman spectroscopy holds great potential for enantioselective detection and may lead to significant advancements in pharmaceutical research and other applications.

ACKNOWLEDGEMENTS

This work was supported by the GACR under the project 23-05197S

REFERENCES

[1] WU, W.; PAULY, M. Chiral plasmonic nanostructures: recent advances in their synthesis and applications. *Materials Advances.* 2022, vol. 3, no. 1, pp. 186-215. Available from: https://doi.org/10.1039/D1MA00915J.

[2] YANG, G.; SUN, L.; TAO, Y.; CHENG, Q.; SUN, X.; LIU, C.; ZHANG, Q. Chiral AuCu heterostructures with site-specific geometric control and tailored plasmonic chirality. *Science China Chemistry.* 2023, vol. 66, no. 11, pp. 3280-3289. Available from: https://doi.org/10.1007/s11426-023-1685-3.

[3] ABDALI, S.; BLANCH, E. W. Surface enhanced Raman optical activity (SEROA). *Chemical Society Reviews.* 2008, vol. 37, vol. 5, pp. 980-992.

[4] KARTAU, M.; SKVORTSOVA, A.; TABOUILLOT, V.; CHAUBEY, S. K.; BAINOVA, P.; KUMAR, R.; BURTSEV, V.; SVORCIK, V.; GADEGAARD, N.; IM, S. W.; et al. Chiral Metafilms and Surface Enhanced Raman Scattering for Enantiomeric Discrimination of Helicoid Nanoparticles. *Advanced Optical Materials.* 2023, vol. 11, vol. 9, 2202991. Available from: https://doi.org/10.1002/adom.202202991.

[5] HU, L.; XI, F.; QV, L.; FANG, Y. Searching the Theoretical Ultimate Limits of Probing Surface-Enhanced Raman Optical Activity. *ACS Omega.* 2018, vol. 3, no. 1, pp. 1170-1177. Available from: https://doi.org/10.1021/acsomega.7b02098.

[6] LEE, H.-E.; KIM, R. M.; AHN, H.-Y.; LEE, Y. Y.; BYUN, G. H.; IM, S. W.; MUN, J.; RHO, J.; NAM, K. T. Cysteine-encoded chirality evolution in plasmonic rhombic dodecahedral gold nanoparticles. *Nature Communications.* 2020, vol. 11, no. 1, 263. Available from: https://doi.org/10.1038/s41467-019-14117-x.

THE EFFECT OF SILVER NANOPARTICLES ON THE MICROALGAE *HAEMATOCOCCUS PLUVIALIS*

[1]Liliana CEPOI, [1]Ludmila RUDI, [1]Tatiana CHIRIAC, [2]Ecaterina PLINGAU, [1]Ana VALUTA, [1]Vera MISCU

[1]Technical University of Moldova, Institute of Microbiology and Biotechnology, Chisinau, Republic of Moldova, liliana.cepoi@imb.utm.md

[2]Moldova State University, Doctoral School of Natural Sciences, Chisinau, Republic of Moldova, kateaplingau@gmail.com

https://doi.org/10.37904/nanocon.2024.5021

Abstract

The green alga *Haematococcus pluvialis* is highly regarded as a producer of astaxanthin. Due to its antioxidant properties, astaxanthin has become a popular nutritional ingredient. Astaxanthin has demonstrated potential pharmacological effects including antidiabetic, anti-inflammatory, and antioxidant activities and cardiovascular and neurological protective properties. Nanoparticles have become integral components of microalgal technologies, serving as stimulators for producing biologically active compounds. The stimulating effect of nanoparticles is based on the induction of oxidative stress. In this study, oxidative stress was induced by silver nanoparticles (AgNPs) and combined with stress from a nutrient-deficient cultivation medium. Citrate-stabilized AgNPs of 10 nm and 20 nm were used. They were applied to the mineral medium from the first day of the cultivation cycle in concentrations ranging from 0.01 mM to 10 mM. The microalgal culture in the red cyst phase was used as the inoculum. The results demonstrated a significant increase of 50-80% in the astaxanthin content in the biomass of the red cysts without altering the biomass content. Additionally, the lipid content in the biomass also increased significantly. The dose/effect relationship between the concentrations of 20 nm AgNPs and astaxanthin values was direct. For 10 nm AgNPs, low concentrations resulted in increased astaxanthin values. Combined stress, with the variable factor being the concentrations of silver nanoparticles, can be proposed as a strategy in the biotechnology of *Haematococcus pluvialis* for astaxanthin production.

Keywords: *Haematococcus pluvialis*, silver nanoparticles, biomass, astaxanthin, lipids

1. INTRODUCTION

The green microalga *Haematococcus pluvialis* is one of the most well-known producers of astaxanthin, a carotenoid with superior antioxidant properties. Astaxanthin has demonstrated anti-inflammatory, antitumor, antidiabetic, and immunomodulatory effects [1]. It is widely used in the pharmaceutical, nutraceutical, cosmetic, and food industries [2]. Technologies applied to *Haematococcus pluvialis* for astaxanthin production include adjustments to cultivation conditions such as pH, temperature, and light intensity and modifications to the mineral composition of the growth medium [3,4]. Nutrient deficiencies and excesses can significantly affect cell biomass productivity and astaxanthin accumulation [3,5]. Studies on the application of nanoparticles in microalgae cultivation technologies show varied results. Nanoparticles can undoubtedly be applied to biomolecule production [6]. For example, various types of nanoparticles have stimulated lipid accumulation in microalgae, demonstrating the potential for biofuel production, while others have influenced the accumulation of photosynthetic pigments and biomass production [7]. Some nanoparticles have demonstrated properties as stimulators of astaxanthin production by the microalga *Haematococcus pluvialis*, significantly increasing biomass content [8]. Silver nanoparticles, at low concentrations, have stimulated carotenoid synthesis in *Haematococcus pluvialis* biomass [9]. Several studies compare the action of nanoparticles on astaxanthin synthesis based on their origin, highlighting this as an essential factor in nanoparticle characteristics [10,11]. Toxic effects on growth and carotenoid accumulation have been observed for various types of nanoparticles, with different forms of oxidative stress being identified [12,13]. In this study, we aimed to identify the effect of AgNPs with sizes of 10 nm and 20 nm, in combination with nutrient deficiency, on astaxanthin accumulation in the biomass of *Haematococcus pluvialis*.

2. EXPERIMENTAL DESIGN

The green microalga strain *Haematococcus pluvialis* CNMN-AV-05, deposited in the National Collection of Nonpathogenic Microorganisms at the Institute of Microbiology and Biotechnology, Technical University of Moldova, was used as the study object. In the experiments, the microalga was cultivated in a mineral medium with the following composition: $NaNO_3$ – 0,3 g/L; KH_2PO_4 – 0,02 g/L; K_2HPO_4 – 0,02 g/L; NaCl – 0,02 g/L; $CaCl_2$ – 0,05 g/L; $MgSO_4 \cdot 7H_2O$ - 0,01 g/L; $ZnSO_4 \cdot 7H_2O$ – 0,0001 g/L; $MnSO_4 \cdot 5H_2O$ – 0,0015 g/L; $CuSO_4 \cdot 5H_2O$ – 0,00008 g/L; H_3BO_3 - 0,0003 g/L; $(NH_4)_6MoO_{24} \cdot 4H_2O$ – 0,0003 g/L; $FeCl_3 \cdot 6H_2O$ – 0,0175 g/L; $Co(NO_3)_2 \cdot 6H_2O$ - 0,0002 g/L; EDTA - 0,0075 g/L, in 100 mL Erlenmeyer flasks containing 50 mL of microalgal culture, at a temperature of 26°C, with constant illumination of 1500 lx and periodic agitation during the first ten days of cultivation. The induction of astaxanthin production was carried out by excessive illumination with an intensity of 3000 lx for 72 hours. Silver nanoparticles (AgNPs) stabilized with citrate, 10 ± 0.2 nm and 20 ± 0.2 nm particle size (TEM) (SIGMA-ALDRICH CHEMIE GmbH, Germany), in various selected concentrations, and were added to the mineral medium. The biomass content collected at the end of the growth phase of red aplanospores was determined spectrophotometrically, with a quantitative calculation based on the calibration curve. The astaxanthin content was determined in the ethanolic extract from pre-treated red aplanospore biomass. For this purpose, native *Haematoccus pluvialis* biomass was subjected to acid hydrolysis with 0.1N HCl at a temperature of 90°C for 10 minutes. The residual acid was removed by washing the biomass with distilled water. The extraction mixture consisted of 10 mg of pre-treated aplanospore biomass and 1 mL of 96% ethanol. Astaxanthin extraction was carried out under continuous agitation for 120 minutes. The absorbance of the ethanolic extract was recorded at 478 nm. The quantitative calculation was based on the calibration curve constructed for synthetic astaxanthin. The lipid content was determined spectrophotometrically with phospho-vanillin reagent. All measurements and determinations were performed in triplicate. Statistical analysis was performed using Microsoft Excel software (Microsoft 365 Excel version 2108). The data were statistically analysed by calculating the arithmetic mean and standard error. The t-Student test was used to determine statistically significant differences between the experimental and control values

3. RESULTS AND DISCUSSION

The mineral medium used for the cultivation of the microalga *Haematococcus pluvialis* CNMN-AV-05 is characterized by a low phosphate content (KH_2PO_4 – 0.02 g/L and K_2HPO_4 – 0.02 g/L), with a P/N ratio of 0.164. In the red aplanospore phase, astaxanthin constitutes 3% of the obtained biomass, with a total biomass production of 2.2-2.4 g/L and a cultivation cycle of 14 days. Under nutrient-deficient conditions, *Haematococcus pluvialis* cells enter a state of moderate stress, triggering protective mechanisms, including synthesizing carotenoids to neutralize free radicals. This represents a survival strategy under stress conditions, stimulating the production of biologically valuable compounds such as astaxanthin [14].

Under such conditions, when the microalgal culture is subjected to moderate stress, the addition of potentially toxic compounds, such as AgNPs, may influence either astaxanthin production or the productivity of the culture. The effect may depend on the nanoparticles' size and/or concentration. The presence of silver nanoparticles in the cultivation medium of the microalga induced changes in the biomass content during the red aplanospore phase. **Figure 1** shows the changes in the biomass content of *Haematococcus pluvialis*, collected in the red phase, due to the action of silver nanoparticles at various concentrations in the cultivation medium.

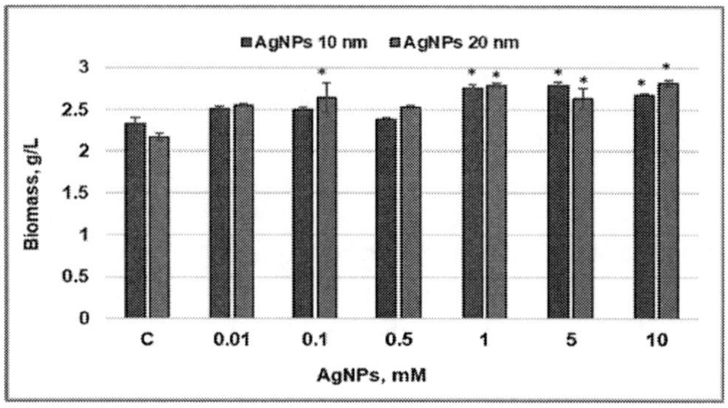

Figure 1 Biomass content (g/L) of *Haematococcus pluvialis*, red aplanospore phase, cultivated in the presence of AgNPs with sizes of 10 nm and 20 nm, C-control; *p < 0.05

The addition of 10 nm AgNPs to the mineral medium of *Haematococcus pluvialis* at concentrations of 1.0, 5.0, and 10 mM led to biomass increases of 18.6% ($p > 0.01$), 19.8% ($p > 0.05$), and 14.9% ($p > 0.05$), respectively. Lower concentrations, ranging from 0.01 to 0.5 mM, resulted in a 7% increase in biomass content. In the case of 20 nm AgNPs, applied at the same concentrations, an increase in biomass accumulation between 17.7% and 29.8% ($p > 0.05$) was observed compared to the control. Higher biomass content values are associated with higher nanoparticle concentrations. Thus, at the applied concentrations, citrate-stabilized silver nanoparticles did not exhibit toxic effects on the *Haematococcus pluvialis* culture, suggesting good tolerance of this microalga to silver nanoparticles in the growth medium.

Figure 2 illustrates the changes in astaxanthin and lipid content in the aplanospore biomass due to the exposure of *Haematococcus pluvialis* culture to AgNPs.

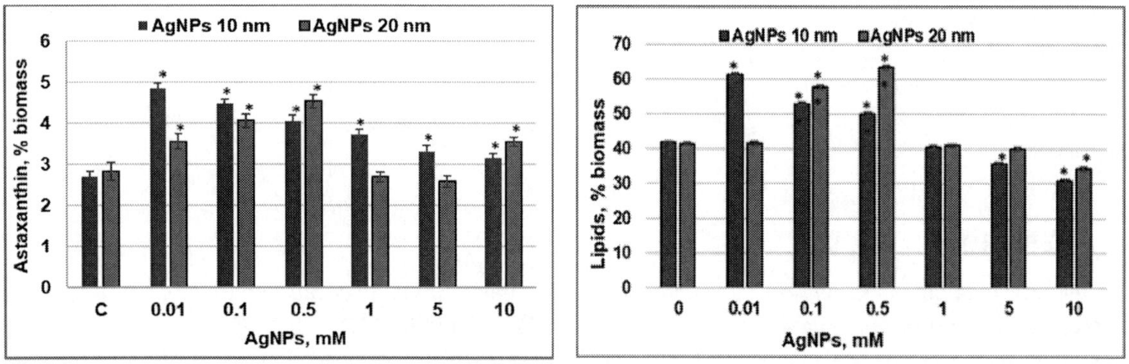

Figure 2 Astaxanthin content, % **(A)** and lipid content, % **(B)**, in the biomass of *Haematococcus pluvialis*, red aplanospore phase, cultivated in the presence of AgNPs with sizes of 10 nm and 20 nm, C-control; *p < 0.01

Figure 2A shows the changes in astaxanthin content in the aplanospore biomass as the result to the exposure of *Haematococcus pluvialis* to AgNPs. AgNPs of 10 nm, applied at concentrations between 0.01 mM and 0.5 mM, led to a significant increase in astaxanthin content in the biomass. The concentration of 0.01 mM AgNPs resulted in an 80% increase in astaxanthin content ($p<0.01$) compared to the control. Concentrations of 0.1 mM and 0.5 mM induced increases in astaxanthin values by 65.9% ($p<0.01$) and 50% ($p<0.01$), respectively. For 10 nm AgNP concentrations of 1.0-10 mM, an increase in astaxanthin of 37.8% ($p<0.01$) to 17% ($p<0.01$) was recorded. The stimulatory effect directly depends on the concentration of nanoparticles in the

Haematococcus pluvialis cultivation medium. The Pearson coefficient r = -0.9976, calculated for the relationship between 10 nm AgNP concentrations and astaxanthin content, shows a strong negative correlation. As the AgNP concentration increases, the astaxanthin content decreases proportionally. For 20 nm AgNPs, a concentration of 0.05 mM led to a 26.8% increase ($p<0.01$) in astaxanthin content in the microalgal biomass. Concentrations of 0.1 and 0.5 mM resulted in pigment content increases of 45% ($p<0.01$) and 62% ($p<0.01$), respectively, compared to the control. Therefore, increasing the concentration of AgNPs in the cultivation medium led to a proportional increase in astaxanthin content in the microalgal biomass, with a strong positive correlation (r = 0.9326). The concentrations of 20 nm AgNPs, 1.0 and 10 mM, considered high, also showed a tendency to increase astaxanthin levels, though the stimulatory effect was moderate. The values obtained at concentrations of 1.0 and 5.0 mM were similar to those of the control sample. An increase of 26.9% ($p<0.01$) in astaxanthin content was observed for the concentration of 10.0 mM AgNPs in the cultivation medium. Thus, *Haematococcus pluvialis* responds in two distinct ways to the presence of AgNPs in the culture medium, depending on the size and concentration of the nanoparticles. Low concentrations of AgNPs stimulate astaxanthin synthesis: for 10 nm AgNPs, the correlation between nanoparticle concentrations and astaxanthin content is inverse; for 20 nm AgNPs, the correlation is direct.

The size and concentration are the determining factors in the influence of nanoparticles on astaxanthin biosynthesis in *Haematococcus pluvialis.* Within the applied concentration limits, no inhibition effect on astaxanthin synthesis was established, suggesting a potential adaptation of the culture to the presence of AgNPs in the microalgal growth medium. As evidence of this statement, the results of changes in lipid content accumulated in the biomass of *H. pluvialis* (**Figure 2B)**. Silver nanoparticles (AgNPs) of 10 nm, applied in the *Haematococcus pluvialis* cultivation medium at low concentrations, stimulated lipid synthesis, and their content in the red aplanospore biomass increased considerably. AgNPs of 10 nm, at concentrations of 0.01 and 0.5 mM, led to a lipid content increase of 53.6% ($p<0.01$) and 24.7% ($p<0.01$), respectively. Nanoparticle concentrations of 1.0 and 5.0 mM generated a neutral response, while at 10 mM AgNP concentration, a 17.3% decrease in lipid content was observed ($p<0.05$). AgNPs of 20 nm, at concentrations of 0.01 and 0.5 mM, resulted in a lipid content increase of 19.6% ($p<0.01$) and 46.9% ($p<0.01$), respectively. Nanoparticle concentrations of 5.0 mM and 10.0 mM decreased lipid content by 14.6% and 16.8% ($p<0.05$) in the microalgal biomass. Lipid accumulation in the experimental variants with increased astaxanthin levels in the biomass is characterized by the same correlations between AgNP concentrations and astaxanthin values. The correlation between the concentrations of 10 nm AgNPs with a stimulatory effect and the lipid content is negative and strong (r=-0.9455). In the case of 20 nm AgNPs, this correlation is positive and very strong (r=0.9959). Lipid synthesis in the *Haematococcus pluvialis* biomass in the presence of AgNPs in the cultivation medium is not determined by astaxanthin accumulation. Increasing AgNP concentrations leads to a decrease in lipid content in the biomass. The correlation between the applied concentrations of 10 nm AgNPs and the lipid content in aplanospore biomass is negative and strong (r=-0.7956). For 20 nm AgNPs, this correlation is also negative and strong (r=-0.7490). It can be assumed that the stimulation of astaxanthin synthesis results from strong oxidative stress, considering lipid content values are decreasing.

The general trend of AgNPs influencing *Haematococcus pluvialis* culture in the growth medium is to reduce lipid synthesis, which generates a stressful environment for the microalga. Higher concentrations of AgNPs disrupt the metabolic processes of the microalga, negatively affecting its ability to accumulate lipids. At the same time, it can be stated that low concentrations of AgNPs have a safe stimulatory effect on the microalgal culture. There is limited evidence indicating a stimulatory effect on the growth of the microalga *Haematococcus pluvialis* due to nanoparticle exposure. Studies on the effects of nanoparticles, such as ZnONPs and TiO_2NPs, on the growth and astaxanthin synthesis by *Haematococcus pluvialis* have shown negative effects, manifested by the inhibition of growth and pigment synthesis [10,12]. Silver nanoparticles have also been studied from the perspective of their toxicity. AgNPs obtained by reducing a biological substrate were toxic to *Haematococcus pluvialis* culture. At concentrations up to 8 mg/L, AgNPs harmed cell morphology, culture productivity, and astaxanthin synthesis. Culture productivity in the green phase was reduced by up to 85% as AgNP concentration increased to 8 mg/L, while the effects on the red aplanospore phase culture were less severe. Additionally, AgNPs reduced the rate of astaxanthin production [11]. A significant increase in astaxanthin production in response to nanoparticles in the cultivation medium was reported due to the application of magnesium aminoclay nanoparticles (MgAC) [8]. A concentration of 1.0 mg/L of MgAC nanoparticles induced a 13.7-fold increase in astaxanthin content in the *Haematococcus pluvialis* microalga cells. Lipid content in the cells increased 13.6 times. The Tris-Acetate-Phosphate mineral medium, specifically for photoautotrophy, with a high phosphate content (KH_2PO_4 – 0.34 g/L and K_2HPO_4 – 1.72 g/L), was used for microalga growth. The astaxanthin increase was calculated by comparison with the control, with the total astaxanthin production being 7.3 ± 0.6 mg/L. Astaxanthin content increased by 40% as a result of applying MgAC nanoparticles. A nutrient-poor mineral medium significantly stimulates astaxanthin production [14]. In this context, AgNPs were added to the *Haematococcus pluvialis* CNMN-AV-03 cultivation medium, which had a reduced mineral content. In the control, without nanoparticle supplementation, astaxanthin production was 60.7 mg/L. Adding a concentration of 0.01 mM of 10 nm AgNPs led to a 97.7% increase in astaxanthin production compared to the control. The combined stimulatory effect of the nanoparticles and the deficient mineral medium was evident. Higher lipid content values were recorded in variants with increased astaxanthin content due to the action of 10 nm and 20 nm AgNPs. A similar result was obtained in experiments using MgAC nanoparticles, which can be evidence of specific nanoparticles' stimulation of astaxanthin synthesis [8]. Nanoparticles are used as stimulators of lipid synthesis in microalgal biomass, and this process is based on the induction of oxidative stress. Citrate-stabilized AgNPs, with sizes of 10 and 20 nm, demonstrated a stimulatory effect on the microalga *Porphyridium cruentum*, manifested by a significant increase in lipid content in the biomass [15]. In the study of the action of AgNPs on the *Haematococcus pluvialis* microalga culture, lipid synthesis induction was a secondary process, with the primary effect being the stimulation of astaxanthin synthesis. The absence of toxicity can be attributed to the stabilization of the nanoparticles, which prevented their aggregation and, consequently, the subsequent damage to cell membranes and the associated consequences of this process

4. CONCLUSION

An important conclusion of this study is that 10 nm and 20 nm citrate-stabilized AgNPs act as stimulators of astaxanthin synthesis in the green alga strain *Haematococcus pluvialis* CNMN-AV-03 when cultivated in a nutrient-poor mineral medium. This microalgal response results from moderate cell stress, which did not lead to reductions in biomass content at the end of the cultivation cycle. The stimulatory effect of silver nanoparticles on astaxanthin synthesis by the *Haematococcus pluvialis* CNMN-AV-03 strain used in this study depends on the size and concentration of the nanoparticles in the cultivation medium. Low nanoparticle concentrations significantly increased astaxanthin content, and no evident toxic effects were observed within the applied concentration range.

ACKNOWLEDGEMENTS

This research was supported by a grant from the National Agency for Research and Development, project number 24.80012.7007.08 SE.

REFERENCES

[1] JANNEL, S., CARO, Y., BERMUDES, M., PETIT, T. Novel Insights into the Biotechnological Production of *Haematococcus pluvialis*-Derived Astaxanthin: Advances and Key Challenges to Allow Its Industrial Use as a Novel Food Ingredient. *Journal of Marine Science and Engineering*. 2020, vol. 8, no.10, 789. Available from: https://doi.org/10.3390/jmse8100789.

[2] PATIL, A.D., KASABE, P.J., DANDG, E.P.B. Pharmaceutical and nutraceutical potential of natural bioactive pigment: astaxanthin. *Nat Prod Bioprospect*. 2022, vol. 12, no. 1, 25. Available from: https://doi.10.1007/s13659-022-00347-y.

[3] LI, F., CAI, M., WU, Y., LIAN, Q., QIAN, Z., LUO, J., ZHANG, Y., ZHANG, N., LI, C. AND HUANG, X. Effects of Nitrogen and Light Intensity on the Astaxanthin Accumulation in Motile Cells of *Haematococcus pluvialis*. *Front. Mar. Sci.* 2022, vol. 9, no. 909237. Available from: https://doi.org/10.3389/fmars.2022.909237.

[4] MOTA, G.C.P., MORAES, L.B.S., OLIVEIRA, C.Y.B., OLIVEIRA, D.W.S., ABREU, J.L, DANTAS, D.M.M., GÁLVEZ, A.O. Astaxanthin from *Haematococcus pluvialis*: processes, applications, and market. *Prep Biochem Biotechnol.* 2022, vol. 52, issue 5, pp.598–609. Available from: https://doi.org/10.1080/10826068.2021.1966802.

[5] de MORAES, L.B.S, MOTA, G.C.P., dos SANTOS, E.P., da SILVA CAMPOS, C.V.F., da SILVA, B.A.B., GÁLVEZ, A.O., de SOUZA BEZERRA, R. *Haematococcus pluvialis* cultivation and astaxanthin production using different nitrogen sources with pulse feeding strategy. *Biomass Conversion and Biorefinery*. 2024, vol.14, pp. 16231–16243. Available from: https://doi.org/10.1007/s13399-023-03824-7.

[6] VARGAS-ESTRADA, L., TORRES-ARELLANO, S., LONGORIA, A., ARIAS, D.M., OKOYE, P.U., SEBASTIAN, P.J. Role of nanoparticles on microalgal cultivation: A review. *Fuel.* 2020, vol. 280, no. 118598. Available from: https://doi.org/10.1016/j.fuel.2020.118598

[7] SENDRA, M, YESTE, M.P, GATICA, J.M., MORENO-GARRIDO, I, BLASCO, J. Direct and indirect effects of silver nanoparticles on freshwater and marine microalgae (*Chlamydomonas reinhardtii* and *Phaeodactylum tricornutum*). *Chemosphere.* 2017, vol. 179, pp. 279–289. Available from: https://doi.org/10.1016/J.CHEMOSPHERE.2017.03.123.

[8] KIM, Y.E., MATTER, I.A., LEE, N.A, JUNG, M., LEE, Y.C., CHOI, S.A., LEE, S.Y., KIM, J. R., OH, Y.K. Enhancement of astaxanthin production by *Haematococcus pluvialis* using magnesium aminoclay nanoparticles. *Bioresource Technology*. 2020, vol. 307, no. 123270. Available from: https://doi.org/10.1016/j.biortech.2020.123270.

[9] CHENG, W.H., SHING, W.L., TIAN, L.M., YOKE, C. M., HOCK, O. G. Effects of Silver Nanoparticles on the Carotenoid Production from *Haematococcus pluvialis*. *EnvironmentAsia.* 2020, vol. 13, no.1, pp. 106-111. Available from: https://doi.org/10.14456/ea.2020.10.

[10] NASRI, N., KEYHANFAR, M., BEHBAHANI, M., DINI, G. Enhancement of astaxanthin production in *Haematococcus pluvialis* using zinc oxide nanoparticles. *Journal of Biotechnology*. 2021, vol. 342, pp. 72-78, Available from: https://doi.org/10.1016/j.jbiotec.2021.10.004.

[11] VENCKUS, P., ENDRIUKAITYTĖ, I., ČEKUOLYTĖ, K., GUDIUKAITĖ, R., PAKALNIŠKIS, A., & LASTAUSKIENĖ, E. Effect of Biosynthesized Silver Nanoparticles on the Growth of the Green Microalga *Haematococcus pluvialis* and Astaxanthin Synthesis. *Nanomaterials*. 2023, vol. 13, no. 1618. pp. 1-14. Available from: https://doi.org/10.3390/nano13101618.

[12] MAHENDRAN, M.S., WONG, L.S., ANTONY DHANAPAL, A.C.T., DJEARAMANE, S. The Effect of Titanium Dioxide Nanoparticles on *Haematococcus pluvialis* Biomass Concentration. *Journal of Experimental Biology and Agricultural Sciences*. 2023, vol. 11, no. 2, pp. 416-422. Available from: https://doi.org/10.18006/2023.11(2).416.422.

[13] WONG, L.S., BATUMALAI, T., CHENG, W.H., ONG, G.H. The effect of copper nanoparticle to astaxanthin content in microalgae. *Transactions on Science and Technology.* 2019, vol. 6, no.1-2, pp. 81-85.

[14] BUTLER, T. O., MCDOUGALL, G. J., CAMPBELL, R., STANLEY, M. S., DAY, J. G. Media screening for obtaining *Haematococcus pluvialis* red motile macrozooids rich in astaxanthin and fatty acids. *Biology*. 2018, vol. 7, no.1, 2. Available from: https://doi.org/10.3390/biology7010002.

[15] RUDI, L., CEPOI, L., CHIRIAC, T., MISCU, V., VALUTA, A., DJUR, S. Effects of citrate-stabilized gold and silver nanoparticles on some safety parameters of *Porphyridium cruentum* biomass. *Front. Bioeng. Biotechnol.* 2023, vol.11, 1224945. Available from: https://doi.org/10.3389/fbioe.2023.1224945.

TRANSPORT AND IMMOBILIZATION OF PHARMACEUTICALS IN HUMIC HYDROGEL

[1]Martina KLUČÁKOVÁ, [1]Petra ZÁVODSKÁ, [1]Jan FUČÍK, [1]Ludmila MRAVCOVÁ

[1]*Brno University of Technology, Faculty of Chemistry, Brno, Czech Republic, EU, klucakova@fch.vutbr.cz*

https://doi.org/10.37904/nanocon.2024.5019

Abstract

Pharmaceuticals can be found in soil and aquatic environments as a result of human activity. They are released into the environment where they undergo soil sorption, photodegradation, and chemical transformation. It is known that their migration ability and toxicity can be significantly affected by interactions with organic matter. The interactions can result in their immobilization in humic structure and decrease in their bioavailability. Humic substances as important constituents of natural organic matter are investigated from point of view of their interactions with different pollutants (including pharmaceuticals) mainly by means of traditional batch adsorption experiments. Our approach is different. Humic substances are used in the form of hydrogel which can be considered as a model system of soil with characteristic content of organic matter. The hydrogel form is advantageous for its defined parameters which can be reflected in initial and boundary conditions necessary for mathematical description of transport in the system. Ibuprofen (sodium salt) was studied from point of view of its transport and immobilization in humic hydrogel by the method of instantaneous planar source. It means that small defined amount of diffusing substance is placed on the circular surface of cylinder-shaped hydrogel (placed in glass tube to achieve one-dimensional diffusion). Diffusion study was complemented by adsorption experiments to compare traditional and our novel approaches. The mobility in humic systems as well as the immobilization ability and their effect on the bioavailability of pharmaceutical are discussed.

Keywords: Ibuprofen, humic acid, hydrogel, diffusion, immobilization

1. INTRODUCTION

The presence of pharmaceuticals in environment gains significant attention and raises health concern due to their adverse effect and environmental risks [1-3]. The fate of pharmaceuticals in the environment mainly depends on their adsorption behaviour in soils and sediments. The adsorption can affect their distribution between aqueous medium (such as soil solution) and solid surfaces, migration, and bioavailability [1,3]. The adsorption degree is influenced by pharmaceutical properties as well as environment characteristics such as temperature, pH, and content of minerals and organic matter [3,4]. Therefore, the adsorption studies are the most extensively used way of the investigation of interactions between pharmaceuticals and soils or sediments. Classical batch experiments [5-7], column experiments [8,9], and monitoring of pharmaceuticals in environment by means of the method of diffusive gradients in thin films (DGT) [10,11] are mostly performed to study of fate and immobilization possibilities of pharmaceuticals in environment.

Our study is focused on the transport of pharmaceuticals in systems containing humic acid as an active substance able to interact with diffusing drug particles. We believe that humic substances as main components of organic matter play important role in the immobilization of pollutants in nature. The significance of humic substances in interactions of pollutants with organic matter and the drug removal from environment and wastewater were reported in many studies, e.g. [1-4,12-15]. Our approach combines interaction (sorption) experiments and diffusion experiments and therefore allows the direct observation of the actual mobility of pollutants affected by their reactivity in the transport through model humic system. This approach was developed on the basis of general mathematical model for the diffusion accompanied by chemical reaction [16,17] and optimized for the application to the transport in humic hydrogels in our previous works, e.g.

[18-20]. In this study, the hydrogel based on the precipitation of humic acids resulting in the gel form prepared in defined shape and size which are necessary for the mathematical description and correctly defined initial and boundary conditions corresponding with our experimental arrangement. Ibuprofen, which was chosen for this study as an example of commonly used drug, is a clinically important pharmaceutical for the treatment of pain and inflammation [21,22]. Its high consumption can result in a contamination of aquatic environment including surface, drainage, and drinking waters [22,23] connected with toxic effects observed on non-target organisms even at low concentrations [24]. Our study is focused on the reactivity mapping of this pharmaceutical in systems containing humic substances, determination of its diffusivity, and immobilization potential of humic substances as active constituents of organic matter.

2. MATERIALS AND METHODS

Ibuprofen sodium salt (CAS 31121-93-4) was purchased from Sigma-Aldrich. Humic acids was extracted from lignite mined in the Czech Republic (Mikulčice in South Moravia) using a mixture of NaOH and $Na_4P_2O_7$ by the procedure described previously [18-20]; for more details of chemical structure and the isolation procedure see previous works [18-20,25].

Humic hydrogel was prepared by dissolving powdered humic acids in 0.5M NaOH and acidifying the solution with HCl to value of pH close to 1. After 24 hours, the gel phase was separated from the solution by repeated centrifugation followed by washing with deionized water [18-20]. The content of dry matter in prepared hydrogel was 15.9 ± 1.1% (wt.).

Diffusion experiments were performed with hydrogels pressed gently into glass tubes (length = 3 cm and diameter = 1 cm). A circular slice of filter paper (diameter = 1 cm) was sunk into a solution of ibuprofen with concentration equal to 232 mg /dm^3 (1 min) and then added to one side of the tube (filled with the hydrogel). The tube was packed with parafilm and aluminium foil to prevent the hydrogel drying. The durations of the diffusion experiments were 24, 48 and 72 h. Then, the hydrogel was sliced into thicknesses of 0.5 – 1 mm and each slice was extracted separately in 10 cm^3 of deionized water. Leachates were analysed using ultra-performance liquid chromatography (UHPLC Agilent 1290 Infinity LC) in tandem with triple quadruple (Bruker EVOQ LC-TQ) with electrospray ionisation (ESI). Chromatographically data resulted in concentration profiles of ibuprofen in humic hydrogels and determination of diffusion coefficient.

Adsorption and desorption experiments were performed in the ratio 1 g of powdered humic acids to 50 cm^3 of drug solution (adsorption) or deionized water (desorption). Initial concentrations of ibuprofen in solutions were in the range between 1 and 10 mg dm^{-3}. The solutions of were mixed with humic acids, stirred for 48 h, centrifuged, and analysed chromatographically as in diffusion experiments. Solid residues were mixed with deionized water, stirred for 48 h, centrifuged, and analysed as in the case of adsorption experiments. The equilibrium concentrations of solutions, adsorbed, and desorbed amounts were determined and used for the determination of adsorption capacity and the determination of free-mobile (leachable) and strongly bound (residual) fractions of ibuprofen in humic acids.

All experiments were triplicated and performed at laboratory temperature (25 ± 1 °C). Data are presented as average values with standard deviation bars.

3. RESULTS AND DISCUSSION

Diffusion experiment was based on the diffusion from instantaneous planar source. It means that a small defined amount of pharmaceutical was placed on the hydrogel surface, and it immediately diffuse into the hydrogel. A mathematical description was based on one-dimensional Fick's law

$$\frac{\partial c}{dt} = D_{\text{eff}} \frac{\partial^2 c}{\partial x^2} \qquad (1)$$

Where c represents the concentration of the diffusing compound in the time t and the position x (the coordinate parallel to the direction of diffusion movement). Where D_{eff} is the so called "effective diffusion coefficient" in which two main effects (tortuous movement of the diffusing matter, chemical interactions in the system) are involved. While solving this partial differential equation, appropriate initial and boundary conditions must be applied according to the particular experimental settings [16-20]. If the diffusion is realised in inert medium, the effective diffusion coefficient includes only the tortuous movement of diffusing particles through the porous hydrogel structure.

A solution of the **Equation 1** for the diffusion from instantaneous planar source can be found in the form valid for the zero initial concentration of diffusing matter in hydrogel

$$c = \frac{n}{S\sqrt{\pi D_{\mathrm{eff}} t}} exp\left(-\frac{x^2}{4D_{\mathrm{eff}}t}\right) \quad (2)$$

Where n stands for the total mass of diffusing compound applied in the form of a narrow pulse and S is the cross-section area available for the transport of the compound.

The **Equation 2** was linearized to calculate the effective diffusion coefficient directly from the experimental concentration profiles:

$$ln\, c = ln \frac{n}{S\sqrt{\pi D_{\mathrm{eff}} t}} - \frac{x^2}{4D_{\mathrm{eff}}t} \quad (3)$$

The example of measured concentration profile of ibuprofen in humic hydrogel is shown in **Figure 1a.** The concentration decreases gradually with the distance from the interface. According to **Equation 3**, the D_{eff} value can be determined from the slope of linear dependence of ln c on x^2 which is usual and commonly used calculation method [16-20]. The effective diffusion coefficient based on the slope of **Equation 3** was determined as $(3.42 \pm 0.04) \times 10^{-10}$ m^2/s. This value involves both the tortuous movement of the diffusing matter and chemical interactions of ibuprofen with humic acids, therefore the D_{eff} value is lower in comparison with its diffusion coefficient in water (5.0×10^{-10} m^2/s [26]). The question is if one of these effects predominates or if they contribute to the decrease in the diffusion coefficient to a similar degree. In consideration of the character of humic hydrogel, we are not able to compare experimental data with similar inert hydrogel. The humic hydrogel is composed of humic acids and water in ideal case, where majority component is water. Our experiences with diffusion of metal ions in humic hydrogel showed that their interactions with humic acids resulted in a noticeable decrease in the diffusion coefficient [16-20]. In previous works [19,27], we studied the effect of selective blocking of carboxylic groups of humic acids as active centres for the interactions on the diffusivity of copper(II) ions in the hydrogels. The decrease was stronger for hydrogels containing more active centres (less blocked functional groups) which corresponded with the hypothesis that the influence of hydrogel reactivity on the diffusion is considerable. Investigation of the diffusion in agarose hydrogel [28] showed that the diffusion coefficients for several drugs (including ibuprofen) were found to be essentially the same as in water. It means that the diffusion in the agarose hydrogel enriched by reactive humic acids should by caused mainly by the interactions of diffusing particles. Our experiences with the diffusion of pharmaceuticals in agarose hydrogels are different [29]. The diffusion coefficients obtained for sulphonamide antibiotics in agarose hydrogel were much lower in comparison with their values in water and an enrichment of hydrogel by humic acids resulted in further changes in their diffusivities.

In consideration to the discrepancy in our and published findings, we decided to investigate the interactions between ibuprofen and humic acids in details. The measured adsorption isotherm for ibuprofen and powdered humic acids is shown in **Figure 1b**. Adsorption experiments provided adsorption capacity of humic acids a_{max} and the ratio between adsorption and desorption rate constants b ($b = k_{ads}/k_{des}$) according to Langmuir isotherm [1,30] frequently used for chemical adsorption

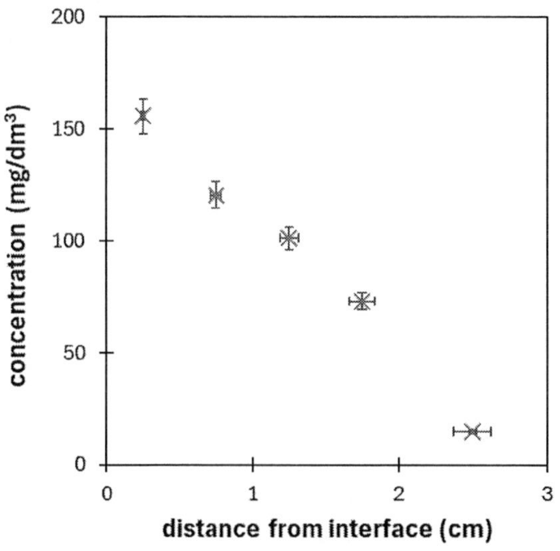

(a)

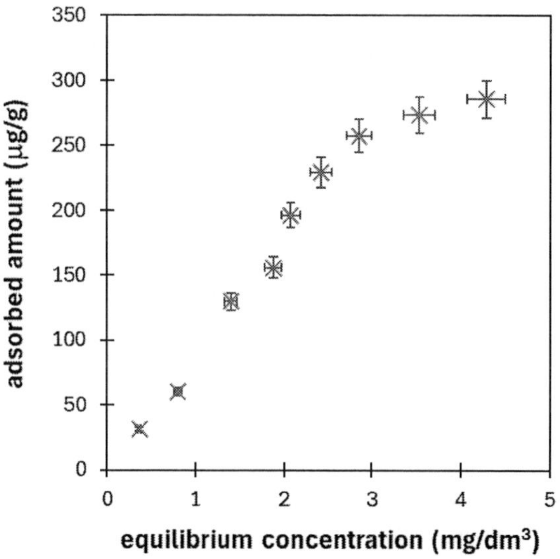

(b)

Figure 1 Concentration profile of ibuprofen in humic hydrogel after 72 h (a) and adsorption isotherm of ibuprofen on powdered humic acids (b).

$$a = a_{max} \frac{bc}{1+bc} \tag{4}$$

where a is adsorbed amount and c is equilibrium concentration. Parameters of adsorption isotherm (a_{max} and b) was determined on the basis of the linearized form of **Equation 4**:

$$\frac{1}{a} = \frac{1}{a_{max}b}\frac{1}{c} + \frac{1}{a_{max}}. \tag{5}$$

The adsorption capacity a_{max} was determined as 460 μg g^{-1} and the ratio b was equal to 0.39 dm^3 g^{-1}. The ratio between adsorption and desorption rate constants was lower than 1 which means that desorption should predominate, and more drug particles should be in mobile form. This was confirmed by the desorption experiments. Average leachability of adsorbed drug into water was 64 % and 36 % remained bound on humic acids. The effect of interactions between ibuprofen and humic acids thus is important but relatively big amount of ibuprofen can remain movable and bioavailable. As we can see in **Equation** 3, the calculation of effective diffusion coefficient can be based also on the intercept of the line where the total mass of diffusing drug applied on the interface is involved. The total mass is in the diffusion fractionated into movable mobile for and immobilized drug bound to humic acids. In reality, the diffusion coefficients based on the slope and intercept of the line cannot be the same as a result of the partial immobilization of diffusing particles and local equilibrium between movable and immobilized particles [16-20]. The difference between D_{eff} values based on the slope: (3.42 ± 0.04) × 10^{-10} m^2/s and the intercept: (4.82 ± 0.21) × 10^{-10} m^2/s can be considered as a degree of the immobilization and effect of the interactions between humic acids and diffusing pharmaceutical.

4. CONCLUSION

In this contribution, the diffusion of ibuprofen in humic hydrogel was studied. The hydrogel was used as a model medium of soil porous system containing reactive organic matter. The hydrogel should contain only humic acids (as a representative of organic matter) and water therefore can be considered as a model system for the investigation of transport of pharmaceutical through soil affected by their partial immobilization. It was found that the effective diffusion coefficient determined for the diffusion of ibuprofen in humic hydrogel is lower in comparison with its diffusion in water which is caused by hydrogel porous structure and interactions in diffusion through reactive medium. Complementary adsorption/desorption experiments confirmed that ibuprofen can be partially immobilized. The average effectivity of adsorption was around 60%. The ratio between adsorption and desorption rate constants based on Langmuir isotherm as well as desorption experiments showed that the desorption predominated, and ibuprofen was leachable from humic acids in large measure (64 %) and a residue 36 % remained bound on humic acids.

REFERENCES

[1] BEHERA, S.K., OH, S.Y., PARK, H.S. Sorptive removal of ibuprofen from water using selected soil minerals and activated carbon. *International Journal of Environmental Science and Technology*. 2012, vol. 9, pp. 85-94. Available from: https://doi.org/10.1007/s13762-011-0020-8

[2] WEBB, S., TERNES, T., GIBERT, M., OLEJNICZAK, K. Indirect human exposure to pharmaceuticals via drinking water. *Toxicology Letters*. 2003, vol. 142, pp. 157-167. Available from: https://doi.org/10.1016/S0378-4274(03)00071-7

[3] BUI, T.X., CHOI, H. Influence of ionic strength, anions, cations, and natural organic matter on the adsorption of pharmaceuticals to silica. *Chemosphere*. 2010, vol. 80, pp. 681-686. Available from: https://doi.org/10.1016/j.chemosphere.2010.05.046

[4] JERMANN, D., PRONK, W., BOLLER, M., SCHAFER, A.I. The role of NOM fouling for the retention of estradiol and ibuprofen during ultrafiltration. *Journal of Membrane Science*. 2009, vol. 329, pp. 75-84. Available from: https://doi.org/10.1016/j.memsci.2008.12.016

[5] THIEBAULT, T., BOUSSAFIR, M. LE FORESTIER, L., LE MILBEAU C., MONNIN, L., GUEGAN, R. Competitive adsorption of a pool of pharmaceuticals onto a raw clay mineral. *RCS Advances* 2016, vol. 6, pp. 65257-65265. Available from: https://doi.org/10.1039/C6RA10655B

[6] KODEŠOVÁ R., GRABIC R., KOČÁREK M., KLEMENT A., GOLOVKO O., FÉR M., NIKODEM A., JAKŠÍK O. Pharmaceutical´s sorptions relative to properties of thirteen different soils. *Science of the Total Environment* 2015, vol. 511, pp. 435-443. Available from: https://doi.org/10.1016/J.SCITOTENV.2014.12.088

[7] WU L., BI E. Sorption of ionic and neutral species of pharmaceuticals to loessial soil amended with biochars. *Environmental Science and Pollution Research* 2019, vol. 26, pp. 35871-35881. Available from: https://doi.org/10.1007/S11356-019-06721-7

[8] BIEL-MAESO M., BURKE V., GRESKOWIAK J., MASSMANN G., LARA-MARTÍN P.A., CORADA-FERNÁNDEZ C. Mobility of contaminants of emerging concern in soil column experiments. *Science of the Total Environment* 2021, vol. 762, pp. 144102. Available from: https://doi.org/10.1016/j.scitotenv.2020.144102

[9] SALVIA M.V., EXPERTON J., GEANDEL C., CREN-OLIVÉ C., VULLIET E. Fate of pharmaceutical compounds and steroid hormones in soil: study of transfer and degradation in soil columns. *Environmental Science and Pollution Research* 2014, vol. 21, pp. 10525-10535. Available from: https://doi.org/10.1007/s11356-014-3038-x

[10] CAO H.M., BU Q.W., LI Q.S., YANG L., TANG J.F., YU G. Evaluation of the DGT passive samplers for integrating fluctuating concentrations of pharmaceuticals in surface water. *Science of the Total Environment* 2024, vol. 926, pp. 172067. Available from: https://doi.org/10.1016/j.scitotenv.2024.172067

[11] LIU X., ZHANG R,L., CHENG H., KHORRAM M.S., ZHAO S.Z., THAM T.T., TRAN T.M., MINH T.B., JIANG B., JIN B., ZHANG G. Field evaluation of diffusive gradients in thin-film passive samplers for wastewater-based epidemiology. *Science of the Total Environment* 2021, vol. 773, pp. 145480. Available from: https://doi.org/10.1016/j.scitotenv.2021.145480

[12] GONCALVES N.P.F., MINELLA M., FABBRI D., CALZA P., MALITESTA C, MAZZOTTA E, PREVOT A.B. Humic acid coated magnetic particles as highly efficient heterogeneous photo-Fenton materials for wastewater treatments. *Chemical Engineering Journal* 2020, vol. 390, pp. 124619. Available from: https://doi.org/10.1016/j.cej.2020.124619

[13] CHRISTL I., RUITZ M., SCHMIDT J.R., PEDERSEN J.A. Clarithromycin and tetracycline binding to soil humic acid in the absence and presence of calcium. *Environmental Science & Technology* 2023, vol. 30, pp. 122420-122436. Available from: https://doi.org/10.1007/s11356-023-30900-2

[14] SUN H., SHI X., MAO J., ZHU D. Tetracycline sorption to coal and soil humic acids: An examination of humic structural heterogeneity. *Environmental Toxicology and Chemistry* 2010, vol. 29, pp. 1934-1942. Available from: https://doi.org/10.1002/etc.248

[15] GU C., KARTHIKEYAN K.G. Sorption of the antibiotic tetracycline to humic-mineral complexes. *Journal of Environmental Quality* 2008, vol. 37, pp. 704-711. Available from: https://doi.org/10.2134/jeq2007.0030

[16] CRANCK J. *The Mathematics of Diffusion*. Oxford: Clarendon Press, 1956.

[17] CUSSLER E.L. *Diffusion: Mass Transfer in Fluid Systems*. Cambridge: Cambridge University Press, 1984.

[18] KALINA, M. et al. Utilization of fractional extraction for characterization of the interactions between humic acids and metals. *Geoderma*. 2013. vol. 207-208, pp. 92-98. Available from: https://doi.org/10.1016/j.geoderma.2013.04.031

[19] KLUČÁKOVÁ, M., KALINA, M. Diffusivity of Cu(II) ions in humic gels - influence of reactive functional groups of humic acids. *Colloids and Surfaces A: Physicochemical and Engineering Aspects A* 2015, vol. 483, pp. 162-170. Available from: https://doi.org/10.1016/j.colsurfa.2015.05.041

[20] KLUČÁKOVÁ M., PEKAŘ M. Study of diffusion of metal cations in humic gels. In *Humic Substances: Nature`s Most Versatile Materials*. New York: Taylor & Francis, 2004, pp. 263-273.

[21] FARRAR H., LETZIG L., GILL M. Validation of a liquid chromatographic method for the determination of ibuprofen in human plasma. *Journal of Chromatography B*. 2002, vol. 780, pp. 341-348. Available from: https://doi.org/10.1016/S1570-0232(02)00543-3

[22] BEIRAGHI A., POURGHAZI K., AMOLI-DIVA M. Mixed supramolecular hemimicelles aggregates and magnetic carrier technology for solid phase extraction of ibuprofen in environmental samples prior to its HPLC-UV determination. *Chemical Engineering Science*. 2014, vol. 108, pp. 103-110. Available from: https://doi.org/10.1016/j.ces.2013.12.044

[23] BENDZ D., PAXÉUS N.A., GINN T.R., LOGE F.J. [2] Occurrence and fate of pharmaceutically active compounds in the environment, a case study: Höje River in Sweden. *Journal of Hazardous Materials*. 2005, vol. 122, pp. 195-204. Available from: https://doi.org/10.1016/j.jhazmat.2005.03.012

[24] SOUSA T.F.A., AMORIM, C.G., MONTENEGRO M.C.B.S.M., ARAÚJO A.N. Cyclodextrin based potentiometric sensor for determination of ibuprofen in pharmaceuticals and waters. *Sensors and Actuators B: Chemical*. 2013, vol. 176, pp. 660-666. Available from: https://doi.org/10.1016/j.snb.2012.09.016

[25] PEURAVUOURI J., ŽBÁNKOVÁ P., PIHLAJA K., 2006. Aspects of structural features in lignite and lignite humic acids. *Fuel Processing Technology*. 2006, vol. 87, pp. 829-839. Available from: https://doi.org/10.1016/j.fuproc.2006.05.003

[26] MENDES F.S., CRUZ C.E.M., MARTINS, R.N., PRATES RAMALHO J.P. MARTINS L.F.G. On the diffusion of ketoprofen and ibuprofen in water: An experimental and theoretical approach. *The Journal of Chemical Thermodynamics*. 2023, vol. 178, pp. 106955. Available from: https://doi.org/10.1016/j.jct.2022.106955

[27] KLUČÁKOVÁ M., KALINA M., SEDLÁČEK P., GRASSET L. Reactivity and transport mapping of Cu(II) Ions in humic hydrogels. *Journal of Soils and Sediments*. 2014, vol. 14, pp. 368-376. Available from: https://doi.org/10.1007/s11368-013-0730-2

[28] UPADRASHTA S.M., HÄGLUND B.O., SUNDELÖF L.O. Diffusion and concentration profiles of drugs in gels. *Journal of Pharmaceutical Sciences. 1993*, vol. 82, pp. 1094-1098. Available from: https://doi.org/10.1002/jps.2600821106

[29] KLUČÁKOVÁ, M.; ZÁVODSKÁ, P. Diffusion of pharmaceuticals in agarose hydrogels enriched by humic acids. *Colloids and Surfaces, A: Physiochemical and Engineering Aspects*. 2023, vol. 673, pp. 131825. Available from: https://doi.org/10.1016/j.colsurfa.2023.131825

THE EFFECT OF MOLECULAR WEIGHT OF POLYCAPROLACTONE IN BLENDED NANOFIBERS ON THE VAGINAL MICROBIOME RESPONSE

[1,2]Karolína MORÁVKOVÁ, [1]Šárka HAUZEROVÁ, [1]Michal ŘEZANKA, [2]Miroslava RYSOVÁ

[1]Technical University of Liberec, Faculty of Science, Humanities and Education, Department of Chemistry, Liberec, Czech Republic, EU, karolina.moravkova@tul.cz

[2]Technical University of Liberec, Institute for Nanomaterials, Advanced Technology and Innovation, Department of Applied Biology, Liberec, Czech Republic, EU, miroslava.rysova@tul.cz

https://doi.org/10.37904/nanocon.2024.5025

Abstract

The health of the vaginal microbiota is a crucial component of an intricate system. Vaginal drug delivery systems play an integral role in women's health by offering targeted and efficient treatment options for various gynaecological conditions, while aiming to minimally impact said bacterial population. We propose using nanofibrous materials based on the polymers polycaprolactone (PCL) and poly(lactic acid) (PLA) as a drug delivery system for vaginal environment due to their favourable properties. The molecular weight of PCL can significantly influence the properties of blend nanofibers, which in turn affects their interaction with the vaginal microbiome. The molecular weight of polymers tends to enhance the mechanical properties, crystallinity and stability of the nanofibers, potentially leading to a more controlled and sustained release of incorporated substances. This controlled release is crucial for maintaining a balanced microbiome, as it may help in the release of antimicrobial agents or probiotics.

A healthy vaginal microbiome is dominated by various types of *Lactobacilli*, which produce lactic acid and antimicrobial compounds. We tested the effect of blend nanofibers with PCL of two molecular weights (Mn 45,000 and 80,000) on the growth rate of three Lactobacilli species: *L. crispatus, L. gasseri, and L. jensenii*. There was no distinguishable effect on the growth rates of *L. crispatus* and *L. gasseri*, but the growth rate of *L. jensenii* was affected by PCL of both molecular weights. This is an important finding for the future use of PCL based nanofibers in vaginal drug delivery.

Keywords: Nanofibers, vaginal, polycaprolactone, microbiome, *Lactobacillus*

1. INTRODUCTION

Material composition of drug delivery systems used in vaginal drug delivery is crucial for ensuring effective and targeted treatment. They need to be biocompatible to avoid causing irritation or adverse reactions, and their properties must support the controlled release of medications at the site of action. Moreover, these materials help maintain an optimal environment for the natural microbiota, which is essential for overall vaginal health. Currently, vaginal drug delivery employs a variety of dosage forms, each tailored to enhance efficacy and user comfort. These include creams, gels, suppositories, tablets, rings, and films. Each form has unique benefits. For example, vaginal rings offer prolonged and controlled drug release, while gels and creams can provide immediate relief. On the other hand, many of these formulations still have disadvantages due to specifics of the vaginal environment. This includes its self-cleaning properties, acidic environment and most importantly vaginal bacterial population, which plays a crucial role in this intricately balanced system. Nanofibers are emerging as a promising system for vaginal drug delivery due to their unique properties [1], and so their testing for interaction with vaginal microbiota is essential.

Lactobacillus crispatus, *Lactobacillus jensenii*, and *Lactobacillus gasseri* are key players in maintaining vaginal health in Caucasian and Hispanic population [2,3]. These beneficial bacteria help maintain an acidic pH, which prevents the overgrowth of harmful microorganisms. They produce lactic acid and hydrogen peroxide, both of which are natural antimicrobials. Additionally, these lactobacilli enhance the mucosal immune response, providing further physiological protection against infections. Each novel material intended for vaginal use must be tested for interaction with these bacteria in order to assess its biocompatibility.

Polycaprolactone (PCL) and poly(lactic acid) (PLA) nanofibers have been widely researched for drug delivery applications due to their biocompatibility and biodegradability [4,6]. The molecular weight of polymeric chains combined with fiber dimeter can impact the bulk mechanical properties of the resulting nanofibers. Higher molecular weight PCL was described to produce nanofibers with increased mechanical strength, crystallinity and slower degradation kinetics, which is beneficial for long-term drug delivery applications. On the other hand, lower molecular weight PCL can result in nanofibers with faster degradation rates, suitable for applications requiring quicker drug release [7]. Lower molecular weight also increases frequency of the chain end groups affecting surface functionality, reactivity and energy, thereby affecting interactions between nanofibers surface and biological environment [8]. The choice of molecular weight depends on the specific therapeutic needs and desired release kinetics. As there is a wide range of potential vaginal application, it is crucial to test the response of vaginal microbiota for both higher and lower molecular weight PCL. For this study, polycaprolactone with molecular weight of 45,000 and 80,000 Da was chosen. Nanofibers were produced by blend electrospinning of the PCL and PLA mixture, tested for their basic morphological and surface properties and then their effect on growth kinetics of three vaginal bacterial strains was assessed for 44 hours. Presence of adhered bacteria was evaluated after the incubation period by electron microscopy.

2. MATERIALS AND METHODS

2.1 Materials

The polymers used for nanofibers preparation were biodegradable polyesters poly(lactic) acid (PLA, Corbion, Purac Biochem bv, NL) and poly-ε-caprolactone (PCL, Sigma-Aldrich, Mw 45,000 and 80,000 Da). The ratio of the polymers in the final spinning solution was 50:50 (wt.). Nanofibrous sheets were prepared from chloroform-based solvent system while utilizing DC needleless electrospinning device Nanospider™ (NS Line 1WS500U, Elmarco) set to production speed 15 mm/min.

2.2 Morphological and wettability analysis

Morphology of the nanofibers was evaluated by scanning electron microscopy (Tescan Vega 3 SEM). The fiber diameters were measured using built in software and expressed as histograms of diameter distribution and mean ± standard deviation. The initial water contact angle of sessile drop was measured on Drop Shape Analyzer DSA30E (Krüss, software 1.16 DSA30) in 5 repeats per sample. Contact angle development was measured after 1, 3 and 5 minutes on selected sample.

2.3 Interaction with vaginal *Lactobacilli*

Lactobacillus strains were *Lactobacillus jensenii* (CCM 7560), *Lactobacillus gasseri* (CCM 7009) and *Lactobacillus crispatus* (CCM 7776). Prior to the experiment, each lactobacilli strain underwent three passages during the exponential growth phase to ensure optimal metabolic activity. All three strains were incubated in De Man, Rogosa and Sharpe broth (MRSB, HiMedia), supplemented with 0.1% L-cysteine (Merck) as a reducing agent, under microaerophilic conditions at 37°C with CO_2 tension.

Materials used for testing interactions with vaginal lactobacilli were cut into 1 cm diameter circles and sterilized using UV light. For the experiment, each sample was immersed into 3.9 ml of bacterial suspension of MRSB and 0.1ml of the third passage of the bacterial strain. Optical density at 600 nm (OD600) was measured at specified intervals (0, 2, 4, 6, 8, 12, 17, 19, 21, 24, 32, and 44 hours) to assess turbidity. An increase in OD600 value reflects a rise in bacterial cell concentration. After 44 hours of incubation, the materials were gently washed with sterile phosphate buffered saline and fixed with 2.5 % paraformaldehyde for 10 minutes. The materials were then assessed by SEM for the presence of adhered bacteria.

3. RESULTS AND DISCUSSION

In this work, two types of biodegradable nanofibrous materials comtaining 50:50 mixture of two polymers – PLA and PCL. The difference between the materials was given by variation in molecular weight of PCL, which was 45,000 and 80,000. Combination of PLA and PCL represent a potentially new material for use in gynaecological applications, and as such, the response of vaginal microbiome to different PCL component needs to be assessed.

The materials were firstly evaluated by SEM for fiber diameter distribution. The PLA/PCL45 fibers had smaller mean fiber diameter (0.76 ± 0.3 µm) compared to the PLA/PCL80 nanofibers. Both showed randomly oriented fibrous structure with smooth surface of nanofibers (**Figure 1**).

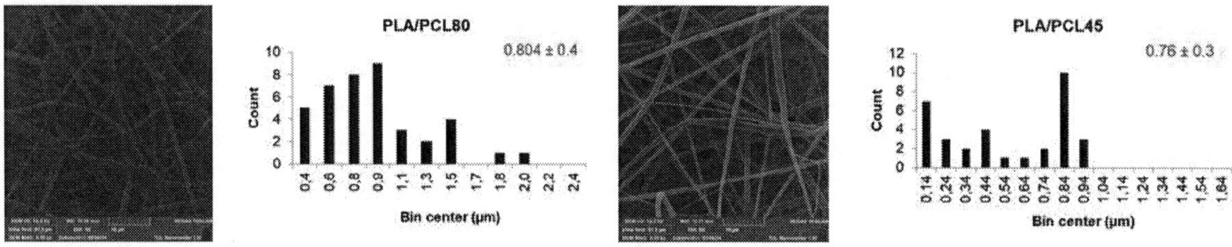

Figure 1 Scanning electron microscope images and histograms of the measured fibers size distribution including mean ± standard deviation values (red).

There was no significant difference between the samples observed with the sessile water drop wettability measurement. Based on the results, both of the materials can be considered hydrophobic, as the initial contact angle (CA) over 90° (**Figure 2a**). Accelerated wettability in time was observed for the PLA/PCL80. The contact angle dropped to 85.26° after 5 minutes while the PLA/PCL45 remained more hydrophobic with CA(m) 96.85° (**Figure 2b**).

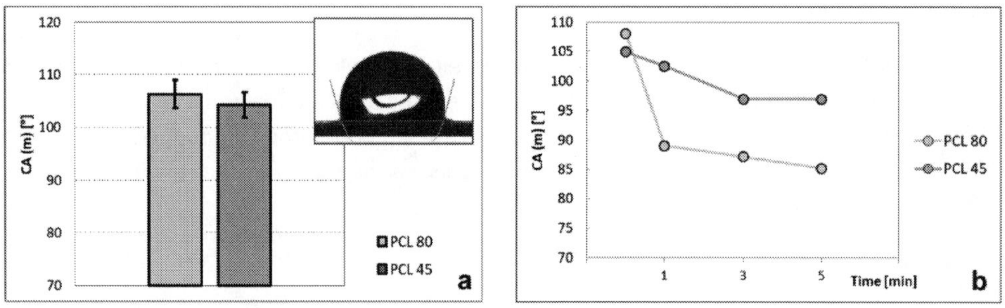

Figure 2 Contact angle of PLA/PCL80 and PLA/PCL45 materials (labeled by their PCL component), a) initial contact angle, b) contact angle after 1, 3 and 5 minutes.

The growth of three Lactobacilli strains was monitored by measuring optical density at a wavelength of 600 nm, where light scattering is referred to as a factor of turbidity, rather than light absorption. Growth of bacterial cells typically progresses through a series of consecutive phases including lag, log and stationary. When plotted, the OD600 data of *L. crispatus*, *L. jensenii* and *L. gasseri* from 44 hours of incubation (**Figure 3**) shows that *L. jensenii* growth curve is flatter indicating slower bacterial cells proliferation. Lactobacilli generally as a genus are considered aerotolerant anaerobes that produce lactic acid through the fermentation of glucose [9]. However, *L. jensenii* is considered to be an anaerobic strain [10], and so the experiment condition which were microaerophilic, were suboptimal for its growth. Nevertheless, the growth kinetics was the same in three consecutive experiments, so it was considered as a normal growth for this type of bacteria.

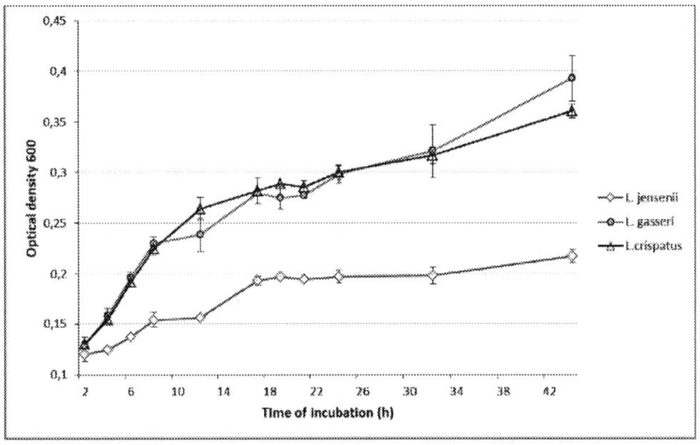

Figure 3 The growth curves of *Lactobacillus crispatus, Lactobacillus jensenii and Lactobacillus gasseri* incubated for 44 hours under microaerophilic conditions at 37 °C.

Coincubation with PLA/PCL materials yielded similar results with *L. crispatus* and *L. gasseri*. There is no difference in growth kinetics with or without contact with either of the tested nanofibers (**Figure 4**).

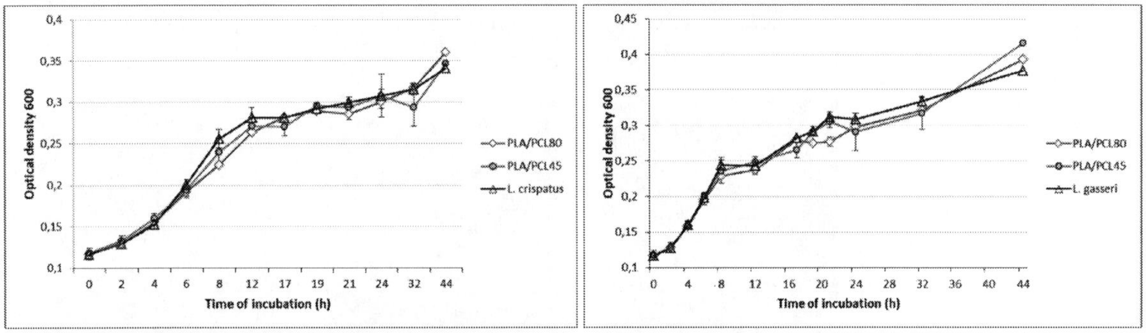

Figure 4 The growth curves of *Lactobacillus crispatus (left) and Lactobacillus jensenii (right)* in contact with tested materials PLA/PCL$_{80}$ and PLA/PCL$_{45}$. Incubation for 44 hours under microaerophilic conditions at 37 °C.

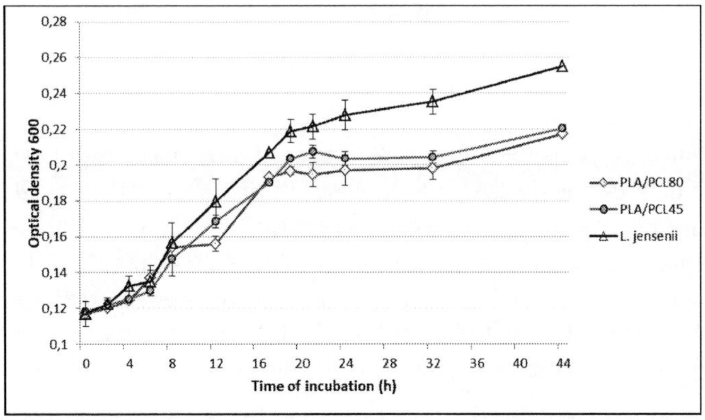

Figure 5 The growth curves of *Lactobacillus gasseri* in contact with tested materials PLA/PCL$_{80}$ and PLA/PCL$_{45}$. Incubation for 44 hours under microaerophilic conditions at 37 °C.

Coincubation of PLA/PCL materials with *Lactobacillus jensenii* resulted in different growth curve compared to positive control (bacterial suspension without any material) (**Figure 5**). The bacteria coincubated with the PLA/PCL materials demonstrated slower growth rate after 20 hours of incubation and reached lower maximum optical density after 44 hours. There is no difference between growth kinetics of *L. jensenii* in contact with either PLA/PCL80 and PLA/PCL45.

The materials were assessed by scanning electron microscopy after coincubation with bacteria (**Figure 6**). There are very few bacteria attached to the fibers after 44 hours, with the exception of PLA/PCL45 and *Lactobacillus jensenii*, which exhibited strings of rod-shaped bacterial cells on its whole surface serving as an evidence of increased adhesion of *L. jensenii* to the surface of PLA/PCL45 nanofibers. This could contribute to *L. jensenii* altered growth in presence of this material, nevertheless, it does not explain the similar trend in PLA/PCL80 material, which also altered the growth without the bacterial adhesion.

Material	*L. jensenii*	*L. gasseri*	*L. crispatus*
PLA/PCL_{80}	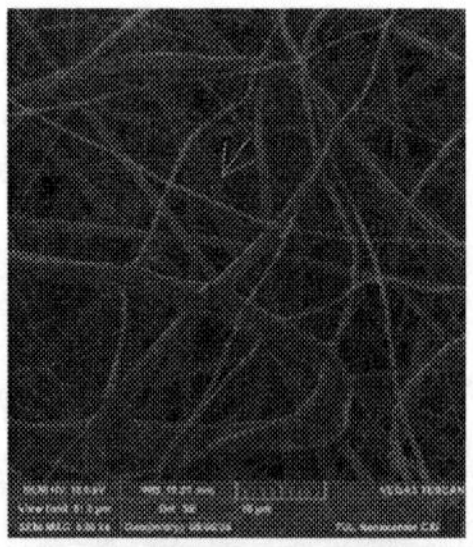		
PLA/PCL_{45}			

Figure 6 SEM images of PLA/PCL80 and PLA/PCL45 nanofibers after 44 hours (microaerophilic conditions, 37 °C) of coincubation with vaginal bacteria. Red arrows indicate bacterial cells attached to the fibers.

4. CONCLUSION

The objective of this study was to evaluate potential impact of molecular weight of PCL in blend PLA/PCL nanofibers on the growth of three vaginal bacterial strains: *Lactobacillus crispatus, Lactobacillus jensenii,* and *Lactobacillus gasseri*. The lower molecular weight (45.000 Da) of PCL resulted in nanofibres with smaller average diameter while maintaining hydrophobicity after 5 minutes (CA 96.95°). The incorporation of lower molecular weight PCL also led to increased adhesion of *Lactobacillus jensenii* with a growth curve comparable to sample containing higher molecular weight PCL (PLA/PCL80). Growth curves of *Lactobacillus crispatus* and *Lactobacillus gasseri* remained unaffected by any of the tested samples.

The obtained results indicate potential of the PLA/PCL nanofibres for applications in vaginal environment stimulation and drug delivery. Further research on biodegradation and biocompatibility is needed.

ACKNOWLEDGEMENTS

This work was supported by the Student Grant Competition of the Technical University of Liberec under the project No. SGS-2023-4406 and by OABi CXI TUL funds (8330/117).

REFERENCES

[1] CHINDAMO, G.; SAPINO, S.; PEIRA, E.; CHIRIO, D.; GALLARATE, M. Recent Advances in Nanosystems and Strategies for Vaginal Delivery of Antimicrobials. *Nanomaterials.* 2021, vol. 11, no. 311. Available from: https://doi.org/10.3390/nano11020311.

[2] SADEGHPOUR HERAVI, F. Host-vaginal microbiota interaction: shaping the vaginal microenvironment and bacterial vaginosis. *Current Clinical Microbiology Report*, 2024, vol. 11, pp 177–19. Available from: https://doi.org/10.1007/s40588-024-00227-8.

[3] CHEE, W.J.Y., CHEW, S.Y., THAN, L.T.L. Vaginal microbiota and the potential of Lactobacillus derivatives in maintaining vaginal health. *Microbial Cell Factories*, 2020, vol. 19, 203. Available from: https://doi.org/10.1186/s12934-020-01464-4.

[4] DUAN, X., CHEN, HI. GUO, C. Polymeric Nanofibers for Drug Delivery Applications: A Recent Review. *Journal of Material Science: Materials in Medicine,* 2022, vol. 33, 78. Available from: https://doi.org/10.1007/s10856-022-06700-4

[5] BHADRAN, A et al. Recent Advances in Polycaprolactones for Anticancer Drug Delivery. *Pharmaceutics*, 2023, vol. 15, no. 7, 1977. Available from: https://doi.org/10.3390/pharmaceutics15071977.

[6] HEDVICAKOVA, V. et al. The Gradual Release of Alendronate for the Treatment of Critical Bone Defects in Osteoporotic and Control Rats. *International Journal of Nanomedicine.* 2023, vol. 18, pp. 541-560. Available from: https://doi.org/10.2147/IJN.S386784.

[7] ALHARBI, N., DARAEI, A., LEE, H., GUTHOLD, M. The effect of molecular weight and fiber diameter on the mechanical properties of single, electro spun PCL nanofibers. *Materials Today Communications*. 2023, vol. 35. Available from: https://doi.org/10.1016/j.mtcomm.2023.105773.

[8] HENDRIKSON, W. J. et al. Influence of PCL molecular weight on mesenchymal stromal cell differentiation. *RSC Advances*. 2015, vol. 5, 54510. Available from: https://doi.org/10.1039/c5ra08048g.

[9] WILSON, W. A., ROACH, P. J., MONTERA, M., BAROJA-FERNANDEZ, E., MUNOZ, F. J., EYDALLIN, G., VIALE, A. M., POZUETA-ROMERO, J., Regulation of glycogen metabolism in yeast and bacteria, *FEMS Microbiology Reviews*. 2010, vol. 34, no. 6, pp. 952–985. Available from: https://doi.org/10.1111/j.1574-6976.2010.00220.x

[10] MEDINA-COLORADO, A. A., VINCENT, K. L., MILLER, A. L., MAXWELL, C. A., DAWSON, L. N., OLIVE, T., KOZLOVA, E.V., BAUM, M.M., PYLES, R.B. Vaginal ecosystem modelling of growth patterns of anaerobic bacteria in microaerophilic conditions. *Anaerobe*. 2017, vol. 45, pp. 10-18. Available from: https://doi.org/10.1016/j.anaerobe.2017.04.014.

NANOBIOSENSOR FOR RAPID DETECTION OF CREATININE CRYSTALS PRESENT IN URINE

[1]Julia WERLE, [1]Jakub PODHAJSKY, [2]Martina STANKOVA, [1]Magdalena FORTOVA, [3]Egon WERLE, [4]Warawan EIAMPHUNGPORN, [1]Karel KOTASKA, [1]Richard PRUSA, [1]Rene KIZEK

[1]Department of Medical Chemistry and Clinical Biochemistry, Second Faculty of Medicine, Charles University in Prague and Motol University Hospital, Prague, Czech Republic, EU

[2]Department of Research and Development, EcoNanoLife, Prague, Czech Republic, EU

[3]Department of Medicine I and Clinical Chemistry, Heidelberg University Hospital, Heidelberg, Germany, EU

[4]Department of Clinical Microbiology and Applied Technology, Faculty of Medical Technology, Mahidol University, Salaya, Nakhon Pathom, Thailand

https://doi.org/10.37904/nanocon.2024.5023

Abstract

Creatine is synthesized primarily in the liver and kidneys from amino acids and is phosphorylated to phosphocreatine, which serves as an energy source in muscle. Creatine is non-enzymatically degraded to creatinine at a constant rate and excreted in the urine. Its formation mainly depends on the muscle mass. Changes in creatine and creatinine levels may be related to various diseases (e.g. kidney disease, rhabdomyolysis etc.). Several methods are used to determine creatinine, such as for example the Jaffe method using the reaction of creatinine with picric acid in an alkaline medium, forming a red-orange complex or the enzymatic Trinder reaction. In some cases, crystals or larger aggregates are present in the urine. Primary diagnosis of such crystals is focused on microscopic evidence of the presence of the structure and may be confirmed by FTIR. The three enzymes, creatinase, creatininase and sarcosine oxidase can also be used for qualitative assessment of creatinine. The aim of this work was to design a SPION nanobiosensor for qualitative proof of the presence of creatinine in a sample. The methodology uses the highly sensitive Red Amplex fluorophore. Absorbance dependence on creatinine concentration read off at the wavelength of 572 nm was linear over a total range of 0 -1000 μM.

Keywords: Sarcosine; urinary crystals; photometric detection; automated analysis; sarcosine oxidase; SPION

1. INTRODUCTION

Creatinine is a metabolic product of creatine phosphate in the muscles, which provides energy to the muscle tissue. Creatinine levels are considered an indicator of renal function, thyroid disorders, and muscle damage. The serum creatinine level ranges between 45-140 μM and is excreted in the urine at 0.8 - 2.0 g/day [1]. The accuracy of serum creatinine concentration measurement affects the estimation of glomerular filtration rate (GFR) [2]. A simplified diagram of creatinine metabolism is shown in **Figure 1**. In addition to the above-mentioned importance of creatinine levels in clinical practice, there are more rarely occurring biologically significant phenomena. Elevated concentrations of low molecular weight substances can lead to the formation of small crystals to relatively large stones present in various organs. The analysis of urinary stones present in the urinary tract is of considerable diagnostic value and may provide significant benefits to the patient. Very few cases of creatinine/creatine crystals in urine have been described so far [3]. A number of analytical approaches (spectrophotometry, chromatography) are available for the determination of creatinine. In routine clinical practice, creatinine determination by non-enzymatic (Jaffé method) and enzymatic methods are used.

The principle of the Jaffé method is the reaction of creatinine with picric acid in an alkaline medium to form an orange-coloured complex. The enzymatic method is based on the conversion of creatinine by creatininase to creatine. Creatine is hydrolysed to sarcosine by creatinase. The sarcosine formed is oxidised by sarcosine oxidase to glycine, formaldehyde and hydrogen peroxide.

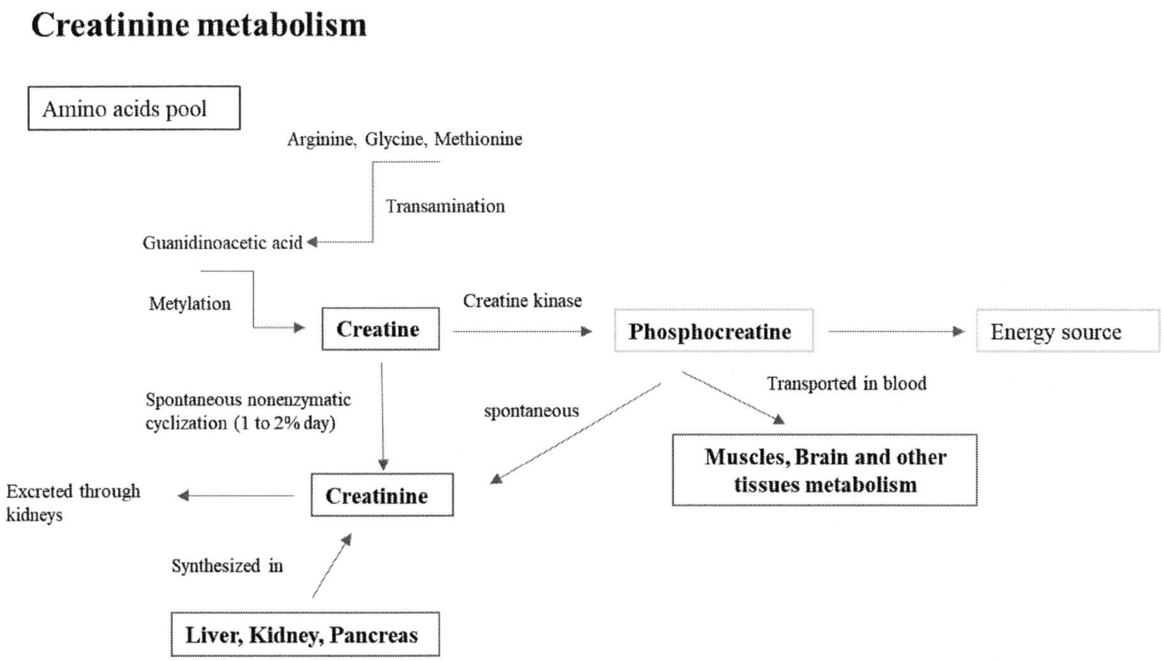

Figure 1 Simplified diagram of the proposed metabolic pathways of creatine and creatinine

It is well known that amino acids are colourless, making their direct photometric determination challenging. A suitable method for analysing amino acids involves their reaction with chromophores to form coloured complexes [4]. The well-known reaction of amino acids with ninhydrin produces a complex ranging from red-violet to blue-violet in colour [5], with the absorbance of these products measured at 570 nm [6-9].

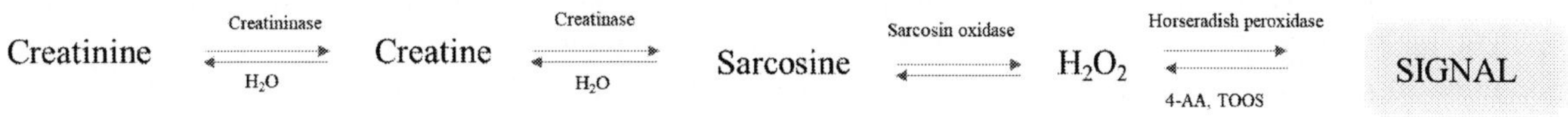

Figure 2 Simplified diagram of the proposed reactions enabling the measurement of creatinine concentration in a biological sample

Figure 2 illustrates a simplified scheme for the detection of creatinine. In the proposed method, creatinine is converted by creatininase into creatine. The subsequent reaction with creatinase leads to the synthesis of sarcosine. Sarcosine oxidase (EC 1.5.3.1) is a specific enzyme that oxidizes only the amino acid sarcosine in the presence of water and oxygen to glycine, producing hydrogen peroxide and formaldehyde as by-products. The resulting hydrogen peroxide can subsequently be quantified using the Trinder reaction, detectable spectrophotometrically at 510 nm. In this reaction, hydrogen peroxide interacts with phenol and 4-aminoantipyrine in the presence of the sodium salt of 3-(N-ethyl-3-methylaniline) propanesulfonic acid (TOPS) and peroxidase, leading to the formation of a quinonimine complex.

Biosensors, comprising a biologically active component, a chemical-physical transducer, and an advanced mathematical evaluation system, are also available and have demonstrated the capability to yield comparable results. The potential of nanotechnology in bioanalytical approaches is substantial. This approach allows for significant enhancements in both the sensitivity and selectivity of analyses. Clinical laboratory medicine demands a high level of precision, often working with minimal sample volumes and requiring the avoidance of unnecessary repetition of tests. Nanotechnology presents a promising solution to these challenges. Numerous types of nanoparticles have been developed, with superparamagnetic iron oxide nanoparticles (SPIONs) being the preferred choice for efficient analyse detection. SPIONs can be easily modified with appropriate biomolecules, enabling the selective and sensitive detection of target analyses [16]. Consequently, this approach is increasingly integrated into rapid point-of-care testing (POCT) devices.

-The objective of this research is to design and develop an advanced nano-biosensor based on superparamagnetic iron oxide nanoparticles (SPIONs), specifically tailored for the precise and efficient detection of creatinine. This work places particular emphasis on refining in-house analytical methodologies to achieve high sensitivity and specificity in creatinine measurement. A critical issue addressed by this study is the current deficiency in rapid screening capabilities within point-of-care (POC) testing systems, which this novel SPION-based nano-biosensor is poised to significantly enhance.

2. MATERIAL AND METHODS

2.1 Material and chemicals

The analytical grade chemicals (ACS) used were purchased from Sigma Aldrich (USA): NaH_2PO_4, Na_2HPO_4, Na_3PO_4, KCl, NaCl, NaOH, HCl, 4-aminoantipyrine, phenol bioextra, 3-(N-ethyl-3-methylanilino)-propanesulfonic acid sodium salt, sarcosine, creatinine, bovine serum albumin, enzymes (horseradish peroxidase, sarcosine oxidase from Bacillus sp.). The disposable plastic tips and microtubes used were of RNA and DNA free quality (Eppendorf, Germany). Disposable 1 cm UV-VIS cuvettes for the single-beam spectrophotometer were purchased from VWR (USA).

2.2 UV-VIS spectroscopy

Spectrophotometric measurements were conducted using a UV-3100 PC spectrophotometer (VWR, USA). For these measurements, plastic cuvettes (UV-VIS, BRAND, USA) were utilized. 500 µL reagent was pipetted into the cuvette followed by 100 µL of sample. The reaction solution was used as blank. Further spectrophotometric analyses were performed on a polystyrene microtiter plate, which had been pre-washed with 18 MΩ water (3 times, 300 µL) at room temperature using a reader (TECAN, Switzerland). 50 µL of sample and 250 µL of reagent were pipetted. Absorbance measurements were taken every minute at 540 nm over a period of 60 minutes. The analysis was conducted using the VARIOSKAN Lux multimode reader (Thermo Scientific, USA). The scanning was performed across a wavelength range of 300 to 800 nm, with 1 nm increments. Data control and collection were managed using the SKANIT RE software (version 6.0.1).

2.3 Fourier-Transform Infrared (FTIR) Spectroscopy

All samples were dehumidified (ECOCELL BMT, Czechia, 130 °C, 24 h) and maintained in a desiccator (silica gel) prior to analysis. The functional group modifications of the samples were characterized using Fourier-transform infrared spectroscopy (Nicolet iS10 Spectrometer) by the potassium bromide (KBr) pellet method. The dried samples (30 mg) were mixed with KBr (500 mg) and pressed under mechanical pressure in mortar (5 min). Spectra were recorded between the wavenumber range of 4000 and 400 cm^{-1}. Software was used to remove noise and smoothening utilizing Savitsky-Golay filter, baseline correction and Fourier transformation.

2.4 Automated analysis

For the chemical analysis of specimens, the Atellica Solutions CH 930 analyser (Siemens, USA) was used for photometric analyses. The urine sediment was analysed using Atellica 1500, which combines the Clinitek

Novus analyser and the Atellica Automated Urinalysis System 800 (Siemens, USA). The semi-automated Aution Eleven AE-4020 (Arkray, Japan) with Aution Sticks 10EA was utilized for the chemical urine analysis with a measurement principle of test strip method and dual-wavelength reflection photometric method. For the detection of ions, an integrated multisensor technology (IMT) on printed electrodes was employed. Osmolality was measured using the OsmoPRO osmometer (Advanced Instruments, USA).

2.5 Reaction procedure

Sample contained 4-AAP (4-aminoantipyrine), sarcosine oxidase, horseradish peroxidase in 0.2 M phosphate buffer (pH 8). After pipetting the sample into the reaction solution, incubation was performed for 30 minutes. Artificial urine: sodium chloride (170 mM), potassium chloride (80 mM), sodium phosphate (39 mM), urea (300 mM), creatinine (18 mM), bovine serum albumin (750 µM) were prepared according to available protocols [18-20].

2.6 Synthesis and modification of SPION

Preparation of SPION nanoparticles and their modification with SOX SPION nanoparticles were prepared according to the procedure published [21]. Shortly, SPIONs nanoparticles were prepared from $Fe(NO_3)_3$ and a solution of borohydride in 3.5% NH_3. The synthesis was carried out stirring at 100 °C for 2 hours. After 2 hours, the mixture was allowed to stir overnight at room temperature. The prepared iron nanoparticles were washed, and their surface was modified by the reaction with $HAuCl_4$, the mixture was stirred for 3 hours. Trisodium citrate was then added, and the mixture was stirred again overnight. The resulting SPIONs nanoparticles were washed, dried and ready for further use. For further modification, the SPIONs nanoparticles were weighed, washed and SOX was attached to them using chitosan and tripolyphosphate.

2.7 Microscopy

Microscopic analysis was performed on an OLYMPUS BX53 microscope paired with a CANON camera and an evaluation program (Quick PHOTO CAMERA 3.2). Dry urine sample (1 mg) was used on a glass slide after intensive grinding. For each magnification (100, 200, 400, and 600 times), ten images were evaluated in both transmitted and polarized lighting. Selected images were recorded in the evaluation program and saved.

2.8 Data treatment and descriptive statistics

The experimental work was carried out in three independent experiments. The analysis of each sample was carried out 5 times. The obtained data are presented as average values. From the proposed study no experimental subjects were excluded from the proposed experimental studies. All the obtained data were stored in the QINSLAB database (Czech Republic). When possible, data was processed and evaluated mathematically and statistically in the QINSLAB database. The exclusion of outliers in the data sets was performed by calculating the Grubbs test. Photographs were processed by programme ColorTest, which assigns intensity to the individual pixels of the studied image in the colour area. For the publication purpose, data were processed using Microsoft (USA).

3. RESULTS AND DISUSSION

The spectrophotometric analysis of biological samples conducted using automated techniques is illustrated in **Figure 3**. In the initial phase of this study, the effect of a substantial creatinine excess in a model serum sample was investigated. Our findings indicate that a creatine concentration of approximately 100 mmol/L induces a signal variation of up to 5%, thereby confirming the robustness of the proposed reaction mechanism (see **Figure 2**). It is well known that haemolysis can affect analytical accuracy, and this was assessed by testing haemolysis levels ranging from 0 to 6000 mg/L. As demonstrated, no significant dependence on haemolysis was observed. The method showed typical concentration-dependence relationships in urine (y = 40.1x - 45.2;

r = 0.9994, LOD = 2, LOQ = 6, QC = 3.4 mmol/L) and serum (y = 71.4x - 71.5; r = 0.9991, LOD = 1.2, LOQ = 3.8, QC = 4.2 µmol/L). Further evaluation of the method was performed using serum samples from 98 subjects, with creatinine concentrations ranging from 10 to 900 µmol/L. The data exhibited a normal distribution with a mean concentration of 174 µmol/L and a median of 91 µmol/L. In urine samples (n = 80), concentrations ranged from 700 to 30,000 mmol/L, displaying a normal distribution with a mean of 9872 mmol/L and a median of 8219 mmol/L. Additionally, this pilot study included a unique cohort of 64 children (age 0–1 year), with creatinine levels in serum ranging from 29 to 79 µmol/L, and a mean and median of 60 and 61 µmol/L, respectively.

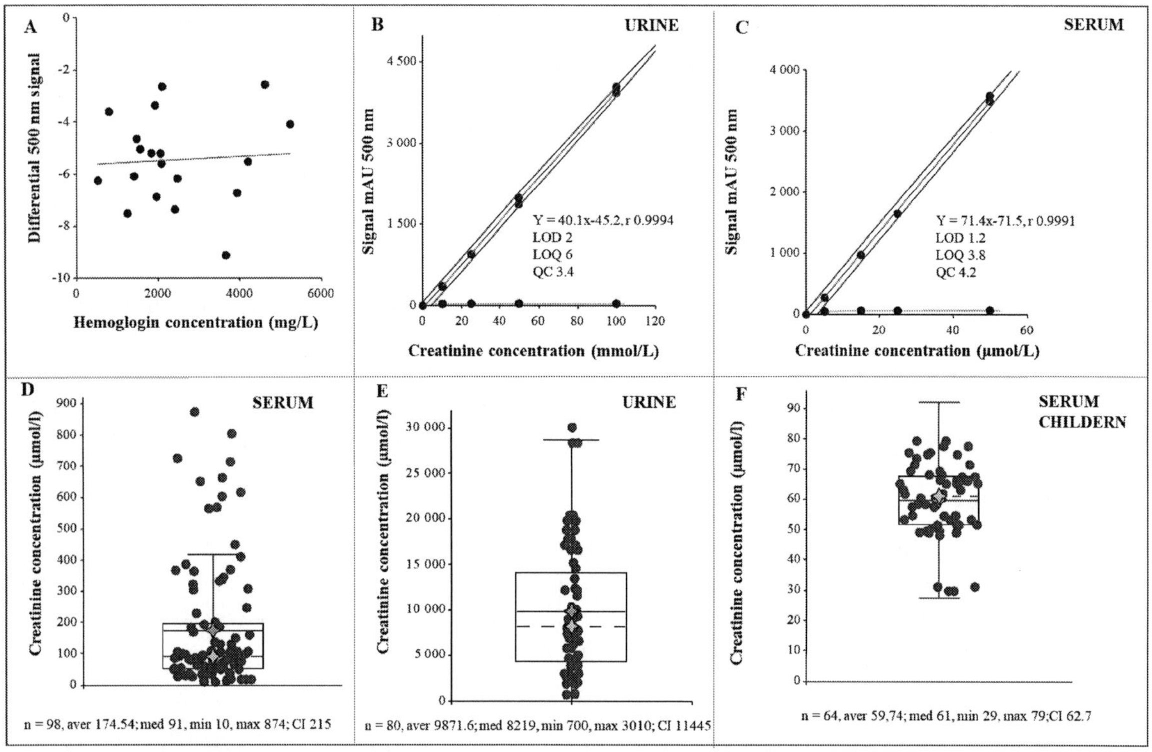

Figure 3 Spectrophotometric analysis of creatinine of biological samples by automated analyser. Study of the effect of erythrocyte haemolysis (haemoglobin 0 - 6000 mg/l) (**A**) on the measured signal of creatinine concentration in serum sample. Dependence of creatinine signal in model serum (0 - 60 µmol/l) (**B**) and in model urine (0 - 100 mmol/l) (**C**). Biological variability (including pathology) of serum creatinine concentration (n = 98) (**D**) and urine **(E)**, group of children (0 - 1 year; n = 64) (**F**).

In our designed concept, chemical synthesis of magnetic particles (SPION) modified with citric acid and subsequently enzymes (creatininase, creatinase, sarcosine oxidase and peroxidase) was carried out. SPION modification was performed according to our previous work with chitosan [17]. Modification was carried out in 0.1 M phosphate buffer (pH 8.0) with added chitosan (8 mg/mL) for 30 min, 100 rpm, 25 C. After that, TPP (0.1 mg/mL) was added to the particles and further incubated for 30 min, 100 rpm, 25 C. Subsequently, the modified particles were washed 3 times (in 0.1 M phosphate buffer, pH 8.0). Based on the results obtained in **Figure 3**, we extended our approach to the analysis of other biological samples (blood, serum, urine, stones). The particles prepared in this way were used to detect the presence of creatinine (**Figure 4**). Attention was focused on the final stage of detection of sarcosine conversion by the enzyme sarcosine oxidase. In the experiment, creatinine was determined by an enzymatic method using 4-aminoantipyrine (4-AAP) and the sodium salt of 3-(N-ethyl-3-methylaniline) propanesulfonic acid (TOPS). The use of these reagents produces a stable mixture with very good analyte sensitivity. Creatinine is converted by creatininase into creatine and the reaction with creatinase leads to the synthesis of sarcosine. Sarcosine oxidase oxidizes sarcosine in the presence of water and oxygen to glycine, producing hydrogen peroxide and formaldehyde as by-products (see **Figure 2**). The product of the reaction is a distinct purple coloured quinonimine complex with an absorption maximum at 546 nm, as shown in **Figure 4C**. Very good responses were obtained in the tested samples even at low creatinine concentrations. The procedure is therefore suitable for further development of POCT technologies.

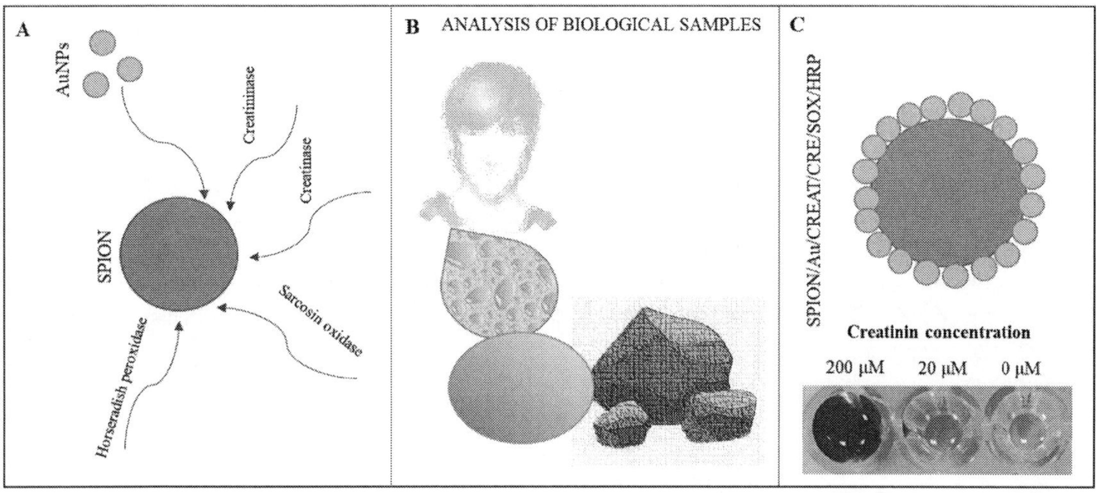

Figure 4 Scheme of SPION modification with gold nanoparticles AuNPs and enzymes sarcosine oxidase, creatinase, creatininase (**A**). Potential use of SPION/Au/CREAT/CRE/SOX/HRP for biological sample analysis (serum, blood, urine, exudates, concrements) (**B**). Color reaction of detected creatinine (200, 20, 0 µM) in 1 mM PBS using modified nanoparticles (**C**)

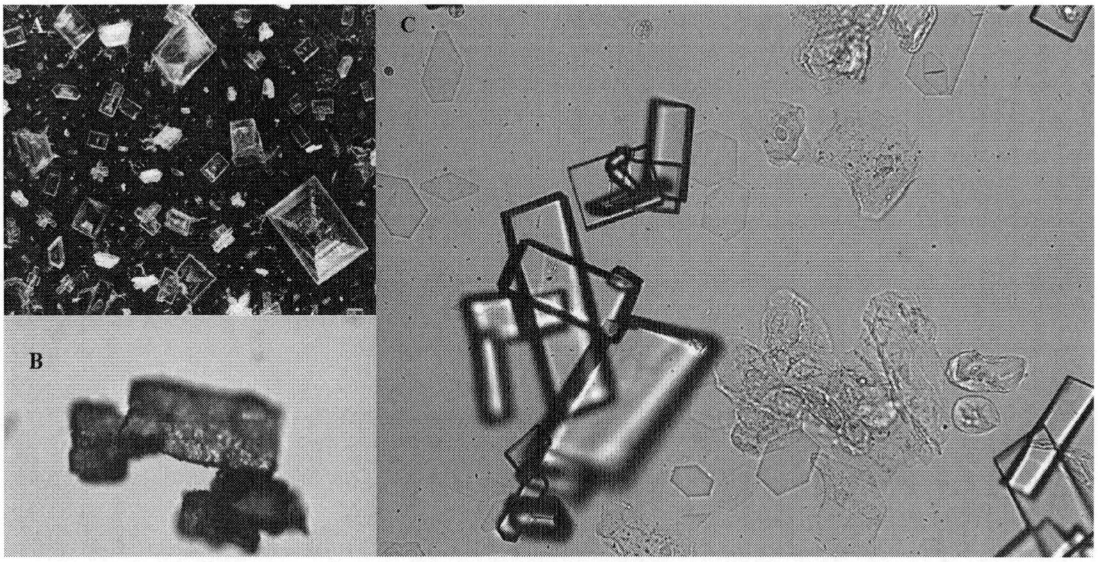

Figure 5 Crystallisation of creatinine (**A**). Polarization microscopy of creatinine crystals (**B**). Light field microscopy of creatinine crystals in a biological urine sample (**C**)

The present study focuses on the application of the proposed concept to the analysis of unique samples, such as crystals. Although it is exceedingly uncommon for creatinine/creatine crystals to manifest, with the majority of cases occurring within the urinary tract [22], recently, we were able to document a case of this phenomenon. By employing a range of analytical techniques, we were able to confirm the presence of these crystals. However, it remains uncertain whether this infrequent occurrence is attributable to the specialisation of the department and whether the requisite tools are accessible in less specialised departments. This hypothesis may be verified by the modified SPIONs developed by our research group. Moreover, our team has also synthesised creatinine and creatine crystals. The structure of creatinine crystals was first elucidated over seven decades ago [23], and it is established that diverse complexes can be formed [24]. The resulting crystals are illustrated in **Figure 5** and were subsequently subjected to a comprehensive analysis. The presence of creatinine was confirmed in the samples using the specified analytical techniques.

4. CONCLUSION

In routine laboratory operations, the limited space and time prevents a more detailed study and analysis of the biological sample. Usually, a rapid and one-time examination is necessary, which is an approach not sufficiently sensitive to detect rare occurrences, such as the presence of creatinine crystals. Our approach involved the design of a rapid procedure for specific identification, which was made possible by the use of nanotechnology. SPION nanoparticles were modified with enzymes for this purpose. This resulted in the development of a highly sensitive and selective tool that is suitable for incorporation into a point-of-care testing (POCT) analyser.

ACKNOWLEDGEMENTS

The work was supported by the SARKOTEST 165/2015 project and the European Technology Platform for Nanomedicine. We would like to thank Tomáš Sedlák, Iva Křížková and Irina Stýborová for their technical assistance.

REFERENCES

[1] KOTASKA, K., PRUSA, R., BUNESOVA, M., CEPOVA, J. *Reference values for biochemical and pharmacological analysis by age, sex, ethnicity, transgender and trimester of pregnancy*. Prague: University Hospital of Motol, 2023, p. 297. ISBN: 9788087347430.

[2] FORTOVA, M., SOPKO, B., FENCL, F., KIZEK, R., KOHUTIAR, M., PRUSA, R. Estimation of glomerular filtration rate from serum creatinine in children. *Prakt. Lek.* 2023, vol. 103, no. 4, pp. 174-179.

[3] LIVROZET, M., ESTEVE, E., DAUDON, M., ZAWORSKI, J., HAYMANN, J.-P., FROCHOT, V., BOFFA, J.-J., BAZIN, D., BENOIST, J.-F., LETAVERNIER, E. The Case | Fluctuating serum creatinine and crystals in urine. *Kidney International.* 2020, vol. 97, no. 6, pp. 1307-1308.

[4] YAMKAMON, V., PHAKDEE, B., YAINOY, S., SUKSRICHAWALIT, T., TATANANDANA, T., SANGKUM, P., EIAMPHUNGPORN, W. Development of sarcosine quantification in urine based on enzyme-coupled colorimetric method for prostate cancer diagnosis. *Excli Journal.* 2018, vol. 17, pp. 467-478.

[5] CERNEI, N., HEGER, Z., KOPEL, P., SKLADANKA, J., ZITKA, O., ADAM, V., KIZEK, R. Isolation of Biogenic Amines Using Paramagnetic Microparticles Off-Line Coupled with Ion Exchange Liquid Chromatography. *Chromatographia.* 2014, vol. 77, no. 21-22, pp. 1451-1459.

[6] YEMM, E. W., COCKING, E. C., RICKETTS, R. E. The determination of amino-acids with ninhydrin. *Analyst.* 1955, vol. 80, no. 948, pp. 209-214.

[7] CESSII, C., PILIEGO, F. The Determination of Amino Sugars in the Presence of Amino Acids and Glucose. *Biochem. J.* 1960, vol. 77.

[8] DENT, C. E. A Study of the Behaviour of some Sixty Amino-acids and other Ninhydrin-reacting Substances on Phenol 'collidine' Filter-paper Chromatograms, with Notes as to the Occurrence of some of them in Biological Fluids. *Biochem. J.* 1948, vol. 43, no. 2, pp. 169-180.

[9] GAITONDE, M. K. A Spectrophotometric Method for the Direct Determination of Cysteine in the Presence of Other Naturally Occurring Amino Acids. *Biochem. J.* 1967, vol. 104, no. 2, pp. 627-633.

[10] JONES, P. F., JOHNSON, K. E. Estimation of Phenols by the 4-Aminoantipyrine Method: Identification of the Colored Reaction Products by Proton Magnetic Resonance Spectroscopy. *Canad. J. Chem.* 1973, vol. 51, pp. 2860-2868.

[11] TRINDER, P. Determination of blood glucose using 4-amino phenazone as oxygen acceptor. *J. Clin. Pathol.* 1969, vol. 22, no. 2, p. 246.

[12] VECERA, M. *Detection and Identification of Organic Compounds.* Boston: Springer, USA, 1971.

[13] WIEWIORKA, O., DASTYCH, M., CERMAKOVA, Z. Trinderova reakce v klinické biochemii - přínosy a limity. *Chem. Listy.* 2017, vol. 111, no. 3, pp. 186-191.

[14] PUNDIR, C. S., KUMAR, P., JAIWAL, R. Biosensing methods for determination of creatinine: A review. *Biosensors & Bioelectronics.* 2019, vol. 126, pp. 707-724.

[15] SHARMA, A. C., JANA, T., KESAVAMOORTHY, R., SHI, L. J., VIRJI, M. A., FINEGOLD, D. N., ASHER, S. A. A general photonic crystal sensing motif: Creatinine in bodily fluids. *Journal of the American Chemical Society,* 2004, vol. 126, no. 9, pp. 2971-2977.

[16] SODIPO, B. K., AZIZ, A. A. Recent advances in synthesis and surface modification of superparamagnetic iron oxide nanoparticles with silica. *Journal of Magnetism and Magnetic Materials.* 2016, vol. 416, pp. 275-291.

[17] LUPPA, P. B., MÜLLER, C., SCHLICHTIGER, A., SCHLEBUSCH, H. Point-of-care testing (POCT): Current techniques and future perspectives. *Trac-Trends in Analytical Chemistry.* 2011, vol. 30, no. 6, pp. 887-898.

[18] PIETRZYNSKA, M., VOELKEL, A. Stability of simulated body fluids such as blood plasma, artificial urine and artificial saliva. *Microchem. J.* 2017, vol. 134, pp. 197-201.

[19] CHUTIPONGTANATE, S., THONGBOONKERD, V. Systematic comparisons of artificial urine formulas for in vitro cellular study. *Anal. Biochem.* 2010, vol. 402, no. 1, pp. 110-112.

[20] SHMAEFSKY, B. R. ARTIFICIAL URINE FOR LABORATORY TESTING. *Amer. Biol. Teacher.* 1990, vol. 52, no. 3, pp. 170-172.

[21] STANKOVÁ, M., RUTTKAY-NEDECKY, B., DOCEKALOVÁ, M., UHLÍROVÁ, D., MIKELOVÁ, R., HOSNEDLOVÁ, B., RUZICKA, J., KIZEK, R. Spectrophotometric Detection of the Amino Acid Sarcosine Using its Hydrolysis by Sarcosine Oxidase. *Chemicke Listy.* 2019, vol. 113, no. 10, pp. 603-609.

[22] LIVROZET, M., ESTEVE, E., DAUDON, M., ZAWORSKI, J., HAYMANN, J. P., FROCHOT, V., BOFFA, J. J., BAZIN, D., BENOIST, J. F., LETAVERNIER, E. Fluctuating serum creatinine and crystals in urine - Excessive dietary creatine intake. *Kidney International.* 2020, vol. 97, no. 6, pp. 1307-1308.

[23] DUPRE, S., MENDEL, H. The Crystal Structure of Creatinine. *Acta Crystallographica.* 1955, vol. 8, no. 6, pp. 311-313.

[24] KARLE, I. L., DRAGONETTE, K. S., BRENNER, S. A. The Crystal and Molecular Structure of Serotonin-Creatinine Sulphate Complex. *Acta Crystallographica.* 1965, vol. 19, pp. 713-716.

INVESTIGATION OF COVALENT BINDING OF GOLD NANOPARTICLES TO CHITOSAN NANOFIBERS USING CELLULOSE AND HYALURONATE DIALDEHYDES

[1]Lukáš MÜNSTER, [1]Alžběta DŮBRAVOVÁ, [1]Vítek HRBÁČEK, [1]Monika MUCHOVÁ, [1,2]Ivo KUŘITKA, [1,3]Petr HUMPOLÍČEK, [1]Jan VÍCHA

[1]*Centre of Polymer Systems, Tomas Bata University in Zlín, Zlín, Czech Republic, EU,* munster@utb.cz

[2]*Department of Chemistry, Faculty of Technology, Tomas Bata University in Zlín, Zlín, Czech Republic, EU,* kuritka@utb.cz

[3]*Department of Fat, Surfactant and Cosmetics Technology, Faculty of Technology, Tomas Bata University in Zlín, Zlín, Czech Republic, EU,* humpolicek@utb.cz

https://doi.org/10.37904/nanocon.2024.5024

Abstract

In this study, the reaction mechanisms of gold nanoparticles (AuNPs) synthesis using dialdehyde cellulose (DAC) and dialdehyde hyaluronate (DAH), and their covalent binding to chitosan nanofibers (CHITs), was investigated. The synthesis uses a redox reaction where dialdehyde polysaccharides are oxidized to dicarboxy polysaccharides and the gold salt precursor is reduced to elemental gold. The formation of the AuNPs-CHIT composite involves Schiff base chemistry, where reactive aldehyde groups of the polysaccharide shell around AuNPs react with chitosan's amine groups, forming pH-labile imine groups, which can be subsequently stabilized using reductive amination. FT-IR and XPS analyses were used to confirm the proposed reaction mechanisms. Next, the catalytic activity of AuNPs synthesized using DAC and DAH was evaluated for the reduction of 4-nitrophenol to 4-aminophenol with sodium borohydride. Rapid conversion rates and high turnover frequency were observed. The morphology and structure of the composites were characterized using transmission electron microscopy (TEM), scanning electron microscopy (SEM), and X-ray diffraction (XRD). These analyses confirmed the sub-10 nm size of the AuNPs and their uniform distribution on chitosan nanofibers. The findings confirm the proposed reaction mechanisms, showcase the morphology and catalytic activity of the prepared nanoparticles and highlight their potential industrial applications. The versatility of this method also opens avenues for further functionalization and broader applications of AuNPs in biomedical fields such as biosensors and drug delivery systems.

Keywords: Gold nanoparticles, dialdehyde cellulose, dialdehyde hyaluronate, chitosan nanofibers, catalysis

1. INTRODUCTION

Gold nanoparticles (AuNPs) are widely studied due to their unique optical, electronic, and catalytic properties, making them valuable for applications in catalysis, drug delivery, and biosensing [1–3]. However, ensuring stable binding of AuNPs to biocompatible substrates such as chitosan nanofibers remains a complex challenge [4]. Chitosan, a natural polymer known for its biocompatibility and antimicrobial properties, is an ideal candidate for such composite materials. This study investigates the covalent binding of AuNPs to chitosan nanofibers (CHITs) using dialdehyde cellulose (DAC) and dialdehyde hyaluronate (DAH) prepared employing simple periodate oxidation (POX), providing a robust method for composite synthesis with enhanced stability and functionality. Dialdehyde polysaccharides (DAPs), such as DAC and DAH, are valuable due to their triple role as reducing and stabilizing agents in nanoparticle synthesis, and as liner for covalent attachment to CHITs [5]. Derived from cellulose and hyaluronic acid, respectively, these materials contain reactive aldehyde (–CHO) groups that readily form Schiff base linkages with the amine (–NH_2) groups of chitosan. This process creates covalent imine bonds, which can be further stabilized via reductive amination, ensuring stable attachment of AuNPs to chitosan nanofibers. This method effectively combines the catalytic properties of AuNPs with the functional benefits of substrates based on natural polymers.

In the synthesis represented in **Figure 1**, DAPs prepared by POX (part A) partially oxidize to dicarboxy polysaccharides, while the gold salt precursor is reduced to elemental gold (part B), forming AuNPs surrounded by a stabilizing polysaccharide shell. The –CHO groups in the shell then bind covalently to chitosan's –NH_2 groups, ensuring robust nanoparticle attachment (part C). This approach provides stability in aqueous environments, a critical factor for many biomedical and industrial applications. Here, the reaction mechanisms of synthesis and covalent binding are studied and confirmed through Fourier-transform infrared spectroscopy (FT-IR) and X-ray photoelectron spectroscopy (XPS), which validates the reduction of Au^{3+} to Au^0 and partial oxidation of DAPs to dicarboxy polysaccharides (DCP), and the formation of imine/secondary amine bonds between –CHO groups of AuNPs shell and chitosan's –NH_2 groups. Morphological analyses using transmission electron microscopy (TEM), scanning electron microscopy (SEM), and X-ray diffraction (XRD) revealed a uniform distribution of sub-10 nm AuNPs on the nanofiber surface, crucial for maximizing catalytic efficiency. The catalytic performance of the synthesized AuNPs was demonstrated in the reduction of 4-nitrophenol to 4-aminophenol, a common catalysis reaction [6], using UV-Vis analysis. The AuDAP-CHIT composites thus should also exhibit excellent catalytic performance suitable for industrial applications.

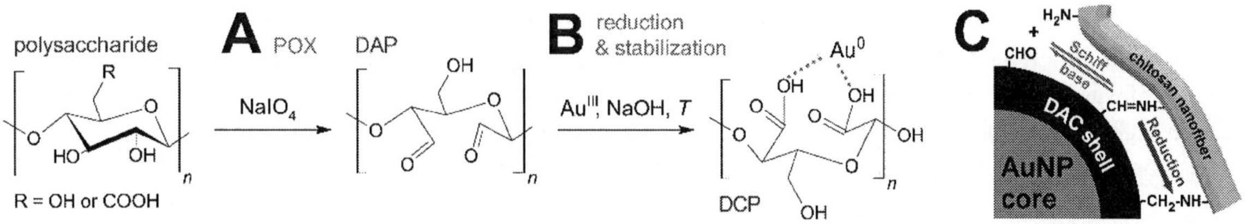

Figure 1 Scheme of polysaccharide POX to DAP (part A), subsequent reduction of gold(III) salt precursor and DAP partial oxidation to DCPs, and representation of covalent nanoparticle binding to CHITs *via* Schiff base chemistry and reductive amination forming AuDAP-CHIT composite (part C).

In conclusion, this study presents a versatile method for covalently binding AuNPs to chitosan nanofibers using DAPs. The composite materials show promising catalytic and biomedical potential, opening new possibilities for functionalized nanomaterials in applications such as biosensing and drug delivery.

2. EXPERIMENTAL PART

2.1 Materials

Cellulose SigmaCell type 20 (CELL, weight-average molecular weight M_w = 76 kDa, dispersity $Đ$ = 4.7, degree of polymerization DP = 468, Sigma Aldrich Co.), sodium hyaluronate (HA, M_w = 396 kDa, $Đ$ = 5.9, DP = 987) (Contipro, Czech Republic), sodium periodate ($NaIO_4$, Penta, Czech Republic), gold(III) chloride trihydrate ($HAuCl_4 \cdot 3H_2O$, Sigma Aldrich Co.), medium molecular weight chitosan (75–85 % deacetylated, viscosity-average molecular weight M_v = 190–310 kDa, Sigma Aldrich Co.), acetic acid (CH_3COOH, Sigma Aldrich Co.) polyethylene oxide (PEO, M_v = 600 kDa, Sigma Aldrich Co.), sodium hydroxide (NaOH, Lachner, Czech Republic), hydrochloric acid (HCl, Penta, Czech Republic), sodium triacetoxyborohydride (STAB, Sigma Aldrich Co.), sodium borohydride ($NaBH_4$, Sigma Aldrich Co.), and 4-nitrophenol (4-NP, Sigma Aldrich Co.) were used as received without further purification. Demineralized water (DEMI) of conductivity < 0.1 µS/cm was used throughout the study.

2.2 Synthesis of DAPs

DAPs with a high degree of oxidation (*DO*) were prepared by POX of CELL and HA using a 1.2 molar excess of $NaIO_4$ at room temperature for 72 and 24 h, respectively [7]. Resulting DAC suspension was purified by 5 cycles of repeated centrifugation and mechanical homogenization, followed by solubilization procedure (i.e, stirring for 2 h at 80 °C under reflux in an oil bath [8]) and final centrifugation to obtain clear DAC solution. Both DAC and DAH solution were then dialyzed (dialysis tubing *MWCO* = 14 kDa) against DEMI for 72 h to remove oxidizing agent residues, flash-frozen, and lyophilized.

2.3 DAPs assisted AuNPs synthesis, AuDAP-CHIT composite preparation

DAPs were dissolved in water (0.6 wt%, V = 9.4 mL, shaken for 24 h at 40 °C), pH was adjusted to 7 using diluted NaOH solution, and DAPs solutions were placed in an oil bath heated to 90 °C for 15 min. Next, 100 µL of stock precursor solution (50 mg/mL of $HauCl_4{\cdot}3H_2O$ in DEMI) was added to hot DAPs solutions and vortexed for 5 s. Subsequently, 500 µL of 0.1 M NaOH solution was added to the reaction mixture, vortexed again, and immediately cooled under tap water. The formation of AuNPs took place instantly after NaOH addition. The cold AuNPs samples were transferred into dialysis tubing (*MWCO* = 14 kDa) and place in acid dialysis medium (diluted HCl solution of pH 3.5) and dialyzed for 2 h. The resulting purified AuDAP samples, designated as AuDAC and AuDAH, were characterized by methods described below. AuDAPs samples were then covalently attached to CHITs (for details on CHITs preparation, see [5]). To each of AuDAP sample, with pH adjusted to 6.5, a 10 % w/v of washed and deprotonated CHIT was added and left to react for 3 h. The resulting AuDAC-CHIT and AuDAH-CHIT samples were thoroughly washed in DEMI, one half of each sample was left in unreduced state (without reductive amination, designated as AuDAP-CHIT_unred samples) and second half was placed in solution containing excess of STAB for 1 h to reduce imine groups formed between DAPs and CHIT, and washed in DEMI (designated as AuDAP-CHIT_red). Both types of samples were flash-frozen, lyophilized, and characterized.

2.4 Characterization of DAPs, AuDAPs, and AuDAPs-CHIT

All synthesized samples, including source materials, were characterized by FT-IR using a Nicolet 6700 spectrometer (Thermo Fisher Scientific, USA) equipped with a diamond crystal in ATR mode and 500–4,000 cm^{-1} wavenumber range. AuDAPs and AuDAPs-CHIT (unred and red) were characterized by XPS using an Axis Ultra DLD spectrometer equipped with a monochromatic Al Kα (*hv* = 1486.7 eV) X-ray source (75 W, 5 mA, 15 kV), using base pressure of 2×10^{-8}, Kratos charge neutralizer system, and standard Shirley background. SEM and TEM analyses of prepared AuDAPs and AuDAPs-CHIT (red) were carried out on a Nova NanoSEM 450 scanning electron microscope (FEI, Czech Republic), and a JEM-2100 transmission microscope (JEOL, Japan), operated at a 5 and 160 kV accelerating voltages, respectively. AuDAPs-CHIT diffractograms were measured using a MiniFlex600 X-ray diffractometer (Rigaku, Japan) in the diffraction 2θ angle range 5–95° (CoKα λ = 1.789 Å, Kβ line filter). Catalytic performance of AuDAPs was determined using 4-NP to 4-AP in $NaBH_4$ model catalytic reaction by monitoring decrease of 4-NP signal at 400 nm in time using a Lambda 1050 UV-Vis spectrometer (Perkin Elmer, USA) and calculations based on the work of Kozuch and Martin [9].

3. RESULTS AND DISCUSSION

The results of FT-IR analysis (**Figure 2**, part A) of all materials used and prepared in this study, as well as XPS analysis of the synthesized AuDAPs and AuDAPs-CHIT (red and unred) (**Figure 2**, parts B–M), are discussed below. After POX, both DAC and DAH IR spectra show typical band at 1730 cm^{-1} (–CHO groups), DAC spectrum shows also increase in absorbance at 885 cm^{-1} (hemiacetal structures). DAPs conjugation with CHITs form AuDAPs-CHIT (unred) samples, in which bands at 1730 (DAC's –CHO) and 1560 cm^{-1} (chitosan's $-NH_2$) disappears evidencing formation of imine bonds (1640 cm^{-1}). Subsequent imines reduction results in formation of secondary amines manifested by band shoulder at 1065 cm^{-1} and weakening of absorbance at 1640 cm^{-1}, which is well in line with previously reported observations [10, 11].

XPS analysis of AuDAPs confirmed proposed reaction mechanisms of AuNPs synthesis and their covalent binding to CHITs. Both DAPs reduce gold salt as evidenced in XPS high-resolution Au 4f scans (**Figure 2**, parts B–C) while being partially oxidized (–CHO to –COOH groups) as noticeable from high-resolution C 1s scans, see **Figure 2**, parts D–E. The comparison of high-resolution C 1s and N 1s scans of unreduced and reduced AuDAPs-CHIT composites (**Figure 2**, parts F–I and J–M showing DAC- and DAH-based composites, respectively) exemplifies presence of imines after conjugation in both material types and their subsequent elimination (AuDAC-CHIT red) or decrease (AuDAH-CHIT red) accompanied by –CHO group reduction as a result of the reductive amination.

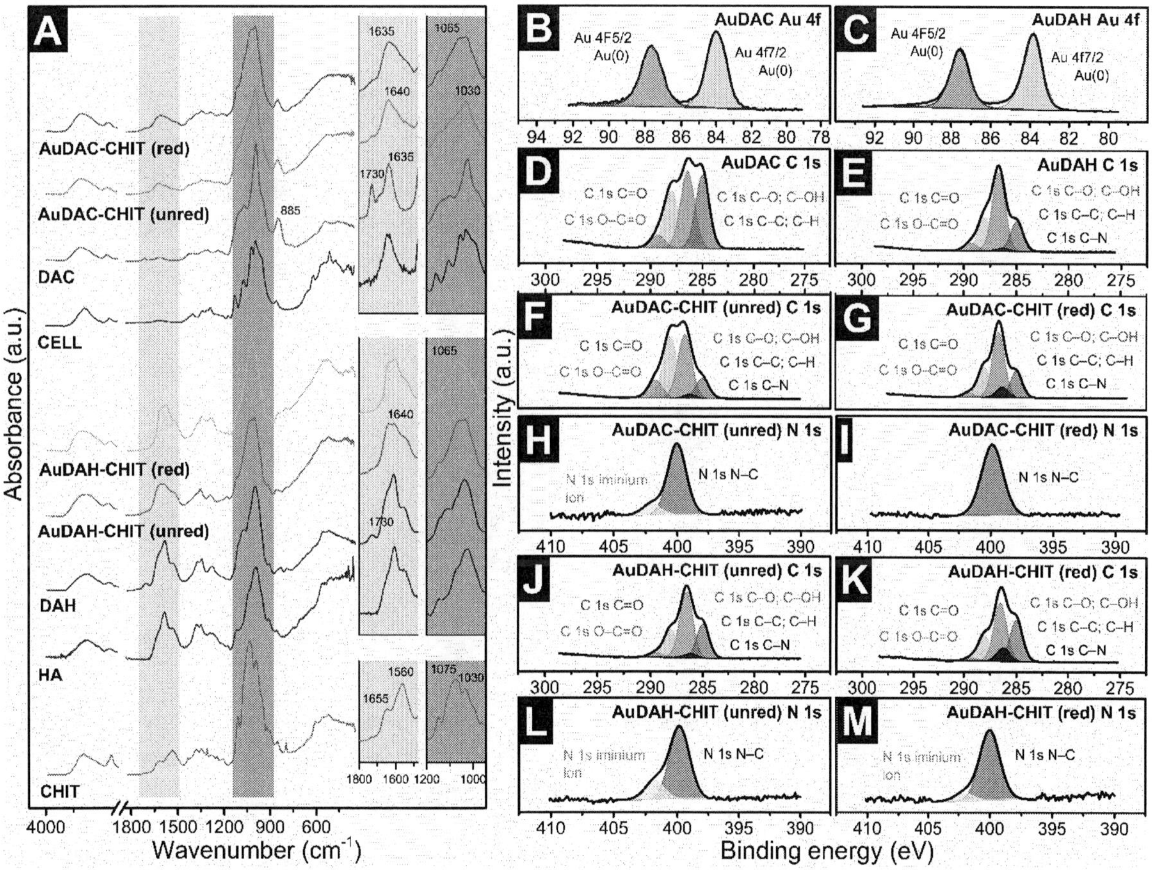

Figure 2 FT-IR analysis of used and resulting materials (part A). XPS analysis of all prepared composites including XPS high-resolution Au f4, C 1s, and N 1s scans.

SEM and TEM analyses showed densely covered CHITs by covalently attached AuDAC (mostly spherical/polyhedral 6.2 ± 2.4 nm nanoparticles) and AuDAH (triangular, bipyramidal, and spherical nanoparticles 9.8 ± 4.1 nm), although AuDAC-CHIT exhibit more uniform and densely surface-conjugated nanoparticles (**Figure 3**, left and center parts). The XRD analysis (**Figure 3**, right part) confirmed presence of Au in cubic crystalline lattice manifested by diffraction peaks at 44.64 (111), 52.04 (200), and 76.67 (220) 2Theta °. The loss of cellulose's crystallinity after POX, manifested by elimination of diffraction peak at 25.37 (-110) 2 Theta °, is well in line with literature [12].

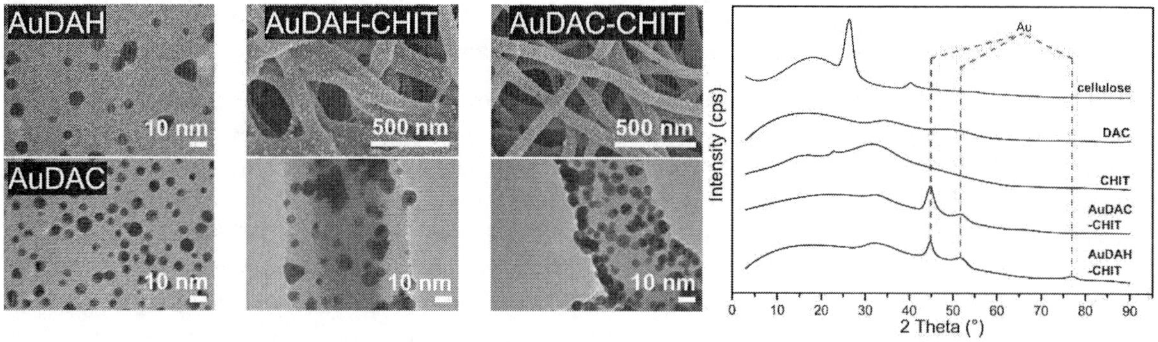

Figure 3 SEM and TEM analyses of the prepared AuDAPs and AuDAPs-CHIT composites (left and center parts, respectively), XRD analysis of used materials and resulting composites (right part).

Catalysis efficacy evaluation, performed using 4-NP to 4-AP reaction in presence of $NaBH_4$ (**Figure 4**, part A) and 2 µg of prepared AuDAPs samples, showed excellent performance with turn over frequencies (TOFs) between 27 and 29 min^{-1} (higher value for AuDAC) while maintaining relatively high catalysis reaction rate constant (k) between 0.65 and 0.7 min^{-1} (**Figure 4**, parts B and C). Catalytic performance was also for AuDAPs-CHIT composites and gold nanoparticles prepared using other DAPs, however, this topic will be discussed in more detail in our upcoming contribution.

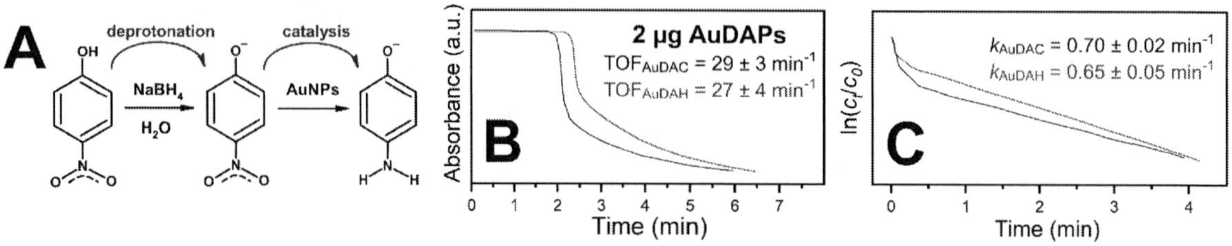

Figure 4 Catalysis efficacy evaluation using 4-NP to 4-AP (part A), calculated turn over frequency (TOF, part B) and catalysis reaction rate constant (k, part C) for both of the prepared AuDAPs samples.

4. CONCLUSION

In conclusion, this study successfully demonstrated the mechanisms of synthesis and covalent binding of gold nanoparticles (AuNPs) to chitosan nanofibers (CHITs) using dialdehyde cellulose (DAC) and dialdehyde hyaluronate (DAH) as both reducing and stabilizing agents and as linker. The synthesis was confirmed through various spectroscopic techniques, and the resulting AuNP-CHIT composites exhibited sub-10 nm nanoparticle size and uniform distribution. These composites demonstrated promising catalytic efficiency in the reduction of 4-nitrophenol, with excellent catalytic performance. The findings highlight the versatility and potential applications of these nanomaterials in biomedical fields such as drug delivery and biosensing.

ACKNOWLEDGEMENTS

L. Münster and J. Vícha gratefully acknowledge the Czech Science Foundation grant 23-07361S. M. Muchová and P. Humpolíček acknowledge the Centre of Polymer Systems Internal Development Project RP/CPS/2024-28/001 financed by the Ministry of Education, Youth, and Sports of the Czech Republic for infrastructural support. V. Hrbáček acknowledges the support by the Internal Grant Agency of TBU project no. IGA/CPS/2024/002. A. Důbravová and I. Kuřitka acknowledge the Centre of Polymer Systems Internal Development Project RP/CPS/2024-28/007 financed by the Ministry of Education, Youth and Sports of the Czech Republic for infrastructural support.

REFERENCES

[1] KUSTOV, L. M. Catalytic properties of supported gold nanoparticles in organic syntheses. *Russian Chemical Bulletin*. 2013, vol. 62, no. 4, pp. 869–877.

[2] COMENGE, J., C. SOTELO, F. ROMERO, O. GALLEGO, A. BARNADAS, T. G.-C. PARADA, F. DOMÍNGUEZ and V. F. PUNTES. Detoxifying Antitumoral Drugs via Nanoconjugation: The Case of Gold Nanoparticles and Cisplatin. *PLoS ONE*. 2012, vol. 7, no. 10, pp. e47562.

[3] LEPINAY, S., A. STAFF, A. IANOUL and J. ALBERT. Improved detection limits of protein optical fiber biosensors coated with gold nanoparticles. *Biosensors and Bioelectronics*. 2014, vol. 52, pp. 337–344.

[4] MATSUMOTO, M., K. KANEKO, M. HARA, M. MATSUI, K. MORITA and T. MARUYAMA. Covalent immobilization of gold nanoparticles on a plastic substrate and subsequent immobilization of biomolecules. *RSC Advances*. 2021, vol. 11, no. 38, pp. 23409–23417.

[5] DŮBRAVOVÁ, A., M. MUCHOVÁ, D. ŠKODA, L. LOVECKÁ, L. ŠIMONÍKOVÁ, I. KUŘITKA, J. VÍCHA and L. MÜNSTER. Highly efficient affinity anchoring of gold nanoparticles on chitosan nanofibers via dialdehyde cellulose for reusable catalytic devices. *Carbohydrate Polymers*. 2024, vol. 323, pp. 121435.

[6] MEJÍA, Y. R. and N. K. REDDY BOGIREDDY. Reduction of 4-nitrophenol using green-fabricated metal nanoparticles. *RSC Advances*. 2022, vol. 12, no. 29, pp. 18661–18675.

[7] MUCHOVÁ, M., L. MÜNSTER, A. VÁVROVÁ, Z. CAPÁKOVÁ, I. KUŘITKA and J. VÍCHA. Comparison of dialdehyde polysaccharides as crosslinkers for hydrogels: The case of poly(vinyl alcohol). *Carbohydrate Polymers*. 2022, vol. 279, pp. 119022.

[8] KIM, U.-J., M. WADA and S. KUGA. Solubilization of dialdehyde cellulose by hot water. *Carbohydrate Polymers*. 2004, vol. 56, no. 1, pp. 7–10.

[9] KOZUCH, S. and J. M. L. MARTIN. "Turning Over" Definitions in Catalytic Cycles. *ACS Catalysis*. 2012, vol. 2, no. 12, pp. 2787–2794.

[10] KIM, U.-J., H. J. KIM, J. W. CHOI, S. KIMURA and M. WADA. Cellulose-chitosan beads crosslinked by dialdehyde cellulose. *Cellulose*. 2017, vol. 24, no. 12, pp. 5517–5528.

[11] GAO, C., S. WANG, B. LIU, S. YAO, Y. DAI, L. ZHOU, C. QIN and P. FATEHI. Sustainable Chitosan-Dialdehyde Cellulose Nanocrystal Film. *Materials*. 2021, vol. 14, no. 19, pp. 5851.

[12] KIM, U.-J., S. KUGA, M. WADA, T. OKANO and T. KONDO. Periodate oxidation of crystalline cellulose. *Biomacromolecules*. 2000, vol. 1, no. 3, pp. 488–492.

ANTIBACTERIAL ACTIVITY OF SILVER NANOPARTICLES BIOSYNTHESIZED USING THE SULFATED POLYSACCHARIDES FROM *ARTHROSPIRA PLATENSIS* BIOMASS

[1]Oxana RADU, [1]Liliana CEPOI, [2]Irina NEGUȚ, [1]Ludmila RUDI, [1]Tatiana CHIRIAC, [2]Gabriela DORCIOMAN, [3]Greta BALAN, [3]Elena ZARICIUC

[1]*Technical University of Moldova, Chisinau, Republic of Moldova,* oxana.radu@sa.utm.md

[2]*National Institute for Laser, Plasma and Radiation Physics, Magurele, Romania, UE,* negut.irina@inflpr.ro

[3]*Nicolae Testemitanu State University of Medicine and Pharmacy, Chisinau, Republic of Moldova,* greta.balan@usmf.md

https://doi.org/10.37904/nanocon.2024.5022

Abstract

Arthrospira platensis (spirulina) is known for its high biological activity, including antioxidant, anti-inflammatory, and antiviral properties. In recent years, spirulina polysaccharides have been actively researched in fields such as tissue engineering, targeted drug delivery, and wound healing. These substances also possess a high reducing capacity, making them suitable matrices for the biosynthesis of metallic nanoparticles from metal ions. For instance, silver, known for its antimicrobial properties, can be biosynthesized as AgNPs on polysaccharide matrices derived from Spirulina biomass, potentially combining the effects of the nanoparticles and the matrix for developing wound coverings. In this study, sulfated polysaccharides were extracted from Spirulina biomass grown under controlled conditions to stimulate carbohydrate accumulation. AgNPs biosynthesis was carried out using polysaccharides solubilized in deionized water at 10 mg/ml and $AgNO_3$, under different process parameters. The nanoparticles obtained were tested for antibacterial properties against *Staphylococcus aureus, Escherichia coli, Pseudomonas aeruginosa*, and antifungal properties against *Candida albicans*. The AgNPs demonstrated significant antimicrobial activity, with MIC and MBC values varying based on the species. These results indicate that AgNPs biosynthesized on spirulina sulfated polysaccharide matrices possess antimicrobial properties, making them promising for developing wound dressing materials.

Keywords: Silver nanoparticles, sulfated polysaccharides, antimicrobial activity

1. INTRODUCTION

It is now well-established that metallic nanoparticles, including silver nanoparticles, can be synthesized in an environmentally friendly manner through biosynthetic procedures using various microorganisms, cell cultures, and biological extracts. The resulting nanoparticles' shape, size, and biological properties vary depending on the matrix used for synthesis [1-3]. Cyanobacterial extracts are being studied due to their high reducing power and the abundance of components that enable the synthesis of zero-valent particles from target ions [4-6]. Most research focuses on using complex cyanobacterial extracts as bionanosynthesis matrices, while specific extracts, such as phycobiliprotein and protein extracts, are applied less frequently [5, 7]. Cyanobacterial polysaccharides can also be utilized in the biosynthesis of silver nanoparticles, resulting in polydisperse nanoparticles coated with a thin layer of polysaccharides [8].

Arthrospira platensis, commonly known as Spirulina, is one of the most intensively studied cyanobacteria due to its unique properties, including its diverse content of compounds with potent therapeutic effects [9]. In addition to its well-documented antioxidant, anti-inflammatory, and immunomodulatory activities, the antimicrobial effects of certain components of Spirulina biomass have also gained recognition in recent years

[10]. In this context, the sulfated polysaccharides of Spirulina, which exhibit inhibitory and biocidal effects against various viruses, bacteria, and fungi, have been particularly studied [11]. Moreover, whole biomass and different Spirulina biomass fractions are considered suitable matrices for the biosynthesis of silver nanoparticles [7, 8, 12].

The potential of biosynthesized silver nanoparticles using sulfated polysaccharide extracts from Spirulina biomass as a biosynthesis matrix for treating various wounds, including infected ones, is significant. A successful combination of Spirulina's sulfated polysaccharides and silver nanoparticles could provide a promising solution for joint use in treating wounds. Therefore, as a first step toward developing active therapeutic agents, we aimed to synthesize silver nanoparticles using sulfated polysaccharide extracts from Spirulina biomass and to test their activity against microorganisms with the potential to infect various skin lesions.

2. EXPERIMENTAL DESIGN

2.1. Silver nanoparticles biosynthesis

The sulfated polysaccharide extract from Spirulina (SPES) was used as a reducing agent for the biosynthesis of AgNPs. The source of SPES was the cyanobacterium *Arthrospira platensis* CNMN-CB-02 (Spirulina) strain from the National Collection of Non-Pathogenic Microorganisms at the Institute of Microbiology and Biotechnology, Technical University of Moldova. The cyanobacterium was cultivated in SP1 medium under 12:12 (L) photoperiod conditions, which ensured the production of up to 26% carbohydrates from the dry Spirulina biomass [13]. SPES was extracted from the native Spirulina biomass using deionized water at a 1:45 (mg/mL) ratio at a temperature of 90°C for 120 minutes. After centrifugation for 10 minutes at 4500 rpm, a 1% cetyltrimethylammonium bromide solution was added to the supernatant to precipitate the polysaccharides. Following centrifugation under the same conditions, the residue was washed with a saturated solution of sodium acetate in 95% alcohol, then with 96% alcohol, solubilized in deionized water at a concentration of 10mg/ml [14]. SPES contains 89.2% carbohydrates and 10.8% peptides.

Silver nitrate ($AgNO_3$) (SIGMA-ALDRICH GmbH, Germany) was used as the precursor substance in the biosynthesis of AgNPs. $AgNO_3$, at a silver concentration of 200 mg, was resuspended in 250 ml Erlenmeyer flasks with 100 ml of a sulfated polysaccharide extract from Spirulina (SPES) at a concentration of 10 mg/ml. The mixture was placed on a WU-4 universal laboratory shaker at 38-40°C and stirred at an oscillation frequency of 200 rpm for 3 hours. As a result, the active AgNPs-SPES mixture was obtained.

2.2. Methods of investigation

The FTIR method (Perkin-Elmer FTIR spectrometer (PerkinElmer Inc., USA) equipped with an air-cooled DTGS (Deuterated Triglycine Sulfate) detector) was used to identify the active chemical groups involved in the biosynthesis of silver nanoparticles.

Scanning electron microscopy was used to visualize the obtained nanoparticles. Scanning Electron Microscopy (SEM) was performed using the Quanta 3D FEG (FEI Company, USA). The operational features of the microscope used in the experiment were as follows: magnification 100,000-250,000x; voltage 30 kV.

The antimicrobial activity of AgNPs-SPES was evaluated against three bacterial strains: *Staphylococcus aureus* (ATCC 25923), *Escherichia coli* (ATCC 25922), *Pseudomonas aeruginosa* (ATCC 27853) and fungal strain *Candida albicans* (ATCC 10231). Determination of the MIC (minimum inhibitory concentration, µg/ml) and MBC (minimum bactericidal concentration, µg/ml) was done using the serial dilutions in liquid broth method. As a pozitive control was used SPES.

3. RESULTS AND DISCUSSION

The formation of nanoparticles was visually monitored by observing changes in the color of the reaction mixture, and their presence was confirmed through scanning electron microscopy (**Figure 1**).

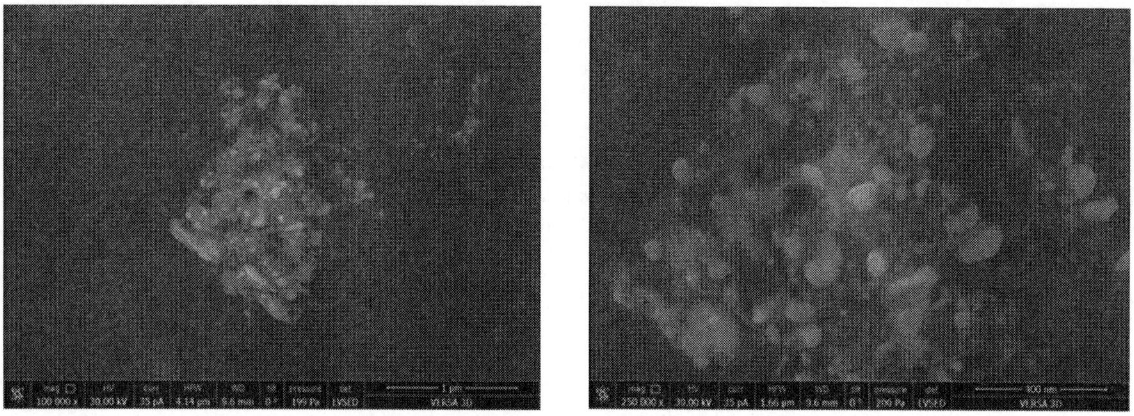

Figure 1 SEM of AgNPs-SPES

Due to their metallic nature, silver nanoparticles have a distinctive shine and are easily visualized against the amorphous polysaccharide matrix. They are spherical or nearly spherical, with sizes ranging from a few nanometers to several tens of nanometers, and are predominantly grouped in clusters.

Table 1 presents the FTIR absorption peaks for the sulfated polysaccharide extract from Spirulina (SPES) and the silver nanoparticles in the extract (AgNPs-SPES).

Table 1 FTIR Absorption peaks for sulfated polysaccharide extract from Spirulina (SPES) and silver nanoparticles synthesized in SPES (AgNPs-SPES)

Material	**Regions**			
	4000-2500, -OH; -NO	**2000-1500, C=O; N-H**	**1500-1000, C-O; C-N; S=O**	**<1000**
SPES	3283, 2971, 2930	1632, 1557	1450, 1406,1395, 1241, 1231,1065, 1037, 1027	919, 901, 893, 798, 696
AgNPs-SPES	3347, 2968, 2934	1634, 1557	1393, 1387, 1352, 1231, 1120, 1066, 1028	923, 828, 776, 715, 699

For the sulfated polysaccharide extract from Spirulina (SPES), the absorption peak at 3283 cm^{-1} suggests the presence of -OH or -NH groups, indicative of hydrogen bonds, reflecting a structure rich in hydrophilic functionalities. The values at 2930 cm^{-1} and 2971 cm^{-1} correspond to C-H stretching vibrations associated with hydrocarbon chains, suggesting the presence of methyl or methylene fragments. The peaks around 1632 cm^{-1} are attributed to C=O stretching vibrations, indicating carbonyl or amide groups. In the 1400-1600 cm^{-1} region, multiple peaks indicate the presence of carboxylate (COO^-) groups or N-H vibrations. In the 1500-1000 cm^{-1} region, the peaks at 1037 cm^{-1}, 1027 cm^{-1}, 1085 cm^{-1}, and 1025 cm^{-1} are typical of C-O stretching vibrations, characteristic of carbohydrates or polysaccharides, which are major components of the extract. A peak at 1120 cm^{-1} indicates S=O groups (from sulfated polysaccharides). The values below 1000 cm^{-1} correspond to molecular backbone vibration modes of organic compounds, indicating deformations characteristic of complex structures such as polysaccharides. For the active AgNPs-SPES mixture, the absorption peak at 3347 cm^{-1} indicates the presence of -OH and -NH groups that remain active. The peak at

1634 cm^{-1} suggests that carbonyl groups (C=O) play a role in the stabilization and potential functionalization of the formed nanoparticles. The 1400-1500 cm^{-1} region changes reflect possible interactions between carboxylate groups from Spirulina and silver ions (Ag^+). These interactions are essential for the ions' reduction process and the nanoparticles' stabilization. Peaks below 1000 cm^{-1} present signals associated with the vibrations of organic compounds and preliminary interactions between silver ions and the biomolecules from Spirulina. Thus, the FTIR spectrum for AgNPs-SPES indicates that in the mixture of silver nitrate and Spirulina extract, the biomolecules act as reducing and stabilizing agents for the silver ions. The spectrum reflects interactions involved in the reduction of Ag^+ and the synthesis of AgNPs, highlighting the structural and chemical changes that occur.

The antimicrobial activity of SPES and AgNPs-SPES is presented in **Table 2**.

Table 2 Antimicrobial activity of sulfated polysaccharide extract from Spirulina (SPES) and silver nanoparticles synthesized in SPES (AgNPs-SPES)

Strain	Product	MIC (µg/mL)	MBC (µg/mL)
Staphylococcus aureus (ATCC 25923)	AgNPs-SPES	0.48	0.48
	SPES	5.00	n/a
Escherichia coli (ATCC 25922),	AgNPs-SPES	0.11	0.22
	SPES	5.00	5.00
Pseudomonas aeruginosa (ATCC 27853)	AgNPs-SPES	0.11	0.22
	SPES	5.00	5.00
Candida albicans (ATCC 10231).	AgNPs-SPES	1.95	3.90
	SPES	5.00	n/a

The sulfated polysaccharide extract from Spirulina exhibits inhibitory action against all studied strains and bactericidal action against *Escherichia coli* (ATCC 25922) and *Pseudomonas aeruginosa* (ATCC 27853). The minimum inhibitory concentrations determined for the extract are the same, constituting 5.0 µg/mL. The silver nanoparticles obtained based on the sulfated polysaccharide extract from Spirulina show high antimicrobial activity against all the studied strains of microorganisms. For instance, the minimum inhibitory concentration for *Staphylococcus aureus* ATCC 25923 was 0.48 µg/mL, the minimum bactericidal concentration for this microbial strain. The AgNPs-SPES exhibited a more pronounced antibacterial activity against the two Gram-negative bacterial strains studied. The minimum inhibitory concentration for *Escherichia coli* ATCC 25922 and *Pseudomonas aeruginosa* ATCC 27853 was 0.11 µg/mL, while the minimum bactericidal concentration was 0.22 µg/mL. Importantly, the biosynthesized nanoparticles were also active against the yeast-like fungus *Candida albicans* ATCC 10231, with a minimum inhibitory concentration of 1.95 µg/mL and a minimum bactericidal concentration of 3.90 µg/mL.

In previous studies, polysaccharide extracts from Spirulina have demonstrated moderate antimicrobial activity against Gram-negative and Gram-positive strains. For example, Spirulina polysaccharides have been reported to have MIC values ranging between 2 and 10 µg/mL for *E. coli* and other Gram-negative bacteria. More recent studies highlight that Spirulina and its components may have bacteriostatic activity, while bactericidal effects are observed at higher concentrations. The MIC values for *E. coli* and *P. aeruginosa* obtained in our study are consistent with those reported in the literature, while the bactericidal activity is significantly higher. We believe this is likely due to applying a specific procedure for growing Spirulina under moderate stress conditions, which resulted in biomass with enhanced properties [13].

Silver nanoparticles are well-documented for their remarkable antimicrobial properties, particularly against Gram-negative bacteria, due to their thinner cell walls and higher nanoparticle permeability. In many studies, the MIC for *S. aureus* ranges between 0.5 and 1 µg/mL, consistent with our results. *E. coli* and *P. aeruginosa* tend to have even lower MICs against AgNPs, ranging from 0.1 to 0.2 µg/mL, aligning with our study's 0.11 µg/mL value. The antifungal activity of silver nanoparticles against *C. albicans* is also well-documented in the literature, with MIC values ranging from 1 to 10 µg/mL, depending on the synthesis method of the nanoparticles. The values obtained in this study, 1.95 µg/mL, fall well within these limits [15-17].

4. CONCLUSION

It has been demonstrated that the sulfated polysaccharide extract from Spirulina biomass is a suitable matrix for the biosynthesis of silver nanoparticles from the $AgNO_3$ precursor. It provides conditions for reducing and stabilizing silver ions formed in solution. The formed nanoparticles aggregate into small clusters on the available polysaccharide matrix.

The antimicrobial test results demonstrate excellent antimicrobial activity for both SPES and AgNPs-SPES, particularly against Gram-negative bacteria and fungi. Compared to the literature, the values obtained are consistent with previously reported results, and AgNPs-SPES shows a substantial improvement in antimicrobial activity compared to SPES. In this context, the results obtained can serve as a basis for the use of the biosynthesized nanoparticles from this study in impregnating wound dressing textiles of various etiologies, allowing the combination of antibacterial effects with the known therapeutic properties of the polysaccharide matrix.

ACKNOWLEDGEMENTS

This work was supported by a grant of the Ministry of Research, Innovation and Digitization, CNCS-UEFISCDI, contract number 12ROMD/20.05.2024, project PN-IV-P8-8.3-ROMD-2023-0060, within PNCDI IV.

REFERENCES

[1] MALIK, M., IQBAL, M.A., IQBAL, Y., MALIK M., BAKHSH, S., IRFAN, S., AHMAD, R., PHAM. P.V. Biosynthesis of silver nanoparticles for biomedical applications: A mini review. *Inorganic Chemistry Communications*. 2022, vol. 145, no. 109980. Available from: https://doi.org/10.1016/j.inoche.2022.109980.

[2] RIZWAN, A., UDDIN R. A review on recent developments in the biosynthesis of silver nanoparticles and its biomedical applications. *MEDICAL DEVICES & SENSORS*. 2021, vol.4, issue 1, no. e10158. Available from: https://doi.org/10.1002/mds3.10158.

[3] QAMER, S., ROMLI, M.H., CHE-HAMZAH, F., MISNI, N., JOSEPH, N.M.S., AL-HAJ, N.A., AMIN-NORDIN, S. Systematic Review on Biosynthesis of Silver Nanoparticles and Antibacterial Activities: Application and Theoretical Perspectives. *Molecules*. 2021, vol. *26*, no. 5057. Avalaible from: https://doi.org/10.3390/molecules26165057.

[4] EL SEMARY, N.A., BAKIR, E.M. Multidrug-Resistant Bacterial Pathogens and Public Health: The Antimicrobial Effect of Cyanobacterial-Biosynthesized Silver Nanoparticles. *Antibiotics*. 2022, vol. *11*, no.1003. Avalaible from: https://doi.org/10.3390/antibiotics11081003.

[5] ISMAIL, G.A., EL-SHEEKH, M.M., SAMY, R.M. GHEDA S.F.Antimicrobial, Antioxidant, and Antiviral Activities of Biosynthesized Silver Nanoparticles by Phycobiliprotein Crude Extract of the Cyanobacteria *Spirulina platensis* and *Nostoc linckia*. *BioNanoSci.* 2021, vol. 11, pp. 355–370. Avalaible from: https://doi.org/10.1007/s12668-021-00828-3.

[6] YOUNIS, N.S., MOHAMED, M.E., E.L., 0SEMARY, N.A. Green Synthesis of Silver Nanoparticles by the Cyanobacteria *Synechocystis* sp.: Characterization, Antimicrobial and Diabetic Wound-Healing Actions. *Mar. Drugs*. 2022, vol. *20*, no. 56. Avalaible from: https://doi.org/10.3390/md20010056.

[7] AMEEN, F., ABDULLAH, M.M.S., AL-HOMAIDAN, A.A., AL-LOHEDAN, H.A., AL-GHANAYEM, A.A., ALMANSOB, A. Fabrication of silver nanoparticles employing the cyanobacterium *Spirulina platensis* and its bactericidal effect against opportunistic nosocomial pathogens of the respiratory tract. *Journal of Molecular Structure*. 2020, vol. 1217, no. 128392. Avalaible from: https://doi.org/10.1016/j.molstruc.2020.128392.

[8] HAMOUDA, R.A., HUSSEIN, M.H., ELHADARY, A.M.A., ABUELMAGD, A. Extruded polysaccharide/protein matrix from *Arthrospira platensis* cultures mediated silver nanoparticles biosynthesis and capping. *Appl Nanosci.* 2020, vol. 10, pp. 3839–3855. Avalaible from: https://doi.org/10.1007/s13204-020-01490-z.

[9] GENTSCHEVA, G., NIKOLOVA, K., PANAYOTOVA, V., PEYCHEVA, K., MAKEDONSKI, L., SLAVOV, P., RADUSHEVA, P., PETROVA, P., YOTKOVSKA, I. Application of *Arthrospira platensis* for Medicinal Purposes and the Food Industry: A Review of the Literature. *Life.* 2023, vol. *13*, no. 845. Avalaible from: https://doi.org/10.3390/life13030845.

[10] BANCALARI, E., MARTELLI, F., BERNINI, V., NEVIANI, E., GATTI, M. Bacteriostatic or bactericidal? Impedometric measurements to test the antimicrobial activity of *Arthrospira platensis* extract. *Food Control.* 2020, vol. 118, no.107380. Avalaible from: https://doi.org/10.1016/j.foodcont.2020.107380.

[11] AI, X., YU, P., LI, X., LAI, X., YANG, M., LIU, F., LUAN, F., MENG, X. Polysaccharides from *Spirulina platensis*: Extraction methods, structural features and bioactivities diversity. *International Journal of Biological Macromolecules.* 2023, vol. 231, no. 123211. Avalaible from: https://doi.org/10.1016/j.ijbiomac.2023.123211.

[12] EL-DEEB, N.M., ABO-ELENEEN, M.A., AWAD, O.A., ABO-SHADY, A.M. *Arthrospira platensis*-Mediated Green Biosynthesis of Silver Nano-particles as Breast Cancer Controlling Agent: *In Vitro* and *In Vivo* Safety Approaches. *Appl Biochem Biotechnol.* 2022, vol. 194, pp. 2183–2203. Avalaible from: https://doi.org/10.1007/s12010-021-03751-1.

[13] CEPOI, L. Environmental and technological stresses and their management in Cyanobacteria. In: *Cyanobacteria From Basic Science to Applications*, Mishra A.K., Tiwari D.N., Rai A.N. eds., 1st edition, Amsterdam. Amsterdam: Elsevier Inc. 2019, pp. 217-244. ISBN 978-012814668-2, 978-012814667-5. Available from: https://doi.org/10.1016/B978-0-12-814667-5.00011-8

[14] CHIKAHAN, R., CHRIASUWAN, N., TRIRATANA, P., LOHA, V., TIA, S., BUNNAG, B. Polysaccharide extraction from *Spirulina* sp. and antioxidant capacity. *International Journal of Biological Macromolecules.* 2013, vol. 58, pp. 73-78. Avalaible from: https://doi.org/10.1016/j.ijbiomac.2013.03.046

[15] SINGH, P., MIJAKOVIC, I. Strong Antimicrobial Activity of Silver Nanoparticles Obtained by the Green Synthesis in *Viridibacillus* sp. Extracts. *Front. Microbiol.* 2022, vol.13, no. 820048. Avalaible from https://doi.org/10.3389/fmicb.2022.820048.

[16] TORMENA, R.P.I., VIEIRA, R.E., MOTA, B.F.O., CHAKER, J.A., FAGG, C.W., FREIRE, D.O., MARTINS, P.M. RODRIGUES DA SILVA, I.C., SOUSA, M.H. Evaluation of the antimicrobial activity of silver nanoparticles obtained by microwave-assisted green synthesis using *Handroanthus impetiginosus* (Mart. ex DC.) Mattos underbark extract. *RSC Advances.* 2020, vol. 10, no. 35, pp. 20676–20681. Avalaible from: https://doi.org/10.1039/d0ra03240a

[17] ATTALLAH, N.G.M., ELEKHNAWY, E. NEGM, W.A., HUSSEIN, I.A., MOKHTAR, F.A., AL-FAKHRANY, O.M. *In Vivo* and *In Vitro* Antimicrobial Activity of Biogenic Silver Nanoparticles against *Staphylococcus aureus* Clinical Isolates. *Pharmaceuticals.* 2022, vol. *15,* no. 194. Avalaible from: https://doi.org/10.3390/ph15020194

SESSION D
MONITORING AND TOXICITY OF NANOMATERIALS

Chairman

Dr. Jan TOPINKA Institute of Experimental Medicine of the CAS, Prague, Czech Republic, EU

THE EFFECTS OF Β-IRRADIATION ON ELECTROSPUN POLYCAPROLACTONE FIBRES

Jaroslav MIKULE[1], Manikandan SIVAN[1], Divyabharathi MADHESWARAN[1,2], Ema CHUDOBOVÁ[1], Šárka HAUZEROVÁ[1], David CHVÁTIL[3], Václav OLŠANSKÝ[3], Luboš BĚHÁLEK[4], Eva KUŽELOVÁ KOŠŤÁKOVÁ[1], David LUKÁŠ[1]

[1]Technical University of Liberec, Faculty of Science, Humanities and Education, Department of Chemistry, Liberec, Czech Republic, EU, jaroslav.mikule@tul.cz

[2]Technical University of Liberec, Faculty of Textile Engineering, Liberec, Czech Republic, EU

[3]Nuclear Physics Institute, Czech Academy of Sciences, Řež, Czech Republic, EU

[4]Technical University of Liberec, Faculty of Mechanical Engineering, Department of Engineering Technology, Liberec, Czech Republic, EU

https://doi.org/10.37904/nanocon.2024.5027

Abstract

Microplastics originating from the degradation of plastic waste are considered a global concern for the environment, with microplastic fibres from synthetic textiles being responsible for the majority of microplastic pollution. Over the past few decades, there has also been a notable increase in the production of synthetic polymeric materials in the form of nanofibers for various applications. Despite this fact, knowledge in the field of aging of nanofibrous materials is very limited. To evaluate the effects of polymer aging, accelerated aging techniques involving high-energy irradiation are commonly used. This work focuses on the impact of β-irradiation on electrospun polycaprolactone nanofibers. Polycaprolactone (PCL) is a semicrystalline biodegradable polyester widely used in medical applications; thus, any potentially formed fiber fragments pose less risk to the environment and health compared to non-biodegradable counterparts. The impact of irradiation was evaluated in terms of the mechanical properties of the prepared nanofibrous mats. Additionally, the effects on fibre morphology and crystallinity were assessed, as well as the impact of irradiation on the biocompatibility of the material.

Keywords: β-irradiation, polymer aging, polycaprolactone, nanofibers

1. INTRODUCTION

Plastics or in another words synthetic polymers found their way to variety of applications since the discovery of first synthetic polymer bakelite in 1909. Wide spread of plastics was boosted due to its unique mechanical and chemical properties, as well as low production costs in comparison to conventional materials [1]. Worldwide production of plastics is still increasing to this day, which goes in hand with production of plastic waste. Around 80 % of plastic waste ends up in landfills or in the environment, thus significantly changing the biosphere [1, 2]. Furthermore, plastic waste in the environment is subjected to numerous forces, which facilitate disintegration into smaller particles, eventually generating secondary microplastics (< 5 mm) and nanoplastics (< 100 nm) [2]. Vast majority of microplastic pollution in aquatic ecosystems is in form of microplastic fibres originating from synthetic textiles [3]. In addition to conventional synthetic microfibrous materials, nanofibers found its way to multiple applications in last decades, which may end up in environment in some form of waste after successfully fulfilling its lifecycle [4]. Despite this fact there is little to no information available about fate of nanofibrous materials, when subjected to environment.

Aging of polymers is understood as a change of mechanical and chemical properties over the time [5]. Polymer aging is complex process influenced by many factors – sunlight exposure, temperature, humidity, external stress, to highlight few [6]. In case of polymers with C-C backbone aging process is initiated by UV irradiation, which causes breakage of C-H bonds within polymeric chain forming free radicals, followed by formation of peroxy radicals in presence of oxygen. Subsequent radical reaction result in further chain scission, crosslinking, branching or formation of oxygen-containing groups. For polyesters UV irradiation results in ester bond cleavage, as well as formation of carboxyl and hydroxyl groups, which further promote hydrolysis of ester bonds in aqueous environment. Thus photo-oxidation results mainly in chain scission in case of polyesters [7].

Standard approaches to study aging process of polymeric materials employ accelerated aging methods. Accelerated aging methods were originally developed to predict lifecycle and time depend performance of polymeric materials within certain application, which is otherwise extremely time-consuming. Since solar irradiation is considered main reason of polymer aging, majority of commonly used accelerated aging methods is based on UV irradiation [8]. Despite this fact, proposed study suggests use of high energy electrons (β-irradiation) to study aging of polymeric nanofibers, which allows even further reduction in time-consuming nature of the experiment in comparison to UV irradiation.

2. MATERIALS AND METHODS

2.1 Materials

Polycaprolactone PURASORB® PC 08 was obtained from Corbion (Netherlands), chloroform and ethanol were purchased from Penta (Czechia). PCL in form of pellets was dissolved in chloroform/ethanol mixture in ratio 8:2 by weight resulting in 16% PCL solution (w/w). The solution was stirred over night at room temperature. Nanofibrous layers were prepared by needless electrospinning device Nanospide™ NS 1WS500U (Elmarco, Czechia) equipped with 200 μm string as spinning electrode using following parameters: spinning electrode voltage of 40 kV, collector voltage of -10 kV, spinning electrode – collector distance of 180 mm, polymeric solution dosing device with aperture of 700 μm, substrate winding speed of 33 mm/min, temperature of 22 °C and relative humidity of 50 % inside the spinning chamber. Formed nanofibers were deposited on polypropylene nonwoven substrate, for purpose of following experiments nanofibrous layer was separated from supporting substrate. Final grammage of PCL nanofibrous layer obtained by needless electrospinning was 33 g/m^2.

2.2 Radiation resistance testing

Electrospun nanofibrous layers were cut into rectangles (70 × 100 mm) and sealed in sterilization packaging material. Afterwards 45 samples were placed into custom made frame to ensure homogenous irradiation of all samples. Sample irradiation was conducted with microtron MT25 located at Nuclear Physics Institute, CAS in Prague. The electron beam with intensity of 16,5 MeV and dose rate of 1,56 Gy/s was used for sample irradiation. Frame with samples was placed on rotating base at distance of 450 mm from microtron outlet, rotation around vertical axis ensured homogenous irradiation of all samples in the frame. The aluminium foil (thickness of 3 mm) was placed at distance of 120 mm from microtron outlet to disperse electron beam, resulting in homogenous field 100 × 100 mm in the target plane allowing for irradiation of entire area of nanofibrous samples. During the irradiation, samples were cooled by stream of cold air to prevent overheating caused by electrons colliding with material. Individual samples were removed from the frame after obtaining desired dose. Experimental setup is depicted in **Figure** 1.

2.3 Morphological and structural analysis

The morphology was analysed using scanning electron microscopy (SEM) Vega 3, Tescan, Czechia). Samples for SEM analysis were coated by gold to ensure surface conductivity using sputter coater Q150R ES (Quorum,

United Kingdom). Fibre diameter was measured by image analysis software ImageJ, average fibre diameter for each sample was determined as average value of 250 measurements from 5 SEM images. The impact of β-irradiation on internal structure of PCL nanofibers was analysed by differential scanning calorimetry (DSC), (DSC 1/700, Mettler Toledo, Switzerland). For each sample 2 measurements were conducted, samples were heated from -20 to 100 °C at a heating rate of 10 °C/min.

2.4 Tensile testing

For tensile testing irradiated and control samples were cut into strips with width of 10 mm. Obtained samples were afterwards attached to paper supporting frame to ensure parallel axial alignment in tensile tester grips. Subsequently supporting paper frame was cut before measurement was performed. Tensile tests were carried out by LabTest 6.0051 (LaborTech, Czechia) universal tensile tester equipped with 10 N (accuracy class: 1 from 0,030 N according to EN ISO 7500-1, accuracy class: 0,5 from 0,1 N according to EN ISO 7500-1) load cell, (AST, Germany), with sampling frequency of 50 Hz. For all samples, tensile tests were performed with initial distance of 20 mm and velocity of 10 mm/min, preload of 0.1 N. Average values of mechanical properties were determined as average value of 10 measurements.

2.5 Cytotoxicity testing

Cytotoxicity testing was conducted following EN ISO 10993-5 (855220) with 3T3 mouse fibroblasts. DMEM High Glucose (Dulbecco's Modified Engles Medium, Merck) with 10% Fetal Bovine Serum (Biosera, Czechia), 1% glutamine (Biosera, Czechia) and 1% antibiotic - Pen / Strep Amphotericin B (Lonza, Switzerland) was used as a cell culture media. From each sample, circle with a diameter of 14 mm was cut out of sample by die and placed in 24 well microtiter plate with 1 ml of media. Samples were incubated with cells for 24 hours in a CO2 incubator at 37°C (Heracell Vios 160i, Thermo Fisher Scientific, USA). For each sample 8 +2 (control) wells were seeded with cells at cell density of $1*10^4$ cells/ml in DMEM. As a negative control DMEM culture media was used, as positive control served 0.5% wt. of triton in DMEM. Seeded cells were incubated in media for 24 hours, afterwards culture media was extracted and 100 µl of media incubated with tested nanofibrous samples was added to seeded cells. After next 24hour incubation cycle cytotoxicity was evaluated by CCK metabolic test (spectrophotometry analysis).

2.6 Statistical evaluation

The statistic evaluation of experiments was performed in MATLAB R2024a software, normality of data was tested by Kolomogorov-Smirnov test, multiple comparisons between groups were performed by Kruskal-Wallis test followed by post hoc Dunnett's test.

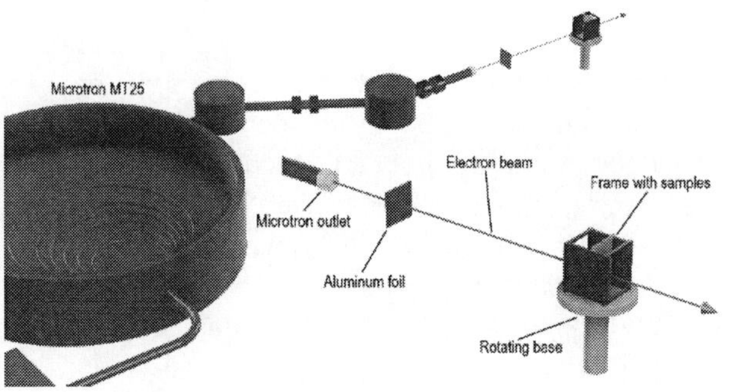

Dose [kGy]	Q [mC]	Irradiation time [s]		Ḋ [Gy/min]
50	16 025	T_1	785	3822
100	32 050	T_1+595	1 380	4348
150	48 075	T_2+591	1 971	4566
200	64 100	T_3+589	2 560	4688
250	80 125	T_4+608	3168	4735
300	96 150	T_5+627	3 795	4743
350	112 175	T_6+569	4 364	4812
400	128 200	T_7+577	4 941	4857

Figure 1 Scheme of experimental setup of β-irradiation process of PCL nanofibers. Table representing irradiation times necessary for each sample (dose) and corresponding dose rates Ḋ.

3. RESULTS AND DISCUSSIONS

3.1 Mechanical properties

Mechanical properties were evaluated by uniaxial tensile test on samples (0–200 kGy). Samples subjected to higher β-irradiation doses (250–400 kGy) were excluded due to severe mechanical degradation, which disallowed attachment of the samples into the testing apparatus. Obtained stress-strain curves are plotted in **Figure 2**. Statistically insignificant differences were observed between sample subjected to dose of 50 kGy and control sample. However, greater doses of resulted in significant decreases in both ultimate tensile strength (up to 70 %), and strain at failure (up to 91 %), see **Figure 3**. These results confirm β-irradiation-induced mechanical deterioration in PCL nanofibers.

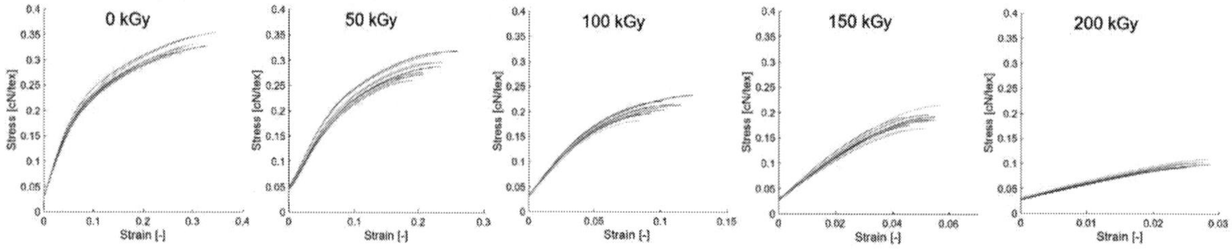

Figure 2 Obtained stress-strain curves for β-irradiated PCL nanofibrous layers.

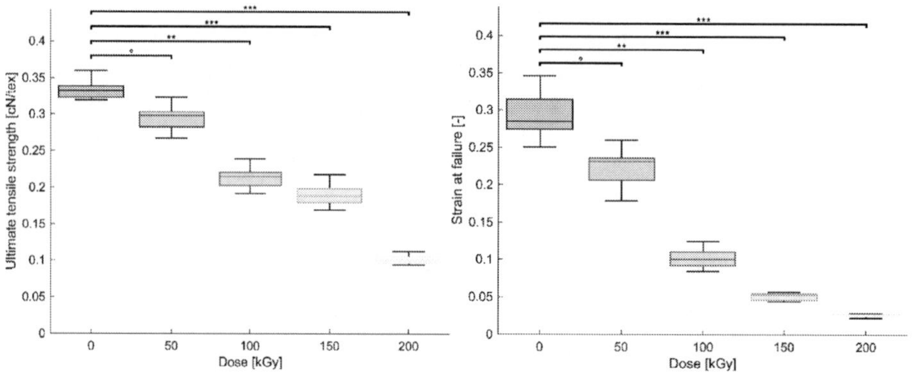

Figure 3 Box plots representing obtained ultimate tensile strength and strain at failure for β-irradiated PCL nanofibres and control sample. The sign ° denotes no statistical significance, *, **, *** denote statistical significance at significance levels of 0.05, 0.01, 0.001 (Kruskal-Wallis).

3.2 Morphological and structural analysis

Scanning electron microscopy revealed no visual differences between irradiated and control samples, see **Figure 5**. All samples exhibited a mix of smooth submicron and micron fibers with sporadic bead defects. Despite mechanical degradation, no fiber fractures or surface cracks were observed, unlike UV-weathered macroscopic polymer materials, as described by White [9]. Average fiber diameter analysis showed statistically significant differences, but no meaningful trend can be observed in statistically significant changes of average fibre diameter with respect to obtained dose, see **Figure 5**. Furthermore, effect size expressed by Cohen's d for all samples except 50 kGy proofed to be small (d < 0.22). Thermal analysis indicated a gradual increase in crystallinity from 59 % up to 68 % (350 kGy), see

Table 3, with a slight drop at 400 kGy, probably due to inappropriate handling with fragile sample. Secondly melting temperature decreased with obtained dose. These observations suggest chain scissions in amorphous regions allowing for secondary crystallization, as previously described by White [9].

Table 3 Thermal analysis data of β-irradiated PCL nanofibers

Dose [kGy]	*0*	*50*	*100*	*150*	*200*	*250*	*300*	*350*	*400*
Enthalpy of crystallite melting – ΔH_m [J/g]	*79.7*	*81.8*	*83.4*	*88.5*	*90.1*	*89.6*	*90.3*	*92.1*	*90.9*
Degree of crystallinity – X_C [%]	*58.8*	*60.4*	*61.6*	*65.3*	*66.5*	*66.2*	*66.7*	*68*	*67.1*
Melting temperature – T_m [°C]	*62.4*	*60.4*	*61.5*	*62.3*	*62*	*60.3*	*59*	*59*	*58.8*

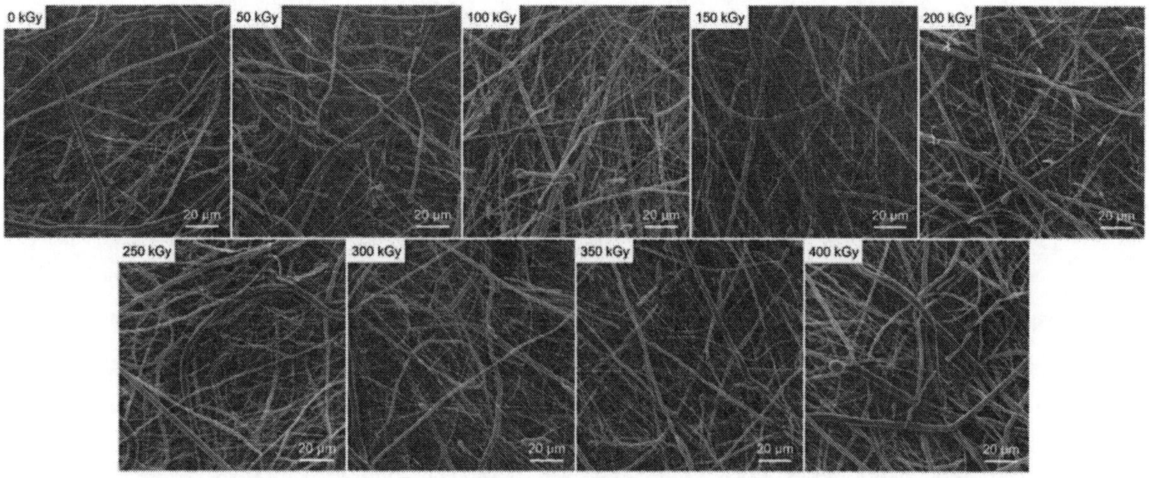

Figure 4 Scanning electron microscope images of PCL fibres irradiated by β-irradiation

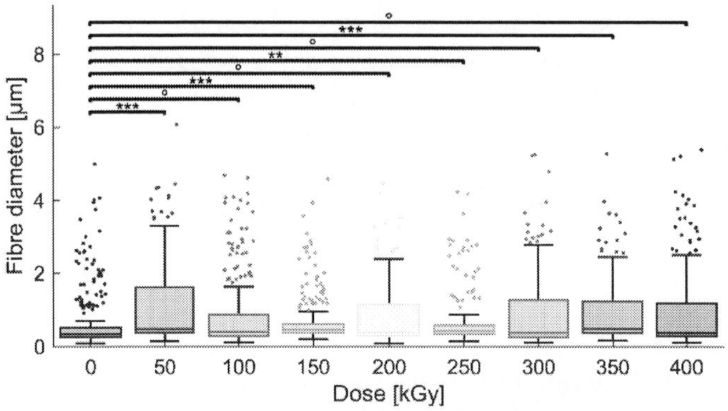

Dose	*Average diameter*	*Cohen's d*
0 kGy	0.70 ± 0.83 μm	-
50 kGy	1.06 ± 1.05 μm	0.378
100 kGy	0.87 ± 1.01 μm	0.182
150 kGy	0.71 ± 0.69 μm	0.018
200 kGy	0.79 ± 0.81 μm	0.116
250 kGy	0.66 ± 0.68 μm	0.048
300 kGy	0.88 ± 0.97 μm	0.202
350 kGy	0.88 ± 0.82 μm	0.220
400 kGy	0.90 ± 1.05 μm	0.214

Figure 5 Box plot representing fibre diameters for β-irradiated samples and control sample. The sign ° denotes no statistical significance, *, **, *** denote statistical significance at significance levels of 0.05, 0.01, 0.001 (Kruskal-Wallis). Table representing average fibre diameter and calculated values of Cohen's d to evaluate size effect against control sample.

3.3 Cytotoxicity

All of the β-irradiated samples were considered sterile, control sample was sterilized by ethylene oxide. None of the tested samples show any toxic effects after incubation in the DMEM culture medium. All tested materials except 300 kGy sample exhibited more than 90 % cell viability, see **Figure 6**. Nevertheless, cell viability is still above toxicity threshold of 70 %, set by used standard. Notably, the cell viability did not decrease further even for samples with higher doses (350 and 400 kGy). Based on obtained data, β-irradiation at doses up to 400 kGy does not compromise biocompatibility of electrospun PCL nanofibres.

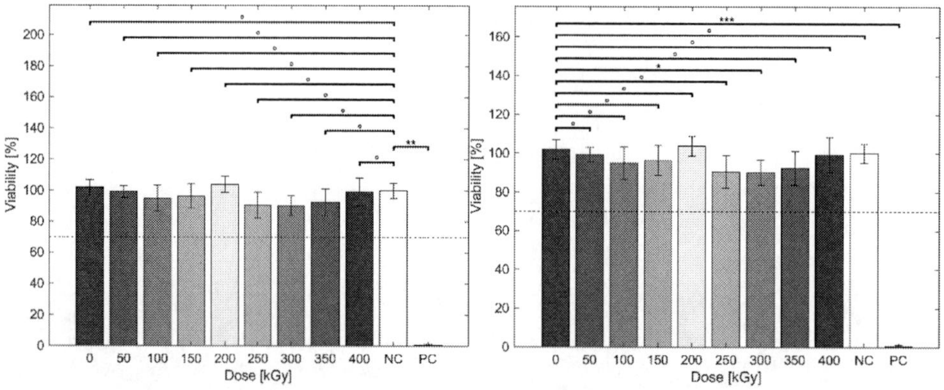

Figure 6 Bar charts representing cell viability of β-irradiated PCL nanofibers. Dashed line denotes viability threshold representing cytotoxicity following EN ISO 10993-5 (855220). The sign ° denotes no statistical significance, *, **, *** denote statistical significance at significance levels of 0.05, 0.01, 0.001 (Kruskal-Wallis).

4. CONCLUSION

This study evaluated the effects of β-irradiation on the morphology, structure, mechanical properties, and biocompatibility of electrospun polycaprolactone nanofibers. PCL was chosen for its biodegradability, biocompatibility, and relevance in biomedical applications. Furthermore, biodegradable material was chosen to lower environmental and health risk connected with potentially harmful products of aging process. Nanofibers were produced via needleless electrospinning, and irradiation was conducted using a microtron MT25 with doses ranging from 0 to 400 kGy in 50 kGy increments. Tensile tests revealed significant reductions in mechanical properties for doses above 50 kGy, confirming β-irradiation as a viable method for accelerated aging studies. SEM showed no visible fiber damage, while thermal analysis highlighted increased crystallinity and reduced melting temperatures due to chain scission in amorphous regions and secondary crystallization. Despite structural and mechanical changes, biocompatibility was unaffected, with all samples showing cell viability above 90 %. This study demonstrates that β-irradiation offers a time-efficient alternative to UV aging, enabling the study of polymer nanofiber aging and its effects on material behaviour in environmental conditions.

ACKNOWLEDGEMENTS

This work was developed thanks to the financial support of the student grant competition project (SGS-2024-4481) from the Technical University of Liberec.

REFERENCES

[1] N. EVODE, S. A. QAMAR, M. BILAL, D. BARCELÓ, AND H. M. N. IQBAL. 'Plastic waste and its management strategies for environmental sustainability'. *Case Stud. Chem. Environ. Eng.* Dec. 2021, vol. 4, p. 100142, Available from: https://doi.org/10.1016/j.cscee.2021.100142.

[2] L. LI *et al.* 'Effective uptake of submicrometre plastics by crop plants via a crack-entry mode'. *Nat. Sustain.* Jul. 2020, vol. 3, no. 11, pp. 929–937. Available from: https://doi.org/10.1038/s41893-020-0567-9.

[3] S. T. L. SAIT *et al.* 'Microplastic fibres from synthetic textiles: Environmental degradation and additive chemical content'. *Environ. Pollut.* Jan. 2021, vol. 268, p. 115745. Available from: https://doi.org/10.1016/j.envpol.2020.115745.

[4] V. S. NARAGUND and P. K. PANDA. 'Electrospun nanofiber-based respiratory face masks—a review'. *Emergent Mater.* Apr. 2022, vol. 5, no. 2, pp. 261–278. Available from: https://doi.org/10.1007/s42247-022-00350-6.

[5] C. MARAVEAS, I. V. KYRTOPOULOS, K. G. ARVANITIS, and T. BARTZANAS. 'The Aging of Polymers under Electromagnetic Radiation'. *Polymers*. Mar. 2024, vol. 16, no. 5, p. 689. Available from: https://doi.org/10.3390/polym16050689.

[6] F. ZHANG, R. YANG, and D. LU. 'Investigation of Polymer Aging Mechanisms Using Molecular Simulations: A Review'. *Polymers*. Apr. 2023, vol. 15, no. 8, p. 1928. Available from: https://doi.org/10.3390/polym15081928.

[7] B. GEWERT, M. M. PLASSMANN, and M. MACLEOD. 'Pathways for degradation of plastic polymers floating in the marine environment'. *Environ. Sci. Process. Impacts*. 2015, vol. 17, no. 9, pp. 1513–1521. Available from: https://doi.org/10.1039/C5EM00207A.

[8] L. WANG, J. ZHANG, W. HUANG, and Y. HE. 'Laboratory simulated aging methods, mechanisms and characteristic changes of microplastics: A review'. *Chemosphere*. Feb. 2023, vol. 315, p. 137744. Available from: https://doi.org/10.1016/j.chemosphere.2023.137744.

[9] J. R. WHITE. 'Polymer ageing: physics, chemistry or engineering? Time to reflect'. *Comptes Rendus Chim.* Oct. 2006, vol. 9, no. 11–12, pp. 1396–1408. Available from: https://doi.org/10.1016/j.crci.2006.07.008.

DEVELOPMENT OF THERANOSTIC FERUMOXIDE AND GADOLINIUM NANOLIPOSOMES FOR CONTROLLED RELEASE OF CYTOKINES

[1,3]Josef SKOPALÍK, [1]Jaroslav PRŮCHA, [3]Tomáš PARÁK, [2]Vratislav ČMIEL, [2]Jiří SEKORA

[1]Faculty of Biomedical Engineering, Czech Technical University in Prague, Czech Republic, EU, skopajos@fbmi.cvut.cz

[2]Faculty of Electrical Engineering and Communication, Brno University of Technology, Brno, Czech Republic, EU

[3]Faculty of Pharmacy, Masaryk University, Brno, Czech Republic, EU

https://doi.org/10.37904/nanocon.2024.5028

Abstract

Several experimental strategies have been formulated to develop theranostic liposomes by incorporating both drugs and various imaging contrast agents.in last decades. Liposome kinetics and cracking could be potentiated by external forces (ultrasound, electro-magnetic, laser induced). Gadolinium is used in many contrast entities, and they could be detectable by various methods of electron microscopy and tomography if the concentration is within a good range of values. Our workplace is working with a wide range of extracellular vesicles, which are natural transporters of regenerative factors. In order to evaluate the Gd-label effectiveness for some specific extracellular vesicles (specific exosomes derived from mesenchymal stromal cells), experimental labeling methods were prepared. We prepared original conjugation methods for Gd-DSPS binding to the surface lipid layer of extracellular vesicles, and evaluation by electron microscopy and DLS methods was conducted. To reveal possible adverse effects of Gd-vesicles that may be initiated after contact with living cells, we also evaluated basic cytotoxicity tests, and we conducted a statistical survey for metal distribution from dermal dressing into different tissues. Results of tests revealed that vesicles before and after Gd-labeling had very similar sizes (115 versus 122 nm) and similar zeta potential. It has been demonstrated that Gd-vesicles are stable particles in the dermis and subdermal region, still detectable after 12 days as compact nanobodies with high contrast under electron microscopy in dermal tissue. A simple dose of 20 μg per 1 cm² on dermal tissue (mixed in a hydrogel drop) showed very good diffusion into the tissues over the next 6 days. The diffused concentration of vesicles is sufficient for the quantification of particles with mass spectrometry after taking a biopsy, but not yet sufficient for the quantification of particles by traditional X-ray or MRI tomography. Gd-release had no adverse toxicity. A modified variant, "Gd + extra sono-stimul," was evaluated at the end of the experiment. There is evidence that sono-stimulation could accelerate the penetration of healing vesicles into the target dermal tissue.

Keywords: Nanovesicles, metal toxicity, medical device, tissue model, liposome kinetic

1. INTRODUCTION

Extracellular vesicles (EVs) are microscopical objects released by almost all cell types, sometimes called as nature liposomes, and there are three main groups of EVs according to their size and genesis: (i) apoptotic bodies (>1,000 nm) released during early apoptosis; (ii) microvesicles (100 to 1,000 nm) formed via outward budding of the plasma membrane; and (iii) exosomes (40 to 100 nm) secreted after fusion of multivesicular bodies with the plasma membrane [1,2]. The third group of EVs, exosomes, mostly contains anti-inflammatory and regenerative cargos (cytokines, enzymes and miRNAs) and they have great potential in modern medicine of metabolic disorders and wounds. However, the current understanding of these all nature EVs, especially their *in vivo* behaviors and distributions, remains inadequate. So, it is necessary to establish marking and detection methods, methods that would be safety for the *in vivo* imaging and tracking of EVs in tissues of animal and human body [3]. Our preliminary result have shown, that labeling by gadolinium could be an optimal and safe way, the initial *in vivo* tests of these methods are based on micromanipulation with EVs lipid bilayer and precise ICP-MS (Inductively coupled plasma mass spectrometry) and MRI detection, and the summary of experiments is presented in this work.

2. METHODS

2.1 Cells and exosomes preparation

The MSCs were isolated from the adipose tissue of three healthy rats, and then expanded in the complete Dulbecco's modified Eagle's medium (5 % FBS) for three weeks, details in [2]. Exosomes were isolated from these adult MSC by combining ultracentrifugation and stimlultaion by AMC generator. The original AMC generator was applied on cell culture 24 hours before ultracentrifugation, generator was designed on our workplace, the aim was to generate sinusoidal induced electric currents with the amplitude modulation and the harmonic carrier frequency of 5,000 Hz and very high current density of induced electrical in cell sample area (above 1 A/m2), details in [4], which accelarate EVs release from cells to the liquid culture medium, ultracentrifugation was used for aseptic separation of exosomes from liuid medium into the 100 sterile microliter vial. Summary of isolation and following Gd-labeling are illustrated on **Figure 1**. EV with Gd-labels were prepared for in vivo application on dermal tissue (back-side of ZDF rats), EVs were not applied in drop of solution, however mixed with hydrogel Olfen (Teva, Czech Rep.), the amount of EV was adjusted to achieve a final concentration of 20 µg of gadolinium per milliliter of gel.

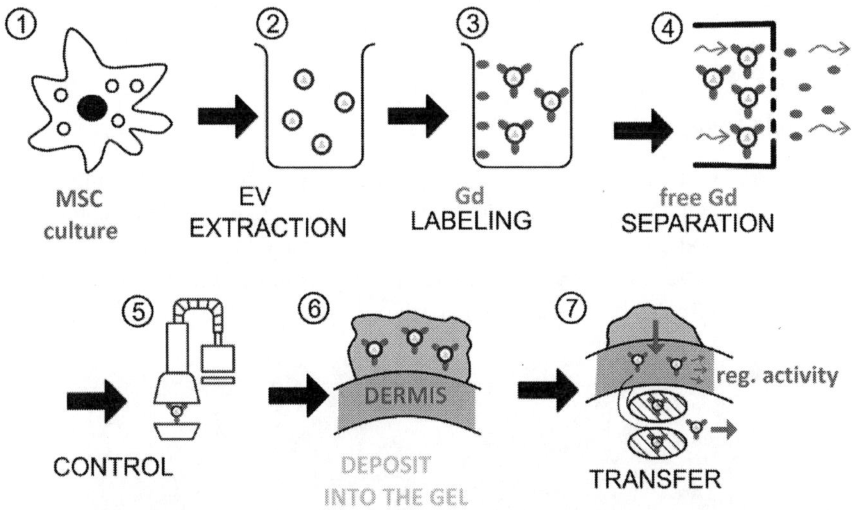

Figure 1 All steps for EV-Gd particles production and application in vivo. Details of step #3,#5 and #7 and their analytical evaluation are described bellow. Step #4 was important for separation of free gadolinium komplex nonincorporated EV, separation separation was realized by a special centrifugation machine ExoDisc (LabSpiner, Korea).

2.2 Exosomes labeling, quantification and transfer

Gadolinium labeled exosome was prepared by lipid insertion following membrane extrusion method [5,6]. In brief, 500 µg DSPE-DOTA-Gd (Avanti Research, USA) were solubilize in 1× DPBS at 65 °C, suspension was cooled down to 25 °C and co-incubated with 500 µg exosome protein for 10 minutes. The exosome and Gd-complex mixture was sonicated in 1ml plastic vials for 1 minutes (5 W, 10s-ON/10s-OFF, original Sono-Ap machine, by CVUT Prague, Czech Rep.). The mixture was extruded trough 100 nm membrane pore size to unify the hydrodynamic size. All other reagents were purchased from Sigma–Aldrich, if not indicated otherwise.

2.3 Determination of EV-Gd particles before *in vivo* application

EVs were applied to grid coated with Formvar (Sigma-Aldrich, Czech Republic). Philips 208 S Morgagni (FEI, Czech Republic) was used for samples visualization (using 7,500× amplification and a quickening voltage of 80 kV). After that, another part of EVs sample were diluted in Milli-Q water. The diluted suspension was placed in a disposable, low volume cuvette with path length 10 mm (Malvern Instruments, UK). For the liposomes size distribution, the Zetasizer Ultra (Malvern Panalytical, UK) operating at detection angle of 173° at room temperature was used.

2.4 In vivo sampling and determination of Gd content using inductively coupled plasma optical emission spectrometer (ICP–OES) and tissue monitoring

Small samples of gel, samples of dermis or internal organs (all compartements, where the high retention is expected, described in **Figure 1**) was collected in different days after gel application on dermis. Concentration of Gd in these samples was measured by ICP–OES Ultima 2 (Horiba Jobin Yvon, France). The Gd standard for ICP (1 g/l) was used as the calibration solution diluted from 0.001 to 50 mg/l. Measurement was carried out in argon flow and using spectral emission characteristics at wavelength of 342 and 246 nm. Gd deposit was recomputed to whole organ mass.

Clinical tomography machines (1,5 T MRI, Bruker) were prepared for experimental scan of EVs deposit in gel and sub-epidermal layers was tested on at day 1, 6 and 12. At the same time, the analysis of viability of dermal cells was conducted on bioptic sample using histological staining and D/L flow cytometry tests..

3. RESULTS

3.1 In vitro characterization

A typical EV-Gd micrograph is shown in **Figure 2**, the exosome with Gd-labels is compared with unlabeled one. As shown in the figure, the most prominent property of both types of EVs is their almost spherical shape.. Unilamellarity of liposomes and homogenity in the size distribution is in correlation with DLS data, the statistical result for hydrodynamic parameters of unconjugated EV (orang) or Gd-conjugated EV (blue) are 115,5± 8,6 nm 122±12,8 nm.The EVs samples had zeta potential values −16,5±0,6 and −22±0,8 mV, respectively.

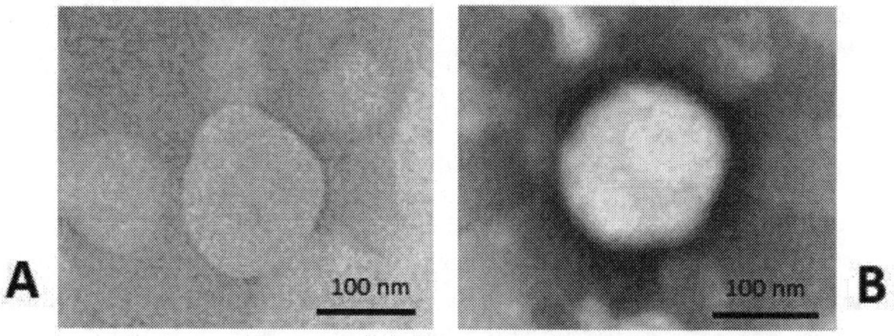

Figure 2 Cryo-TEM images: (**A**) control EV, (**B**) EV after labeling by gadolinium.

3.2 In vivo characterization

The results ICP-MS determination of EV redistribution in different compartment of animal body are summarized on **Figure 4-A**. Very interesting is **Figure 4-B**, which represents modified experimental setup, where the EV was applied under additinal ultrasound stimuls by Sono-Ap machine (CVUT Prague, Czech Rep.; direct contact of sonotrode, 20 minutes in the morning and in the evening at first day of experiment, animal under anesthesia)

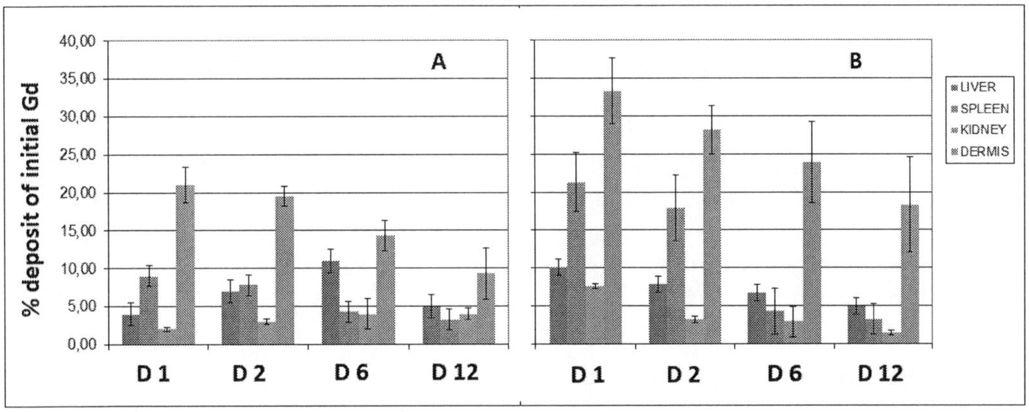

Figure 3 Redistribution of EV-Gd complex and their realtive concentration in different days after initial drop of gel to dermal area. Result derive from ICP-MS determination, absolute mass of Gd for all organs was recomputed as % from initial Gd-cargo of EVs mixed in initial gel drop. Error bars represent the mean standard deviation of 3 samples.

The last experimental step was imaging the gel-drop (20 µg of Gd per 0.5 ml of hydrogel) using an MRI tomography. A concentrated deposit of EV-Gd in the gel can be detected on day 1 (**Figures 4**). In the following days, however, the lower concentrations in the dermis and tissues deeper in the body are not easily detectable in day 6 or day 12 (scan not shown). Improving the contrast of the vesicles and/or the sensitivity of the tomograph will therefore be yet another task for our development in future.

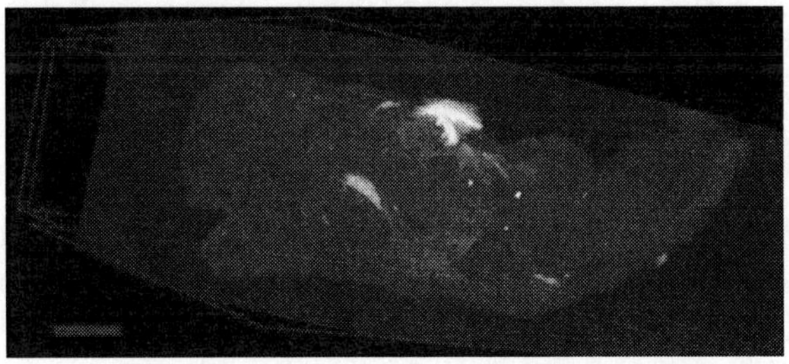

Figure 4 Contrast gel drop containing EV-Gd particles on dermis, scanned by MRI (1.5T). Bar = 1 mm.

Results and their impact can be presented in 4 basic points: (1) The method based on exosomes isolation and Gd-labeling is effective and generates safety particles carrying important pro-regenerative cytokines, no significant effects for viability of cell inside dermis was detected. (2) A gel body was used as a scaffold, from which exosomes diffuse successfully. Unlike intravenous administration of exosomes, the gel body used ensures concentration in the dermal target area, there is evidence from Fig.3 that 20% of initial deposit was transferred to dermis during the first 48 hours; if sono-stimulation is used, this ratio may increase to 33%. (3) Using of the sono-stimulation the longtime concentration of Gd-labels a in internal organs (mainly spleen and liver) can be significantly increased in the first 6 days, however the concentration is relatively well shifting back to under 1 µg/g of tissue after day 12. The clearance of Gd from the the internal organ seems as very effective in all variants of application. (4) The release of cytokines in target dermis was detected by western blot (data not shown), the delivery by used variant of exosomes is effective. Higher local concentration of exosomes and cytokines inside target tissue are a harbinger of a better pharmacological effect.

4. CONCLUSION

EV vesicles have traditionally a problem with the absence of contrast and the possibility of determining the concentration in the target tissue [7,8]. The presented method of gadolinium labeling proved to be effective and safe, it does not significantly change the physical properties of the nature vesicles. Our test of Gd-EVs in vivo approved, that the diffused concentration of vesicles is sufficient for quantification of particles with a mass spectrometry in different dermal and internal location of body (by taking a microbiopsy), but not yet for quantification of particles by traditional X-ray or MRI tomography. And this is a shared unresolved issue in a number of recent publications looking for multimodal high-resolution detection methods [9,10]. We did not detect Gd-induced adverse toxicity. Modified variant "Gd + extra sono-stimul" was also evaluated, this variant have shown that sono-stimulation could accelerate the penetration of vesicles into the target dermal tissue. Production Gd-labeled vesicles is chance to precise supply and precise monitoring of specific curative agents for variety of dermal pathology.

ACKNOWLEDGEMENTS

The project was supported by Technology Agency of the Czech Republic within the execution of the projects FW01010106 "Development of new generation medical devices using the translational medicine and physical interventions principle".

REFERENCES

[1] ZHAO, J., DING, Y., HE, R., HUANG, K., LIU, L., JIANG, C., LIU, Z., WANG, Y., YAN, X, MA, Z., YI, X. Dose-effect relationship and molecular mechanism by which BMSC-derived exosomes promote peripheral nerve regeneration after crush injury. Stem. Cell Res. Ther. 2020, vol. 11, p. 360.

[2] ALMERIA, C., WEISS, R., ROY, M., TRIPISCIANO, C., KASPER, C., WEBER, V., EGGER, D. Hypoxia conditioned mesenchymal stem cell-derived extracellular vesicles induce increased vascular tube formation in vitro. Front. Bioeng. Biotechnol. 2019, vol. 7.

[3] LIANG, Y., DUAN, L., LU, J., XIA, J. Engineering exosomes for targeted drug delivery. Theranostics. 2021, vol. 11, no. 7, pp. 3183-3195.

[4] PRŮCHA, J., SKOPALÍK, J., JUSTAN, I., PARÁK, T., GABRIELOVÁ, E., HÁNA, K., NAVRÁTIL, L. High inductive magnetic stimuli and their effects on mesenchymal stromal cells, dendritic cells, and fibroblasts. Physiol. Res. 2019, vol. 68 (Suppl. 4).

[5] ARYAL, S., STIGLIANO, C., KEY, J., RAMIREZ, M., ANDERSON, J., KARMONIK, C., FUNG, S., DECUZZI, P. Paramagnetic Gd(3+)-labeled red blood cells for magnetic resonance angiography. Biomaterials. 2016, vol. 98, pp. 163-170.

[6] PITCHAIMANI, A., NGUYEN, T.D.T., WANG, H.W., BOSSMANN, S.H., ARYAL, S. Design and characterization of gadolinium infused theranostic liposomes. RSC Adv. 2016, vol. 6, pp. 36898-905.

[7] WAWROWICZ, K., DURAK-KOZICA, M., MARZEJON, M., TRUSIAK, M., STĘPIEŃ, E. Extracellular vesicles produced in hollow fiber bioreactor as potential drug delivery systems–cell uptake tracking with label-free optical diffraction tomography. Stem Cell Res Ther, 2021, vol. 12, no.1, p.127

[8] DURAK-KOZICA, M., BASTER, Z., KUBAT, K., STĘPIEŃ, E. 3D visualization of extracellular vesicle uptake by endothelial cells. Cell. Mol. Biol. Lett. 2018, vol. 23, p. 57.

[9] ROY, S., GU, J., XIA, W., MI, C., GUO, B. Advancements in manganese complex-based MRI agents: Innovations, design strategies, and future directions. Drug Discov. Today. 2024, vol. 29, no. 9, p. 104101.

[10] CHARCZUK, N., TARGOŃSKA, S., ZÁKUTNÁ, D., WATRAS, A., PATEJ, A., WIGLUSZ, R. J. Europium(III) and Gadolinium(III) co-doped nanohydroxyapatite with enhanced photoluminescence as potential multimodal bioimaging agent. Ceram. Int. 2024, vol. 50, no. 9, part A, pp. 14601-14613.

SESSION E
ADVANCED METHODS OF PREPARATION AND CHARACTERIZATION OF NANOMATERIALS

Chairmen

Dr. Petr KLAPETEK	Czech Metrology Institute, Brno, Czech Republic, EU
Dr. Oldrich SCHNEEWEISS	Institute of Physics of the CAS, Prague, Czech Republic, EU
Dr. Michael SOLAR	Czech Technical University in Prague, Czech Republic, EU

XPS CHARACTERIZATION OF Mo-Nb MIXED-OXIDE NANOROD ARRAYS VIA THE ANODIZING OF THIN Al/Mo-Nb-ALLOY BILAYERS ON SUBSTRATES

[1]Maria BENDOVA, [1,2]Jan PRASEK, [1]Alexander MOZALEV

[1]*Department of Microelectronics, Faculty of Electrical Engineering and Communication, Brno University of Technology, Brno, Czech Republic, EU, bendova@vutbr.cz, mozalev@vutbr.cz*

[2]*CEITEC – Central European Institute of Technology, Brno University of Technology, Brno, Czech Republic, EU, prasek@vutbr.cz*

https://doi.org/10.37904/nanocon.2024.4993

Abstract

Molybdenum oxides form a large group of materials resulting from the ability of Mo to possess different formal oxidation states and local coordinations, which may give rise to mixed-valency oxides indispensable for energy storage, gas sensing, electrochromic or catalytic applications. Nanostructuring and doping molybdenum oxides with foreign elements or mixing with similar or dissimilar metal oxides may substantially advance the properties of the individual oxide, although such an approach typically requires exceptionally high temperatures, high vacuum, and high-budget equipment. In this work, we present an alternative approach to creating arrays of molybdenum oxide mixed with niobium oxide nanostructures via a simple, self-organized, room-temperature electrochemical anodizing of thin Al layers superimposed on thin Mo-Nb metal-alloy layers of variable component concentrations prepared by the magnetron sputter-deposition. Anodizing was carried out in oxalic acid solution, forming porous anodic alumina (PAA), followed by PAA-assisted oxidation of the alloy underlayers, and finishing with selective dissolution of the PAA overlayers. Detailed X-ray photoelectron spectroscopy (XPS) analysis of Mo 3d and Nb 3d narrow-scan spectra revealed that the highest amount of Mo in the nanostructures' surface was 79 at.% (Mo+Nb=100 at.%). The dominating molybdenum-oxide component comprised Mo cations with various oxidation states (Mo^{6+}, Mo^{5+}, Mo^{4+}, and Mo^{3+}), whereas the niobium-oxide inclusions were mostly Nb_2O_5. The success of preparing and analyzing the Mo-Nb mixed-oxide nanostructures is vital for understanding the formation-structure-morphology relationship and exploring the functional properties of the novel nanoarrays.

Keywords: Anodic oxidation, metal oxides, molybdenum, niobium, X-ray photoelectron spectroscopy

1. INTRODUCTION

Molybdenum oxides (MoO_x) reveal diverse electronic properties attributed to various formal oxidation states and local coordination of Mo cations, giving rise to mixed-valency oxides [1]. Molybdenum oxides have been applied for constructing energy storage and conversion, electrochromic, gas sensing, and photo- and electrocatalytic devices [2]. For metal oxides in general and MoO_x specifically, nanostructuring to produce nanowires, nanorods, or nanotubes substantially improves the oxide functional properties [3]. This is due to an increased surface-to-volume ratio and/or shorter carrier transport pathways compared with bulk materials [4]. Moreover, doping or mixing molybdenum oxides with similar or dissimilar metal oxides may substantially advance the properties of the individual oxides, although associated techniques typically require exceptionally high temperatures, high-vacuum conditions, and high-budget equipment.

An alternative approach for creating MoO_x-based nanostructures self-organized on a substrate could be based on the PAA-assisted anodizing of valve metals, as was reported for Ta, Nb, W, Ti, Hf, Zr, or some alloys [5]. This results in the formation of short oxide nanoprotrusions of the underlying metal in the PAA barrier layer.

Further re-anodizing to a higher voltage extends the oxide protrusions into the PAA nanopores, forming metal-oxide nanorods following the morphology of the PAA layer. A layer of molybdenum alloy must be used instead of pure molybdenum to obtain the mixed-oxide nanorods. The oxide mixing should be advantageous since there have been no reports on a successful PAA-assisted anodizing of pure molybdenum due to the peculiar electrochemical behavior of molybdenum and the instability of molybdenum oxides in aqueous electrolytes [1,6]. We have selected Nb as the alloying element because it behaves electrochemically as a typical valve metal, and its oxide has excellent chemical stability [5,7]. In this work, we have anodized an Al/Mo-Nb-alloy bilayer in an oxalic acid solution to prepare novel MoO_x-Nb_2O_5 nanostructure arrays. The chemical composition of nanoarrays, mainly the Mo and Al contents and the stoichiometry of the oxides, has been investigated by X-ray photoelectron spectroscopy (XPS).

2. EXPERIMENTAL PART

2.1 Sample preparation

The precursor thin-film Al/Mo-Nb-alloy bilayer was prepared on an oxidized Si wafer by the magnetron co-sputtering of a 200-nm thick Mo-Nb alloy layer followed by sputtering of a 700-nm thick Al layer using Mo (99.95%), Nb (99.95%), and Al (99.999%) targets. 1 cm × 1 cm samples were cut from the wafer and galvanostatically anodized at a steady-state voltage of 50 V in 0.6 M oxalic acid at room temperature. For selected samples, re-anodizing was performed in a borate buffer (0.5 M H_3BO_3/0.05 M $Na_2B_4O_7$) to a more anodic voltage of 180 V. The PAA overlayer was partially or fully selectively etched away in CrO_3/H_3PO_4-based etchant [5] to obtain the 'PAA-residues' or 'PAA-free' samples. Field-emission SEM, utilizing an InBeam detector for secondary electrons, was used to examine the morphology of the anodic films.

2.2 XPS investigation

XPS analysis was conducted in a Kratos Axis Ultra DLD spectrometer using a monochromatic Al Kα source. The X-ray emission power was 75 W with a 15 kV accelerating voltage focused to 300 μm × 700 μm. The emitted electrons were detected at fixed pass energies of 160 eV for the survey spectra and 20 eV for the high-resolution spectra. The Kratos charge neutralizer system was used for all specimens.

The spectra were analyzed using CasaXPS. GL(30) profiles were used for all components except the metallic core lines of Mo 3d and Nb 3d, for which asymmetric profiles in the form of LA(1.1,2,2) and LA(2,3.3,12) were applied, respectively. A standard Shirley background was used. Spectra from all samples were charge-corrected to give the adventitious C 1s spectral component (C−C, C−H) binding energy of 284.8 eV. The deconvolution of C 1s spectra was performed as described elsewhere [5,8].

3. RESULTS AND DISCUSSION

3.1 Film formation and morphology

The precursor Al/Mo-Nb-alloy bilayer was initially characterized by XPS to obtain the Nb content after selectively removing the Al overlayer in 1 wt.% NaOH solution. The presence of C, O, Mo, Nb, and Al was confirmed on the alloy surface. From the narrow scan Mo 3d and Nb 3d spectra, the Nb content was calculated to be 19 at.% (when Mo+Nb=100 at.%). Therefore, the alloy layer will be referred to as the Mo-19Nb and the initial bilayer as the Al/Mo-19Nb.

The PAA-assisted anodizing and re-anodizing of the Mo-19Nb layer, followed by selectively etching the PAA layer, resulted in an array of well-aligned oxide nanorods, as confirmed by SEM observations (**Figure 1**). The rods are about 400 nm long, anchored to the substrate via a compact bottom-oxide layer (**Figure 1a,b**). The lower parts of the rods (originally residing in the PAA barrier layer) are slightly narrower than their upper parts, which are ~40 nm in diameter (**Figures 1c,d**). Therefore, the alloying of Mo with a small amount of Nb enabled stabilization of the growing oxide, preventing it from dissolving, which resulted in a successful PAA-assisted formation of MoO_x-based nanorods.

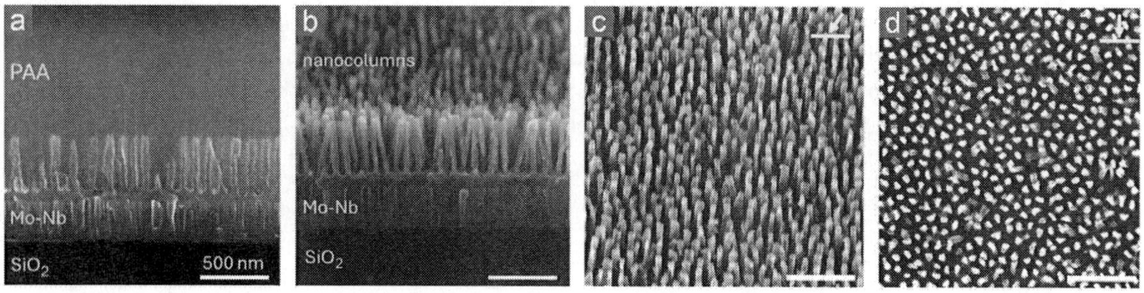

Figure 1 Cross-sectional, tilted, and top SEM views of (a) PAA-embedded and (b–d) PAA-free re-anodized anodic film prepared from the Al/Mo-19Nb bilayer

Based on the SEM observation, we have developed a schematic model describing the formation process for the anodized and re-anodized MoO_x-based nanofilms (**Figure 2**). The migration rates of Mo^{n+} and Nb^{n+} cations likely differ, and therefore, a compositional gradient along the nanorods is expected, as indicated by the colors in the schematics. For XPS analysis, we selected three films made from the Al/Mo-19Nb bilayer: the anodized PAA-free (**Figure 2e**), the re-anodized PAA-free (**Figure 2f**), and the re-anodized PAA-residues (**Figure 2g**) films. The third option was added to mask the bottom oxide layer and thus to distinguish XPS signals originating from various parts of the re-anodized film.

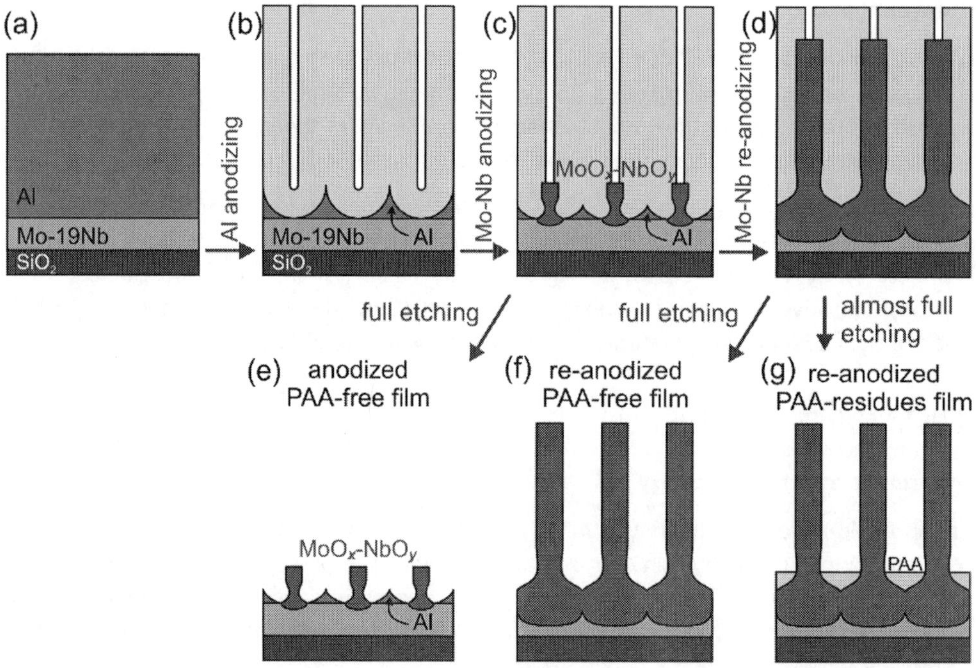

Figure 2 Schematic processes for forming MoO_x-based nanoarrays: (a) the sputter-deposition of an Al/Mo-19Nb bilayer onto an oxidized Si wafer, (b) the anodizing of the Al layer at 50 V to form a PAA film, (c) the anodizing of the Mo-19Nb alloy layer through the PAA barrier layer, and (d) the re-anodizing of the Mo-19Nb layer to 180 V; (e) the anodized PAA-free, (f) the re-anodized PAA-free, and (g) the re-anodized PAA-residues films

3.2 Chemical composition

XPS examination was performed on the anodized PAA-free, re-anodized PAA-free, and re-anodized PAA-residues films made from the Al/Mo-19Nb bilayer to estimate the film surface layers' chemical composition and bonding states. C, O, Mo, Nb, Al, P, and Cr were identified in the survey spectra; the last two elements were present due to the etching in the CrO_3/H_3PO_4-based solution. Narrow-scan C 1s, O 1s, Mo 3d, Nb 3d, and Al 2p spectra were collected to perform quantitative analysis and reveal the bonding states. The experimental and fitted Mo 3d, Nb 3d and Al 2p spectra are shown in **Figure 3**.

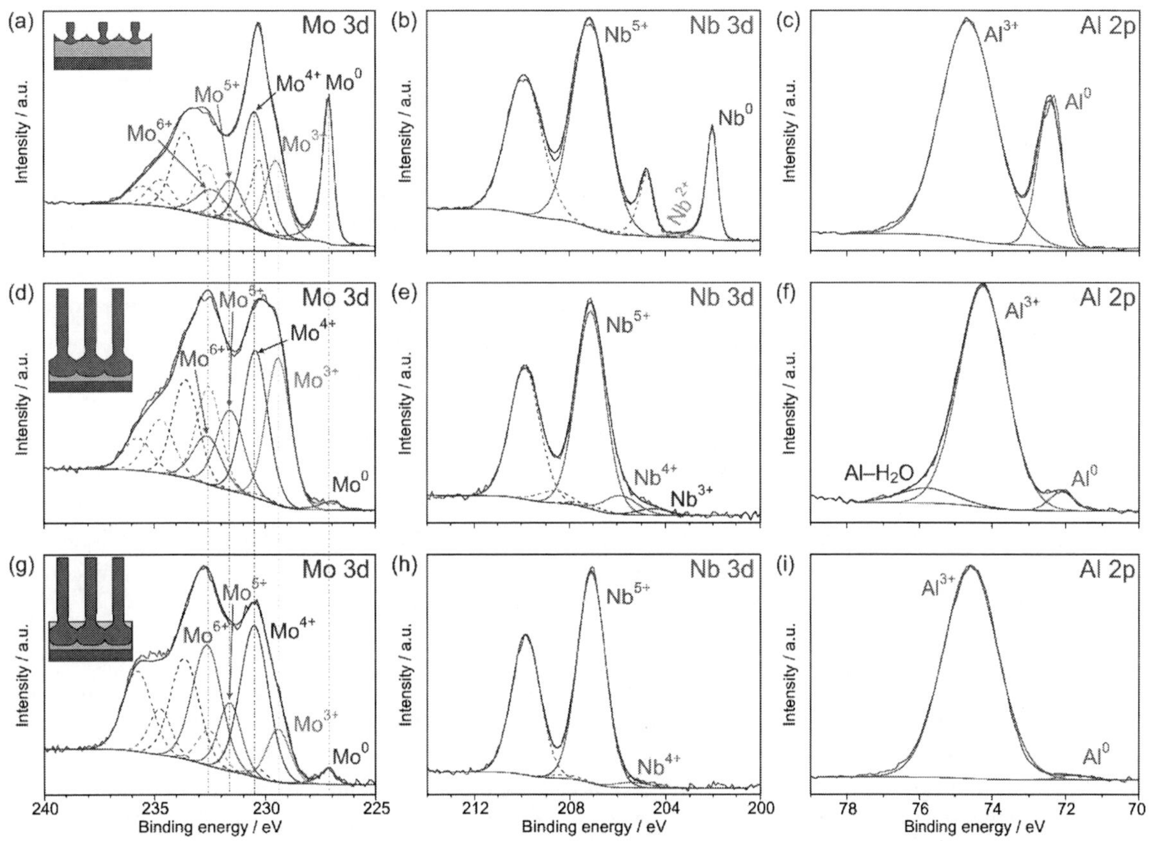

Figure 3 Experimental and fitted XPS spectra of the surface of the (a–c) anodized PAA-free, (d–f) re-anodized PAA-free, and (g–i) re-anodized PAA-residues anodic films

The *Mo 3d spectra* (**Figures 3a,d,g**) are fitted with several constrained doublets (Mo $3d_{5/2}$ and $3d_{3/2}$ with a fixed peak-separation energy of 3.13 eV and with a fixed intensity ratio of 3:2 [9,10]). All the films reveal the presence of four Mo^{n+} cations, with n = 6, 5, 4, or 3, in various ratios and a doublet of Mo^0. The Mo $3d_{5/2}$ positions agree with the literature values [9–11] and are consistent between the samples (as indicated by the vertical lines across the spectra), although the separation between the doublets is relatively small (~1.0 eV) and the assignment of the doublets to the specific oxidation states of Mo is, to some extent, questionable [9–11]. The high amount of Mo^0 in the anodized film (**Figure 3a**) is explained by the Mo-19Nb residues around the nanoprotrusions, whereas in the re-anodized films (**Figures 3d,g**), such metallic residues are not expected. When subtracting the Mo^0 contributions, an average calculated oxidation state of Mo in MoO_x in the anodic parts of the films is 4.2, 4.1, and 4.7 for the anodized PAA-free, re-anodized PAA-free, and re-anodized PAA-residues films, respectively. Clearly, the PAA-residues sample contains more oxidized MoO_x. This means the rod's surface contains more oxidized MoO_x compared with the bottom oxide.

The *Nb 3d spectra* (**Figures 3b,e,h**) are fitted with doublets of appropriately constrained peaks (Nb $3d_{5/2}$ and $3d_{3/2}$ with a fixed peak separation energy of 2.75 eV and with a fixed intensity ratio of 3:2 [5]). All the spectra contain a high-intensity doublet at the highest binding energy (with Nb $3d_{5/2}$ at 207.1±0.1 eV), assigned to Nb^{5+} cations in Nb_2O_5 [5]. The anodized sample (**Figure 3b**) reveals a substantial amount of Nb^0 [5] due to the metallic residues. Furthermore, minor amounts of suboxides Nb^{4+} (9%) and Nb^{3+} (3%) are identified in the re-anodized PAA-free film (**Figure 3e**). The PAA-residues sample has even fewer suboxides (**Figure 3h**). Therefore, similarly to MoO_x, the NbO_x in the nanorods is more oxidized than the bottom oxide layer.

After subtracting the metallic contributions, the fitted Mo 3d and Nb 3d narrow-scan spectra were also used to calculate the Mo content (when at.%(Mo+Nb)=100%) in the surface layer of the anodic parts of the films. Mo content decreases from 79 through 69 to 60 at.% for the anodized PAA-free, re-anodized PAA-free, and re-anodized PAA-residues films, respectively, whereas the initial Mo-19Nb alloy has 81 at.% Mo. This finding means that the surface layer of the anodic films contains both MoO_x and NbO_x in appreciable amounts; however, the amount of Mo is lower in the re-anodized samples than in the initial alloy, and it is substantially lower in the PAA-residues film. Thus, the rods contain less Mo than the bottom oxide, which implies that the Mo^{n+} cations migrate slower than the Nb^{n+} cations under the high electric field during the anodizing and re-anodizing processes.

The *Al 2p spectra* (**Figures 3c,f,i**) are fitted with a few singlet peaks. The highest-intensity peak is associated with Al_2O_3 [5]. In addition, a narrow Al^0 peak residing at 71.8–72.5 eV is noted in all the spectra. A relatively broader low-intensity peak in the spectrum of the re-anodized PAA-free sample, shifted to +1.6 eV from the Al_2O_3 position, is attributed to hydrated Al_2O_3 [12]. After subtracting the metallic contributions originating from the metallic residues surrounding the nanorods (and also the corresponding native oxide formed on the aluminum [13]), an approximate amount of Al in the surface layer of the anodic films is obtained (providing that at.%(Al+Nb+Mo)=100%): 68, 79, and 90 at.% Al for the anodized PAA-free, re-anodized PAA-free, and re-anodized PAA-residues films, respectively. An amount of Al up to 80% has been reported for PAA-assisted nanofilms made from various precursor metals. The higher amount of Al in the PAA-residues film is explained by underetched alumina between the lower parts of the oxide nanorods.

In summary, the surface layer of the anodic films contains MoO_x with an average oxidation state of Mo ~4.2, mixed with Nb_2O_5, substoichiometric niobium oxides, and Al_2O_3. The amount of Mo is lower than in the initial Mo-19Nb alloy. The surface of the oxide nanorods, compared with the whole surface of the re-anodized anodic film, has a relatively smaller amount of Mo, suggesting a slower migration of Mo^{n+} cations compared with Nb^{n+} cations, and comprises more oxidized Mo and Nb oxides, whereas the bottom oxide layer is composed of more reduced Mo and Nb oxides. The oxide nanorods' inner part is probably alumina-free mixed Mo-Nb anodic oxide [5].

4. CONCLUSIONS

In the present work, we report, for the first time, the electrochemical PAA-assisted synthesis of a MoO_x-based nanorod array made from the Al/Mo-19Nb-alloy bilayer. The alloying of Mo with a relatively low amount of Nb in the precursor material enabled the field-assisted growth of chemically stable mixed-oxide nanorods. The XPS analysis revealed that the rods' surface has a complex chemical composition comprising a dominating amount of MoO_3, Mo_2O_5, MoO_2, and Mo_2O_3 mixed with a minor amount of Nb_2O_5, NbO_2, and Nb_2O_3. Due to the interaction of growing Mo-Nb anodic oxide with the PAA cell walls, the rods' surface also contains Al_2O_3. A compositional difference between the various parts of the anodic film was suggested based on the XPS analysis.

Mixing molybdenum oxides with minor amounts of niobium oxides, likely at the nanoscale level, is expected to substantially impact the oxides' electrochemical, electronic, and electro-optical properties. With a focus on energy storage applications [6,7], the nanoarrays developed here will be the subject of future investigation.

ACKNOWLEDGEMENTS

The research leading to these results was supported by GACR grant no. 23-07848S. We acknowledge the use of CzechNanoLab Research Infrastructure supported by MEYS CR (LM2018110).

REFERENCES

[1] de CASTRO, I.A., DATTA, R.S., OU, J.Z., CASTELLANOS-GOMEZ, A., SRIRAM, S., DAENEKE, T., KALANTAR-ZADEH, K. Molybdenum oxides – From fundamentals to functionality. *Advanced Materials*. 2017, vol. 29, 1701619, pp. 31.

[2] CONCEPCIÓN, O., de MELO, O. The versatile family of molybdenum oxides: synthesis, properties, and recent applications. *Journal of Physics: Condensed Matter*. 2023, vol. 35, 143002, pp. 26.

[3] DEVAN, R.S., PATIL, R.A., LIN, J.-H., MA, Y.-R. One-dimensional metal-oxide nanostructures: recent developments in synthesis, characterization, and applications. *Advanced Functional Materials*. 2012, vol. 22, pp. 3326–3370.

[4] OSTERLOH, F.E. Inorganic nanostructures for photoelectrochemical and photocatalytic water splitting. *Chemical Society Reviews*. 2013, vol. 42, pp. 2294–2320.

[5] MOZALEV, A., VÁZQUEZ, R.M., BITTENCOURT, C., COSSEMENT, D., GISPERT-GUIRADO, F., LLOBET, E., HABAZAKI, H. Formation–structure–properties of niobium-oxide nanocolumn arrays via self-organized anodization of sputter-deposited aluminum-on-niobium layers. *Journal of Materials Chemistry C*. 2014, vol. 2, pp. 4847–4860.

[6] ELKHOLY, A.E., DUIGNAN, T.T., SUN, X., ZHAO, X.S. Stable α-MoO_3 electrode with a widened electrochemical potential window for aqueous electrochemical capacitors. *ACS Applied Energy Materials*. 2021, vol. 4, pp. 3210–3220.

[7] SHEN, F., SUN, Z., HE, Q., SUN, J., KANER, R.B., SHAO, Y. Niobium pentoxide based materials for high rate rechargeable electrochemical energy storage. *Materials Horizons*. 2021, vol. 8, pp. 1130–1152.

[8] PAYNE, B.P., BIESINGER, M.C., McINTYRE, N.S. X-ray photoelectron spectroscopy studies of reactions on chromium metal and chromium oxide surfaces. *Journal of Electron Spectroscopy and Related Phenomena*. 2011, vol. 184, pp. 29–37.

[9] SPEVACK, P.A., McINTYRE, N.S. Thermal reduction of MoO_3. *Journal of Physical Chemistry*. 1992, vol. 96, pp. 9029–9035.

[10] BIESINGER, M.C. *X-ray Photoelectron Spectroscopy (XPS) Reference Pages, Molybdenum* [online]. 2009–2021 [viewed: 2024-09-17]. Available from: http://www.xpsfitting.com/search/label/Molybdenum

[11] OKONKWO, I.A., DOFF, J., BARON-WIECHEC, A., JONES, G., KOROLEVA, E.V., SKELDON, P., THOMPSON, G.E. Oxidation states of molybdenum in oxide films formed in sulphuric acid and sodium hydroxide. *Thin Solid Films*. 2012, vol. 520, pp. 6318–6327.

[12] BIESINGER, M.C. *X-ray Photoelectron Spectroscopy (XPS) Reference Pages, Aluminum* [online]. 2009–2021 [viewed: 2024-09-16]. Available from: http://www.xpsfitting.com/search/label/Aluminum.

[13] STROHMEIER, B.R. An ESCA method for determining the oxide thickness on aluminum alloys. *Surface and Interface Analysis*. 1990, vol. 15, pp. 51–56.

ADVANCED PREPARATION AND CHARACTERIZATION OF PEDOT-BASED NONWOVEN MATERIALS FOR WEARABLE ELECTRONICS

[1]Emanuele Alberto SLEJKO, [1]Giovanni CARRARO, [2]Dario CAVALLO, [2]Roberto SPOTORNO, [1]Marco SMERIERI

[1]CNR-IMEM – Institute of Materials for Electronics and Magnetism, National Research Council of Italy, Genoa, Italy, EU

[2]Department of Chemistry, University of Genoa, Genoa, Italy, EU

https://doi.org/10.37904/nanocon.2024.4992

Abstract

Poly(3,4-ethylenedioxythiophene)-polystyrene sulfonate (PEDOT-PSS) nanofiber-based materials represent advanced solutions for application in wearable electronics due to their biocompatibility, lightweight, permeability and improved sensitivity related to their large surface area. Since PEDOT is a conductive polymer, the electrospinning process for manufacturing thin, flexible nonwovens made of homogeneous nanofibers is not well-studied yet. In this research, we have investigated the fabrication and characterization of electrospun conductive nanofibers composed of varying concentrations of PEDOT-PSS mixed with poly(ethylene oxide) (PEO). The electrospinning technique was employed to produce these conductive nonwovens, and their morphological properties were evaluated using scanning electron microscopy (SEM). To enhance the water resistance of nanofibers, a thermal treatment was applied to remove excess PEO. The resulting samples demonstrated significant improvements in water stability, whereas untreated samples dissolved in pure water almost immediately. Electrical characterization of conductive nonwovens was performed, revealing resistivity in several hundreds of Ohm•m. To further investigate the electrical response and electrochemical behavior of the material when in contact with specific compounds, we conducted electric impedance spectroscopy (EIS) and cyclic voltammetry (CV). The EIS measurements provided insights into the impedance characteristics, while the CV analysis highlighted the occurrence of reversible faradaic reactions. Our findings suggest that the optimized PEDOT-PSS/PEO nanofibers exhibit promising electrical and electrochemical properties, making them suitable for applications in flexible electronics and wearable sensors. This study contributes to the understanding of the relationship between the fabrication process, structural morphology, and functional performance of electrospun conductive nonwovens with the potential to significantly impact the wearable technology field.

Keywords: Nonwovens, conductive polymers, electrospinning, flexible electronics

1. INTRODUCTION

The production of nanofibers has become a highly active area of research due to their unique properties and broad applications in various fields such as biomedicine, energy storage, textiles and flexible electronics, and environmental science [1–3]. Nanofibers are characterized by their high surface area-to-volume ratio, flexibility, and the ability to be tailored for specific applications, making them particularly versatile materials for diverse industrial and scientific uses. Among the different techniques for nanofiber fabrication, electrospinning has emerged as one of the most widely used and effective methods due to its scalability, versatility, and ability to produce nanofibers with customizable properties. Electrospinning involves the application of a high-voltage electric field to a polymer solution or melt, which causes the formation of a thin jet that elongates and solidifies into continuous nanofibers as the solvent evaporates and the polymer solidifies [4]. These nanofibers typically have diameters ranging from a few nanometers to several micrometers, exhibiting excellent uniformity and purity. The primary advantage of electrospun nanofibers lies in their high surface area-to-volume ratio, which enhances their interaction with surrounding environments. Additionally, the properties of electrospun nanofibers can be fine-tuned by adjusting various parameters, including polymer concentration, solvent type, applied voltage, and collection distance [5]. The scalability and cost-effectiveness of electrospinning further enhance its industrial viability, as the process is relatively simple yet capable of producing high-quality, uniform nanofibers on a large scale. This has made electrospinning an attractive technique for various fields, particularly those requiring large-scale nanofiber production.

In recent years, conductive polymers have also gained significant attention due to their combination of electrical conductivity, mechanical flexibility, and processability [6]. These properties open up new possibilities for the development of advanced electronic devices and energy storage systems. Among the various conductive polymers, materials such as polyaniline, polypyrrole, and polythiophenes have been extensively studied. However, one of the most promising conductive polymers is poly(3,4-ethylenedioxythiophene) (PEDOT), which stands out for its exceptional conductivity, environmental stability, and ease of synthesis. PEDOT belongs to the polythiophene family and is characterized by a π-conjugated structure that imparts high electrical conductivity. The conductivity of PEDOT, measured at approximately 4380 S/cm, approaches that of indium tin oxide, a commonly used conductive material in electronic applications [7]. Unlike many other conductive polymers, PEDOT remains stable when exposed to air, moisture, and UV radiation, making it particularly well-suited for applications requiring long-term durability, such as organic electronic devices and biosensors. In addition to its stability, PEDOT's electrical and optical properties can be finely tuned through chemical doping and structural modifications. In recent years, PEDOT has found widespread use in a variety of fields, including organic electronics, energy storage, biomedicine, and wearable technology [1,2,8]. PEDOT's compatibility with flexible substrates and its biocompatibility make it an attractive candidate for flexible electronics, implantable medical devices, and bioelectronic interfaces.

Electrospinning has proven to be a valuable technique for fabricating PEDOT nanofibers, as it allows the production of nanofibers with high surface area-to-volume ratios, precise control over fiber diameter and alignment, and the ability to create complex three-dimensional structures. By electrospinning PEDOT, researchers can create bioactive interfaces with enhanced conductivity, mechanical properties, and biocompatibility, which are critical for the development of advanced biomedical devices such as biosensors, neural implants, and drug delivery systems. Significant advancements have been made in the fabrication of PEDOT nanofibers, particularly in terms of improving their conductivity and mechanical properties. Early techniques involved electrospinning insulating polymers followed by vapor-phase polymerization of PEDOT, resulting in high conductivity and resistive heating capabilities [9]. In the 2010s, researchers introduced new techniques such as oxidative polymerization and blending PEDOT with other polymers for solution-based electrospinning. These methods have further enhanced the structural stability, conductivity, and flexibility of PEDOT nanofibers [10].

In this paper, we investigate the spinnability conditions of PEDOT solutions for the fabrication of nonwoven materials. Furthermore, we characterize the electrochemical response and stability of nanofiber-based nonwovens for potential applications as wearable sensors and biosensors. In fact, one of the major limitations in the production of electrospun PEDOT-based nanofibers is related to the conductive nature of PEDOT. Therefore, being able to identify the best processing conditions for the reliable and repeatable production of nanofibers is of upmost importance in terms of industrial adoption of these materials.

2. MATERIALS AND METHODS

PEDOT:PSS PH1000 water dispersion at 1.3% (from Heraeus) was mixed with Polyethylene oxide 600 kDa (PEO, from Sigma Aldrich). The amounts of reagents used for the preparation of PEDOT-PEO starting solutions are reported in **Table 1**.

Table 1 Preparation of starting solutions made of PEDOT:PSS and PEO in deionized water

ID	Volume of PEDOT:PSS solution (ml)	Mass of PEO (mg)	Volume of deionized water (ml)	PEDOT Concentratio (mM)	PEO Concentratio (μM)
1	10.0	0.6	0.0	23.2	91.7
2	5.0	0.6	5.0	15.5	61.1
3	5.0	0.3	0.8	20.0	79.0
4	5.0	0.3	1.6	17.6	69.4
5	2.2	0.2	6.5	5.9	38.3
6	11.0	0.6	6.0	15.0	61.5
7	7.5	0.4	0.8	21.1	76.0
8	6.6	0.6	9.0	9.8	64.6
9	9.0	0.3	1.0	20.9	50.0
10	4.0	0.4	2.0	15.5	111.0

A Linari RT Advanced electrospinner has been used for the fabrication of nonwoven samples. Using the roll-to-roll configuration, 4 ml of PEDOT-PEO starting solution have been loaded in a plastic syringe, with a working distance between Al foil collector and syringe needle of 15 cm. The voltage has been set to 17 kV, and the flow rate to 0.1 ml/h. After 6 hours, the electrospinner has been stopped and the sample has been left overnight to completely dry before characterization.

2.1 Characterization techniques

Rheological analysis was performed using an DSR500-CP-4000-plus rheometer (Lamy rheology, France). The starting solutions were subjected to increasing and decreasing shea rate ramps from 10 to 500 s^{-1} for the evaluation of the viscosity. Images of nonwoven samples after the electrospinning fabrication have been collected using a scanning electron microscope (Zeiss Germany). Electrochemical characterization was performed in a naturally aerated aqueous 0.1 M sodium chloride solution (VWR, AnalaR NORMAPUR, purity >99.5%) at room temperature. Experiments were performed using a proprietary three-electrode cell suitable for exposing a fixed area of 1 cm2 of a sample. The cell configuration consisted of a sample connected as a working electrode (WE), a reference electrode (RE) of Ag/AgCl (KCl 3 M), and a platinum wire as a counter electrode (CE) in a volume of 0.250 L of test solution under static conditions. The electrochemical characterization has been carried out using a Palmsens4 potentiostat (Palmsens BV, The Netherlands).

3. RESULTS AND DISCUSSION

3.1 Solution Optimization

Figure 1a shows the results of the electrospinning tests with the PEDOT-PEO prepared starting solutions. As can be seen, when the concentration of PEO and PEDOT is relatively low the nanofiber production is successful, while if the concentrations are increased the spinning process fails. The spinnability region of starting solutions is limited on the right side due to high concentrations of PEO, which causes a marked increase of viscosity; while on top side it is limited by high concentrations of PEDOT which, being a conductive polymer, causes the electrospinning process to fail due to the disruption of the electric field between needle and collector. **Figures 1b and c** show the quality of obtained nanofibers, with a uniform large area and a mean fibers diameter of 180 ±40 nm. The rheological behavior of starting solution is typical pseudoplastic, with a decrease of viscosity with increase shear rate, as shown in **Figure1d**. Using small needle size with the same flow rate, therefore, is beneficial as the viscosity of the solution is reduced and the spinning process is promoted. Therefore, the spinnability region can be extended towards the right side of the PEO concentration recurring to smaller needles and high local fluxes, an interesting observation for increasing the output production (and, therefore, the process yield) of PEDOT-based nanofibers in industrial environments.

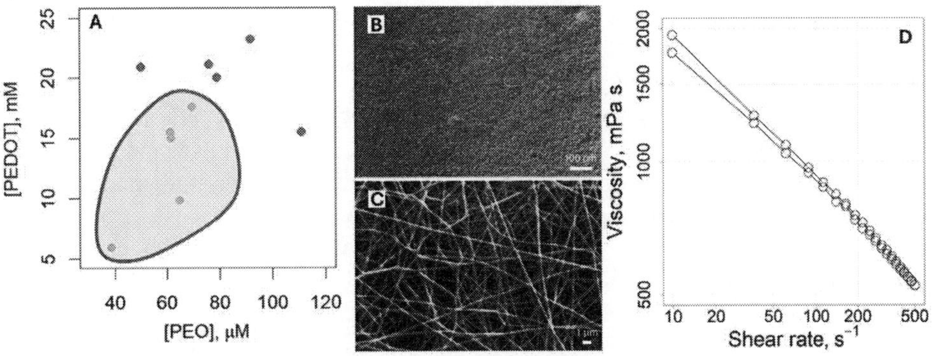

Figure 1 Characterization of the PEDOT-PEO nonwoven samples. A) spinnability as a function of PEDOT and PEO concentrations in the starting solution. B) 250X and C) 10000X SEM images of the nonwoven samples, SE 20 kV and WD 8 mm. D) Viscosity as a function of the shear rate for solution number 4 in **Table 1**.

3.2 Electrochemical Characterization

Cyclic voltammetry is a widely used technique for studying the charge/discharge process of PEDOT by measuring the current as a function of applied potential at a fixed scan rate (mV·s^{-1}). Throughout the repeated charge/discharge process, the nonwoven material undergoes physical changes, such as swelling, shrinkage, and cracking, due to the reaction of charge carriers with the conductive polymer. The ability to maintain its mechanical integrity and preserve the cyclic voltammetry profile (**Figure 2a**) without degradation over successive testing indicates the durability of the material during measurements in aqueous media, for example for sensing applications.

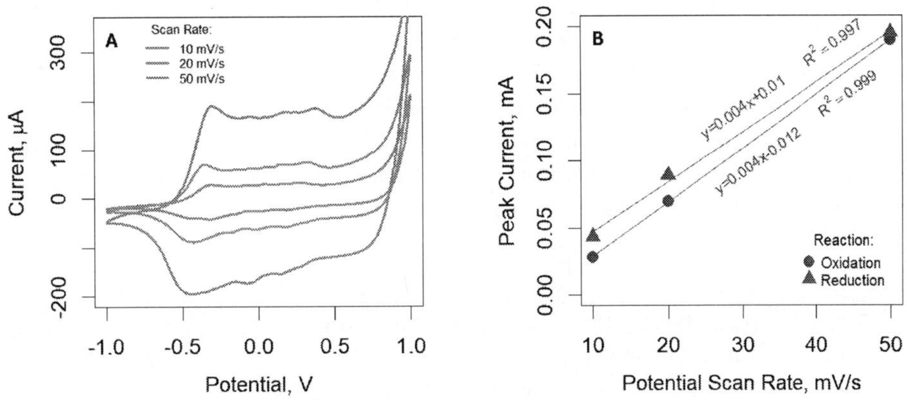

Figure 2 A) Cyclic voltammetry at different scan rate and B) linear trend of the absolute value of main current peaks

In a cyclic voltammetry experiment, the presence of symmetric peaks in the oxidizing and reducing regimes suggested the occurrence of reversible electrochemical reactions. By plotting the intensity of the main current peak as a function of the potential scan rate (**Figure 2b**) a linear relationship has been obtained, indicating the presence of a surface-controlled process, rather than a diffusion-controlled one. This behavior is characteristic of a reaction where the redox species are adsorbed on the electrode surface rather than being freely diffusing in the bulk solution. In fact, when a redox species is adsorbed on the electrode surface, the number of available molecules for electron transfer is constant and does not depend on their diffusion from the solution to the electrode. In this case, the peak current is proportional to the scan rate, as the faster the scan rate, the more electrons are transferred per unit time. The observed linear relationship between peak current and scan rate indicates that the rate-limiting process is electron transfer at the electrode surface or adsorption/desorption of the species. Since the redox species are adsorbed, the process does not rely on diffusion from the bulk solution, and the current increases directly with the rate of potential change (scan rate). This is consistent with the very nature of nonwoven samples: nanofibers display a large surface-to-volume ratio, meaning that conducting species can easily move within the materials structure and interact at the surface of nanofibers, therefore diffusion does not represent the limiting step in the process.

4. CONCLUSION

The electrospinning experimental campaign revealed that successful nanofiber production occurs at lower concentrations of PEO and PEDOT, while higher concentrations hinder the process due to increased viscosity and disruption of the electric field. The obtained nanofibers showed uniformity with an average diameter of 180 ± 40 nm, and the starting solution exhibited pseudoplastic behavior, with viscosity decreasing as shear rate increased. By using smaller needle sizes and higher local fluxes, the spinnability region can be extended, offering a promising approach to enhance PEDOT-based nanofiber production for industrial applications. Furthermore, CV is a valuable tool for assessing both the mechanical and electrical stability of PEDOT nonwoven samples, making it particularly useful in evaluating the long-term performance and durability of nonwovens. The observed linear trend in peak current as a function of scan rate indicates that the cyclic voltammetry experiment is based on a surface-confined redox process, where the limiting process is electron transfer of the adsorbed species, not diffusion from the bulk solution. These findings indicate the PEDOT-based nonwoven samples are viable materials for the development of biological sensors and wearable electronics.

ACKNOWLEDGEMENTS

Funded by the European Union—NextGenerationEU and by the Ministry of University and Research (MUR), National Recovery and Resilience Plan (NRRP), Mission 4, Component 2, Investment 1.5, project "RAISE—Robotics and AI for Socio-economic Empowerment" (ECS00000035). E.A. Slejko, G. Carraro, and M. Smerieri are part of RAISE Innovation Ecosystem. We also acknowledge financial support under the National Recovery and Resilience Plan (NRRP), Mission 4, Component 2, funded by the European Union – NextGenerationEU - Project Title "SMART non-woven fabrics for multifUnctional wearable haPtic devices (SMART_UP)" - CUP B53D23008810006.

REFERENCES

[1] YI, H.; PATEL, R.; PATEL, K.D.; BOUCHARD, L.S.; JHA, A.; PERRIMAN, A.W.; PATEL, M. Conducting Polymer-Based Scaffolds for Neuronal Tissue Engineering. *Journal of Materials Chemistry B.* 2023, vol. 11, pp. 11006–11023. Available from: https://doi.org/10.1039/d3tb01838e.

[2] SEITI, M.; GIURI, A.; CORCIONE, C.E.; FERRARIS, E. Advancements in Tailoring PEDOT: PSS Properties for Bioelectronic Applications: A Comprehensive Review. *Biomaterials Advances.* 2023, vol. 154, pp. 213655.

[3] KEIROUZ, A.; WANG, Z.; REDDY, V.S.; NAGY, Z.K.; VASS, P.; BUZGO, M.; RAMAKRISHNA, S.; RADACSI, N. The History of Electrospinning: Past, Present, and Future Developments. *Advanced Materials Technologies.* 2023, vol. 8, pp. 2201723. Available from: https://doi.org/10.1002/admt.202201723.

[4] LONG, Y.-Z.; YAN, X.; WANG, X.-X.; ZHANG, J.; YU, M. *Electrospinning: Nanofabrication and Applications*. Norwich: William Andrew Publishing, 2019.

[5] VERPOORTEN, E.; MASSAGLIA, G.; CIARDELLI, G.; PIRRI, C.F.; QUAGLIO, M. Design and Optimization of Piezoresistive PEO/PEDOT:PSS Electrospun Nanofibers for Wearable Flex Sensors. *Nanomaterials.* 2020, vol. 10, pp. 2166. Available from: https://doi.org/10.3390/nano10112166.

[6] SLEJKO, E.A.; CARRARO, G.; HUANG, X.; SMERIERI, M. Advances in the Fabrication, Properties, and Applications of Electrospun PEDOT-Based Conductive Nanofibers. *Polymers.* 2024, vol. 16, pp. 2514. Available from: https://doi.org/10.3390/polym16172514.

[7] KIM, N.; KEE, S.; LEE, S.H.; LEE, B.H.; KAHNG, Y.H.; JO, Y.; KIM, B.; LEE, K. Transparent Electrodes: Highly Conductive PEDOT:PSS Nanofibrils Induced by Solution-Processed Crystallization. *Advanced Materials* 2014, vol. 26, pp. 2109–2109. Available from: https://doi.org/10.1002/adma.201470088.

[8] BABAIE, A.; BAKHSHANDEH, B.; ABEDI, A.; MOHAMMADNEJAD, J.; SHABANI, I.; ARDESHIRYLAJIMI, A.; MOOSAVI, S.R.; AMINI, J.; TAYEBI, L. Synergistic Effects of Conductive PVA/PEDOT Electrospun Scaffolds and Electrical Stimulation for More Effective Neural Tissue Engineering. *European Polymer Journal.* 2020, vol. 140, pp. 110051. Available from: https://doi.org/10.1016/j.eurpolymj.2020.110051.

[9] LAFORGUE, A.; ROBITAILLE, L. Production of Conductive PEDOT Nanofibers by the Combination of Electrospinning and Vapor-Phase Polymerization. *Macromolecules.* 2010, vol. 43, pp. 4194–4200.

[10] PARK, H.; LEE, S.J.; KIM, S.; RYU, H.W.; LEE, S.H.; CHOI, H.H.; CHEONG, I.W.; KIM, J.-H. Conducting Polymer Nanofiber Mats via Combination of Electrospinning and Oxidative Polymerization. *Polymer.* 2013, vol. 54, pp. 4155–4160. Available from: https://doi.org/10.1016/j.polymer.2013.05.072.

LASER-MEDIATED SYNTHESIS OF METAL NANOPARTICLES BY OPTICAL VORTEX BEAM

[1,2]Sabrin ABDALLAH, [1,2]Ondřej HAVELKA, [1,2]Jan BRAUN, [1]Rafael TORRES-MENDIETA

[1]Institute for Nanomaterials, Advanced Technologies and Innovation, Technical University of Liberec, Liberec, Czech Republic, EU, sabrin.abdallah@tul.cz, rafael.torres@tul.cz

[2]Faculty of Mechatronics, Informatics and Interdisciplinary Studies, Technical University of Liberec, Liberec, Czech Republic, EU

https://doi.org/10.37904/nanocon.2024.5030

Abstract

Controlling nanoparticles (NPs) morphology in laser-mediated synthesis remains a significant challenge. This study introduces an innovative approach for synthesizing metal NPs using an Optical Vortex Beam (OVB) instead of the conventional Gaussian Beam (GB) in femtosecond laser ablation in liquids (fs-LAL). The primary objective is to achieve NPs shape modification through an environmentally friendly method, eliminating the need for chemical reagents and producing ligand-free NPs suitable for catalytic and biomedical applications. The transformation from GB to OVB was accomplished using a series of waveplates that first converted the linearly polarized GB to a circularly polarized beam and subsequently to a spiral phase wavefront beam, known as OVB. A key innovation of this study is the design of a 3D-printed ablation cell, which generates a thin liquid layer and ensures continuous liquid flow over the metal target, optimizing the laser-ablation setup. This optimization not only enhances productivity but also significantly reduces operation time. Results indicate that using OVB provides superior control over NPs morphology compared to GB, resulting in NPs with improved optical properties, as confirmed by Scanning Transmission Electron Microscopy (STEM) and UV-Vis spectroscopy. This approach holds potential for nanocatalysis, offering a cleaner and more efficient method for NPs synthesis.

Keywords: Laser-mediated synthesis, metal nanoparticles, optical vortex beam, morphology, set-up optimization

1. INTRODUCTION

Laser ablation in liquids (LAL) has emerged as a novel and sustainable method for producing metal nanoparticles (NPs) of high purity. [1] However, a key challenge preventing LAL from reaching its full potential is the precise control of NPs morphology, including size, shape, and surface chemistry. [2, 3] Since the physico-chemical and optical properties of NPs are closely linked to their morphology, fine-tuning these characteristics is essential for optimizing their performance.

Plasmonic NPs (PNPs) like gold (Au) and silver (Ag) are particularly well-studied in this regard due to their unique optical properties, especially the localized surface plasmon resonance (LSPR), which can be enhanced through controlling their shape and structure. [4, 5] This ability to manipulate LSPR makes PNPs highly valuable in applications like sensing, drug delivery, and cancer therapy. [6–9]

As a result, controlling NPs morphology has been extensively studied through various synthetic routes. However, with the growing focus on environmentally friendly practices that align with the principles of green chemistry, LAL has gained significant attention. [10] By using light instead of hazardous chemicals to produce ligand-free NPs, LAL eliminates the need for chemical reagents and stabilizers. This makes laser-generated NPs particularly suitable for applications in fields such as catalysis, photonics, and biomedicine, where ultra-pure NPs are highly desirable. [11]

Although spherical NPs are the most commonly produced in LAL, the versatility of this method- ranging from the choice of target materials and liquid mediums to the modulation of laser parameters such as fluence, wavelength, pulse duration, and beam structure- has led to the exploration of various techniques for shape modification. [1, 12] Techniques such as altering solvents or applying external forces like electric fields, magnetic fields, or heat have been investigated. [13–17] However, these studies, predominantly utilize nanosecond lasers, require lengthy processes (such as suspension aging), result in modest production yield (100–150 mg/L), and generate bi-modal size distributions and/or various structures in a single experiment with some structures exceeding the 100 nm mark. Furthermore, the introduction of surfactants for shape modification contradicts the fundamental objective of LAL, which is to generate pure NPs.

Despite the numerous studies published on this topic, the employment of structurally-modified laser beams remains unexplored. Among the different possibilities for laser beam manipulation, generating an optical vortex beam (OVB) seems the most promising.[18] Due to the helical phase fronts and orbital angular momentum, OVB can capture and set particles into rotation around the beam axis which proved beneficial in various applications like optical tweezers development or quantum entanglement. [19] OVBs have been used for the creation of spiral-like structures on the surface of metals and for trapping and moving particles in liquids and on the surface of materials.[20, 21] In this context, using an OVB instead of a Gaussian Beam (GB) introduces a potentially direct and facile approach for the generation of morphologically modified NPs by LAL.

Moreover, this study considers the importance of the laser ablation setup in ensuring the delivery of a high intensity beam to the target's surface, as well as the experiments' reproducibility. Recent studies indicate that a thin liquid layer on the target's surface can prevent energy loss and enhance productivity up to 15 times compared to the conventional setup. [1, 22] Hence, the design of a 3D printed ablation cell that generates a thin liquid layer and enables a continuous liquid flow over the target's surface is implemented.

In summary, this study aims to compare the effects of GB and OVB on the morphology of Au and titanium (Ti) NPs synthesized by femtosecond LAL (fs-LAL) in a continuous liquid flow setup. By understanding how OVB influences NPs morphology across metals with different ablation thresholds (0.2 J/cm^2 for Au and 0.05 J/cm^2 for Ti) [23, 24], we hope to establish a new methodology for producing shape-controlled NPs that can be tuned for specific applications in catalysis, biomedicine, and beyond.

2. METHODOLOGY

Drawing inspiration from a water wall, the design of an ablation cell for the generation of a thin liquid layer was modeled in Autodesk Fusion 360 software and printed on an i3 MK3S+ 3D printer (Original Prusa) with PETG filament. To generate a continuous liquid flow, a peristaltic pump with 4/6mm silicon tubes operating at 200 ml/min was employed. The ablation cell was attached to two motorized linear stages (8MT193-100, Standa) working at 10 mm/s to ensure consistent ablation along the foil's surface by a 2-axis controller (8SMC5-USB-B9-2, Standa). For the ablation process, a metal foil was placed into the slit of the 3D printed cell as depicted in **Figure 1**, wherein a thin liquid layer was generated by a set of nozzles.

The ablation of an Au or a Ti foil (> 99.99% purity for Au and > 99.7% purity for Ti, Sigma Aldrich) was carried out by a fs-laser (Onefive Origami XP-S, NKT Photonics) operating at 5.1 W, 400 fs, with the wavelength centered at 1030 nm and a repetition rate of 1 MHz. The 3 mm linearly polarized Gaussian Beam (at FWHM) was converted into a circularly polarized GB by a quarter-waveplate (QWP, ThorLabs) placed at 45 °. An S-waveplate (SWP, WOPhotonics) positioned after the QWP converted the circularly polarized GB into an OVB (**Figure 1**). Subsequently, the OVB was expanded by a Galilean telescope (5X) prior to tight focusing with a 25.4 mm plano-convex lens to reach the highest beam intensity on the foil's surface. For experiments employing only GB, the converters were removed from the set-up. The foils were irradiated for 3 h, wherein 40 ml of demiH_2O (18.2 MΩ·cm), and later colloid, was pumped by the peristaltic pumps through a closed-loop system. After irradiation, the colloids were centrifuged for 20 min at 14500 rpm (MiniSpin plus centrifuge,

Eppendorf) to collect the produced NPs in a single 1.5 mL Eppendorf tubes. The samples were stored in a cool and dark place for later characterization.

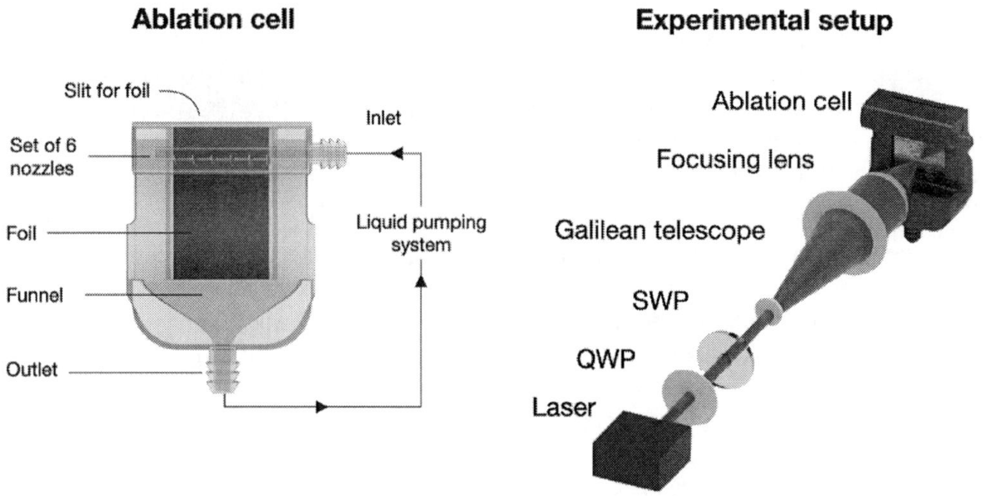

Figure 1 Ablation cell model in Fusion360 with key features explained (left) and the experimental setup for laser ablation of an Au foil with an optical vortex beam (right).

Each sample, labeled based on the type of foil and beam employed, was analyzed by an Ultraviolet–Visible (UV–Vis) spectroscope (Optima 2100Dv, Perkin Elmer) in the range of 220 – 950 nm with a 1 nm resolution to determine the NPs optical properties. Moreover, Scanning Transmission Electron Microscopy (STEM) images were acquired by a focused ion beam microscope (HELIOS 5 PFIM CXe, ThermoFisher Scientific) at 30 kV equipped with a STEM detector to assess the shape and size of the NPs. A drop of the sample was deposited on a formvar/carbon copper grid and dried at room temperature. The size analysis was performed by manually measuring the diameter of 500 NPs.

3. RESULTS AND DISCUSSION

Notably, the visual appearance of the colloids immediately after synthesis revealed significant differences between the samples. The Au-GB colloid exhibited a reddish-purple color, while the Au-OVB colloid displayed a bluish-purple color. Previous studies suggest that Au NPs typically appear red due to their spherical shape, but variations in NPs size and shape can alter the colloid's color. Specifically, Au nanorods, which possess two plasmonic peaks, tend to exhibit a blue coloration, as an additional absorbance peak emerges in the red or near-infrared region. This suggests that the different optical properties observed in our samples could be attributed to variations in NPs morphology.

Thus, based on these observations and the information obtained from the absorbance spectra we can gain an understanding of the shape and size of the NPs, especially in the case of Au NPs. As displayed in **Figure 2**, sample Au-GB has a plasmonic peak centered at 526 nm suggesting the presence of spherical NPs. In contrast, for sample Au-OVB there is a red-shift of the plasmonic peak to 546 nm suggesting the formation of larger spherical NPs or possibly a collective behavior of smaller NPs that are in proximity and acting as a single entity. Moreover, this sample has an additional peak around 633 nm which is an indication of the formation of non-spherical NPs like nanoflowers [25] or nanostars [26], or some degree of agglomeration. However, in both Au spectra a more intense peak is visible at smaller wavelengths (~220 nm). Based on theoretical calculation obtained from the open access Mie calculator from Prof. Lucien Saviot [27] this would suggest the presence of predominantly small NPs, below 10 nm.

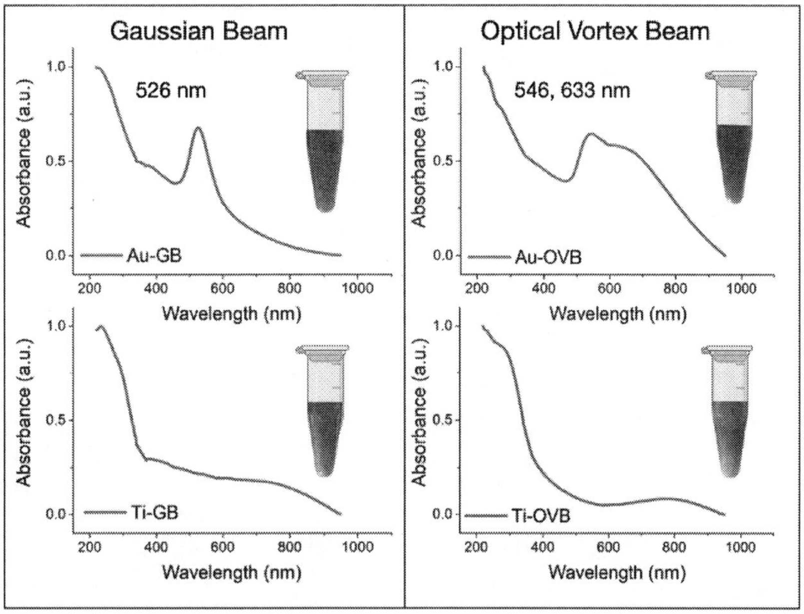

Figure 2 Absorbance spectra of Au and Ti NPs synthesized by a GB and an OVB in the wavelength range of 220 – 950 nm (normalized). The inset images depict the color of each colloid.

In the case of Ti NPs, Ti-GB colloid appeared to have a bluish-grey color, while Ti-OVB colloid appeared grey. In contrast to Au, Ti oxidizes more readily, thus, it is most likely the case that we obtained Ti oxide NPs since the ablation of a Ti foil was carried out in an aqueous environment. Typically, TiO_2 colloid appears white or blue. Sample Ti-GB has an absorbance peak at 230 nm accompanied by a long, broad tail spreading all the way to the near infra-red region which could suggest a wide range of NPs sizes. In the case of Ti-OVB there is an additional peak at 300 nm, which could indicate the presence of larger NPs.

According to the STEM images in **Figure 3**, although both Au-GB and -OVB samples contain spherical NPs, they also contain non-spherical NPs with diameters below 10 nm. This explains the plasmonic behavior observed in the UV region of the spectrum evaluated by the UV-Vis spectroscopy. The Au-GB sample contains predominantly small NPs with a diameter distribution centered at 8.9 nm. Additionally, this sample is characterized by “nanowebs” that create elongated necks linking together NPs. On the other hand, Au-OVB NPs have a smaller diameter centered at 6.9 nm. More importantly, in this case, one can observe elongated NPs that are separated and not forming the nanowebs as in the Au-GB sample. This can be attributed to the fact that the intensity distributions of the GB and OVB are not comparable. GB intensity is distributed in a circle with the highest intensity in the center, in contrast, OVB's intensity is distributed in a ring. Thus, each beam delivers different energy to the foil's surface. Even when using the maximum power, we are ablating at a lower fluence with an OVB (excluding the absorption and scattering of the beam along its path by the optical elements). According to Kabashin et al. [28] who performed fs-LAL of Au in water employing various fluences, such small NPs are observed at a lower fluence (<100 J/cm^2). This phenomenon relates to the different mechanisms of NPs formation in LAL as previously explained by Shih et. al. [29] whom, through molecular dynamic calculations could determine that the immediate laser-matter interaction, preceding the ablation process, enables the solid evaporation forming a solid-liquid mixing region which rapidly condensates forming small NPs. In contrast, the larger particles are formed during the nucleation process after ablation. Thus, according to these studies we can infer that the laser intensity reduction excreted by the OVB favors the solid evaporation process, and with it, small NPs formation. However, as Au holds a massive thermal conductivity (318 W/m·K), compared to Ti (11.4 W/m·K), their NPs formation mechanisms may differ.[30] In particular, when experiencing condensation, higher thermal conductivity leads to rapid cooling and with this, the inability to reach a stable spherical shape, which does not occur with Ti that seems to hold sufficient energy to coalesce until the spheres are formed. When coalescence takes place after the ablation, a high thermal conductivity would be detrimental towards sphere formation, especially when using energies close to the solid targets' ablation threshold, indicating minimal material detachment intervening in the NPs formation leading to metastable shapes as those found in Au that seem to be a middle way to spheres. Moreover, since the OVB carries orbital angular momentum, it can move material immediately after detachment from the solid target. This motion of the detached Au pieces, combined with their high thermal conductivity, appears to influence the coalescence process, and thus, non-spherical particles are formed. In contrast, the GB does not immediately move the detached Au pieces, allowing them to melt and form nanowebs.

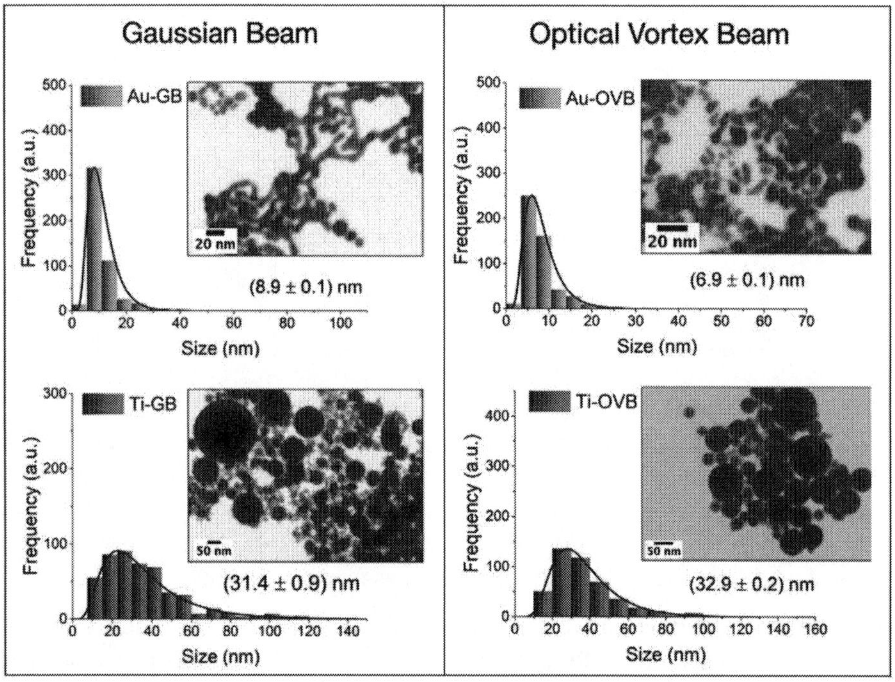

Figure 3 Size distributions of Au and Ti NPs synthesized by a GB and an OVB with log-normal fitting. STEM images of the NPs characteristic of each sample are included.

4. CONCLUSION

In conclusion, this study demonstrates significant differences in the optical and morphological properties of Au and Ti NPs synthesized using GB and OVB in LAL. The Au-OVB sample exhibited a bluish-purple color and a red-shifted plasmonic peak, suggesting the formation of non-spherical NPs, such as nanostars or nanoflowers, while the Au-GB sample displayed a reddish-purple color indicative of spherical NPs. STEM images confirmed these variations in Au-OVB, producing smaller, elongated NPs, however, Au-GB images revealed the formation of nanowebs in addition to larger spherical NPs. These differences are attributed to the unique energy distribution of the OVB, which promotes smaller NPs formation through solid evaporation mechanisms. In the case of Ti, both samples displayed characteristics of Ti oxide, with Ti-GB showing a wider range of NPs sizes and Ti-OVB forming larger particles. The findings highlight the influence of thermal conductivity and beam type on NPs formation, particularly in the case of Au, where high thermal conductivity combined with OVB's orbital angular momentum reduces coalescence and favors non-spherical shapes. While this study enhances understanding of NPs synthesis via LAL, further research is needed to optimize laser parameters and explore how these mechanisms can be applied to other materials.

ACKNOWLEDGEMENTS

The research presented in this article was supported by the Internal Grant of the Technical University of Liberec (SGS-2024-3424).

REFERENCES

[1] ZHANG, Dongshi, Bilal GÖKCE and Stephan BARCIKOWSKI. Laser Synthesis and Processing of Colloids: Fundamentals and Applications. *Chemical Reviews*. 2017, vol. 117, no. 5, pp. 3990–4103. ISSN 0009-2665, 1520-6890.

[2] AMANS, David, Weiping CAI and Stephan BARCIKOWSKI. Status and demand of research to bring laser generation of nanoparticles in liquids to maturity. *Applied Surface Science*. 2019, vol. 488, pp. 445–454. ISSN 01694332.

[3] AMENDOLA, Vincenzo and Moreno MENEGHETTI. What controls the composition and the structure of nanomaterials generated by laser ablation in liquid solution? *Phys. Chem. Chem. Phys.* 2013, vol. 15, no. 9, pp. 3027–3046. ISSN 1463-9076, 1463-9084.

[4] PETRYAYEVA, Eleonora and Ulrich J. KRULL. Localized surface plasmon resonance: Nanostructures, bioassays and biosensing—A review. *Analytica Chimica Acta*. 2011, vol. 706, no. 1, pp. 8–24. ISSN 00032670.

[5] ZHANG, Junxi and Lide ZHANG. Nanostructures for surface plasmons. *Advances in Optics and Photonics*. 2012, vol. 4, no. 2, 157. ISSN 1943-8206.

[6] PLOU, Javier et al. Prospects of Surface-Enhanced Raman Spectroscopy for Biomarker Monitoring toward Precision Medicine. *ACS Photonics*. 2022, vol. 9, no. 2, pp. 333–350. ISSN 2330-4022, 2330-4022.

[7] VERSIANI, Alice F. et al. Gold nanoparticles and their applications in biomedicine. *Future Virology*. 2016, vol. 11. no. 4, pp. 293–309. ISSN 1746-0794, 1746-0808.

[8] BUCHARSKAYA, Alla B. et al. Photothermal and Photodynamic Therapy of Tumors with Plasmonic Nanoparticles: Challenges and Prospects. *Materials*. 2022, vol. 15. no. 4, 1606. ISSN 1996-1944.

[9] TAN, Peng, HeSheng LI, Jian WANG a Subash C.B. GOPINATH. Silver nanoparticle in biosensor and bioimaging: Clinical perspectives. *Biotechnology and Applied Biochemistry*. 2020, vol. 68, no. 6, pp. 1236–1242. ISSN 0885-4513, 1470-8744.

[10] HAVELKA, Ondřej et al. On the Use of Laser Fragmentation for the Synthesis of Ligand-Free Ultra-Small Iron Nanoparticles in Various Liquid Environments. *Nanomaterials*. 2021, vol. 11, no. 6, 1538. ISSN 2079-4991.

[11] ETTEL, David et al. Laser-synthesized Ag/TiO nanoparticles to integrate catalytic pollutant degradation and antifouling enhancement in nanofibrous membranes for oil–water separation. *Applied Surface Science*. 2021, vol. 564, 150471. ISSN 01694332.

[12] SUBHAN, Abdul, Abdel-Hamid Ismail MOURAD and Yarub AL-DOURI. Influence of Laser Process Parameters, Liquid Medium, and External Field on the Synthesis of Colloidal Metal Nanoparticles Using Pulsed Laser Ablation in Liquid: A Review. *Nanomaterials*. 2022, vol. 12, no. 13, 2144. ISSN 2079-4991.

[13] BESNER, Sébastien, Andrei V. KABASHIN, Françoise M. WINNIK and Michel MEUNIER. Synthesis of Size-Tunable Polymer-Protected Gold Nanoparticles by Femtosecond Laser-Based Ablation and Seed Growth. *The Journal of Physical Chemistry C*. 2009, vol. 113, no. 22, pp. 9526–9531. ISSN 1932-7447, 1932-7455.

[14] FRIAS BATISTA, Laysa M. et al. Kinetic Control of $[AuCl_4]^-$ Photochemical Reduction and Gold Nanoparticle Size with Hydroxyl Radical Scavengers. *The Journal of Physical Chemistry B*. 2019, vol. 123, no. 33, pp. 7204–7213. ISSN 1520-6106, 1520-5207.

[15] LIN, X. Z. et al. Synthesis of CuO Nanocrystals and Sequential Assembly of Nanostructures with Shape-Dependent Optical Absorption upon Laser Ablation in Liquid. *The Journal of Physical Chemistry C*. 2009, vol. 113, no. 40, pp. 17543–17547. ISSN 1932-7447, 1932-7455.

[16] TSUJI, Takeshi et al. Morphological changes from spherical silver nanoparticles to cubes after laser irradiation in acetone–water solutions via spontaneous atom transportation process. *Colloids and Surfaces A: Physicochemical and Engineering Aspects*. 2017, vol. 529, pp. 33–37. ISSN 09277757.

[17] PIOTTO, Valentina, Lucio LITTI and Moreno MENEGHETTI. Synthesis and Shape Manipulation of Anisotropic Gold Nanoparticles by Laser Ablation in Solution. *The Journal of Physical Chemistry C*. 2020, vol. 124, no. 8, pp. 4820–4826. ISSN 1932-7447, 1932-7455.

[18] RUBINSZTEIN-DUNLOP, Halina et al. Roadmap on structured light. *Journal of Optics*. 2017, vol. 19, no. 1, 013001. ISSN 2040-8978, 2040-8986.

[19] SHEN, Yijie et al. Optical vortices 30 years on: OAM manipulation from topological charge to multiple singularities. *Light: Science & Applications* [online]. 2019, vol. 8, no. 1, 90. ISSN 2047-7538. Available from: doi:10.1038/s41377-019-0194-2

[20] LEHMUSKERO, Anni et al. Laser Trapping of Colloidal Metal Nanoparticles. *ACS Nano*. 2015, vol. 9, no. 4, pp. 3453–3469. ISSN 1936-0851, 1936-086X.

[21] TOYODA, Kohei et al. Using Optical Vortex To Control the Chirality of Twisted Metal Nanostructures. *Nano Letters*. 2012, vol. 12, no. 7, pp. 3645–3649. ISSN 1530-6984, 1530-6992.

[22] MESSINA, Gabriele C. et al. Pulsed laser ablation of a continuously-fed wire in liquid flow for high-yield production of silver nanoparticles. *Phys. Chem. Chem. Phys.* 2013, vol. 15, no. 9, pp. 3093–3098. ISSN 1463-9076, 1463-9084.

[23] KRÜGER, Jörg, Daniela DUFFT, Robert KOTER and Andreas HERTWIG. Femtosecond laser-induced damage of gold films. *Applied Surface Science*. 2007, vol. 253, no. 19, pp. 7815–7819. ISSN 01694332.

[24] WANG, Liming et al. Surface chemistry of gold nanorods: origin of cell membrane damage and cytotoxicity. *Nanoscale*. 2013, vol. 5, no. 18, 8384. ISSN 2040-3364, 2040-3372.

[25] WOŹNIAK, Anna et al. Size and shape-dependent cytotoxicity profile of gold nanoparticles for biomedical applications. *Journal of Materials Science: Materials in Medicine*. 2017, vol. 28, no. 6, 92. ISSN 0957-4530, 1573-4838.

[26] MINATI, Luca, Filippo BENETTI, Andrea CHIAPPINI and Giorgio SPERANZA. One-step synthesis of star-shaped gold nanoparticles. *Colloids and Surfaces A: Physicochemical and Engineering Aspects*. 2014, vol. 441, pp. 623–628. ISSN 09277757.

[27] SAVIOT, Lucien. Mie scattering calculator. *French National Centre for Scientific Research*. [viewed 2024-01-10]. Available from: https://saviot.cnrs.fr/mie/index.en.html

[28] KABASHIN, A. V. and M. MEUNIER. Synthesis of colloidal nanoparticles during femtosecond laser ablation of gold in water. *Journal of Applied Physics*. 2003, vol. 94, no. 12, pp. 7941–7943. ISSN 0021-8979, 1089-7550.

[29] SHIH, Cheng-Yu et al. Two mechanisms of nanoparticle generation in picosecond laser ablation in liquids: the origin of the bimodal size distribution. *Nanoscale*. 2018, vol. 10, no. 15, pp, 6900–6910. ISSN 2040-3364, 2040-3372.

[30] TOULOUKIAN, Y S, R W POWELL, C Y HO and P G KLEMENS. *Thermophysical properties of matter - the TPRC data series. Volume 1. Thermal conductivity - metallic elements and alloys. (Reannouncement). Data book*. 1970.

OPTICAL SPECTRA OF THIN INTRINSIC AND AL DOPED NANOCRYSTALLINE ZNO THIN LAYERS DEPOSITED BY PULSED LASER DEPOSITION ON FUSED SILICA GLASS SUBSTRATES

Zdenek REMEŠ[1], Neda NEYKOVA[1,2], Naini JAIN[2], Rupendra Kumar SHARMA[2], Jakub HOLOVSKÝ[1,2], Egor UKRAINTSEV[2], Julia MIČOVÁ[3] and Hua Shu HSU[4]

[1]*Czech Academy of Sciences, Institute of Physics, Praha, Czech Republic, EU, remes@fzu.cz,*

[2]*Czech Technical University, Faculty of Electrical Engineering, Prague, Czech Republic, EU, neykoned@fel.cvut.cz, sharmrup@fel.cvut.cz, jainnain@fel.cvut.cz, jakub.holovsky@fel.cvut.cz, ukraiego@fel.cvut.cz*

[3]*Slovak Academy of Sciences, Institute of Chemistry, Bratislava, Slovakia, EU, chemjumi@savba.sk*

[4]*National Pingtung University, Department of Applied Physics, Taiwan, hshsu@mail.nptu.edu.tw*

https://doi.org/10.37904/nanocon.2024.5029

Abstract

Zinc oxide (ZnO) is a low-cost and environmentally friendly material with unique optical properties and a variety of nano and microstructures imposing challenges for energy conversion, scintillators, photocatalytic wastewater treatment, electrochemical energy storage, or sensing applications. In this work, the nominally undoped and Al-doped nanocrystalline ZnO thin layers were pulsed laser deposited (PLD) on fused silica glass. The samples were characterized by photothermal deflection (PDS) and photoluminescence (PLS) spectroscopy. Excitons were observed at low temperature in both samples at about 3.4 eV. The intrinsic samples show the red PL in near infrared region at about 1.9 eV related to zinc vacancies. The results show that optical properties can be evaluated from optical spectra of very thin nanocrystalline ZnO layers with the thickness below 100 nm.

Keywords: Pulse laser deposition, wide band gap semiconductors, optical properties, doping

1. INTRODUCTION

Zinc oxide is a low-cost and environmentally friendly material with unique optical properties and a variety of nano and microstructures imposing challenges for energy conversion, scintillators, photocatalytic wastewater treatment, electrochemical energy storage or sensing applications [1-3]. A number of typical dopant elements such as F, B, Al, Ga, In and Sn have been used so far to produce conducting ZnO films. It is worth underlining that among all the group III elements, Al is a cheap, abundant and non-toxic material [4]. Thus Al-doped ZnO films are prominent, low-cost substitutes for indium tin oxide films as transparent conducting films in the photovoltaic and sensor applications [5-7].

In this paper, we use the PLD setup [8] to deposit very thin nanocrystalline intrinsic and Al-doped ZnO films on fused silica glass substrates to prove that the optical parameters can be evaluated even from the optical spectra of very thin polycrystalline ZnO films with thickness below 100 nm.

1.1 Pulse laser deposition (PLD)

Polycrystalline ZnO and Al-doped ZnO layers were prepared on fused silica substrates by pulsed laser deposition (PLD) setup from TSST B.V. equipped with KrF (λ = 248 nm) excimer laser COMPex 50. After reaching a base pressure of 10^{-4} Pa, oxygen was introduced with process pressures of 10 Pa with a flow rate of 10 cm^3/min. The substrates were placed into a rotating holder (5 rpm). The excimer laser was focused to a spot of the area of 2 mm^2 and set up at the repetition rate of 20 Hz for target ablation. For the deposition from the stoichiometric targets, the pulse energy (measured inside the chamber) was kept at about 24 mJ giving the fluence of 1.2 J/cm^2, respectively. The number of pulses was 20 000 (sample A) and 40 000 (sample B) throughout the experiment, yielding in both cases layers with thickness about 80 nm, see **Table 1**. After the deposition, samples were cooled inside the chamber. We previously discovered a strong effect of the atmosphere during this cool-down step [10].

Table 1 The nanocrystalline ZnO films deposited by PLD on fused silica glass substrates, including the thickness (*d*), the dielectric constant (ε_∞), the parameters of the Lorenc (A, E_0 and E_1) and Drude models (E_p and E_2) as well as the optical absorption coefficient at 1 eV (α)

Sample	*d* (nm)	ε_∞	*A*	E_0(eV)	E_1(eV)	E_p(eV)	E_2(eV)	α (cm^{-1})
A (undoped)	85	3.32	1.21	3.61	0.17	0.18	0.59	< 10
B (Al-doped)	77	3.28	1.41	3.82	0.05	0.39	0.11	730

1.2 Photothermal deflection spectroscopy (PDS)

The transmittance, reflectance and absorptance spectra were measured simultaneously in the 300–1400 nm spectral range by photothermal deflection spectroscopy (PDS) setup with 150 W Xe lamp, SpectraPro-150 monochromator (150-mm focal length, f/4-aperture, slits 1/1mm) equipped with two gratings: a UV holographic (1200 /mm) and a ruled (600 /mm) blazed at 500 nm. The spectral resolution was 5 nm with the UV holographic grating and 10 nm with the ruled grating. Samples were immersed into liquid (Florinert FC72) to measure the relative temperature of the illuminated sample independently for selected photon energies using deflection of probe laser beam. The spectra were spectrally calibrated by measuring PDS of a black carbon sample.

1.3 Photoluminescence spectroscopy (PLS)

PL spectra were measured at low temperature in the spectral range 360-700 nm with a spectral resolution of 2 nm using a closed He-circle cryostat OptistatDry (Oxford Instruments, UK) and a focused UV LED (1 mW Thorlabs M340F3) equipped with narrow bandpass (BP340) and long pass (LP350) optical filters, operating in ac mode at a frequency of 307 Hz. The emitted light was collected and focused onto the 1 mm wide input slit of double gratings SPEX 1672 monochromator (220-mm focal length, f/4-aperture, 1200/mm grating, 1/1 mm slits, spectral resolution 2 nm) equipped with Peltier cooled red sensitive photomultiplier (PMT) and a current preamplifier with 10^5 V/A transimpedance (Ametex 5182) to be measured at selected photon energies independently by a lock-in amplifier (Ametex 5105) with ac noise level about 10 pA (1 μV). The whole setup was spectrally calibrated with Oriel #63358 Quartz Tungsten halogen lamp and converted from wavelength to energy scale for quantitative analysis.

2. RESULTS AND DISCUSSION

Figure 1 shows the transmittance (T), reflectance (R) and absorptance (A) spectra of the Al-doped ZnO layer (sample B). Absorptance spectrum was measured using PDS and put into the absolute scale using 1-*R*-*T* spectrum. It should be noted that unlike PDS, which measures *A* directly, the 1-*R*-*T* spectrum cannot be used to evaluate *A* below 1%. The dielectric function ε (the complex index of refraction $n + ik$) as a function of energy *E* and the film thickness *d* were evaluated from the reflectance spectra using Lorenc and Drude models [10] included in the commercial *FimWizard*$^{(c)}$ software (https://sci-soft.com/product/film-wizard/), see Equation 1 and **Table 1**.

$$\varepsilon(E) = (n + ik)^2 = \varepsilon_\infty \left(1 + \frac{A^2}{E_0^2 - E^2 + iE_1^2} + \frac{E_p^2}{E(E - iE_2)}\right) \tag{1}$$

The fit of *R* spectra is shown in **Figure 2**. **Table 1** also shows the shifts of the central position of the Lorenc peak (E_0) and the plasma energy (E_p) as expected. Once the index of refraction $n(E)$ and the film thickness d were known, the optical absorption coefficient $\alpha(E)$ was evaluated from PDS spectra independently at each measured photon energy E [11], see **Figure 4**.

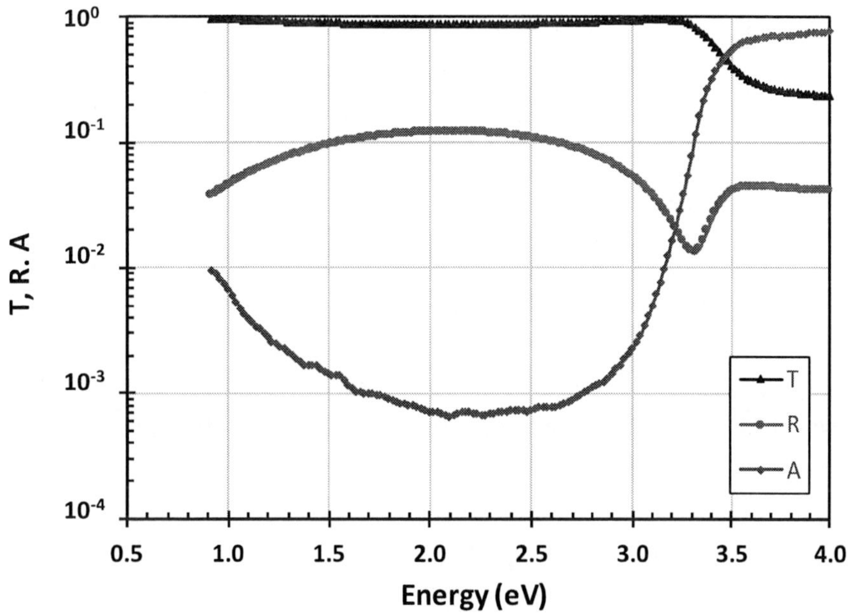

Figure 1 The transmittance (*T*), reflectance (*R*) and absorptance (*A*) spectra of the Al-doped ZnO layer (sample B)

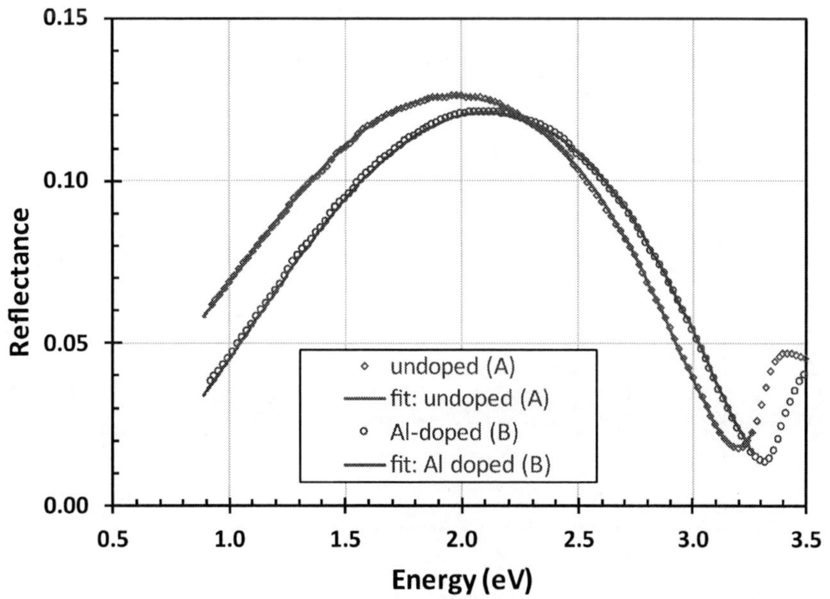

Figure 2 The measured (points) and fitted (solid line) reflectance spectra of the undoped (sample A) and Al-doped (sample B) ZnO layers

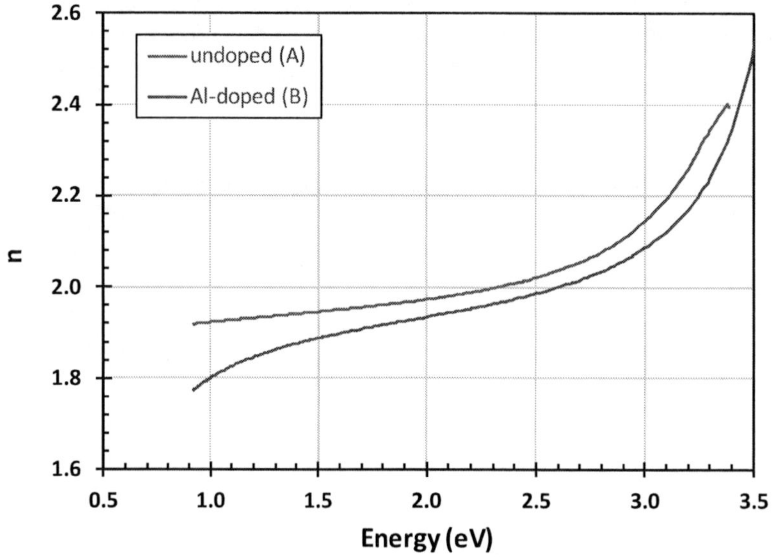

Figure 3 The spectra of the index of refraction of the nominally undoped (A) and Al-doped (B) ZnO layers

The direct bad gap of ZnO [12] is not clearly visible in **Figure 4** because of the Urbach edge [13], localized states below the optical absorption edge related to deep defects and the free carrier absorption in IR region in the case of Al-doped sample. The perturbation imposed by Al atom incorporation leads to the atomic relaxation. This phenomenon influences atomic rearrangement near Al impurity and generates a free electron in the conduction band, which can be considered as a large radius electron polaron increasing the n-type electrical conductivity in the crystal in agreement with the known experimental data [14]. The Burstein–Moss shift, in which the apparent band gap of a semiconductor is increased because of populated states close to the conduction band is also observed [15].

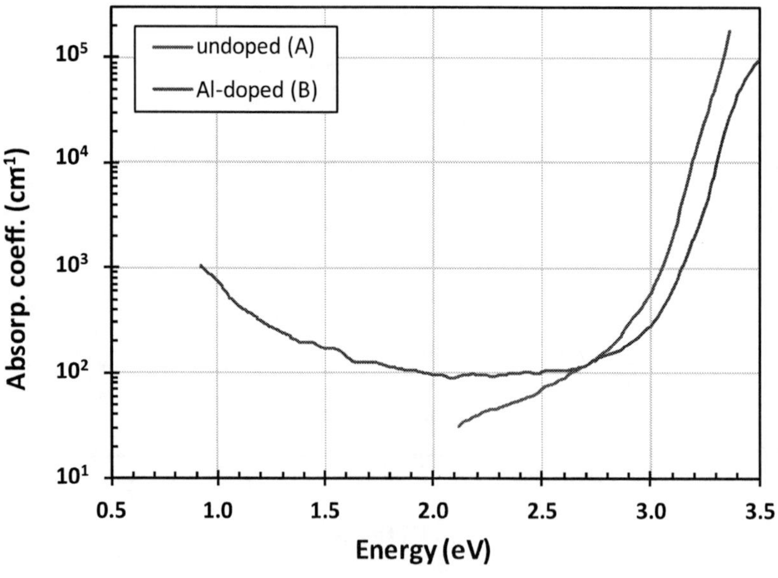

Figure 4 Optical absorption coefficient spectra of the intrinsic and Al-doped ZnO layers. The optical absorptance of the undoped ZnO was below 2.1 eV too low to evaluate the absorption coefficient

There was no measurable PL at room temperature. **Figure 5** shows the PL spectra of the intrinsic and Al doped ZnO layers measured at low temperature under UV LED excitation (340 nm, 1 mW). Excitons observed in both samples at low temperature at about 3.4 eV are more pronounced in Al doped ZnO layer. The intrinsic sample shows the vacancy related PL in near infrared region at about 1.9 eV as it was explained in our previous paper [16].

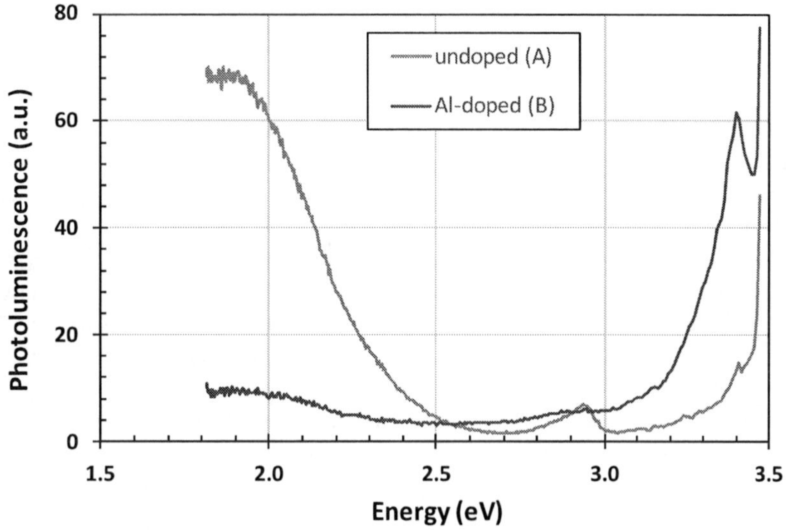

Figure 5 Photoluminescence spectra of the intrinsic (sample A) and the Al-doped ZnO layers (sample B) measured at low temperature (nominally 4 K) under UV LED excitation

3. CONCLUSION

The thin intrinsic and Al-doped nanocrystalline ZnO layers were deposited by PLD on fused silica substrates. The effect of Al doping on the optical properties was investigated. The complex index of refraction and the film thickness were evaluated from the reflectance spectra using Lorenc and Drude models and the commercial *FimWizard* software. The optical absorption coefficient was evaluated from PDS spectra independently. Excitons were observed in photoluminescence spectra at low temperature in both samples at about 3.4 eV. The PLS in near infrared region at about 1.9 eV is related to vacancies. The results show that optical properties can be evaluated from optical spectra of very thin nanocrystalline ZnO layers with the thickness below 100 nm.

ACKNOWLEDGEMENTS

The work was supported by the Czech Science Foundation project 24-10607J, Ministry of Education, Youth and Sports of Czech Republic (MEYS) Danube project 8X23025, the Mobility Plus project SAV-23-13 (CAS-SAS-2022-08) and by Operational Programme Johannes Amos Comenius financed by European Structural and Investment Funds and the Czech Ministry of Education, Youth and Sports Energy conversion and storage – CZ.02.01.01/00/22_008/0004617.

Conceptual design and project supervision (Z. Remeš, J. Mičová, H.S. Hsu, E. Ukraintsev), ZnO deposition growth (N. Neyková, N. Jain, R. K. Sharma), PL measurements (Z. Remeš), interpretation (J. Holovský) .

REFERENCES

[1] KLINGSHIRN, C. F., ed. *Zinc Oxide: from fundamental properties towards novel applications*. B.m.: Springer, 2010. Springer series in materials science, 120. ISBN 978-3-642-10576-0.

[2] KLINGSHIRN, Claus, J. FALLERT, H. ZHOU, J. SARTOR, C. THIELE, F. MAIER-FLAIG, D. SCHNEIDER a H. KALT. 65 years of ZnO research - old and very recent results. *physica status solidi (b)* 2010, vol. 247, no. 6, pp. 1424–1447. Available from: https://doi.org/10.1002/pssb.200983195

[3] KLINGSHIRN, Claus F. *Semiconductor optics*. 3rd ed. Berlin New York: Springer, 2007. ISBN 978-3-540-38347-5.

[4] GÜRBÜZ, Osman, İsmail KURT, Serkan ÇALIŞKAN a Sadık GÜNER. Influence of Al concentration and annealing temperature on structural, optical, and electrical properties of Al co-doped ZnO thin films. *Applied Surface Science* 2015, vol. 349, pp. 549–560. Available from: https://doi.org/10.1016/j.apsusc.2015.04.233

[5] HAO, Ming, Xianhang LU, Fei SUN, Yunhe FU, Yaoshuai BA, Yuanhua XIE a Kun LIU. Optoelectronic effect of Al-based buffer layers on Al-doped ZnO thin transparent conductive films on flexible substrates. *Optical Materials* 2024, vol .157, 116179. Available from: https://doi.org/10.1016/j.optmat.2024.116179

[6] BENAMARA, Majdi, Pedro RIVERO-ANTÚNEZ, Hassen DAHMAN, Manel ESSID, Souhir BOUZIDI, Marc DEBLIQUY, Driss LAHEM, Víctor MORALES-FLÓREZ, Luis ESQUIVIAS, José P. B. SILVA a Lassaad EL MIR. Selective and rapid detection of acetone using aluminum-doped zno-based sensors. *Journal of Sol-Gel Science and Technology* 2023, vol. 108, no. 1, pp. 13–27. Available from: https://doi.org/10.1007/s10971-023-06197-5

[7] YOO, Ran, Andreas T. GÜNTNER, Yunji PARK, Hyun Jun RIM, Hyun-Sook LEE a Wooyoung LEE. Sensing of acetone by Al-doped ZnO. *Sensors and Actuators B: Chemical*. 2019, vol. 283, pp. 107–115. Available from: https://doi.org/10.1016/j.snb.2018.12.001

[8] HOLOVSKÝ, Jakub, Eva HORYNOVÁ, Lukáš HORÁK, Katarína RIDZOŇOVÁ, Zdeněk REMEŠ, Lucie LANDOVÁ a Rupendra Kumar SHARMA. Pulsed laser deposition of high-transparency molybdenum oxide thin films. *Vacuum* 2021, vol. 194, 110613. Available from: https://doi.org/10.1016/j.vacuum.2021.110613

[9] HORYNOVÁ, Eva, Oleksandr ROMANYUK, Lukáš HORÁK, Zdeněk REMEŠ, Brianna CONRAD, Amalraj PETER AMALATHAS, Lucie LANDOVÁ, Jana HOUDKOVÁ, Petr JIŘÍČEK, Tomáš FINSTERLE a Jakub HOLOVSKÝ. Optical characterization of low temperature amorphous MoOx, WOX, and VOx prepared by pulsed laser deposition. *Thin Solid Films*. 2020, vol. 693, 137690. Available from: https://doi.org/10.1016/j.tsf.2019.137690

[10] TOMPKINS, Harland G. *Spectroscopic ellipsometry and reflectometry: a user's guide*. New York: Wiley, 1999. ISBN 0-471-18172-2.

[11] FOWLES, Grant R. *Introduction to modern optics*. 2nd ed., Dover ed. New York: Dover Publications, 1989. ISBN 978-0-486-65957-2.

[12] PANKOVE, Jacques I. *Optical processes in semiconductors*. Unabridged republication, with slight corr. Mineola [NY]: Dover, 1975. ISBN 978-0-486-60275-2.

[13] URBACH, Franz. The Long-Wavelength Edge of Photographic Sensitivity and of the Electronic Absorption of Solids. *Physical Review* 1953, vol. 92, no. 5, pp. 1324–1324. Available from: https://doi.org/10.1103/PhysRev.92.1324

[14] MALDONADO, Frank a Arvids STASHANS. Al-doped ZnO: Electronic, electrical and structural properties. *Journal of Physics and Chemistry of Solids*. 2010, vol. 71, no. 5, pp. 784–787. Available from: https://doi.org/10.1016/j.jpcs.2010.02.001

[15] BURSTEIN, Elias. Anomalous Optical Absorption Limit in InSb. *Physical Review*. 1954, vol. 93, no. 3, pp. 632–633. Available from: https://doi.org/10.1103/PhysRev.93.632

[16] BURYI, M., V. BABIN, Y.-Y. CHANG, Z. REMEŠ, J. MIČOVÁ a D. ŠIMEK. Influence of precursor age on defect states in ZnO nanorods. *Applied Surface Science* 2020, vol. 525, 146448. Available from: https://doi.org/10.1016/j.apsusc.2020.146448

STRUCTURE COMPATIBILITY OF GRAPHITE AND PHASES FORMED FROM ALUMINUM OR MAGNESIUM PHYLLOSILICATES

[1]Jonáš MOLEK, [1]Jonáš TOKARSKÝ

[1]*VSB - Technical University of Ostrava, Ostrava, Czech Republic, EU, jonas.molek.st@vsb.cz, jonas.tokarsky@vsb.cz*

https://doi.org/10.37904/nanocon.2024.5031

Abstract

Calcination of phyllosilicate/organics nanocomposites in inert atmosphere leads to in situ formation of graphitic carbon layers on silicate phases. Electrically conductive material (ceramics) is thus obtained. Despite an intensive research in this area, attention has not yet been paid to carbon/silicate interfaces in these systems. This study is focused on structure compatibility of graphitic carbon and the most abundant phases formed by transformation of aluminum phyllosilicates or magnesium phyllosilicates, i.e. cristobalite, mullite, forsterite, and protoenstatite. The aim is to find whether the phases can or cannot have a significant influence on the graphitic structure formation, and if the type of phyllosilicate (aluminum or magnesium) must be taken into account. Structure compatibility is determined by original method calculating the overlaps of atomic pairs carbon/another element. Carbon atoms lie in graphite(001) plane, atoms of another element lie in the given phase(hkl) plane. Different crystallographic planes of phases formed from the original phyllosilicates exhibit various structure compatibility with the graphite. The average number of overlaps of individual graphite(001)/phase(hkl) systems revealed the highest structure compatibility for mullite (M) followed by forsterite (F), protoenstatite (P) and cristobalite (C), namely M(102) > F(001) > P(100) > C(100). This sequence suggests that aluminum phyllosilicates, from which M is formed, could be a more suitable input component compared to magnesium phyllosilicates, from which F and P are formed. The lower number of overlaps for C compared to other phases indicates its small contribution to the overall structure compatibility.

Keywords: Structure compatibility, graphitic carbon, silicate, aluminosilicate, magnesium silicate

1. INTRODUCTION

Graphitic carbon can be prepared by pyrolysis of macromolecules/phyllosilicate nanocomposites [1-17]. The phyllosilicate component is transformed into cristobalite and, depending on the type of phyllosilicate, also into mullite (Al-rich) or forsterite and/or protoenstatite (Mg-rich) [5,9,15]. While possible applications of these pyrolyzed nanocomposites have been investigated (electrical conductivity [6,7,12,13], hydrogen storage [9], wastewater treatment [10] or coatings absorbing solar radiation [11] to name a few), the study of the structure is usually focused only on the determination of the formed graphitic carbon. Moreover, the carbon is often liberated from the pyrolyzed nanocomposite by etching [1-5,10,17]. Little is known about the mutual interactions of the graphite with different crystallographic planes of the inorganic phases in these pyrolyzed nanocomposites. The question of the interaction remains at the edge of interest in the field of research on these nanocomposites. Our previous work revealed that the determination of mutual interaction of two phases from optimized molecular model can be successfully replaced by a simple method quantifying the interaction by the number of overlapping atoms of two crystallographic planes, each of which belongs to one of the phases [18]. The higher the number of overlaps, the more the given crystallographic planes are structurally compatible [18]. In this study, the method was used to determine the structure compatibility of the basal graphite(001) plane and various crystallographic planes of cristobalite, mullite, protoenstatite, and forsterite. Knowledge of the compatibility of in situ prepared graphitic carbon and silicate phases can be useful for the preparation of electrically conductive ceramics from these pyrolyzed nanocomposites.

2. MATERIALS AND METHODS

2.1 Structure compatibility

Determining the structure compatibility of two phases is based on a purely geometric point of view [18]. Two crystallographic planes, each from one phase, are placed in the same geometric plane so that one atom of each plane lies at the origin of the coordinates. In this study, one crystallographic plane was always (001) plane of graphite (G), the other crystallographic plane was (hkl) plane of cristobalite (C), mullite (M), forsterite (F) and protoenstatite (P). The calculation is performed using an algorithm that quantifies the amount of overlaps of carbon atoms from G(001) plane with another elements from C(hkl), M(hkl), F(hkl) or P(hkl) planes. An overlap is considered a state when the centers of atoms of such a pair are at a mutual distance equal to or less than 1 Å. This value is the average of the covalent radii of considered elements, i.e. 1.21, 1.41, 0.66, 1.11, 0.73 Å for Al, Mg, O, Si, C, respectively [19]. As the area of both crystallographic planes increases, the number of overlaps increases as well, while the number of overlaps relative to the unit area converges. Our earlier investigations showed the area of 100×100 Å to be optimal [18]. The two (hkl) planes are not static during the process of calculating structure compatibility. The in-plane rotation of one plane with respect to the other has a fundamental effect on the number of overlaps. Therefore, the structure compatibility is calculated as an average number of overlaps reflecting the individual values of the number of overlaps at all mutual in-plane orientations. Our earlier investigations showed the rotation step of 1° to be optimal [18]. Each input data file describing the distribution of atoms in a given (hkl) plane consists of at least one line with six Cartesian coordinates: position of the center of given atom (x, y), direction (dx1, dy1) of copying the atom to create a row, and direction (dx2, dy2) of copying an entire row to create the plane. In the case of more complex patterns, the input data may consist of more than one line of coordinates. Due to the periodicity of the atomic arrangement, these coordinates allow creating a given (hkl) plane of arbitrary dimensions.

2.2 Input data

Unit cells of G (a = b = 2.456 Å, c = 6.696 Å, $\alpha = \beta = 90°$, $\gamma = 120°$), C (a = b = 4.978 Å, c = 6.948 Å, $\alpha = \beta = \gamma = 90°$), M (a = 7.663 Å, b = 5.767 Å, c = 7.492 Å, $\alpha = \beta = \gamma = 90°$), F (a = 4.754 Å, b = 10.197 Å, c = 5.981 Å, $\alpha = \beta = \gamma = 90°$), and P (a = 9.250 Å, b = 8.780 Å, c = 5.320 Å, $\alpha = \beta = \gamma = 90°$) [20,21] were cut along (hkl) planes selected according to the flatness and density of occupation of atoms [18]. These (hkl) planes served as a source of input data used for the structure compatibility calculation. Input data of the same (hkl) plane but with different atoms at the origin of coordinates are distinguished by I and II (see an example in **Figure 1a**).

Many (hkl) planes are not completely planar and contain atoms slightly deviated from the central plane. Strict requirement of selecting only atoms lying in a geometrically perfect plane would mean the creation of input data with an unrealistically low density of atoms. Therefore, atoms distant in the perpendicular direction from the central plane by no more than the covalent atomic radius (1.21, 1.41, 0.66, 1.11 Å for Al, Mg, O, Si, respectively [19]) were included in the input data. The (hkl) planes gradually filled with such atoms are marked with letters a,b,c (see an example in **Figure 1b**). In a case of the (hkl) planes without perfect planarity, deviations from the planarity were calculated as the perpendicular distances of atoms from the central geometric plane defined by three other atoms of a given (hkl) plane. Different arrangements of atoms in parallel (hkl) planes of the same phase, whose mutual distance (perpendicular to the plane) is greater than the covalent radius of the atoms, are marked with numbers 1-3 (see an example in **Figure 1c**). For completeness, the elements present are also listed in the designation of each (hkl) plane. Unit cells and input data are provided in Supplementary material [22].

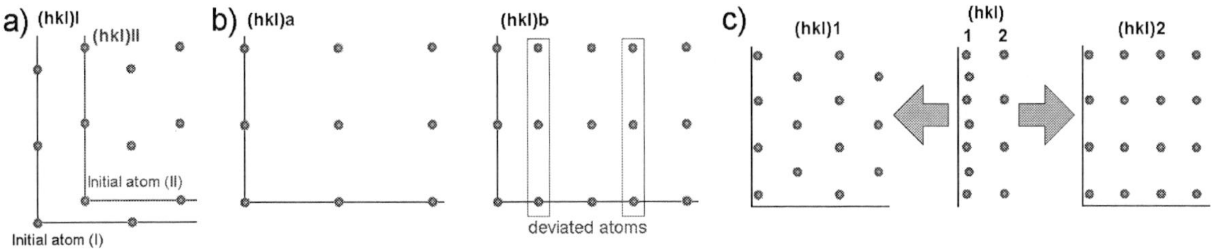

Figure 1 a) Top view of the same (hkl) plane with different atoms at the origin of coordinates (initial atoms); b) Top view of the (hkl) plane gradually filled with atoms deviated from the planarity; c) Side view of two parallel (hkl) planes and top view of each of these (hkl) planes.

3. RESULTS AND DISCUSSION

The descending order of 43 average numbers of overlaps for G(001) plane on M(hkl), F(hkl), P(hkl), and C(hkl) planes show generally the highest values, i.e. the best structure compatibility, for M(hkl) planes (**Figure 2**). The highest and the second highest value is achieved for M(102)SiAlO1 (2482.006 ± 25.048) and M(010)SiAlO (2087.061 ± 32.127) planes, respectively. All numerical data are provided in Supplementary material [22]. In the group of phases formed from magnesium phyllosilicates, F(001)MgSiO (1984.681 ± 21.764) and P(100)O (1545.942 ± 19.360) planes on the 3rd and 8th position, respectively, exhibit the highest average number of overlaps (**Figure 2**). These values represent ~80% a ~62%, respectively, of the 1st value achieved for the M(102)SiAlO1 plane. The C(hkl) planes generally show the lowest average numbers of overlaps and from two-thirds are only found in the second half of all 43 values (**Figure 2**). The highest value (1428.375 ± 43.963) at the 14th position achieved for the C(100)SiO plane represents ~58% of the first value achieved for the M(102)SiAlO1 plane. Based on the average number of overlaps, the degree of structure compatibility of the studied phases with the G(001) plane can be expressed as follows: M > F > P > C.

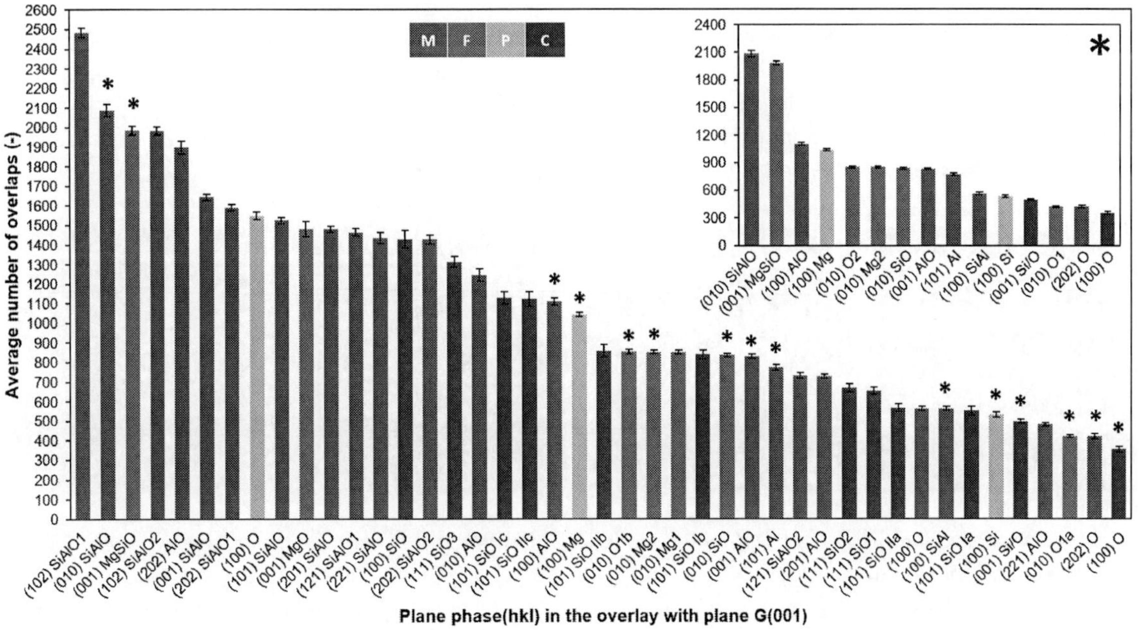

Figure 2 Descending order of the average number of overlaps for the G(001) plane on all M(hkl), F(hkl), P(hkl), and C(hkl) planes. Planar (hkl) planes formed by atoms lying in one geometric plane are marked with an asterisk. The inset compares the average numbers of overlaps for these planes only.

If only planar planes are considered, this order does not change (see the inset in **Figure 2**). The M phase still shows the highest average number of overlaps (2087.061 ± 32.127), although the average number of overlaps for the F(001)MgSiO plane begin to approach it much closer in this case (1984.681 ± 21.764) than when all (hkl) planes are compared regardless of planarity.

Figure 3 shows that the dependence of average numbers of overlaps is generally not directly proportional to the deviation from planarity. The reason for the higher values of the average number of overlaps is therefore not the increased density of atoms in non-planar (hkl) planes after the inclusion of slightly deviated atoms, but better structural compatibility based on a more similar arrangement of atoms in the G(001) plane and the given phase(hkl) planes (whether planar or not). All numerical data are provided in Supplementary material [22].

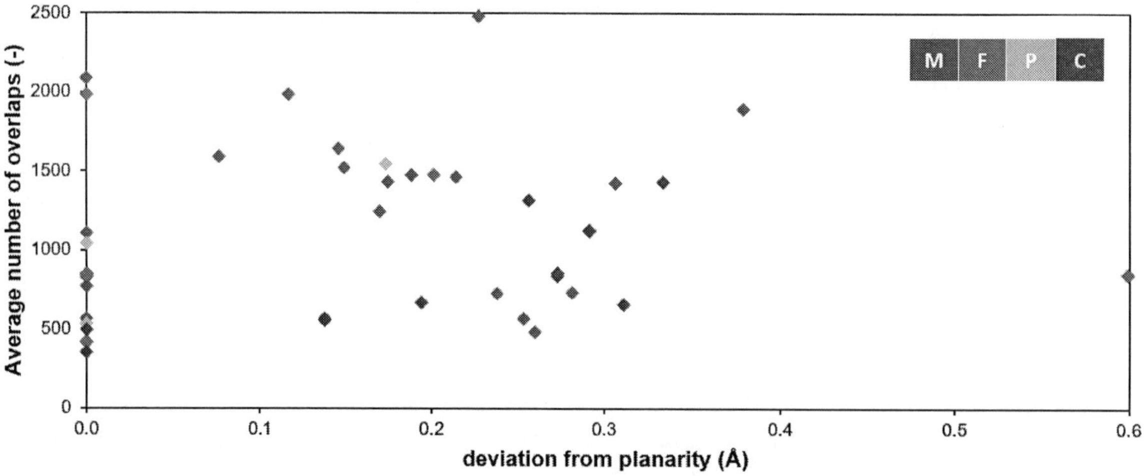

Figure 3 Dependence of the average numbers of overlaps on deviations from ideal planarity. For perfectly planar (hkl) planes, the deviation is zero.

In addition to the numerical values of the number of overlaps, the structure compatibility calculation also provides a visualization of these overlaps for each G(001) in-plane orientation on each phase(hkl) plane. Visualizations of the orientations leading to the highest number of overlaps for each of the M, F, P and C phases can be seen in **Figure 4**. For clarity, only smaller regions (50×50 Å) are shown. Full-size visualizations of the 100×100 Å areas are provided in the Supplementary material [22].

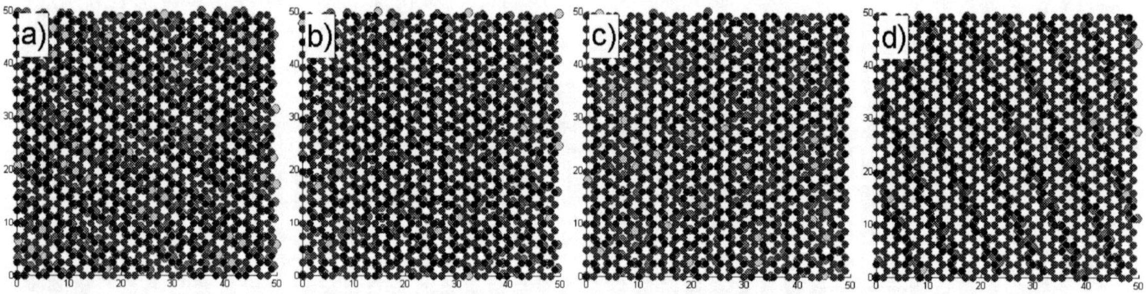

Figure 4 Visualization of the G(001) orientations leading to the highest number of overlaps on each of the phases: (a) M(102)SiAlO1, (b) F(001)MgSiO, (c) P(100)O, (d) C(100)SiO. Color legend: blue – non overlapping atoms in G(001) plane, black - overlapping atoms in G(001) plane, green – non overlapping atoms in phase(hkl) plane, red – overlapping atoms in phase(hkl) plane.

Distinct differences in the distribution of G(001) and phase(hkl) overlapping atoms are apparent. In the case of M(102)SiAlO1, a relatively regular pattern of overlaps can be observed, with the overlaps filling a significant part of the area leaving only tiny regions (~ 5×3 Å) of incompatibility (**Figure 4a**). In the case of F(001)MgSiO and P(100)O, the overlaps form an irregular network of interconnected meshes, denser for F(001)MgSiO and sparser for P(100)O (compare **Figures 4b** and **4c**). In the case of C(100)SiO, the overlaps form narrow parallel bands with "empty" regions approximately 5 Å wide (**Figure 4d**). An overview of the visualizations for the other phase(hkl) planes (see the Supplementary material [22]) shows that the least discontinuous distribution of overlaps is achieved for G(001) on the M(hkl) and F(hkl) planes.

4. CONCLUSION

The results achieved for the G(001) plane on the (hkl) planes of M, F, P and C phases formed by heat treatment of Al-rich and Mg-rich phyllosilicates revealed the best structure compatibility for G(001) and M(hkl). This confirms the empirical experience that nanocomposites containing graphitic carbon can be successfully prepared in situ by thermal treatment of organics with montmorillonite (Al-rich phyllosilicate), which is most often used for this purpose [1-5,8,10,11,13-15]. Relatively high numbers of overlaps found for some F(hkl) and P(hkl) planes suggests that Mg-rich phyllosilicates may also be suitable for this purpose. This is also supported by visualizations of the G(001) on those M, F, P, and C planes that lead to the maximum number of overlaps for each of the phases. The distribution of overlaps more or less over the entire area found for M, F, and P shows these planes as more suitable substrates for continuous growth of G(001) compared to the C plane. Future research will be focused on the correlation of obtained data with real samples prepared in situ by heat treatment of organics not only in the presence of phyllosilicates, but also in the presence of pure M, F, P, and C phases.

ACKNOWLEDGEMENTS

This work was supported by The Ministry of Education, Youth and Sports of the Czech Republic from the projects SGS (SP2024/041) and MATUR - Materials and Technologies for Sustainable Development (CZ.02.01.01/00/22_008/0004631).

REFERENCES

[1] SONOBE, N., KYOTANI, T., HISHIYAMA, Y., SHIRAISHI, M., TOMITA, A. Formation of highly oriented graphite from poly(acrylonitrile) prepared between the lamellae of montmorillonite. *J. Phys. Chem.* 1988, vol. 92, pp. 7029–7034. https://doi.org/10.1021/j100335a037

[2] SONOBE, N., KYOTANI, T., TOMITA, A. Carbonization of polyacrylonitrile in a two-dimensional space between montmorillonite lamellae. *Carbon*. 1988, vol. 26, pp. 573-578. https://doi.org/10.1016/0008-6223(88)90158-3

[3] KYOTANI, T., SONOBE, N., TOMITA, A. Formation of highly orientated graphite from polyacrylonitrile by using a two-dimensional space between montmorillonite lamellae. *Nature*. 1988, vol. 331, pp. 331-333. https://doi.org/10.1038/331331a0

[4] SONOBE, N., KYOTANI, T., TOMITA, A. Carbonization of polyfurfuryl alcohol and polyvinyl acetate between the lamellae of montmorillonite. *Carbon*. 1990, vol. 28, pp. 483-488. https://doi.org/10.1016/0008-6223(90)90042-W

[5] WINANS, R.E., CARRADO, K.A. Novel forms of carbon as potential anodes for lithium batteries. *Journal of Power Sources*. 1995, vol. 54, pp. 11-15. https://doi.org/10.1016/0378-7753(94)02032-X

[6] SANDÍ, G., WINANS, R.E., SEIFERT, S., CARRADO, K.A. In situ SAXS studies of the structural changes of sepiolite clay and sepiolite-carbon composites with temperature. *Chemistry of Materials*. 2002, vol. 14, pp. 739-742. https://doi.org/10.1021/cm010627w

[7] FERNÁNDEZ-SAAVEDRA, R., ARANDA, P., RUIZ-HITZKY, E. Templated synthesis of carbon nanofibers from polyacrylonitrile using sepiolite. *Advanced Functional Materials.* 2004, vol. 14, pp. 77-82. https://doi.org/10.1002/adfm.200305514

[8] RUIZ-HITZKY, E., DARDER, M., FERNANDES, F.M., ZATILE, E., PALOMARES, F.J., ARANDA, P. Supported graphene from natural resources: easy preparation and applications. *Advanced Materials.* 2011, vol. 23, pp. 5250-5255. https://doi.org/10.1002/adma.201101988

[9] ANDRADE, P.F., AZEVEDO, T.F., GIMENEZ, I.F., SOUZA FILHO, A.G., BARRETO, L.S. Conductive carbon–clay nanocomposites from petroleum oily sludge. *Journal of Hazardous Materials.* 2009, vol. 167, pp. 879-884. https://doi.org/10.1016/j.jhazmat.2009.01.070

[10] DING, W., WEI, Z., CHEN, S., QI, X., YANG, T., HU, J., WANG, D., WAN, L.-J., ALVI, S.F., LI, L. Space-confinement-induced synthesis of pyridinic- and pyrrolic-nitrogen-doped graphene for the catalysis of oxygen reduction. *Angewandte Chemie International Edition*. 2013, vol. 52, pp. 11755-11759. https://doi.org/10.1002/anie.201303924

[11] RUIZ-GARCIA, C., PÉREZ-CARVAJAL, J., BERENGUER-MURCIA A., DARDER, M., ARANDA, P., CAZORLA-AMORÓS, D., RUIZ-HITZKY, E. Clay-supported graphene materials: application to hydrogen storage. *Physical Chemistry Chemical Physics.* 2013, vol. 15, pp. 18635-18641. https://doi.org/10.1039/c3cp53258e

[12] ZHANG, Y., OUYANG, J., YANG, H. Metal oxide nanoparticles deposited onto carbon-coated halloysite nanotubes. *Applied Clay Science*. 2014, vol. 95, pp. 252-259. http://dx.doi.org/10.1016/j.clay.2014.04.019

[13] MICÓ VICENT, B., LÓPEZ, M., BELLO, A., MARTÍNEZ, N., MARTÍNEZ VERDÚ, F., Optimum multilayer-graphene-montmorillonite composites from sugar for thermosolar coatings formulations. *Journal of Solar Energy Engineering*. 2017, vol. 139, pp. 1-7. https://doi.org/10.1115/1.4035757

[14] WANG, J., XU, Y., DING, B., CHANG, Z., ZHANG, X., YAMAUCHI, Y., WU, K.C.W. Confined self-assembly in two-dimensional interlayer space: Monolayered mesoporous carbon nanosheets with in-plane orderly arranged mesopores and a highly graphitized framework. *Angewandte Chemie International Edition*. 2018, vol. 57, pp. 2894-2898. https://doi.org/10.1002/anie.201712959

[15] TOKARSKÝ, J., MAMULOVÁ KUTLÁKOVÁ, K., PEIKERTOVÁ, P., ŘEHÁČKOVÁ, L., KORMUNDA, M., MATĚJKOVÁ, P., ŠTUDENTOVÁ, S., KULHÁNKOVÁ, L. Polypyrrole/montmorillonite and polypyrrole/ghassoul intercalates as a source of graphite and multi-layer graphene: preparation of nanocomposites exhibiting strongly anisotropic electrical conductivity. *Materials Research Bulletin.* 2021, vol. 142, pp. 111429. https://doi.org/10.1016/j.materresbull.2021.111429

[16] YALCINKAYA, E., BASKAN-BAYRAK, H., OKAN B.S. Growing of 2D/3D graphene structures on natural substrates from aromatic plastic wastes by scalable thermal-based upcycling process with a comparative CO_2 footprint analysis. *Sustainable Materials and Technologies.* 2023, vol. 37, pp. e00687. https://doi.org/10.1016/j.susmat.2023.e00687

[17] BARRA, A., LAZAR, O., MIHAI, G., BRATU, C., RUIZ-GARCIA C., DARDER, M., ARANDA, P., ENACHESCU, M., NUNES, C., FERREIRA, P., RUIZ-HITZKY E. Graphene-like materials supported on sepiolite clay synthesized at relatively low temperature. *Carbon.* 2024, vol. 218, pp. 118767. https://doi.org/10.1016/j.carbon.2023.118767

[18] TOKARSKÝ, J., MOLEK, J. Overlap and rotate - a simple method for predicting out-of-plane and in-plane orientations of heteroepitaxial thin films. *Surfaces and Interfaces*. 2024, vol. 46, pp. 104129. https://doi.org/10.1016/j.surfin.2024.104129

[19] CORDERO, B., GÓMEZ, V., PLATERO-PRATS, A.E., REVÉS, M., ECHEVERRÍA, J., CREMADES, E., BARRAGÁN, F., ALVAREZ, S. Covalent radii revisited. *Dalton Transactions*. 2008, vol. 21, pp. 2832-2838.

[20] WYCKOFF, R.W.G., *Crystal Structures, second ed., Volume 1*, New York: John Wiley & Sons, Inc., 1963.

[21] WYCKOFF, R.W.G., *Crystal Structures, second ed., Volume 4, Miscellaneous Inorganic Compounds, Silicates, and Basic Structural Information*, New York: John Wiley & Sons, Inc., 1968.

[22] Data collection. *Dataset for Structure compatibility of graphite and phases formed from aluminum or magnesium phyllosilicates.* [online]. 2024. [viewed: 2024-09-27]. Available from: https://doi.org/ 10.5281/zenodo.11574074

AC ELECTROSPINNING: DEVICE FOR CONTROLLED PRODUCTION OF PLANAR, SINGLELAYER NANOFIBER PLUME

[1]Ondrej FRIEDRICH, [1]Jan VALTERA, [1]Milos LICEK

[1]*Technical University of Liberec, Faculty of Mechanical Engineering, Department of Textile Machine Design, Liberec, Czech Republic, EU,* ondrej.friedrich@tul.cz

https://doi.org/10.37904/nanocon.2024.5032

Abstract

The work focuses on the development of a linear electrode for the production of flat nanofibrous material by spinning a polymer solution using the AC electrospinning technique. This method is characterized by high productivity and the absence of an electrically active collector, which predisposes it to the production of advanced nanofibrous structures that will find application in medical and technical applications such as filtration or tissue engineering.

Current spinning electrodes for AC electrospinning are characterized by a spinning surface with an overcritical electric field strength, from which polymer jets and subsequent nanofibers are created in all directions. A spatial, hollow nanofibrous plume is created by the natural periodic interconnection of fibers in the area of the virtual collector. The transport and deposition of such a nanofibrous structure on the supportive spunbond is unstable, when due to the air emerging from the inner space with higher concentration of the residual solvent, folds and other defects are created.

By the geometrically modified electrodes with the help of simulation tools for electric field analysis, it is possible to efficiently modify the geometry and the working surface of the electrode so that the nanofibers are formed only in a single plane, thereby eliminating the negative effects of the spatial nanofibrous plume.

Based on the analyses, electrodes capable of continuous operation producing nanofibers in the single plane and in the desired direction producing a planar nanofiber structure are developed and tested.

With the use of these advanced devices, it is possible to produce more homogeneous surface layered nanofiber materials and advanced nanofiber structures for use especially in medical applications.

Keywords: AC electrospinning, polymeric nanofibers, flat nanofibrous layer, electric field analysis, spinning electrode

1. INTRODUCTION

One of the essential added values of nanofibrous materials is their high specific surface area, which significantly increases their reactivity and efficiency in many applications. Due to their high porosity while maintaining a small pore size, nanofibrous layers can be advantageously used for filtration. Many nanofibrous materials are also biocompatible, predisposing them to biomedical applications. Thanks to these properties, nanofibrous materials are used in biomedicine [1], [2] and tissue engineering [3], filtration [4] and in energy as components of advanced batteries and fuel cells [5]. With continued intensive development, the application of nanomaterials continues to expand [6].

There are several technologies for the production of polymeric nanofibers, such as drawing [7] or forcespinning [8], but electrospinning in a direct electric field (DC electrospinnig) [9] is the most used one in industrial production. Electrospinning in an alternating electric field (AC electrospinning) is a new, highly productive technology for the production of polymer nanofibers, which does not require a counter-electrode typical for direct current spinning, thus opening the way for the production of unique nanofiber structures [8].

Electrospinning technique is based on the difference in electric potentials in the order of tens of kilovolts. In DC electrospinning, the difference is ensured by the presence of a charged counter-electrode. In AC electrospinning technology, an electric field is created between the electrode and the so-called virtual collector. A virtual collector is formed in the air in the vicinity of the electrode due to a periodical change in its polarity and, according to the theory of AC spinning, is formed by charged ions of the air environment and charged segments of nanofibers. The nanofibers are then always created in the direction of the gradient of the electric field from the spinning electrode to the virtual collector to its entire surroundings.

In the area of the virtual collector, fibers moving from the electrode slow down and their mechanical interconnection occurs. The structure of interconnected nanofibers floating above the electrode is called a nanofiber plume. It is possible to manipulate the plume further, for example by placing it on a spunbond or wrapping it on a yarn for the production of a composite yarn with a nanofiber coating [10].

Electrodes used for AC electrospinning, such as the rod-like electrode [10] or the cable electrode, are designed especially towards high productivity. From their working surface with a supercritical electric field strength, nanofibers are created in all directions, which after their connection leads to the formation of a hollow or otherwise closed nanofiber plume.

When the hollow plume is deposited on the spunbond, especially with rod-like electrodes, two layers are applied at once. Leaking air from the space between them then causes irregular folds and other errors in the laying of the plume. The inner space may also contain an increased concentration of vaporized solvent of the polymer solution, which may be the cause of defects in the produced fibers.

2. THEORY OF AC ELECTROSPINNIG

In order to initiate the process of nanofibers spinning on the electrode, a high electric field strength must be applied to the polymeric solution, which will cause the surface tension to be overcome and form Taylor cones. This is determined by the difference in electric potential - the magnitude of the spinning voltage, the relative permittivity of the environment and the geometry of the electrode parts. As the radius of curvature decreases, the concentration of electric charge on the surface of the electrode increases and, as a result, the strength of the electric field is highest in the region of the apexes. This physical principle is used in the design of the spinning area of the electrode. Spinning of the polymeric solution occurs only if the limiting critical value of the electric field strength is exceeded. Not all nanofibers that are created by the effect of the electric field have sufficient energy to connect with the plume. Due to gravity and electrical forces, some of the fibers fall back onto the electrode and into its surroundings. These residual fibers, if deposited on the dry surface of the electrode, gradually change its geometry and can lead to defects in the produced material or to a complete interruption of spinning. Therefore, the design of the electrode for continuous operation must appropriately prevent contamination of the key parts of the electrode with nanofibers. Among the simplest solutions is redissolving of the incident nanofibers in excess of polymeric solution.

The spinning area of the electrode can be closed, without ends (ring, circle), or open, with ends (linear electrode). The closed construction of the spinning region has an indisputable advantage in the homogeneity of the electric field in its vicinity, thanks to which it produces a homogeneous nanofibrous structure around its entire area. However, a nanofibrous plume in the shape of a hollow tube is created. The open spinning region generally has a different overall curvature at its ends than in the part in between. Therefore, a nanofibrous plume with a slightly different structure or different specific weight is formed at the ends.

3. DESIGN OF THE ELECTRODES

Two designs of spinning electrodes for AC electrospinning producing a planar nanofibrous plume were designed. The first, the disc electrode (see **Figure 1.A**) is a suitable design adaptation of the linear electrode so that a continuous homogeneous supply of the polymeric solution and cleaning of the spinning area can be simply provided. It consists of a thin disk rotating around its horizontally oriented axis. In the lower part, where the strength of the electric field is reduced thanks to the frame, the outer edge of the disc is cleaned and a polymeric solution is applied to it, while in the upper part fibers are formed. Thanks to spinning from the thin edge of the disc, a planar nanofibrous plume is formed. Its width is proportional to the diameter of the disc and is limited for use on small widths of the produced material, approximately up to 500 mm.

For the production of material with a larger width, theoretically unlimited, a cable electrode is suitable (see **Figure 1.B**). It consists of a wire or cable wound between a pair of pulleys. The polymeric solution can be applied to the cable while pulleys are going through the solution or with a nozzle. The strength of the electric field in the vicinity of the cable is supercritical and spinning takes place here.

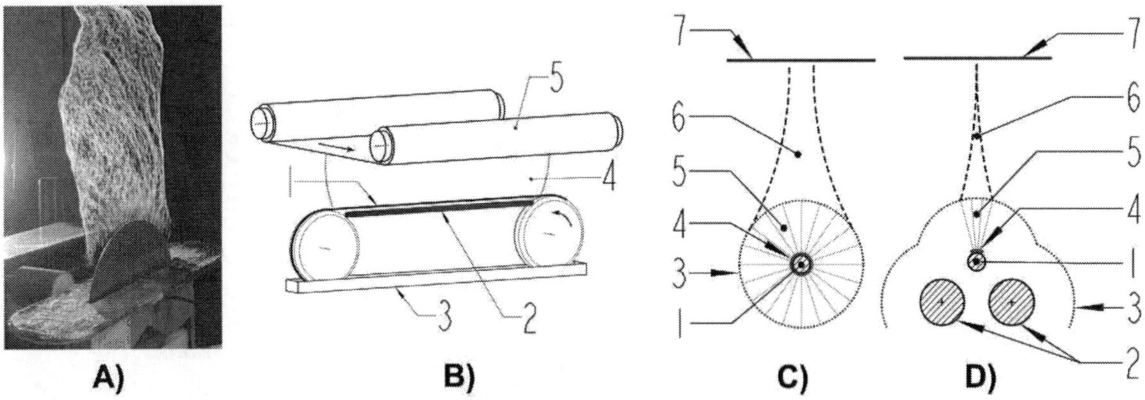

Figure 1 A) Disc-like electrode, **B)** Modified cable electrode: Cable (1), shielding elements (2), reservoir with polymeric solution (3), nanofibrous plume (4), spunbond (5). Schematic representation of spinning of nanofibers on unsuppressed **(C)** and partially-suppressed cable electrode **(D)** in section: cable with polymeric solution (1), shielding elements (2), area of virtual collector (3), area with supercritical strength of electric field (4), nanofibers (5), nanofibrous plume (6) and spunbond (7).

From the point of view of the size of the nanofiber plume, it is a hybrid type showing the characteristics of all the mentioned variants. In general view, this is an electrode with an open spinning area, the nanofibrous plume is flat with edges above the pulleys on both sides. However, in a detailed view of the section of the spinning area (see **Figure 1.C**), the area with supercritical strength of the field on the surface of the cable is closed, spinning runs homogeneously in the entire surroundings, and there is a hollow, closed space between the plume in the area of virtual collector and the spinning electrode. The interconnected nanofibrous plume does not have a clearly defined direction of motion and goes most often up due to external suction and the weak effect of the electric wind.

The aim of modifying the cable electrode is to reduce the strength of the electric field by means of shielding elements on part of its surface, as shown in **Figure 1.D**, so that the character of fiber spinning is similar to a disk electrode. Shielding elements made of electrically conductive material are connected at the same electric potential as the spinning electrode (see **Figure 3.B**). Thanks to this arrangement, the strength of the electric field on the spinning electrode surface exceeds the supercritical value only on a certain section. The nanofiber plume has thus a narrow planar character with a preferred direction of movement.

4. VERIFICATION OF DESIGN WITH PARTIALY SUPPRESSED CABLE ELECTRODE

Simulation tools for electric field analysis are used to study the effect of design and geometric arrangement on the size and shape of the electrode spinning region. To define the task, it is necessary to make several simplifying assumptions. The task is simplified to a static one, so instead of an alternating voltage, a direct current voltage with the same effective value is considered on the electrode. To specify the boundary condition, the theory of a virtual collector is used, which surrounds the spinning electrode with a constant distance, and a voltage of 0 V is assumed here. It is observable during spinning as a region with rapid decrease of nanofibers, and its distance is dependent, among other things, on the frequency of the spinning voltage waveform.

Numerical simulations were performed in the ANSYS Electronics Desktop 2019 program. The boundary condition of the effective electrode voltage U_{ef} was considered to be 35 kV. The distance of the virtual collector from the surface of the electrode, or each element with the boundary condition U_{ef}, was considered to be 50 mm. The model was built using symmetry as a half 2D problem (see **Figure 2** and **Figure 3**).

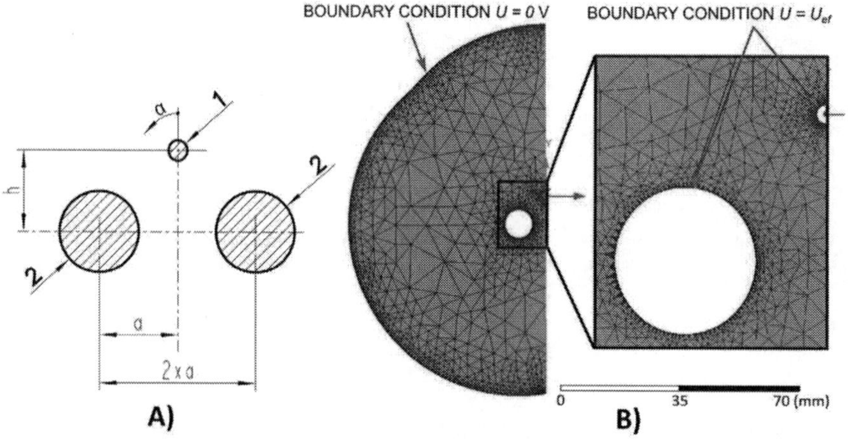

Figure 2 A) Schematic representation of cable electrode (1) with shielding elements (2) in section. **B)** Model of the air environment around the electrode, the shielding elements and the virtual collector with the maximum mesh size: near the electrode - 0.001 mm, near the shielding elements - 0.01 mm.

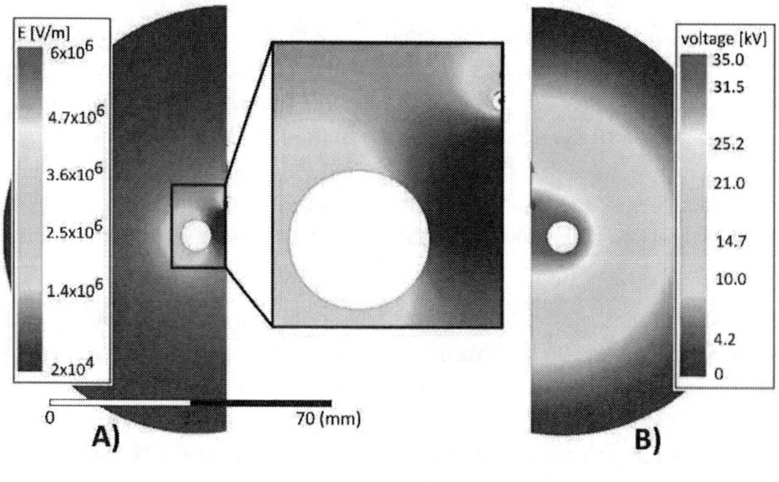

Figure 3 Distribution of electric field strength **(A)** and voltage **(B)** around the electrode with shielding elements according to Fig. 2.A. Applied effective voltage is 35 kV.

A total of 64 variants of the dimensional arrangement of the electrode and shielding elements were simulated in order to find a suitable solution so that spinning occurs only on the part of the electrode facing away from the shielding elements and at the same time so that the strength of the field increase to the supercritical value is as steep as possible. The parameters of the spacing of the shielding elements and their distance from the electrode according to **Figure 2.A** were analyzed in the range a: 8-12 mm and h: 2-10 mm. The diameter of the wire electrode was 1 mm and the diameter of the shielding elements was 8 mm, when the guaranteed subcritical strength of the electric field is experimentally verified, so that there are no electric discharges or possibly spinning on them. Based on experience and comparing experiments with simulations, the critical value of the electric field strength was set at 3000 kV/m.

5. RESULTS AND DISCUSSION

In the performed 2D planar analysis, the value of the electric field strength on the surface of the spinning cable part around the entire perimeter was evaluated (see **Figure 3**). For the selected variants the field strength value is shown in the **Figure 4**. The processing of the obtained data and selection of the best dimensional arrangement was carried out according to the criterion of the highest steepness of the increase in the electric field strength in the region of the transition over the critical value for spinning. From the point of view of the spinning process and fiber formation, a constant value over the spinning area with a jump increase and decrease at the ends is suitable, which is what the design tries to approach.

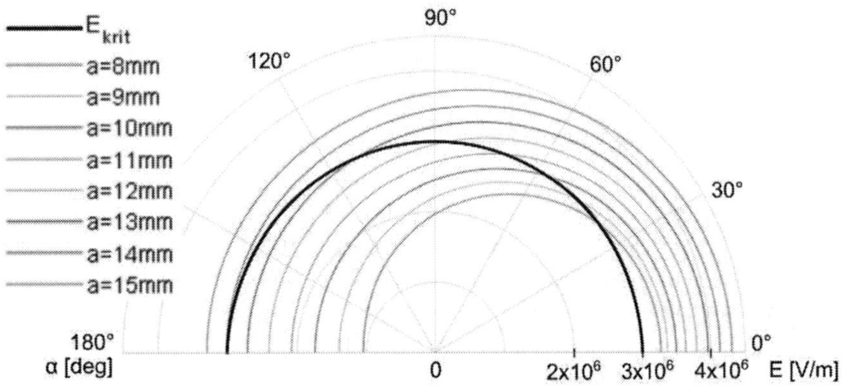

Figure 4 Electric field strength on the electrode surface as a function of the alpha angle according to Fig. 2.A for selected variants. The black line shows the critical electric field strength.

The obtained dependences of the electric field strength on the electrode surface were interleaved with a polynomial curve of degree 9 and then derivative to determine the dimensional arrangement with the highest steepness of the strength of the field increase according to the highest value of the derivative. Based on this analysis, the dimensional option with parameters a=8 mm and h=8 mm was selected as the most suitable for the design.

6. CONCLUSION

Due to its simple design and easy supply of polymer solution, the disc-like electrode is a suitable adaptation of the linear electrode for laboratory purposes and for the production of small-width layers. It forms a singlelayer planar nanofibrous siding with a clear preferred direction of advancement upward from the electrode.

The designed modified cable electrode retains the similar character of nanofiber plume produced by the disc electrode and additionally allows for increasing the fiber production and width of the material produced. Thanks to additional shielding elements, it produces a low thickness nanofibre plume with a preferred direction of motion.

The shielding elements of the cable electrode shall be made of conductive material and placed close to the electrode according to simulation results. For tests, these may be stationary metal rods, but for long-term operation, continuous cleaning is required. A variant with shielding elements as circular belts rewinded together with the cable electrode was already tested. A layer of polymer solution is applied to the belts enabling continuous dissolution of the residual fibers. The polymer solution also provides the electrical charge to the shielding belts, however, due to the subcritical field strength, the spinning does not occur on their surface.

Due to the preferred direction of the motion of the nanofiber plume, the partially suppressed cable electrode does not require such a high level of additional suction, which reduces the energy demands on the device, but due to the spinning only on a part of the surface of the electrode, its productivity is reduced.

The designed electrodes are capable of producing a nanofibrous material with the specific property of a single layer of a planar nanofibrous plume and thus have the potential to be used in the production of specific multilayer materials or even linear nanofiber structures.

ACKNOWLEDGEMENTS

This publication was written at the Technical University of Liberec as part of the project with the support of the Specific University Research Grant 2024-5443.

REFERENCES

[1] Dos SANTOS D. M., CORREA D. S., MEDEIROS E. S., OLIVEIRA J. E., MATTOSO L. H. C. "Advances in Functional Polymer Nanofibers: From Spinning Fabrication Techniques to Recent Biomedical Applications," *ACS Appl. Mater. Interfaces*. 2020, vol. 12, no. 41, pp. 45673–45701.

[2] OMIDIAN H., GILL E. J. 'Nanofibrous Scaffolds in Biomedicine', *J. Compos. Sci.* 2024, vol. 8, no. 7, p. 269. Available from: https://doi.org/10.3390/jcs8070269.

[3] HUANG T., ZENG y., LI C., ZHOU Z., XU J., WANG L., YU D. WANG K. „Application and Development of Electrospun Nanofiber Scaffolds for Bone Tissue Engineering", *ACS Biomater. Sci. Eng*. 2024, vol. 10, no. 7, p. 4114–4144. Available from: https://doi.org/10.1021/acsbiomaterials.4c00028.

[4] YU Z., FAN T., LIU Y., LI L., LIU J., YANG B., RAMAKRISHNA S., LONG Y. „Efficient air filtration through advanced electrospinning techniques in nanofibrous Materials: A review", *Separation and Purification Technology*. 2024, vol. 349, p. 127773. Available from: https://doi.org/10.1016/j.seppur.2024.127773.

[5] THIELKE M. W., GUZMAN S. L., TAFOYA J. P. V., TAMAYO E. G., HERAZO C. I. C., HOSSEINAEI O., SOBRIDO A. J. "Full Lignin-Derived Electrospun Carbon Materials as Electrodes for Supercapacitors," *Front. Mater.* 2022, vol. 9. Available from: https://doi.org/doi.org/10.3389/fmats.2022.859872

[6] LIU H., GOUGH C. R., DENG Q., GU Z., WANG F., HU X. "Recent Advances in Electrospun Sustainable Composites for Biomedical, Environmental, Energy, and Packaging Applications," *Int. J.Mol. Sci.* 2020 vol. 21, no. 11, p. 4019.

[7] XING X., WANG Y., LI B. „Nanofibers drawing and nanodevices assembly in poly(trimethylene terephthalate)", *Opt. Express*. 2008, vol. 16, no. 14, p. 10815. Available from: https://doi.org/10.1364/OE.16.010815.

[8] PADRON S., FUENTRES A., CARUNTU D., LOZANO K. 'Experimental study of nanofiber production through forcespinning'. *Journal of Applied Physics*. 2013, vol. 113, no. 2, p. 024318. Available from: https://doi.org/10.1063/1.4769886

[9] AL-DHAHEBI A. M., LING j., KRISHNAN S. G., YOUSEFZADEH M., ELUMALAI N. K., SAHEED M. S. M., RAMAKRISHNA S., JOSE R."Electrospinning research and products: The road and the way forward," *Appl. Phys. Rev.* 2022, vol. 9, no. 1, p. 011319.

[10] VALTERA J., KALOUS T., POKORNY P., BATKA O., BILEK M., CHVOJKA J., MIKES P., KOSTAKOVA E. K., ZABKA P., ORNSTOVA J., BERAN J., STANISHEVSKY A., LUKAS D. 'Fabrication of dual-functional composite yarns with a nanofibrous envelope using high throughput AC needleless and collectorless electrospinning', *Sci Rep.* 2019, vol. 9, no. 1, p. 1801. Available from: https://doi.org/10.1038/s41598-019-38557-z.